Jim C. DeLoach

Frank J. Ambrosio

Lab Manual: A Troubleshooting Approach

to accompany

DIGITAL SYSTEMS: PRINCIPLES AND APPLICATIONS

Eleventh Edition

By Ronald J. Tocci, Neal S. Widmer, & Gregory L. Moss

Prentice Hall

Boston Columbus Indianapolis New York San Francisco Upper Saddle River
Amsterdam Cape Town Dubai London Madrid Milan Munich Paris Montreal Toronto
Delhi Mexico City Sao Paulo Sydney Hong Kong Seoul Singapore Taipei Tokyo

Editorial Director: Vern Anthony
Acquisitions Editor: Wyatt Morris
Editorial Assistant: Yvette Schlarman
Director of Marketing: David Gesell
Marketing Manager: Harper Coles
Marketing Assistant: Crystal Gonzales
Senior Managing Editor: JoEllen Gohr
Project Manager: Rex Davidson

Senior Operations Supervisor: Pat Tonneman
Operations Specialist: Laura Weaver
Art Director: Diane Ernsberger
Cover Designer: Integra
Printer/Binder: R.R. Donnelley
Cover Printer: R.R. Donnelley
Composition: Aptara Inc.

MAX+PLUS® II software screens are reprinted courtesy of Altera Corporation.
Altera is a trademark and service mark of Altera Corporation in the United States and other countries. Altera products are the intellectual property of Altera Corporation and are protected by copyright laws and one or more U.S. and foreign patents and patent applications.

The data sheets included in this book have been reprinted with the permission of the following companies:
All materials reproduced courtesy of Analog Devices, Inc.
All materials reproduced courtesy of Fairchild, a Schlumberger Company.
All materials reprinted by permission of Intel Corporation. Intel Corporation assumes no responsibility for any errors which may appear in this document, nor does it make a commitment to update the information contained herein.
All materials reprinted with permission of National Semiconductor Corp.
Reprinted by permission of Texas Instruments Inc.

Copyright © 2011, 2007, 2004, 2001, 1998 Pearson Education, Inc., publishing as Prentice Hall, 1 Lake Street, Upper Saddle River, New Jersey, 07458. All rights reserved. Manufactured in the United States of America. This publication is protected by Copyright, and permission should be obtained from the publisher prior to any prohibited reproduction, storage in a retrieval system, or transmission in any form or by any means, electronic, mechanical, photocopying, recording, or likewise. To obtain permission(s) to use material from this work, please submit a written request to Pearson Education, Inc., Permissions Department, 1 Lake Street, Upper Saddle River, New Jersey 07458.

Many of the designations by manufacturers and seller to distinguish their products are claimed as trademarks. Where those designations appear in this book, and the publisher was aware of a trademark claim, the designations have been printed in initial caps or all caps.

10 9 8 7 6 5 4 3 2

Prentice Hall
is an imprint of

www.pearsonhighered.com

ISBN-13: 978-0-13-512395-9
ISBN-10: 0-13-512395-X

To my wife, Margaretta, whose patience and encouragement have been unending; to our son and daughter, John and Cheryl; and to Steve, Debbie, Sara, Justin, Heather, Stevie, and Sarah.

Jim C. DeLoach

To my wife, Ana, for her patience and understanding, and to the most wonderful son in the world, Filip.

Frank J. Ambrosio

PREFACE

This manual is designed to provide practical laboratory experience for students of digital electronics. There are 42 TTL and CMOS experiments that are coordinated with the companion text, *Digital Systems: Principles and Applications,* Eleventh Edition, by Ronald Tocci, Neal Widmer, and Gregory Moss. The sequence of the experiments follows reasonably well with the main text. In addition to the TTL/CMOS experiments, we have included a set of 25 experiments covering the new HDL topics in the text. These are found in the Supplemental Experiments section of this lab manual.

The TTL experiments focus on digital fundamentals, design, and troubleshooting. There is enough material to cover a two-semester laboratory course stressing design, troubleshooting, or even a combination of the two. It is not intended that all experiments be completed or that every experiment be done in its entirety. Instructors will probably want to make some experiments optional for their students.

The troubleshooting exercises take the student from digital IC fault recognition (Experiments 3 and 6) to troubleshooting combinatorial circuits (Experiment 14), flip-flops (Experiment 19), counters (Experiment 25), and systems (Experiment 35). There are also exercises using a logic analyzer (Experiments 31, 32, and 40). The logic analyzer is discussed in Appendix B of this manual. Logic probes and logic pulsers are discussed in Experiment 3 and are needed in most of the troubleshooting exercises.

The design exercises include simplification exercises (Experiment 8), a majority tester (Experiment 13), a programmable counter (Experiment 18), a system clock, variable timer, and serial interface with handshake (Experiment 28), a word recognizer (Experiment 36), a synchronous counter (Experiment 40), and a programmable function sequencer (Experiment 42). Also included in the exercises is an example of a control waveform generator, a synchronous data transmission system, a frequency counter, a 2-digit digital voltmeter, and a data bus system.

Each experiment begins with a set of stated objectives, text references, and required equipment, followed by a procedure for meeting each objective. Most experiments specify logic circuits needed to perform the experiment, whereas a few require that the student design and draw the circuit(s). Almost all experiments end with a set of review questions for the student to complete.

The ICs given in the parts list have been checked for availability. All TTLs and CMOS ICs are currently active. All parts can be obtained from Electronix Express. A complete kit of ICs is available as Kit #32JDDIGSYS1.

The HDL experiments are designed to supplement the TTL exercises, although some instructors may want to use the HDL experiments exclusively. Like their TTL counterparts, there are enough experiments to cover a two-semester course in Digital Electronics. Again, the instructor might want to select parts of the various exercises for the student to do as some of the experiment material may be too lengthy for students to do in a single lab session.

The supplemental experiments assume that the students have access to a PC running MAX+PLUS® II software (Altera Corporation), whereas student access to a CPLD board such as the DeVry University eSOC board, Altera University Board, or R.S.R. Electronics PLDT-2 (or equivalent board with an EPM7128SLC84 CPLD) is optional.

Appendix A covers several topics that should be helpful to students. Included in the material are general hints on proper breadboarding, digital troubleshooting, and information on typical digital system and IC faults. Students are encouraged to become familiar with the contents of the appendix early, even though serious reading of much of the information on troubleshooting and digital faults can be deferred until it is needed.

Appendix B covers the basic principles of logic analyzers. The operation of a typical logic analyzer, the Tektronix Model 7D01, is also discussed. Most of the material covered here can be transferred to another logic analyzer, provided that the instructor explains the differences between that logic analyzer and the 7D01.

The circuits in the experiments will work with most digital trainers. At minimum, the trainer should have the following features:

1. Buffered LED monitors (8)
2. Toggle switches (8)
3. Normally HIGH debounced pushbutton switch (1)
4. Normally LOW debounced pushbutton switch (1)
5. TTL-compatible clock, 1 Hz-100 kHz (1)

Occasionally, additional LED monitors and toggle and pushbutton switches will be required. Figure P-1 shows a circuit for wiring an additional Light Emitting Diode (LED) monitor. Figure P-2 shows a circuit for converting the output of a conventional square wave generator (SWG) so that it is compatible with TTL levels. Figure P-3 shows a circuit for obtaining both normally HIGH and normally LOW debounce switches.

This manual covers 42 ICs, most of which should be readily available. LS-TTL ICs may be substituted for all TTL ICs except the 7476. The data sheets for all digital ICs are found either in Appendix C of this manual or on the World Wide Web.

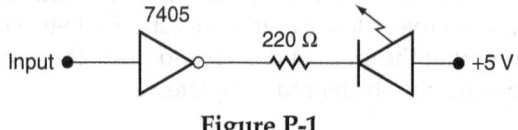

Figure P-1

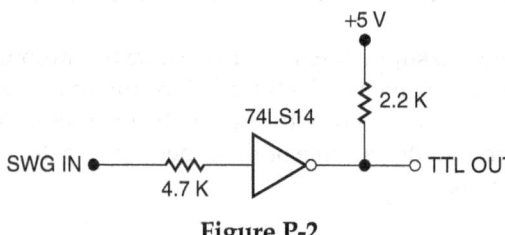

Figure P-2

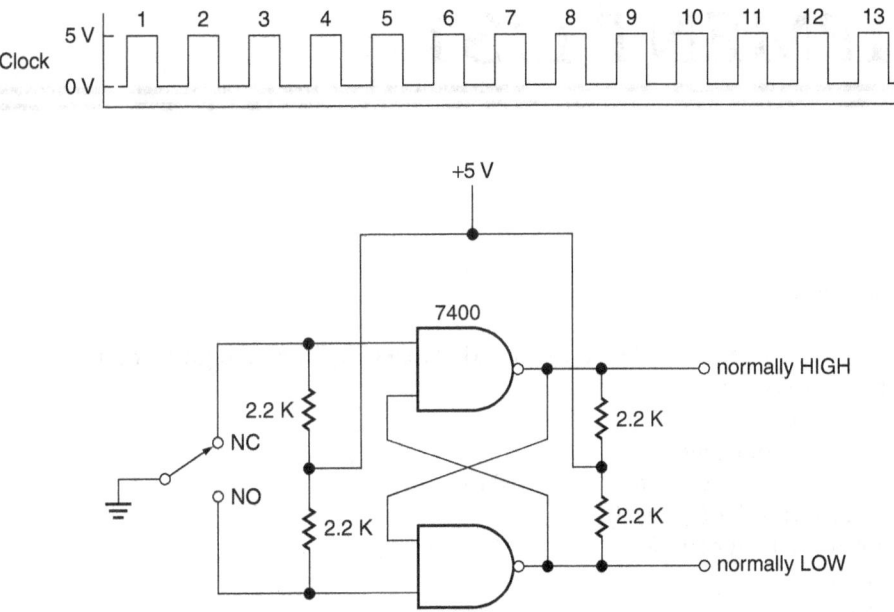

Figure P-3

A list of all required and optional integrated circuits, parts, and laboratory equipment is found in the Equipment List that follows the Preface.

Appendix D covers the availability of toggle switches, pushbutton switches, and LEDs, and their pin assignments, provided by the three CPLD boards listed in the supplemental experiments. It also gives the general programming and downloading procedure that will be mentioned in these experiments. For more information on the CPLD boards and the MAX+PLUS® II software, contact information appears in the Equipment List.

The authors thank several firms for providing up-to-date information that was helpful in preparing this manual, and for permission to use the data sheets that appear in this manual:

Analog Devices, Inc.
Altera Corporation
Elenco Electronics, Inc.
Fairchild Semiconductor, Inc.
Intel Corporation
Motorola Semiconductors
National Semiconductor Corporation
R.S.R. Electronics, Inc.
Texas Instruments Incorporated

Jim C. DeLoach
Frank J. Ambrosio

EQUIPMENT LIST

Laboratory Equipment

1) Power supplies: 0 to +5 V, 50 mA, regulated; 0 to +15 V, 50 mA, regulated; 0 to −15 V, 50 mA, regulated; 6.3 VAC, 60 Hz
2) Dual trace oscilloscope
3) Storage oscilloscope (*optional*)
4) Pulse or square wave generator (SWG), 0 to 1 MHz
5) Time interval counter (*optional*)
6) Volt-ohm-milliameter (VOM)
7) Logic probe
8) Logic pulser
9) Logic analyzer (*optional*)
10) Digital voltmeter (DVM) or Digital Multimeter (DMM)
11) DeVry University Board eSOC with EPM7128SLC84 CPLD (or Altera University Board with EPM7128SLC84 CPLD, R.S.R. PLDT-2 or any other equivalent board) (*optional*)
12) Desktop computer with minimum of INTEL 486/66 CPU and 8 MB RAM (*optional*)
13) MAX+PLUS®II software (Altera Corporation) (*optional*)

For more information and pricing on the CPLD boards and software, contact the following vendor:

Electronix Express, A Division of RSR Electronics, Inc.
Phone: 1-800-972-2225; In New Jersey: 732-381-8020
FAX: 732-381-1006; 732-381-1572
Web Address: www.elexp.com

Digital Integrated Circuits

Quantity	IC	Description
2	7400	Quad 2-input NAND
2	7402	Quad 2-input NOR
2	7404	Hex INVERTERS
2	7405	Hex Open Collector INVERTERS
1	7406	Hex INVERTERS with Open Collector Outputs
2	74LS08	Quad 2-input NAND
1	7410	Triple 3-input NAND
1	74F10	Triple 3-input NAND
2	74LS10	Triple 3-input NAND
1	74LS11	Triple 3-input AND
1	7414	Hex Schmitt-trigger INVERTERS
1	74LS27	Triple 3-input NOR
1	7432	Quad 2-input OR
1	74LS42	1-of-10 DECODER
2	7447A	BCD to Seven-segment DECODER/DRIVER
4	74LS74A	Dual D-type FLIP-FLOP
1	74LS75	4-bit Bistable LATCH
4	74LS76	Dual J-K FLIP-FLOP
1	74LS85	4-bit MAGNITUDE COMPARATOR
1	74LS86	Quad 2-input XOR
2	74LS90	Decade ripple COUNTER
2	74LS93	Binary ripple COUNTER
2	74121	Monostable MULTIVIBRATOR
1	74LS125A	Quad Bus BUFFER w/3-state Outputs
1	74LS138	3-to-8 DECODER/MULTIPLEXER
1	74LS139	Dual 1 of 4 DECODER
1	74HCT147	CMOS 10 TO 1 Priority ENCODER
1	74HCT151	CMOS 1 OF 8 MULTIPLEXER
3	74LS163A	4-bit Binary COUNTER
3	74HCT173	Quad D FLIP-FLOPS
1	74LS174	Hex D FLIP-FLOP
1	74HC192	CMOS Synchronous BCD COUNTER
2	74LS193	Synchronous 4-bit Binary COUNTER
1	74LS194A	4-bit Bidirectional SHIFT REGISTER
1	74HC195	4-BIT Parallel Access REGISTER
1	74LS273	Eight bit LATCH
1	74LS283	4-bit Binary ADDER
2	74LS293	4-bit Binary COUNTER
1	GAL16V8	PLD
1	2732	4k x 8 EPROM
1	4016B	CMOS TRANSMISSION Gate
1	4023B	CMOS Triple 3-input NAND Gate
2	2114A	1k x 4 Static RAM
1	NE555	TIMER
2	DAC0808	8-bit D/A Converter (or use MC1408)
1	LM324	Op-Amp

Linear Integrated Circuits		
Quantity	IC	Description
1	NE555	Timer
2	DAC0808	8-bit digital-to-analog converter (same as MC1408)
2	MC1408	8-bit digital-to-analog converter (same as DAC0808)
1	LM324	Operational Amplifier

Resistors	Qty	Capacitors	Qty	Diodes	Qty
150 ohm	14	270 F	1	1N457	1
1 k-ohm	1	330 pF	1	1N914	2
2.2 k-ohm	2	560 pF	1	4.7 V zener	1
6.8 k-ohm	1	680 pF	1	LEDs	3
10 k-ohm	1	0.01 µF	1	FND507 7-segment	
18 k-ohm	1	1 µF	1	display units	2
33 k-ohm	1	100 µF	1		
180 k-ohm	1				

Potentiometers	Qty	Switches	Qty
1 K-ohm		SPDT	10
(ten-turn recommended)	2	Pushbutton	6
5 k-ohm			
(ten-turn recommended)	2		

Miscellaneous
Decimal or hexadecimal keyboard
SK-10 circuit board

NOTE: ICs and other components listed are available in bulk or in prepackaged kits from Electronix Express, a division of RSR Electronics, Telephone 1-800-972-2225.

CONTENTS

Preface v

EXPERIMENT	1	Preliminary Concepts 1
EXPERIMENT	2	Logic Gates I: OR, AND, and NOT 7
EXPERIMENT	3	Troubleshooting OR, AND, and NOT Gates 13
EXPERIMENT	4	Basic Combinatorial Circuits 27
EXPERIMENT	5	Logic Gates II: NOR and NAND 33
EXPERIMENT	6	Troubleshooting NOR and NAND Gates 39
EXPERIMENT	7	Boolean Theorems 47
EXPERIMENT	8	Simplification Using Boolean Theorems 51
EXPERIMENT	9	DeMorgan's Theorems 55
EXPERIMENT	10	The Universality of NAND and NOR Gates 61
EXPERIMENT	11	Implementing Logic Circuit Designs 67
EXPERIMENT	12	Exclusive-OR and Exclusive-NOR Circuits 71
EXPERIMENT	13	Designing with Exclusive-OR and Exclusive-NOR Circuits 77
EXPERIMENT	14	Troubleshooting Exclusive-OR and Combinatorial Circuits 81
EXPERIMENT	15	Flip-Flops I: Set/Clear Latches and Clocked Flip-Flops 85
EXPERIMENT	16	Flip-Flops II: D Latch; Master/Slave Flip-Flops 93
EXPERIMENT	17	Schmitt Triggers, One-Shots, and Astable Multivibrators 99
EXPERIMENT	18	Designing with Flop-Flop Devices 107
EXPERIMENT	19	Troubleshooting Flip-Flop Circuits 115
EXPERIMENT	20	Binary Adders and 2's Complement System 123
EXPERIMENT	21	Asynchronous IC Counters 131
EXPERIMENT	22	A BCD Counter 137
EXPERIMENT	23	Synchronous IC Counters 143
EXPERIMENT	24	IC Counter Application: Frequency Counter 151

EXPERIMENT 25	Troubleshooting Counters: Control Waveform Generation 157	
EXPERIMENT 26	Shift Register Counters 163	
EXPERIMENT 27	IC Registers 167	
EXPERIMENT 28	Designing with Counters and Registers 173	
EXPERIMENT 29	TTL and CMOS IC Families – Part I 177	
EXPERIMENT 30	TTL and CMOS IC Families – Part II 187	
EXPERIMENT 31	Using a Logic Analyzer 195	
EXPERIMENT 32	IC Decoders 207	
EXPERIMENT 33	IC Encoders 215	
EXPERIMENT 34	IC Multiplexers and Demultiplexers 221	
EXPERIMENT 35	Troubleshooting Systems Containing MSI Logic Circuits 231	
EXPERIMENT 36	IC Magnitude Comparators 235	
EXPERIMENT 37	Data Busing 239	
EXPERIMENT 38	Digital-to-Analog Converters 249	
EXPERIMENT 39	Analog-to-Digital Converters 255	
EXPERIMENT 40	Semiconductor Random Access Memory (RAM) 261	
EXPERIMENT 41	Synchronous Counter Design 271	
EXPERIMENT 42	Programmable Function Sequencer 277	

SUPPLEMENTAL EXPERIMENTS 283

EXPERIMENT S1	Logic Gates: OR, AND, and NOT	285
EXPERIMENT S2	Basic Combinatorial Circuits	291
EXPERIMENT S3	Logic Gates: NOR and NAND	303
EXPERIMENT S4	Boolean Theorems	311
EXPERIMENT S5	Simplification Using Boolean Theorems	323
EXPERIMENT S6	DeMorgan's Theorems	333
EXPERIMENT S7	Implementing Logic Gates and Circuits Using VHDL	343
EXPERIMENT S8	Implementing Logic Circuit Designs	351
EXPERIMENT S9	Exclusive-OR and Exclusive-NOR Circuits	361
EXPERIMENT S10	Designing with Exclusive-OR and Exclusive-NOR Circuits	369
EXPERIMENT S11	Implementing Exclusive-OR and Exclusive-NOR Circuits Using VHDL	387
EXPERIMENT S12	Latches and D-Type Flip-Flops	395
EXPERIMENT S13	J-K and T-Type Flip-Flops	415
EXPERIMENT S14	Flip-Flop Applications	429
EXPERIMENT S15	Implementing Flip-Flops and Flip-Flop Devices with VHDL	447
EXPERIMENT S16	Binary Adders and 2's Complement System	457
EXPERIMENT S17	Asynchronous Counters	471
EXPERIMENT S18	Synchronous Counters	483
EXPERIMENT S19	BCD Counters	497
EXPERIMENT S20	Shift Register Counters	507
EXPERIMENT S21	IC Registers	517
EXPERIMENT S22	One-Shots, Counters, and Registers with VHDL	531
EXPERIMENT S23	IC Decoders and Encoders	545
EXPERIMENT S24	IC Multiplexers and Demultiplexers	555
EXPERIMENT S25	VHDL State Machines	565
PROJECT SP1	Implementing a Simple Frequency Counter	579
Appendix A	Wiring and Troubleshooting Digital Circuits	589
Appendix B	Logic Analyzers	597
Appendix C	Manufacturers' Data Sheets	613
Appendix D	Using Quartus® II and Programming PLD Boards with Quartus® II Projects	641

Experiment 1

Name _____

PRELIMINARY CONCEPTS

OBJECTIVES

1. To observe differences between analog and digital devices.
2. To learn binary-to-decimal conversions.
3. To learn decimal-to-binary conversions.
4. To investigate basic pulse characteristics.

TEXT REFERENCES

Read sections 1.1 through 1.6 and section 2.5. Also read Appendix A of this manual.

EQUIPMENT NEEDED

Components
74HCT 147;
7404 IC;
74LS42 IC;
4 toggle switches;
4 LED monitors;
1 k-ohm resistors (9).

Instruments
volt-ohm-milliameter (VOM);
digital multimeter (DMM);
pulse or square wave generator (SWG);
oscilloscope;
0–5 volt DC power supply.

DISCUSSION

In this experiment, you will discover the major difference between analog and digital quantities by using both an analog and a digital measuring device to measure the output voltage of a power supply. Keep in mind that analog measuring devices display their measurements continuously on a scaled meter, which can be difficult to read with precision. On the other hand, digital devices display their measurements in steps using digits and therefore are capable of being read with much greater precision.

You will also observe how decimal quantities are represented in binary and, conversely, how binary numbers are represented by a decimal number by constructing special circuits called decimal-to-binary encoders and binary-to-decimal decoders. If this is your first experience working with integrated circuits (ICs), you should not be intimidated by your lack of knowledge of what is inside the IC. Concentrate on the experiment at hand, that is, to show that each decimal digit (0–9) has a unique binary representation, and that these same binary numbers have a unique decimal representation.

Finally, you will discover some of the characteristics of pulses and learn how to use an oscilloscope to measure these characteristics.

PROCEDURE

a) *Analog vs. digital:* Turn on the power supply and set it to +5 volts. Set the VOM to measure +5 volts DC, and connect the VOM probes to the output of the power supply, being careful to observe correct polarity. Note the smooth swing of the VOM needle as it swings from zero toward +5 volts. Note also that the amount of deflection of the needle is determined by, and therefore proportional to, the voltage at the VOM's probes. Record the value indicated by the VOM: __5.9__.

Now set the DMM to measure +5 volts DC, and touch the probes to the output of the power supply, being careful to observe correct polarity. Note that the readout displays its measurement in one or more values before it settles near +5 volts. In other words, it displays increasing values in a number of steps. Finally, note that voltage is represented by digits. Record the readout value: __5.01__.

b) *Decimal-to-binary:* In this step, a 74HCT IC and a 7404 IC will be combined to convert decimal digits to binary coded decimal (BCD). Refer to the data sheet for the 74HCT 147 and 7404, and draw the pin layout diagram for each:

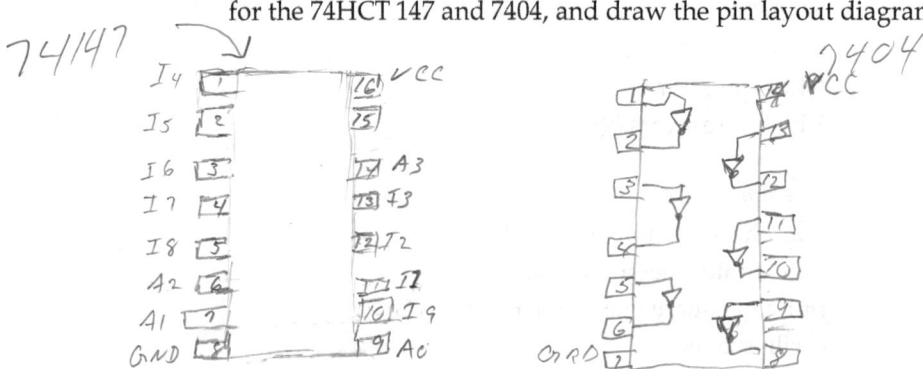

Use the 74HCT 147 and 7404 and connect the circuit in Figure 1-1 according to the procedure outlined in Appendix A. There are nine inputs to this circuit (I_1–I_9), each

PROCEDURE

3

SW = I
A = 0

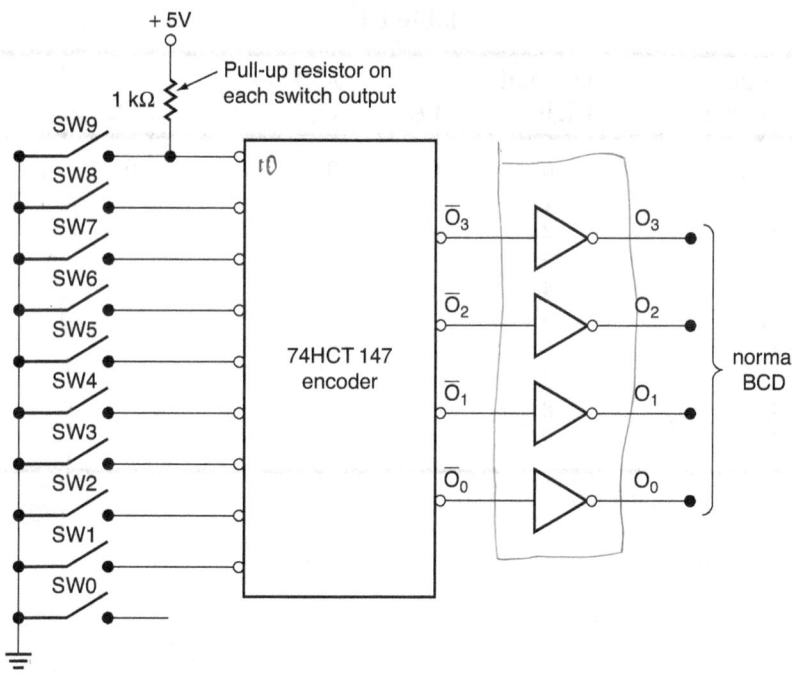

Figure 1-1

representing a decimal digit (1–9, respectively) and each connected through an spdt switch to either ground (input activated) or V_{cc} (input-deactivated, or normal). If nine spdt input switches are not available, leave all input connections open, then connect a single wire to ground. You will use this wire to activate each input as is called for below by touching the wire to the appropriate input pin on the 74HCT 147.

The BCD output of this circuit will be monitored by connecting an LED monitor, DMM, or VOM to the outputs of the circuit labelled O_0–O_3. If LED monitors are used to monitor the output, you will interpret a lighted LED as a binary 1 and an unlighted LED as a binary 0. If a DMM or VOM is used, interpret each output as follows:

$$0 \text{ to } +0.8 \text{ volt} = \text{binary } 0$$
$$+2 \text{ to } +5 \text{ volts} = \text{binary } 1$$

The Least Significant Bit (LSB) of the output is O_0; the Most Significant Bit (MSB) is O_3. The output of this circuit is normally 0000 when none of the 74HCT 147 inputs are activated. Turn on the power supply and check to see that this is the case. If so, then proceed with the experiment; otherwise, turn off the power supply, check your circuit wiring, correct any faults, and repeat this step.

Once you have verified that the circuit is initially working correctly, activate each of the inputs, one at a time, and record the BCD output observed in Table 1-1 on page 4.

Table 1-1

Input Activated	Decimal Digit	BCD Output			
		O_0	O_1	O_2	O_3
None	0	0	0	0	0
I_1	1	0	0	0	1
I_2	2	0	0	1	0
I_3	3	0	0	1	1
I_4	4	0	1	0	0
I_5	5	0	1	0	1
I_6	6	0	1	1	0
I_7	7	0	1	1	1
I_8	8	1	0	0	0
I_9	9	1	0	0	1

c) *Binary-to-decimal:* Refer to the data sheet for the 74LS42 IC, and draw the pin layout diagram:

Turn the power supply off, and disassemble the circuit used in step b. Use the 74LS42, and connect the circuit in Figure 1-2, which is a circuit for converting BCD to decimal. Connect a separate spdt switch to each of the inputs A, B, C, and D of the 74LS42 so that each can be switched to ground (binary 0) or V_{cc} (binary 1). Input A is LSB, and input D is MSB. To monitor the outputs, connect an LED monitor, DMM, or VOM to each 7442 output O_0–O_9, representing decimal digits 0–9, respectively. One (and only one) output will be activated (indicated by an unlighted LED or a reading of near 0 volts DC) for each combination of input switch conditions. Use the binary equivalences given in step b on the previous page to interpret the input and output levels of the 74LS42.

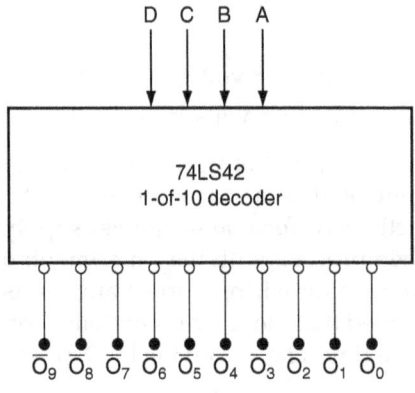

Figure 1-2

PROCEDURE

Set all of the input switches to the ground position (all inputs binary 0). Monitor the O_0 output (pin 1 of the 74LS42): it should be activated. Now check all of the other outputs O_1–O_9 (pins 2–7 and 9–11, respectively), and verify that they are all inactive (all LED monitors connected to these outputs should be lighted or show voltage readings 2.4 V). If this is not the case, recheck the input switches. If the switches are all set to ground position, turn the power supply off and recheck your circuit wiring. If a fault is found, correct it, and repeat this step. If you have problems getting the right results, call your instructor.

Refer to Table 1-2, and toggle the input switches so that each BCD number from 0001 to 1001 is entered into the 74LS42; recording the output conditions for each BCD number in the table.

d) *Pulse characteristics:* Set up an oscilloscope as follows:

DC vertical input at 1 volt/division

Horizontal sweep rate at 1 millisecond/division

Table 1-2

BCD Input				74LS42 Output Conditions										Decimal Digit
D	C	B	A	O_0	O_1	O_2	O_3	O_4	O_5	O_6	O_7	O_8	O_9	
0	0	0	0	0	1	1	1	1	1	1	1	1	1	0
0	0	0	1	1	0	1	1	1	1	1	1	1	1	1
0	0	1	0	1	1	0	1	1	1	1	1	1	1	2
0	0	1	1	1	1	1	0	1	1	1	1	1	1	3
0	1	0	0	1	1	1	1	0	1	1	1	1	1	4
0	1	0	1	1	1	1	1	1	0	1	1	1	1	5
0	1	1	0	1	1	1	1	1	1	0	1	1	1	6
0	1	1	1	1	1	1	1	1	1	1	0	1	1	7
1	0	0	0	1	1	1	1	1	1	1	1	0	1	8
1	0	0	1	1	1	1	1	1	1	1	1	1	0	9

Auto triggering

Negative trigger slope

Set up a pulse, function, or square wave generator to 200 Hz. The output of the generator should be TTL compatible (i.e., LOW level of pulse should be 0 to +0.8 volt and HIGH level of pulse should be +2.0 to +5.0 volts). Connect the output of the generator to both the vertical and external trigger inputs of the scope. Adjust the generator frequency until the oscilloscope display shows two repetitions of the generated pulse. Sketch the display on Timing Diagram 1-1.

Adjust the vertical position control so that the leading edge of a positive-going pulse intersects a convenient vertical graticule at 2.5 volts (50% point). Measure the distance between this point and the 50% point of the trailing edge

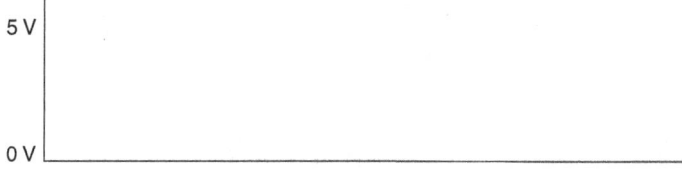

Timing Diagram 1-1

along the horizontal graticule. Record your result: _____. Multiply the distance observed by the setting on the horizontal sweep control. Record your calculation: _____. The calculated result is the *pulse duration* (t_p). Now measure the distance between the leading edges of two consecutive pulses, and convert the distance into time by multiplying by the horizontal sweep setting. Record your computation: _____. This measurement is called the *period* (T) of the waveform. The duty cycle of the waveform can now be calculated using the formula:

$$\text{Duty Cycle} = \frac{(t_p)}{T} \times 100\%$$

Record your computation for the duty cycle: _____.

Set the pulse generator to 1 kHz, and adjust the oscilloscope to display the leading edge of a positive-going pulse. Use the highest horizontal speed possible (magnified, if available). Adjust the oscilloscope's positioning controls so that the 0.5 V (10%) point of the pulse intersects a convenient vertical graticule. Estimate the distance between this point and a vertical line passing through both the 4.5 V (90%) point of the pulse and the main horizontal graticule. Multiply your measurement by the horizontal sweep speed, and record your computation: _____. This is the pulse's *rise time* (t_r). Repeat this procedure on the trailing edge of the pulse. Record your measurement: _____. This is the *fall time* of the pulse (t_f).

One more important pulse characteristic is *propagation delay*. This measurement will be performed in a future experiment.

e) *Review:* This concludes the exercises on preliminary concepts of digital systems. To test your understanding of the principles covered by this experiment, complete the following statements:

1. Based on your observations [analog, *digital*] voltmeters are the easiest to read because voltage is represented by _____. [Analog, Digital] representations are continuous (smooth) while [analog, *digital*] representations are discrete (in steps).
2. In this experiment, a binary 0 is represented by *High* volts. An LED monitor is [*unlighted*, lighted] when indicating a binary 0.
3. In this experiment, a binary 1 is represented by *Low* volts. An LED monitor is [unlighted, lighted] when indicating a binary 1.
4. When input I_5 of the 74HCT147 IC is activated, the outputs of the IC will indicate a binary _____.
5. A 74LS42 IC has a BCD input of 0110. Its output will indicate a decimal *6*.
6. For a series of pulses, T = 8 microseconds and t_p = 3 microseconds. The duty cycle of the waveform is *37.5* %.

$$\frac{t_p = 3}{T = 8} \times 100$$

Experiment 2

Name _____

LOGIC GATES I: OR, AND, AND NOT

OBJECTIVES

1. To investigate the behavior of the OR gate.
2. To investigate the behavior of the AND gate.
3. To investigate the behavior of the NOT gate (inverter).

TEXT REFERENCES

Read sections 3.1 through 3.5.

EQUIPMENT NEEDED

Components
7404 IC;
74LS11 IC;
7432 IC;
4 toggle switches;
1 LED monitor.

Instruments
VOM;
0–5 volt DC power supply;
logic probe (optional).

DISCUSSION

In general, logic gates have one or more inputs and only one output. The gates respond to various input combinations, and a truth table shows this relationship between a gate's input combinations and its output. The truth table for a particular gate explains how the circuit behaves under normal conditions. Familiarization with a logic gate's truth table is essential to the technologist or technician before he or she can design with or troubleshoot it.

In this experiment, three logic gates are covered: the OR, AND, and NOT gates. The OR operation can be summarized as follows:

1) When any input is 1, the output is also 1.
2) When all inputs are 0, the output is also 0.

The AND operation can be summarized similarly:

1) When any input is 0, the output is also 0.
2) When all inputs are 1, the output is also 1.

Finally, the NOT operation is said to be complementary. In other words:

1) If the input is 0, the output is 1.
2) If the input is 1, the output is 0.

You should recall that the logic levels, 0 and 1, have voltage assignments. For TTL circuits, a logic 0 can be anywhere from 0 V to +0.8 V, and a logic 1 is in the range of +2.0 V to +5.0 V.

PROCEDURE

a) *The OR gate:* Refer to the data sheet for the 74LS32 IC and draw its pin layout diagram:

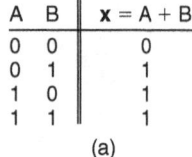

(a)

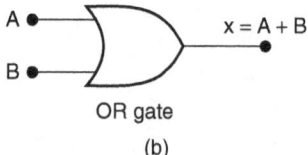

OR gate

(b)

Figure 2-1

PROCEDURE

b) Figure 2-1 shows the logic symbol and truth table for the OR logic gate. The 74LS32 contains four of these gates. Wire one of them as follows:

1) V_{cc} to +5 volts; GND to power ground.
2) Inputs A and B to separate toggle switches.
3) The output to the VOM.

Table 2-1

Data Switches A	B	VOM (Volts)	Logic Level (0/1)
0	0	0.15	0
0	1	4.40	1
1	0	4.40	1
1	1	4.40	1

c) You will now verify the OR operation by setting inputs A and B to each set of logic values listed in the truth table of Figure 2-1, recording the output voltage observed, and converting the output voltage to a logic level. Use

$$0\,V - 0.8\,V = 0 \text{ and } 2\,V - 5\,V = 1$$

for the conversions and record your observations in Table 2-1.

d) Disconnect the VOM from the circuit, and use an LED monitor to observe the output. Repeat step c using the conversion rule LED OFF (unlighted) = 0 and LED ON (lighted) = 1.

e) Disconnect one of the inputs, and set the remaining one to 0. Is the output level 0 or 1? Based on your observation and knowledge of the OR operation, what level does the unconnected input act like? __1 High__

f) *The AND gate:* Refer to the data sheet for the 74LS11 IC, and draw its pin diagram:

g) Figure 2-2 shows the logic symbol and truth table for the three input AND logic gate. The 74LS11 contains three 3-input AND gates. Wire one of the AND gates as follows:

1) V_{cc} to +5 V and GND to power ground.
2) Inputs A, B, and C to toggle switches.
3) Output to an LED monitor.

A	B	C	x = ABC
0	0	0	0
0	0	1	0
0	1	0	0
0	1	1	0
1	0	0	0
1	0	1	0
1	1	0	0
1	1	1	1

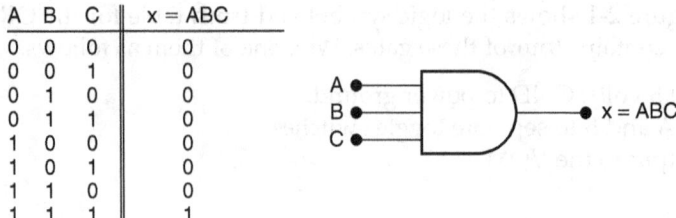

Figure 2-2

Table 2-2

Data Switches			Output LED Monitor (On/Off)	Output Logic Level (0/1)
A	B	C		
0	0	0	OFF	0
0	0	1	OFF	0
0	1	0	OFF	0
0	1	1	OFF	0
1	0	0	OFF	0
1	0	1	OFF	0
1	1	0	OFF	0
1	1	1	ON	1

h) You will now verify the AND operation by setting toggle switches A, B, and C to each set of input values of the truth table in Figure 2-2 and recording the output level observed on the LED monitor using Table 2-2.

i) Disconnect input A from the toggle switch, and set inputs B and C to 1. Note the logic level indicated by the LED monitor. Based on your observation, what logic level does the unconnected input act like? __High__

j) *The NOT gate:* Mount a 7404 IC on the circuit board. Refer to the data sheet for the 7404, and draw its pin layout diagram:

PROCEDURE

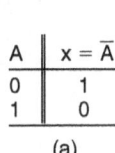
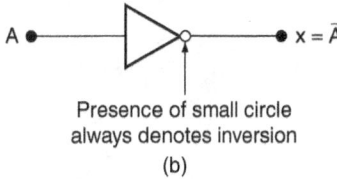

(a) (b)

Figure 2-3

Table 2-3

Data Switch A	Output LED Monitor (On/Off)	Output Logic Level (0/1)
0	on	1
1	oFF	0

k) Figure 2-3 shows the logic symbol and truth table for the NOT gate, also commonly called an inverter. The 7404 IC contains six inverters. Connect one of the inverters as follows:

1) V_{cc} to +5 V and GND to power ground.
2) Input A to a toggle switch.
3) Output to an LED monitor.

l) You will now verify the NOT operation given by the truth table in Figure 2-3 and record your observations in Table 2-3.

m) *Review:* This concludes the investigation of basic logic gate operation. To test your understanding of the logic gates, complete the following statements:

1. The output of an OR gate is LOW only when _All are off_.
2. The output of an AND gate is _oFF_ whenever any input is LOW.
3. The output of an inverter is always _Inverse_ the input.
4. Using the results obtained in step h, one could conclude that to use a three-input AND gate as a two-input gate, one input should be connected to [V_{cc}, GND].
5. If an OR gate input were accidentally shorted to V_{cc}, the output of the gate would always be _High_, no matter what level the other input level might be.

Experiment 3

Name _____

TROUBLESHOOTING OR, AND, AND NOT GATES

OBJECTIVES

1. To test the operation of a logic probe.
2. To test the operation of a logic pulser.
3. To troubleshoot OR, AND, and NOT gates with pulser and probe.

TEXT REFERENCES

Read sections 4.9 through 4.13 and Appendix A of this manual.

EQUIPMENT NEEDED

Components
7404 IC;
74LS08 IC;
7432 IC;
4 toggle switches.

Instruments
DMM or VOM;
0–5 volt DC power supply;
oscilloscope;
logic probe;
logic pulser.

DISCUSSION

Most digital systems incorporate numerous IC logic gates into their circuitry. IC gates are very reliable, but like all electronic devices they fail, and when they do fail, they must be isolated and replaced. Troubleshooting gates and their faults is essential to learning how to troubleshoot a digital system. The techniques employed at the gate level can be used to troubleshoot larger devices.

In this experiment, you will learn to use a logic probe and a logic pulser together to "discover" some typical gate faults. These two tools are probably the most popular of all digital test equipment. They are easy to use, small, and lightweight. They get their power from the power supply of the device being serviced. Figure 3-1 shows a logic probe, and Figure 3-2 shows a logic pulser. Descriptions of these devices can be found in Appendix A of this manual.

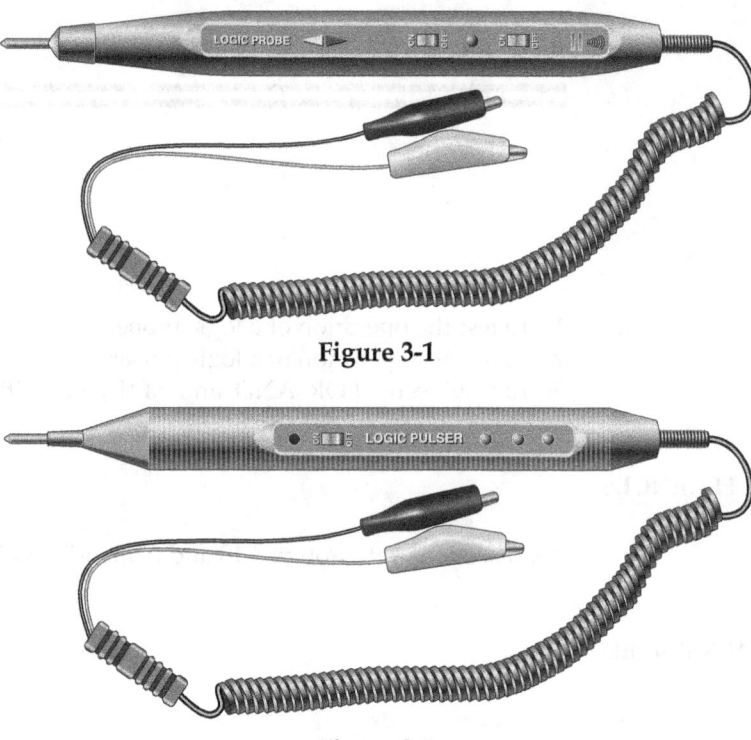

Figure 3-1

Figure 3-2

LOGIC PROBES

A logic probe is a small, hand-held instrument used to indicate the logic level at a point in a digital circuit. It is capable of indicating a logic 0, logic 1, and a level floating between logic 0 and 1. In many cases it is capable of detecting the presence of high speed pulses. Figure 3-3 shows a typical application of the logic probe, that is, static testing a logic gate by monitoring its output while the gate's inputs are switched through their various combinations.

LOGIC PULSERS

A logic pulser generates digital pulses. Like the logic probe, the pulser uses the logic power supply to get its own power. The tip of the pulser is placed on a circuit node where an injected pulse is desired. The pulser senses the logic state of the

PROCEDURE

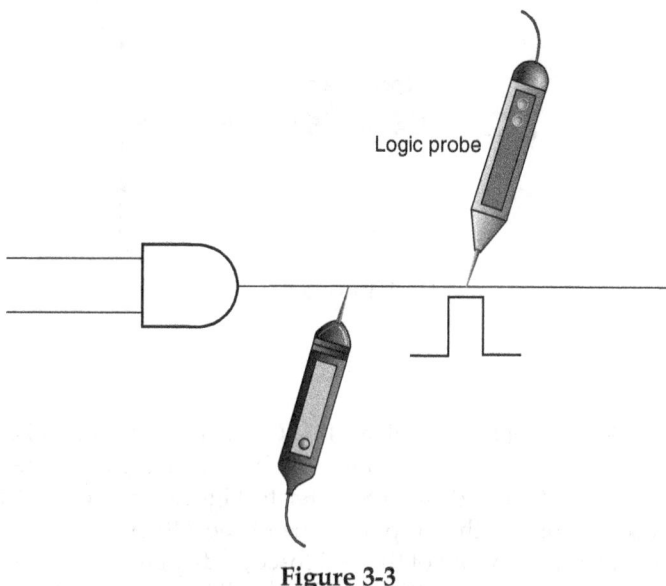

Figure 3-3

node and generates a pulse that will attempt to drive the node to the opposite state. This is a valuable aid in troubleshooting, since it permits the triggering of gates and other devices without removing them from their circuits.

PROCEDURE

a) *Testing the logic probe:* First, make sure the lab power supply is OFF, then set the TTL/CMOS switch to TTL, and connect the logic probe power leads to the logic power supply of the device under test. For this experiment, we will be troubleshooting TTL IC gates, so the logic power supply will be our 0–5 volt DC lab power supply. Make sure to connect the red clip to +5 volts and the black clip to power supply ground. Turn the lab power supply ON. If your probe has a single LED to indicate logic levels, the LED should be blinking to indicate that the level at the probe tip is floating. (If your probe has two LEDs to represent the two logic levels, both should be out.) If you have a DMM available, connect its leads between the logic power ground and the probe tip. The DMM should read a value between 1.3 and 1.5 volts. This value is called by various names—"indeterminate," "bad," "invalid," and "floating." All refer to the fact that if a voltage value falls into a range of greater than 0.8 volts and less than 2.0 volts, the value cannot represent a logic 0 or a logic 1. These values all have a tolerance of 20%.

Refer now to Figure 3-4. Turn the power supply off and connect a 1 k-ohm linear potentiometer and DMM to the power supply as shown. Set the potentiometer so that the DMM reads 0 volts. Touch the probe to the center tap, and turn the power supply ON. Its LED(s) should indicate logic 0. Now slowly turn the potentiometer away from ground and toward V_{cc}. When the LED(s) indicate a floating level, you have reached the top end of the range for logic 0. Now slowly continue to turn the potentiometer toward V_{cc}, this time until the LED(s) indicate a logic 1. This is the bottom end of the range for logic 1.

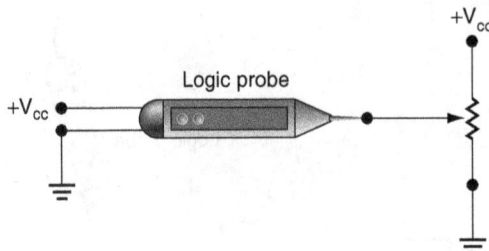

Figure 3-4

b) *Testing the logic pulser:* Install a 7404 IC, a 74LS08 IC, and a 7432 IC on the circuit board. Connect V_{cc} and ground to each IC. For this part of the experiment, we will use only the 7404 and 74LS08. Refer to Figure 3-5. Connect one vertical input of the oscilloscope to the output of one of the 7404 inverters. Select a horizontal graticule near the bottom of the oscilloscope display, and adjust the trace of the oscilloscope to that graticule. This marking will be your LOW reference.

Connect the black power connector of the pulser to the logic power supply ground and the red to V_{cc}. Place the pulser tip on the inverter output that is being observed. Adjust the oscilloscope timing until a stable pulse is displayed. You should note that the pulse is a positive-going pulse.

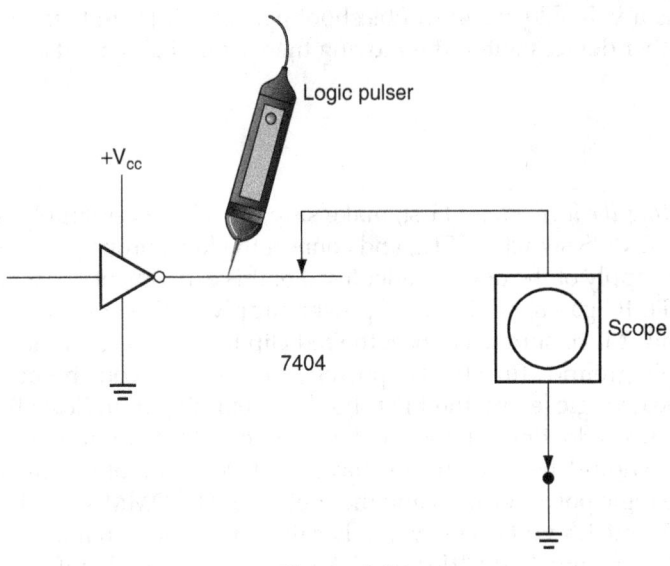

Figure 3-5

Refer now to Figure 3-6. Place the oscilloscope input on the output of one 74LS08 AND gate. Adjust the oscilloscope trace to a suitable horizontal graticule near the top of the screen. This will be your HIGH reference. Touch the pulser tip to the AND output being observed, and adjust the oscilloscope timing until a stable pulse is displayed. You should note that the pulse is negative-going. This confirms that the pulser will attempt to drive a node to its opposite state.

PROCEDURE

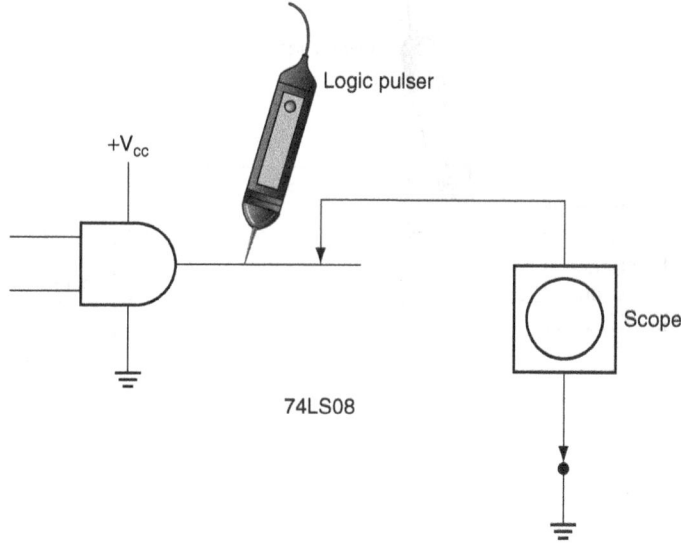

Figure 3-6

Finally, refer to Figure 3-7. Connect the oscilloscope input to V_{cc}. Adjust the oscilloscope trace to a convenient HIGH reference graticule. Touch the pulser tip to V_{cc}. You should note that the pulser cannot drive V_{cc} to a logic 0. Now connect the oscilloscope input to ground, and adjust the oscilloscope trace to a convenient LOW reference graticule. Touch the pulser to ground, and note that it cannot drive ground to a logic 1.

c) *Using the logic probe and logic pulser to troubleshoot gates:* In what follows, the probe-pulser combination will be used to aid in locating certain types of faults of IC gates. We will not look at every possible type of fault, but we can simulate most of the common ones and develop a procedure for testing for them. The faults will be shaded in the figures.

A word of caution: Some of the fault simulations call for shorting gate inputs together and outputs together. **Never** short inputs together that are all connected to your trainer's data switches. **Never** short a trainer data switch directly to ground or to V_{cc}. **Never** short any gate output to ground or to V_{cc}. **Do** exercise care in following the instructions below and elsewhere in this manual. Voltages at shorted LS-TTL outputs cannot be measured with any consistency, so you will not test the 74LS08 with this condition.

OR Gates

1) Refer to Figure 3-8, and wire a 7432 OR gate as shown. Recall that when all inputs to an OR gate are the same, the output is the same as the inputs. Touch the probe tip to one of the inputs and the pulser to the other input. You should observe that the pulsing on one of the inputs is detected by the probe on the other. Normally, unless the two inputs are purposely wired together, you should not observe this condition.

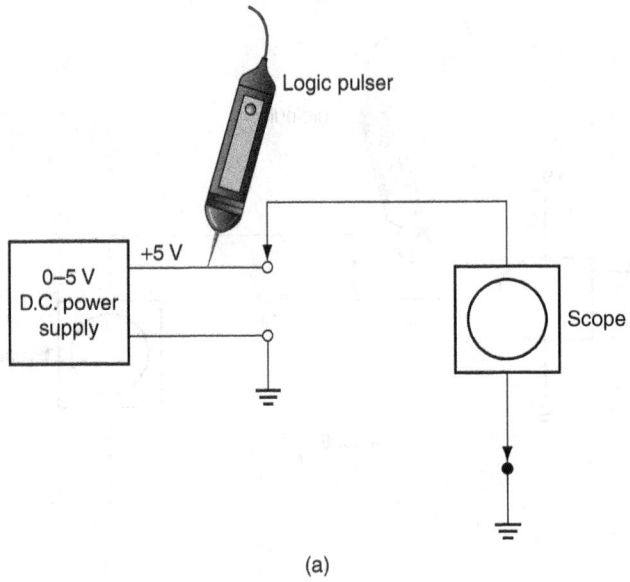

(a)

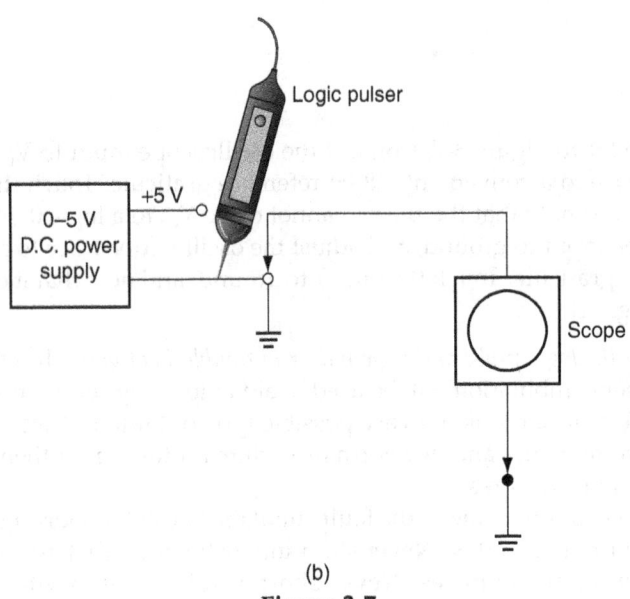

(b)

Figure 3-7

2) Refer to Figure 3-9, and wire a 7432 OR gate as shown. Connect point A to a data switch, but leave points B and C disconnected. Place the probe tip on point C, and observe that the output is HIGH. Toggle the data switch, and observe that it has no effect on the output. Now place the probe tip on point D. You should observe that the level there is floating. This tells you that either the input (point D) has an external open or an internal open. To verify that the open is external, place the pulser tip to point D and the probe tip to point C. Set the data switch at A to LOW. You should observe that the probe now indicates a pulsing condition. If the open had been internal, you would still see a HIGH at point C.

PROCEDURE

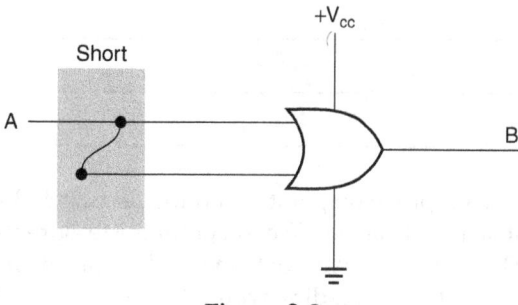

Figure 3-8

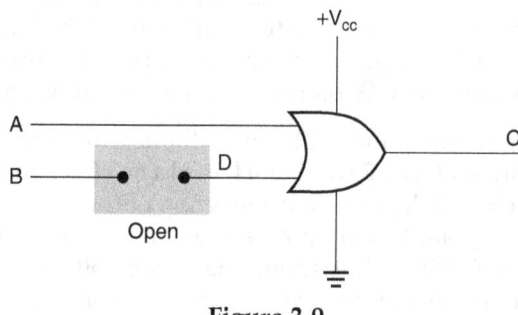

Figure 3-9

3) Refer to Figure 3-10, and wire three 7432 OR gates as shown. Note that a short between the outputs of gate 3 and gate 2 has been wired purposely. Connect data switches to inputs A, B, C, and D. Set switches A and B to LOW and switches C and D to HIGH. Place the tip of the logic probe on output E. You should observe that the level there is LOW. The HIGH output of gate 2 has no effect on the output of gate 1. This is because one of gate 1's inputs is LOW, and the LOW output of gate 3 is pulling the level at the other input of gate 1 down to LOW. Now toggle switch B several times while observing the output at E with the logic probe. You should observe that the output at E toggles also.

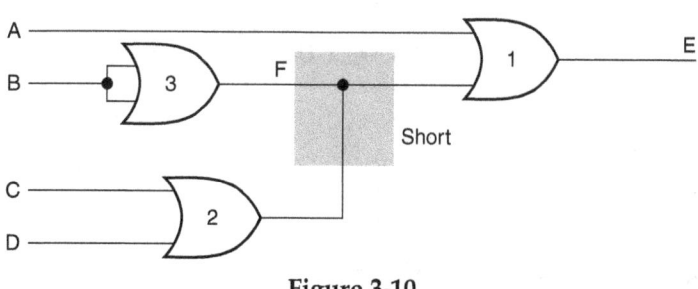

Figure 3-10

Set switches C and D to LOW. Now toggle switch B several times, again monitoring output E with the logic probe. You should observe that the logic probe indicates a constant LOW. Why?

_____

_____

_____

Place the tip of your probe on point F, and toggle switch B several times. You should observe that switch B has no effect on point F. Measure the voltage at point F with the DMM. The voltage at that point should be near the midrange value for a LOW, which is 0.4 V. A typical reading would be about 0.38 V. Remove the short between the outputs of gate 2 and gate 3, and measure the voltage at point F. It should now measure less than 0.4 V, typically 0.15 V. Set switch B to HIGH and measure the voltage at point F with the DMM. Record this voltage: _____.

Finally, replace the short between the outputs of gate 2 and gate 3. Place the tip of the logic probe on the output of gate 1 and the tip of the logic pulser on point F. You should note that the logic probe indicates a pulsing output at E. Thus, the logic pulser can overcome a LOW produced by the output of a gate.

4) Refer to Figures 3-11 and 3-12. You will not wire these circuits. In step 3 above, when the output of gate 2 was HIGH, it had no effect on the output of gate 3. Switch B could be toggled, producing a toggling output at E if switch A was set to 0. A short to V_{cc} at point F could not be overcome by the output of gate 3. So point F would be stuck HIGH. Measuring the actual voltage level at point F can help you determine whether or not a circuit node is shorted to V_{cc}, since the outputs of TTL gates are normally lower than V_{cc}. A logic pulser cannot overcome a short to V_{cc}.

If point F measures 0 V, then it is likely that point F is shorted to ground. Also, a logic pulser cannot overcome this type of short.

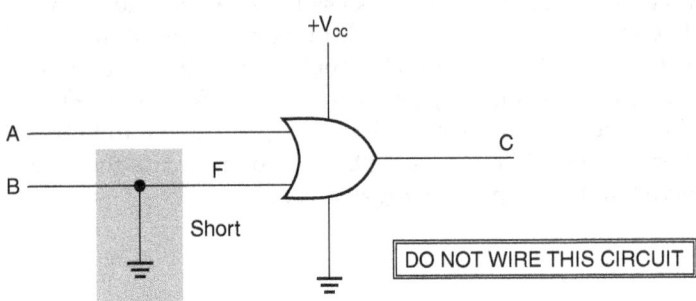

Figure 3-11

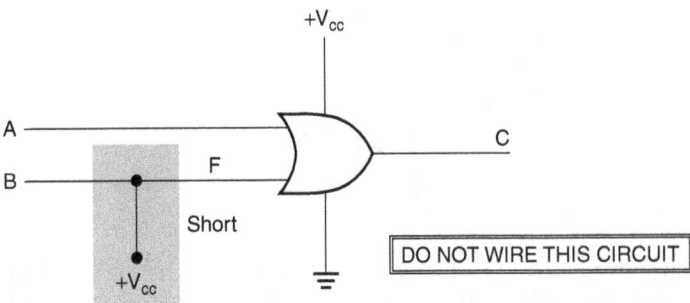

Figure 3-12

PROCEDURE

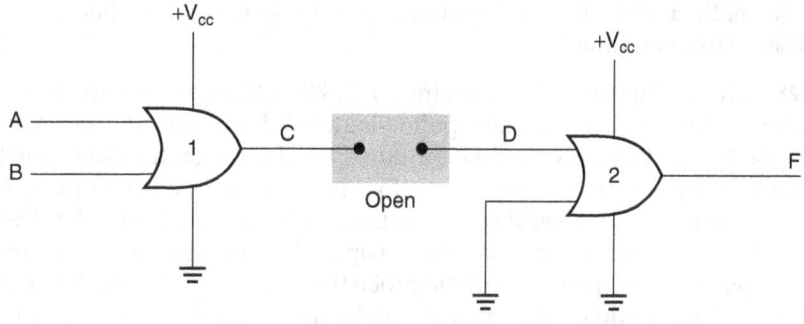

Figure 3-13

5) Refer to Figure 3-13, and wire two 7432 OR gates as shown. Leave points A and B disconnected. We will assume that gate 2 is functioning normally. Place your probe tip on point D. It should indicate a floating level. This is symptomatic of an open in the output circuit of gate 1. The question is whether it is an external or internal open. Place the probe on the output pin of gate 1. If it is HIGH, then there is an external open somewhere between the IC and point C. If the output pin is floating, then an internal open in gate 1's output circuit could be the cause.

AND Gates

1) Refer to Figure 3-14, and wire a 74LS08 AND gate as shown. Recall that when all inputs to an AND gate are the same, the output is the same as the inputs. Touch the probe tip to one of the inputs and the pulser to the other input. You should observe that the pulsing on one of the inputs is detected by the probe on the other.

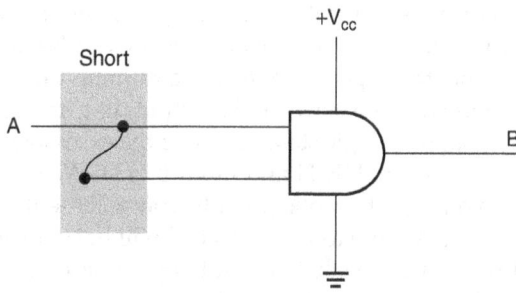

Figure 3-14

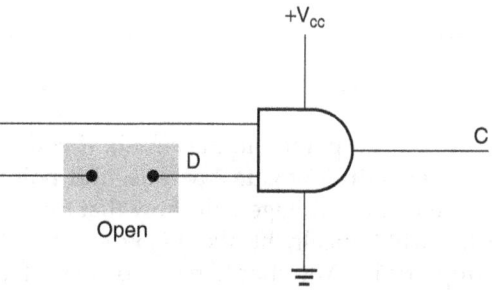

Figure 3-15

Normally, unless the two inputs are purposely wired together, you should not observe this condition.

2) Refer to Figure 3-15, and wire a 74LS08 AND gate as shown. Connect point A to a data switch, but leave points B and C disconnected. Place the probe tip on point C, and observe that the output is HIGH. Toggle the data switch, and observe that the output acts normally. Now place the probe tip on point D. You should observe that the level there is floating. This tells you that either the input (point D) has an external open or an internal open. To verify that the open is external, place the pulser tip on point D and the probe tip on point C. Set the data switch at A to HIGH. You should observe that the probe now indicates a pulsing condition. If the open had been internal, you would still see a HIGH at point C.

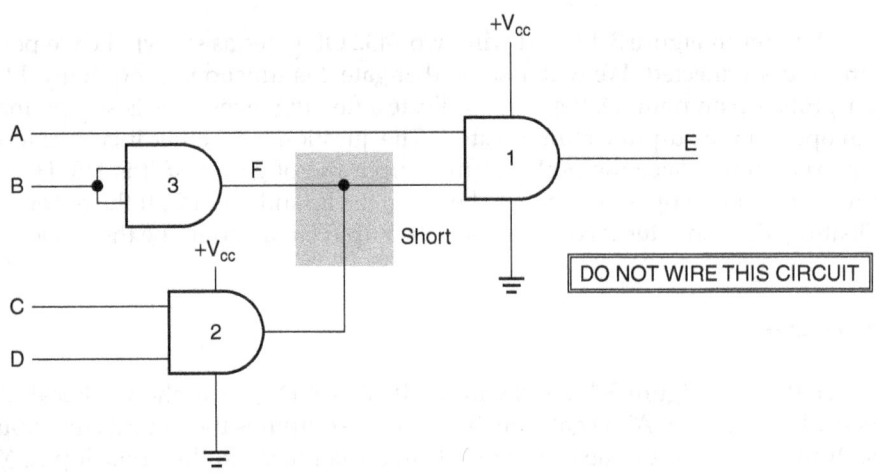

Figure 3-16

3) Refer to Figure 3-16, but do not wire this circuit. The shorted outputs cause higher than specified current in any gate, and consequently more heat, but LS logic has a lower output current specification and is more likely to be destroyed by the excessive current. From your knowledge of how the logic pulser and probe work, you should be able to predict the expected outcome assuming that the circuit was connected. Note that a short is shown between the outputs of gates 2 and 3, so they could try to drive point F to opposite logic states, causing the excessive current mentioned. Assume inputs A, C, and D are HIGH and input B is LOW, and that the circuit has this short in place. If you were to probe point E, you will see that it would be LOW because point F will be LOW due to gate 3. Gate 2 will have no effect on the output. If you toggle input B, you will see that output E toggles also. Can you explain why?

Assume that at least one input to gate 2 is LOW. If you toggle B, the output at E will remain LOW. Why?

_____

_____

_____

Assume you now probed point F in the circuit. If you toggle B, you will see that it has no effect on the output of gate 3 because this point is being held LOW by gate 2. A measurement of the voltage will show that this voltage is not as LOW as you might expect: It will typically be about 0.39 V. If the short is removed, the LOW voltage will drop to 0.1 V. You should be suspicious of a shorted node if you observe this condition.

PROCEDURE

Now, wire the circuit using only gates 1 and 3 and without including the short. Place the tip of the logic probe on the output of gate 1 and the tip of the logic pulser on point F. You should note that the logic probe indicates a pulsing output at E. Thus, the logic pulser can overcome a LOW produced by the output of a gate.

4) Refer to Figures 3-17 and 3-18. You will not wire these circuits. In step 3, when the output of gate 2 was HIGH, it had no effect on the output of gate 3. Switch B could be toggled, producing a toggling output at E. A short to V_{cc} at point F could not be overcome by the output of gate 3, so point F would be stuck HIGH. Measuring the actual voltage level at point F can help you determine whether or not a circuit node is shorted to V_{cc}, since the outputs of TTL gates are normally lower than V_{cc}. A logic pulser cannot overcome a short to V_{cc}.

If point F measures 0 V, then it is likely that point F is shorted to ground. A logic pulser cannot overcome this type of short either.

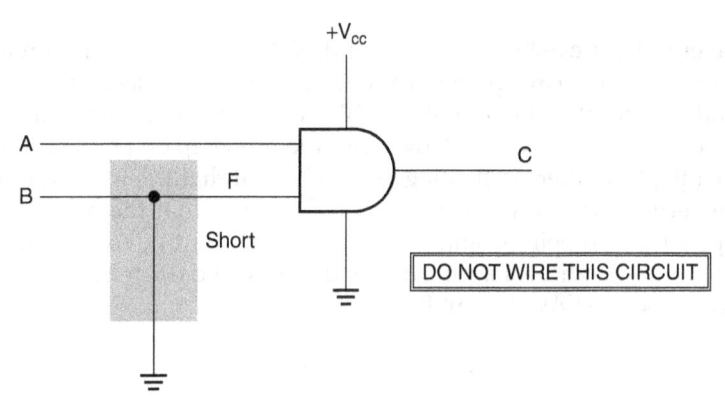

Figure 3-17

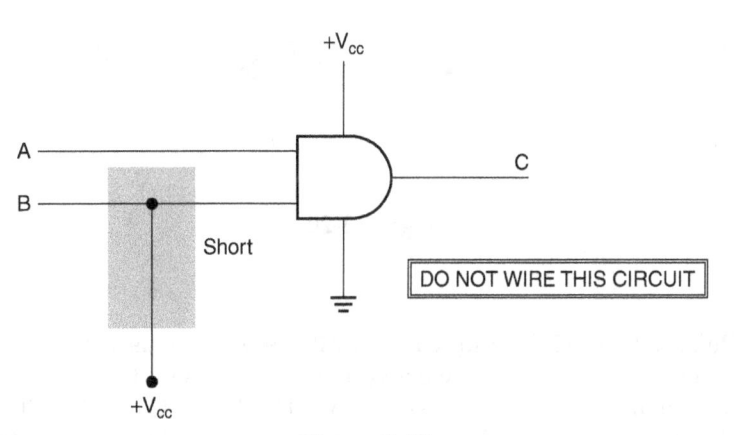

Figure 3-18

5) Refer to Figure 3-19, and wire two 74LS08 AND gates as shown. Leave points A and B disconnected. We will assume that gate 2 is functioning normally. Place your probe tip on point D. It should indicate a floating level. This is symptomatic of an open in the output circuit of gate 1. The question is whether it is an external or internal open. Place the probe on the output pin of gate 1. If it is HIGH, then there is an external open somewhere between the IC and point C. If the output pin is floating, then an internal open in gate 1's output circuit could be the cause.

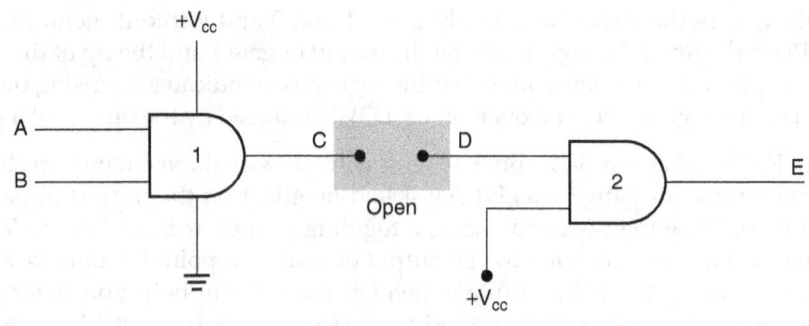

Figure 3-19

NOT Gates

1) Refer to Figure 3-20, and wire a 7404 NOT gate as shown. Connect point A to a data switch, but leave points B and C disconnected. Place the probe tip on point C, and observe that the output is LOW. Toggle the data switch, and observe that it has no effect on the output. Now place the probe tip on point C. You should observe that the level there is floating. This tells you that the input (point C) has either an external open or an internal open. To verify that the open is external, place the pulser tip to point C and the probe tip on point B. You should observe that the probe now indicates a pulsing condition. If the open had been internal, you would still see a LOW at point B.

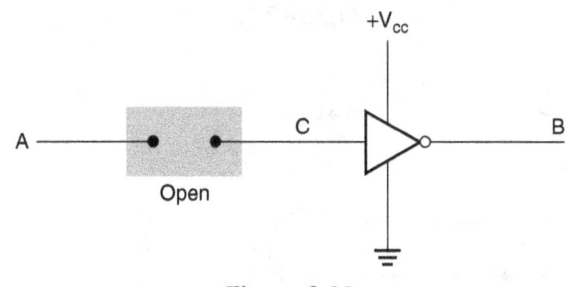

Figure 3-20

2) Refer to Figure 3-21, and wire two 7404 NOT gates as shown. Note that a short between the outputs of the two inverters has been wired purposely. Connect data switches to inputs A and C. Set switch A to HIGH and C to LOW. Place the tip of the logic probe on output B. You should observe that the level there is LOW. The HIGH output of gate 2 has no effect on the output of gate 1. This is because the LOW output of gate 1 is pulling the level of the output of gate 2 down to LOW. Now toggle switch A several times while observing the output at B with the logic probe. You should observe that the output at B toggles also.

Set switch C to HIGH. Now toggle switch A several times, again monitoring output B with the logic probe. You should observe that the logic probe indicates a constant LOW. Why?

_____

_____

PROCEDURE

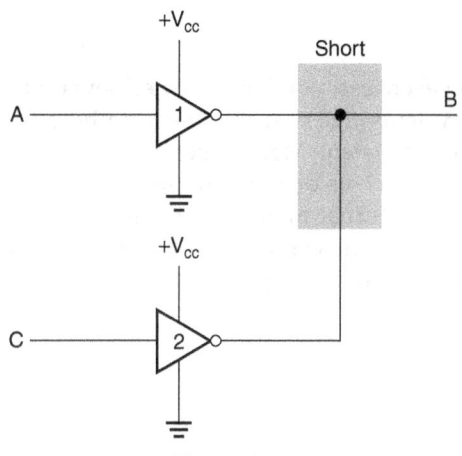

Figure 3-21

Measure the voltage at point B with the DMM. The voltage at that point should be near the midrange value for a LOW, which is 0.4 V. A typical reading would be about 0.39 V. Remove the short between the outputs of gate 1 and gate 2, and measure the voltage at point B. It should now measure considerably less than 0.4 V, typically 0.10 V. Set switch A to LOW and measure the voltage at point B with the DMM. Record this voltage: _____

Finally, replace the short between the outputs of gate 2 and gate 3. Place the tips of the logic probe and the logic pulser on point C. You should note that the logic probe indicates a pulsing output at E. Thus, the logic pulser can overcome a LOW produced by the output of a gate.

3) Refer to Figures 3-22 and 3-23. Do not wire these circuits. As with the AND and OR gates above, shorts to V_{cc} and ground can be found by making a voltage measurement at the point that is stuck. You should never measure exactly V_{cc} or ground unless there is a short to one or the other.

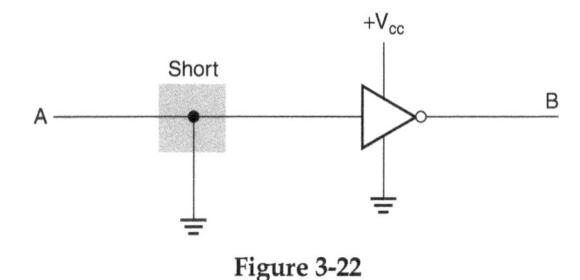

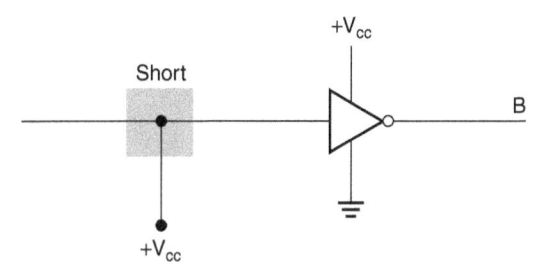

Figure 3-22

Figure 3-23

Summary

In the preceding exercises, you have learned several techniques for isolating problems in digital systems. These techniques can help you to isolate the problems to a small area in the system. Many times, you will hit it lucky and isolate one or more defective ICs in sockets by using these techniques. Other times, you will isolate several defective ICs that must be desoldered, removed, tested, and then replaced. Since digital systems can contain hundreds of ICs (and thousands of gates!), these techniques are extremely important in the field.

Experiment 4

Name _____

BASIC COMBINATORIAL CIRCUITS

OBJECTIVES

1. To investigate the behavior of simple logic circuits constructed from a logic diagram containing AND and OR gates.
2. To investigate the use of parentheses in Boolean expressions.
3. To demonstrate the use of a logic probe in evaluating a logic circuit.
4. To implement a logic circuit from a Boolean expression.
5. To practice troubleshooting on a simple combinatorial circuit using a logic probe.

TEXT REFERENCES

Read sections 3.6 through 3.8. Also review Appendix A of this manual.

EQUIPMENT NEEDED

Components
7404 IC;
74LS 11 IC;
7432 IC;
1 LED monitor;
4 toggle switches.

Instruments
0–5 volt DC power supply;
logic probe.

DISCUSSION

In Experiment 2, you learned the characteristics of three of the fundamental logic gates: the AND, OR, and NOT. In Experiment 3, you learned how to recognize and troubleshoot some of their faults. In this experiment, you will combine these gates into logic circuits, investigate their behavior, and troubleshoot them. In future experiments, we will see that these so-called combinatorial circuits are often manufactured as ICs and become part of digital systems, so this experiment will help to prepare us for dealing with these larger ICs when we encounter them.

PROCEDURE

a) *AND/OR combinations*: Construct the circuit of Figure 4-1. Connect toggle switches to circuit inputs A, B, and C. Monitor the output x of the circuit with an LED monitor. For each input combination in Table 4-1, observe the output state of the LED monitor and record the state.

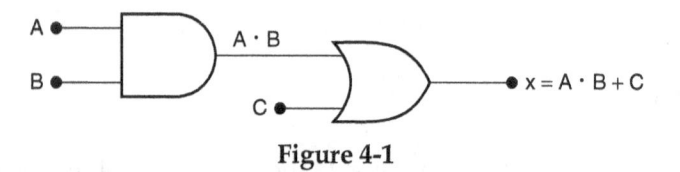

Figure 4-1

Table 4-1

Toggle Switches			Output LED Monitor (On/Off)	Output Logic Level (0/1)
A	B	C		
0	0	0		
0	0	1		
0	1	0		
0	1	1		
1	0	0		
1	0	1		
1	1	0		
1	1	1		

b) Connect the circuit of Figure 4-2. Connect toggle switches to inputs A, B, and C, and observe the output of the circuit with an LED monitor. For each input combination in Table 4-2, observe the output state of the LED monitor, and record the state in the table.

You should have obtained different results for Tables 4-1 and 4-2. This demonstrates experimentally that the two Boolean expressions AB + C and (A + B)C are not the same. Recall that the AND operation must be performed before the OR in an expression containing both operations EXCEPT when parentheses are used. When parentheses are used, the operation (or expression) inside the parentheses is performed first. In the expression (A + B)C, the parentheses enclose the output of the OR gate used as an input to an AND gate. Parentheses are

PROCEDURE

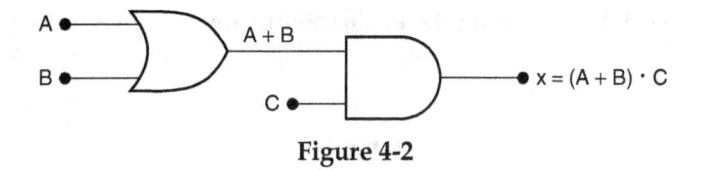

Figure 4-2

$$x = (A + B) \cdot C$$

Table 4-2

Toggle Switches			Output LED Monitor (On/Off)	Output Logic Level (0/1)
A	B	C		
0	0	0		
0	0	1		
0	1	0		
0	1	1		
1	0	0		
1	0	1		
1	1	0		
1	1	1		

not necessary in the expression A + BC, but if they were, the expression would be A + (BC). In this case, the parentheses enclose the output of an AND gate that is used as one input to an OR gate.

c) *Circuits with inverters:* In this step, you will construct a circuit containing all of the basic logic gates covered thus far. You will then use a logic probe to evaluate the circuit at each gate output for all input combinations.

Draw a logic diagram for the expression

$$x = \overline{A}BC(\overline{A + D})$$

in the space provided below.

Complete Table 4-3 using the evaluation procedure given in section 3.7 of the text.

Table 4-3

Toggle Switches				Outputs		
A	B	C	D	$\overline{ABC}$	$\overline{A+D}$	$\overline{ABC}(\overline{A+D})$
0	0	0	0	1	1	1
0	0	0	1	1	0	0
0	0	1	0	1	1	1
0	0	1	1	1	0	0
0	1	0	0	1	1	1
0	1	0	1	1	0	0
0	1	1	0	1	1	1
0	1	1	1	1	0	0
1	0	0	0	1	0	0
1	0	0	1	1	0	0
1	0	1	0	1	0	0
1	0	1	1	1	0	0
1	1	0	0	1	0	0
1	1	0	1	1	0	0
1	1	1	0	0	0	0
1	1	1	1	0	0	0

d) Have the instructor verify your table before proceeding to the next step.

e) Review the operating instructions for your logic probe.

Construct the circuit you drew in step c. Connect toggle switches to circuit inputs A, B, C, and D. Next verify with the logic probe all output entries you made earlier. Write the values you obtain with the probe next to the values in the table. If you note any differences, you have probably wired the circuit incorrectly. If so, find and correct the wiring error(s) before proceeding with the next step.

f) *Troubleshooting:* Have the instructor or lab assistant introduce a fault into your circuit, then troubleshoot the circuit by repeating the circuit evaluation performed in step e with the logic probe. Use Table 4-4 to record your test results. If you spot the fault visually, ignore it. Use your test results to try and isolate the fault. When you think you have correctly isolated the fault, have the instructor or lab assistant verify it.

Repeat this step as many times as your schedule will permit by having your lab partner or another student introduce a fault while you are not looking.

g) *Review:* In Experiment 2, you learned how to test the operation of logic gates. In this experiment, you expanded your capabilities to include the construction and testing of simple logic circuits composed of the basic logic gate. You verified that parentheses are required in a Boolean expression whenever the expression could be interpreted incorrectly, and that a logic circuit can be constructed directly from a Boolean expression. You also practiced troubleshooting a simple combinatorial circuit. To test your understanding of the principles covered in this experiment, complete the following statements:

1. In the expression ABC + DE, the _____ operation is performed first and the _____ operation is performed last.

PROCEDURE

Table 4-4

Toggle Switches				Outputs		
A	B	C	D	$\overline{ABC}$	$\overline{A+D}$	$\overline{ABC}(\overline{A+D})$
0	0	0	0			
0	0	0	1			
0	0	1	0			
0	0	1	1			
0	1	0	0			
0	1	0	1			
0	1	1	0			
0	1	1	1			
1	0	0	0			
1	0	0	1			
1	0	1	0			
1	0	1	1			
1	1	0	0			
1	1	0	1			
1	1	1	0			
1	1	1	1			

2. In the expression (A + C)BD, the expression inside the parentheses is one input to a _____-input _____ gate.
3. The correct operation of a combinatorial circuit is given in a _____.
4. Tracing logic levels through a logic circuit is equivalent to _____ a Boolean expression for given input values.
5. If the circuit used in step f were to be completely contained in one IC, you would evaluate the _____ of the IC.
6. The troubleshooting technique employed in step f is referred to as [static, dynamic] testing.
7. Draw a logic diagram for the expression (AC + B$\overline{C}$) + A$\overline{B}$C.

Experiment 5

Name _____

LOGIC GATES II: NOR AND NAND

OBJECTIVES

1. To investigate the behavior of the NOR gate.
2. To investigate the behavior of the NAND gate.

TEXT REFERENCE

Read section 3.9.

EQUIPMENT NEEDED

Components
7400 IC;
7402 IC;
7404 IC;
74LS08 IC;
7432 IC;
4 toggle switches;
1 LED monitor.

Instruments
0–5 volt DC power supply;
pulse or square wave generator;
oscilloscope.

DISCUSSION

In Experiment 2, you learned the characteristics of three of the fundamental logic gates: the AND, OR, and NOT. You will now be introduced to two of the remaining logic gates: the NAND and NOR. The NAND and NOR gates are nothing more than inverted AND and OR gates, respectively. That is important, but not the most important thing. The fact that a NAND or a NOR can be used to create all other gates is important, because this fact has made them more popular in use than the others.

PROCEDURE

a) *The NOR gate:* Figure 5-1 shows the logic symbol for a two-input NOR gate and its truth table. Examine the truth table, and familiarize yourself with the NOR operation.

b) Refer to the data sheet for a 7402 IC, and draw the pin layout diagram:

c) Mount a 7402 on the circuit board. Connect V_{cc} to +5 V and GND to power ground. Connect toggle switches to inputs A and B of one of the 7402 NOR gates. Observe the output of the gate with an LED monitor.

d) Set the toggle switches to each input combination listed in Table 5-1, observe and record the output state of the LED monitor in the table.

Table 5-1

Data Switch A	B	Output LED Monitor (On/Off)	Output Logic Level (0/1)
0	0		
0	1		
1	0		
1	1		

Verify that your results agree with the truth table in Figure 5-1.

e) Disconnect input B from the toggle switch. Set toggle switch A alternately to 0 and 1, and observe the effect on the output. Based on your observations, the disconnected NOR input acts like a ___ input level.

PROCEDURE

Figure 5-1

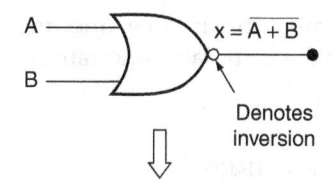

A	B	OR $A+B$	NOR $\overline{A+B}$
0	0	0	1
0	1	1	0
1	0	1	0
1	1	1	0

$x = \overline{A+B}$ (Denotes inversion)

f) Connect a pulse generator set to 1 kHz to input B. Disconnect the LED monitor from the NOR output, and connect one of the vertical inputs of the oscilloscope in its place. Connect the other vertical input to the output of the pulse generator, and trigger on this channel. Set input A alternately to 0 and 1, and observe the effect on the output. Sketch the waveform displayed on the oscilloscope for both settings of switch A using Timing Diagrams 5-1 and 5-2.

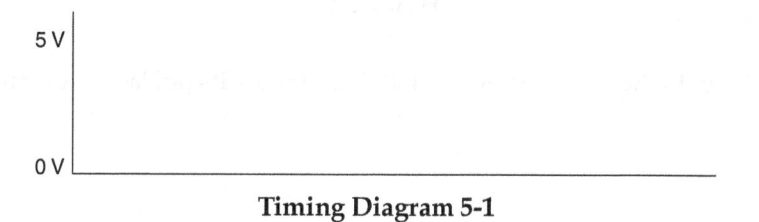

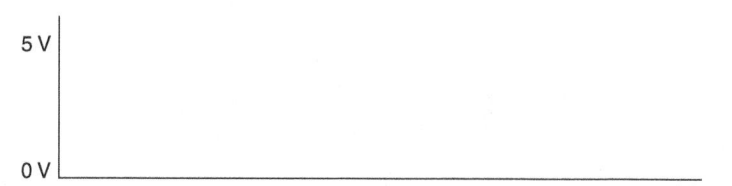

Timing Diagram 5-1

Timing Diagram 5-2

g) *NOR equivalent:* Disconnect the toggle switches from the NOR gate but do NOT remove the 7402 from the board. Mount a 7432 IC and a 7404 IC on the circuit board. Connect V_{cc} to +5 V and GND to power ground for each IC. Connect toggle switches to inputs A and B of one of the 7432 OR gates. Connect the OR gate output to a 7404 inverter as illustrated in Figure 5-2. Observe the output of the inverter with an LED monitor.

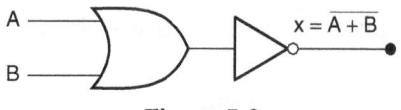

Figure 5-2

$x = \overline{A+B}$

h) Repeat step d. Do not record the output values, but do verify that your results are the same as those recorded for the NOR gate.

i) *The NAND gate:* Figure 5-3 shows the logic symbol, truth table, and an equivalent circuit for a two-input NAND gate. Examine the truth table and familiarize yourself with the NAND operation.

		AND	NAND
A	B	AB	$\overline{AB}$
0	0	0	1
0	1	0	1
1	0	0	1
1	1	1	0

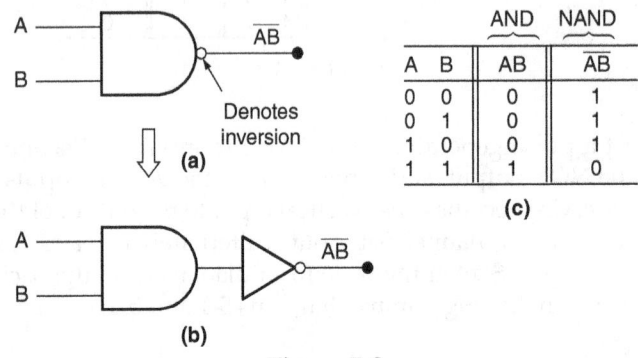

Figure 5-3

j) Refer to the data sheet for a 7400 IC and draw its pin layout diagram:

k) Mount a 7400 IC on the circuit board. Connect V_{cc} to +5 V and GND to power ground. Connect toggle switches to inputs A and B of one of the NAND gates and an LED monitor to its output.

l) Set switches A and B to each input combination listed in Table 5-2, and record your observations of the output monitor in the table.

Verify that your results agree with the truth table in Figure 5-3.

Table 5-2

Data Switch A	B	Output LED Monitor (On/Off)	Output Logic Level (0/1)
0	0		
0	1		
1	0		
1	1		

PROCEDURE

m) Disconnect the toggle switch from input B of the NAND gate. What will the state of output x be when A = 0? ___. When A = 1? ___. Verify your results.

n) Connect the pulse generator to input B and set the generator to 1 kHz.

Remove the output connection to the LED monitor, and connect one of the vertical inputs of the oscilloscope in its place. Connect the output of the pulse generator to the other vertical input of the oscilloscope, and trigger internally using this channel.

o) Set input A alternately to 0 and 1, and observe the effect on the output. Draw the waveforms displayed on the oscilloscope for each setting of A using Timing Diagrams 5-3 and 5-4.

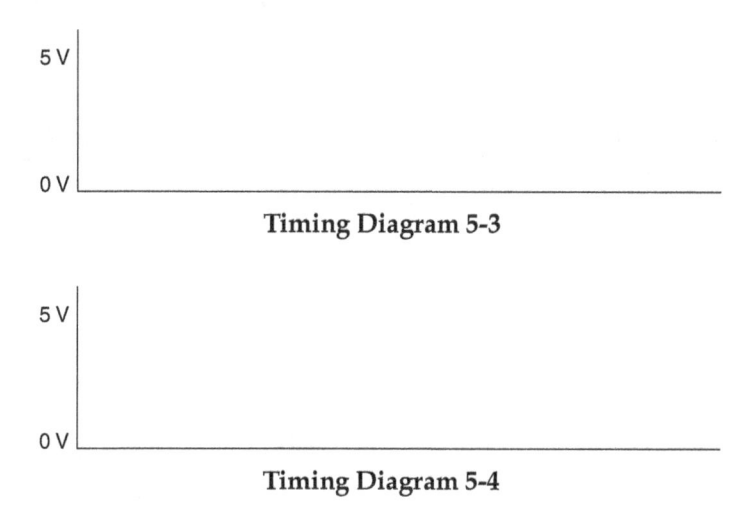

Timing Diagram 5-3

Timing Diagram 5-4

p) *NAND equivalent:* Connect a 74LS08 AND gate and a 7404 inverter so that the wiring agrees with the alternate NAND circuit of Figure 5-3. Connect toggle switches to inputs A and B of the AND gate and an LED monitor to the output of the inverter.

q) Verify that the alternate circuit does perform the NAND operation.

r) *Review:* This concludes the investigation of the NOR and NAND logic gates. To test your understanding of these gates, complete the following statements:

1. A difference between the NOR and NAND operations is that a ____ at any one NOR input results in a LOW output, while a ____ at any one NAND input results in a HIGH output.
2. A gate is said to be *enabled* if a signal at one of its inputs is permitted to pass to the output of the gate. The signal may be inverted or noninverted, depending on the gate's function. For a two-input NOR gate to be enabled, one of its inputs must be ____; the signal at the other input will be [inverted, noninverted] at the output. For a two-input NAND to be enabled, one input must be ____; the signal at the other input will be [inverted, noninverted] at the output.
3. A gate is said to be *inhibited* if the gate's output remains at a constant level regardless of any signals applied at its inputs. To inhibit a NOR gate, a ____ must be applied to one of its inputs. For a NAND gate to be an inhibitor, one input must be ____.
4. The NOR operation can be performed by a(n) _____ gate whose output is _____.
5. The NAND operation can be performed by a(n) _____ gate whose output is _____.

Experiment 6

Name _____

TROUBLESHOOTING NOR AND NAND GATES

OBJECTIVE

To troubleshoot NOR and NAND gates with pulser and probe.

REFERENCES

Read Appendix A of this manual. Also review Experiment 3.

EQUIPMENT NEEDED

Components
7400 IC;
7402 IC;
4 toggle switches.

Instruments
DMM or VOM;
0–5 volt DC power supply;
oscilloscope;
logic probe;
logic pulser.

DISCUSSION

This experiment duplicates part of Experiment 3 but, of course, with NAND and NOR gates. While these exercises should be used to verify what has already been learned in the previous experiments, you should note the distinct differences in troubleshooting NAND and NOR gates as opposed to AND and OR gates. For example, AND and OR gates act like voltage follower circuits (or buffers) whenever their inputs are shorted. You will discover in this exercise that NAND and NOR gates behave like inverters when their inputs are all shorted.

**Using the Probe and Pulser
to Troubleshoot NAND and NOR Gates**

In what follows, the probe-pulser combination will be used to aid in locating certain types of faults of IC NAND and NOR gates. We will not look at every possible type of fault, but we can simulate most of the common ones and develop a procedure for testing for them. The faults will be shaded in the figures.

A word of caution: Some of the fault simulations call for shorting gate inputs together and outputs together. **Never** short inputs together that are all connected to your trainer's data switches. **Never** short a trainer data switch directly to ground or to V_{cc}. **Never** short any gate output to ground or to V_{cc}. **Do** exercise care in following the instructions below and elsewhere in this manual. Do *not* substitute LS-TTL ICs for TTL in this experiment, since shorting LS-TTL outputs together for more than one second is not recommended. Voltages at shorted LS-TTL outputs cannot be measured with any consistency.

PROCEDURE

NOR Gates

1) Refer to Figure 6-1, and wire a 7402 NOR gate as shown. Recall that when all inputs to a NOR gate are the same, the output is the inverse of one of the inputs. Touch the probe tip to one of the inputs and the pulser to the other input. You should observe that the pulsing on one of the inputs is detected by the probe on the other. Normally, unless the two inputs are purposely wired together, you should not observe this condition.

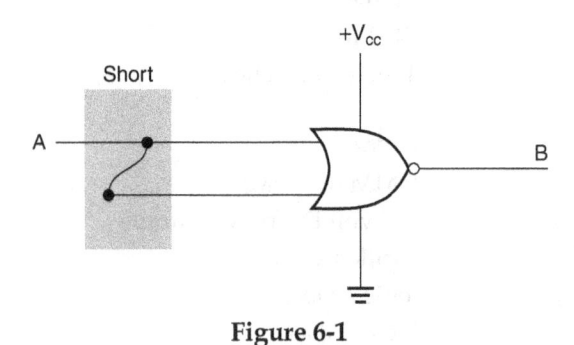

Figure 6-1

2) Refer to Figure 6-2, and wire a 7402 NOR gate as shown. Connect point A to a data switch, but leave points B and C disconnected. Place the probe tip on point C, and observe that the output is LOW. Toggle the data switch and observe that it has no effect on the output. Now place the probe tip on point D. You should observe that the level there is floating. This tells you that the input (point D) has

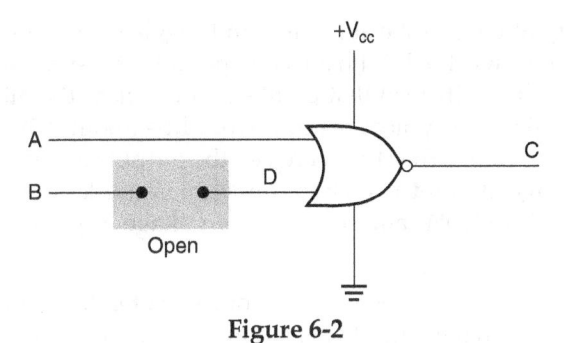

PROCEDURE

Figure 6-2

either an external open or an internal open. To verify that the open is external, place the pulser tip on point D and the probe tip on point C. Set the data switch at A to LOW. You should observe that the probe now indicates a pulsing condition. If the open had been internal, you would still see a LOW at point C.

3) Refer to Figure 6-3, and wire three 7402 NOR gates as shown. Note that a short between the outputs of gate 3 and gate 2 has been wired purposely. Connect data switches to inputs A, B, C, and D. Set switch A to LOW and B to HIGH and switches C and D to LOW. Place the tip of the logic probe on output E. You should observe that the level there is HIGH. The HIGH output of gate 2 has no effect on

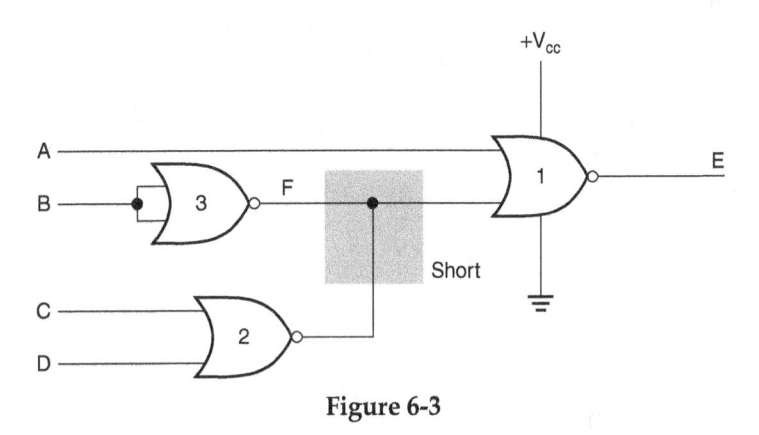

Figure 6-3

the output of gate 1. This is because the LOW output of gate 3 is pulling the level at one input of gate 1 down to LOW.

Now toggle switch B several times while observing the output at E with the logic probe. You should observe that the output at E toggles also.

Set switches C and D to HIGH. Now toggle switch B several times, again monitoring output E with the logic probe. You should observe that the logic probe indicates a constant HIGH. Why?

_____

_____

_____

Place the tip of your probe at point F, and toggle switch B several times. You should observe that switch B has no effect on point F. Measure the voltage at point F with the DMM. The voltage at that point should be near the midrange value for a LOW, which is 0.4 V. A typical reading would be about 0.38 V for a 7402. Set switch B to HIGH, remove the short between the outputs of gate 2 and gate 3, and measure the voltage at point F. It should now measure less than 0.4 V, typically 0.15 V. Set switch B to LOW, and measure the voltage at point F with the DMM. Record this voltage: _____.

Finally, replace the short between the outputs of gate 2 and gate 3. Place the tip of the logic probe on the output of gate 1 and the tip of the logic pulser on point F. You should note that the logic probe indicates a pulsing output at E. Thus the logic pulser can overcome a LOW produced by the output of a gate.

4) Refer to Figures 6-4 and 6-5. You will not wire these circuits. In step 3, when the output of gate 2 was HIGH, it had no effect on the output of gate 3. Switch B could be toggled, producing a toggling output at E if switch A was set to 0. A short to V_{cc} at point F cannot be overcome by the output of gate 3, so point F would be stuck HIGH. Measuring the actual voltage level at point F can help you determine whether or not a circuit node is shorted to V_{cc}, since the outputs of TTL gates are normally lower than V_{cc}. A logic pulser cannot overcome a short to V_{cc}.

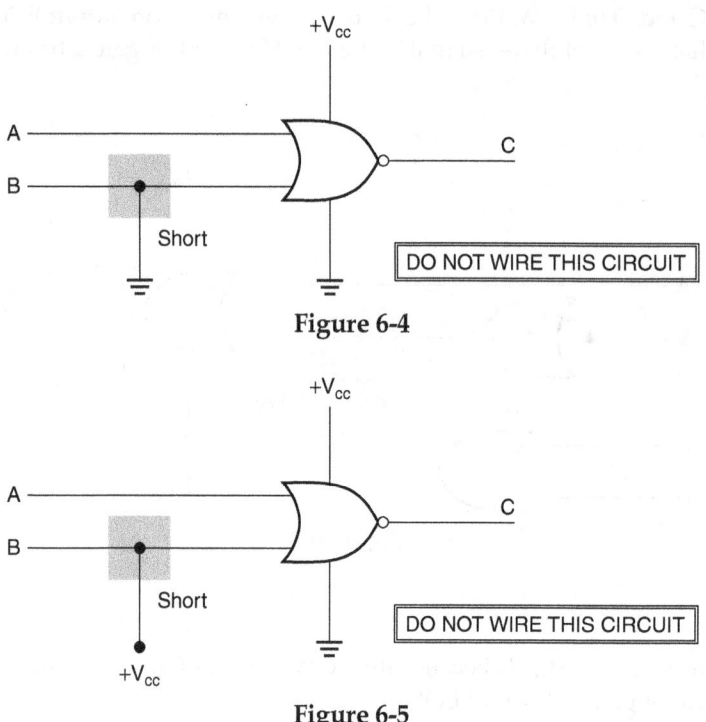

Figure 6-4

Figure 6-5

If point F measures 0 V, then it is likely that point F is shorted to ground. A logic pulser cannot overcome this type of short either.

5) Refer to Figure 6-6, and wire two 7402 NOR gates as shown. Leave points A and B disconnected. We will assume that gate 2 is functioning normally. Place your probe tip on point D. It should indicate a floating level. This is symptomatic of an open in the output circuit of gate 1. The question is whether it is an external or internal open. Place the probe on the output pin of gate 1. If it is LOW, then there is an external open somewhere between the IC and point C. If the output pin is floating, then an internal open in gate 1's output circuit could be the cause.

PROCEDURE

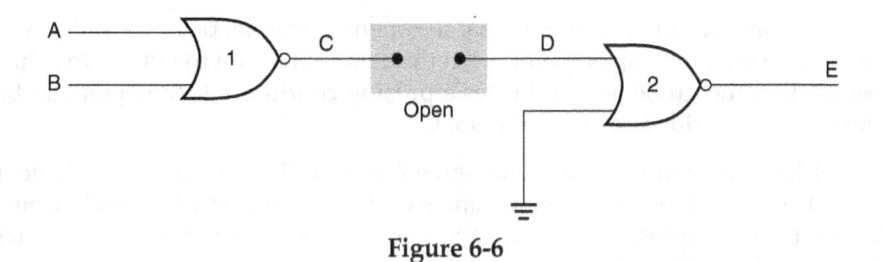

Figure 6-6

NAND Gates

1) Refer to Figure 6-7, and wire a 7400 NAND gate as shown. Recall that, when all inputs to a NAND gate are the same, the output is the same as the inputs. Touch the probe tip to one of the inputs and the pulser to the other input. You

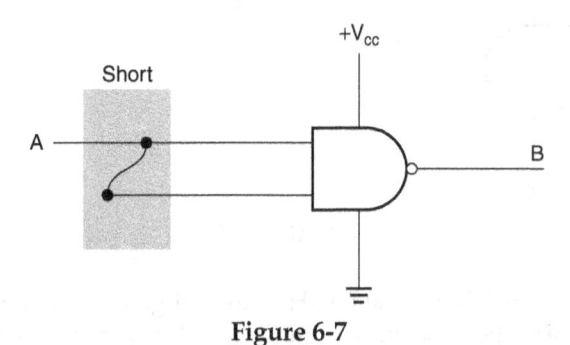

Figure 6-7

should observe that the pulsing on one of the inputs is detected by the probe on the other. Normally, unless the two inputs are purposely wired together, you should not observe this condition.

2) Refer to Figure 6-8, and wire a 7400 NAND gate as shown. Connect point A to a data switch, but leave points B and C disconnected. Place the probe tip on point C and observe the output. Toggle the data switch, and observe that the output acts normally. Now place the probe tip on point D. You should observe that the level there is floating. This tells you that the input (point D) has either an external

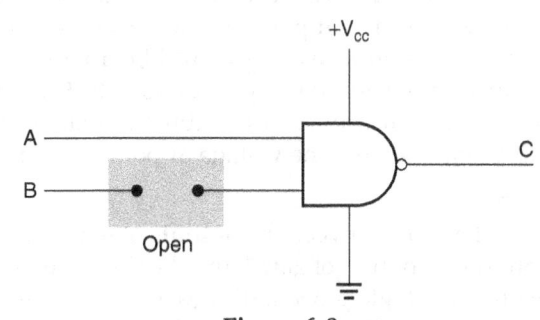

Figure 6-8

open or an internal open. To verify that the open is external, place the pulser tip on point D and the probe tip on point C. Set the data switch at A to HIGH. You should observe that the probe now indicates a pulsing condition. If the open had been internal, you would see a LOW at point C.

3) Refer to Figure 6-9, and wire three 7400 NAND gates as shown. Note that a short between the outputs of gate 3 and gate 2 has been wired purposely. Connect data switches to inputs A, B, C, and D. Set switches A and B to HIGH and switches C and D to LOW. Place the tip of the logic probe on output E. You should observe

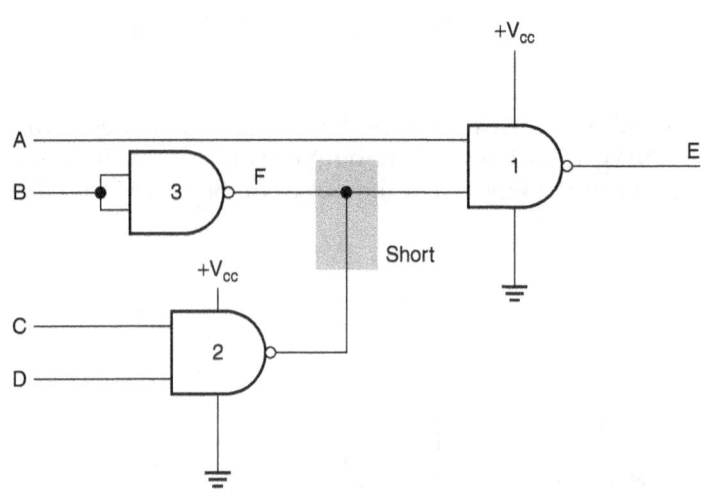

Figure 6-9

that the level there is HIGH. The HIGH output of gate 2 has no effect on the output of gate 1. This is because one of gate 1's inputs is HIGH, and the LOW output of gate 3 is pulling the level at the other input of gate 1 down to LOW. Now toggle switch B several times while observing the output at E with the logic probe. You should observe that the output at E toggles also.

Set switches C and D to HIGH. Now toggle switch B several times, again monitoring output E with the logic probe. You should observe that the logic probe indicates a constant HIGH. Why?

_____

_____

_____

Place the tip of your probe on point F, and toggle switch B several times. You should observe that switch B has no effect on point F. Measure the voltage at point F with the DMM. The voltage at that point should be near the midrange value for a LOW, which is 0.4 V. A typical reading would be about 0.39 V for the 7400. Remove the short between the outputs of gate 2 and gate 3, and measure the voltage at point F. It should now measure considerably less than 0.4 V, typically 0.10 V. Set switch B to LOW, and measure the voltage at point F with the DMM. Record this voltage: _____.

Finally, replace the short between the outputs of gate 2 and gate 3. Place the tip of the logic probe on the output of gate 1 and the tip of the logic pulser on point F. You should note that the logic probe indicates a pulsing output at E. Thus, the logic pulser can overcome a LOW produced by the output of a gate.

PROCEDURE

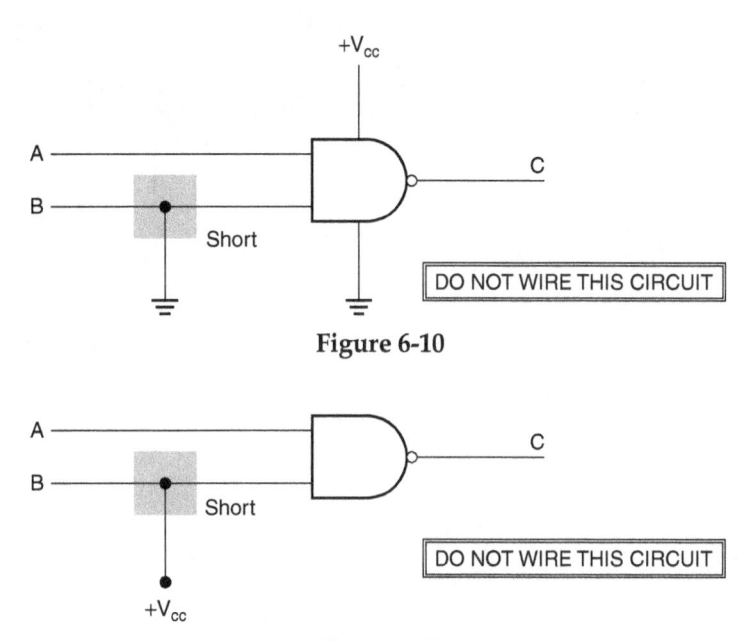

Figure 6-10

DO NOT WIRE THIS CIRCUIT

Figure 6-11

DO NOT WIRE THIS CIRCUIT

4) Refer to Figures 6-10 and 6-11. You will not wire these circuits. In step 3, when the output of gate 2 was HIGH, it had no effect on the output of gate 3. Switch B could be toggled, producing a toggling output at E. A short to V_{cc} at point F could not be overcome by the output of gate 3, so point F would be stuck HIGH. Measuring the actual voltage level at point F can help you determine whether or not a circuit node is shorted to V_{cc}, since the outputs of TTL gates are normally lower than V_{cc}. A logic pulser cannot overcome a short to V_{cc}.

If point F measures 0 V, then it is likely that point F is shorted to ground. A logic pulser cannot overcome this type of short either.

5) Refer to Figure 6-12, and wire two 7400 AND gates as shown. Leave points A and B disconnected. We will assume that gate 2 is functioning normally. Place your probe tip on point D. It should indicate a floating level. This is symptomatic of an open in the output circuit of gate 1. The question is whether it is an external or internal open. Place the probe at the output pin of gate 1. If it is LOW, then there is an external open somewhere between the IC and point C. If the output pin is floating, then an internal open in gate 1's output circuit could be the cause.

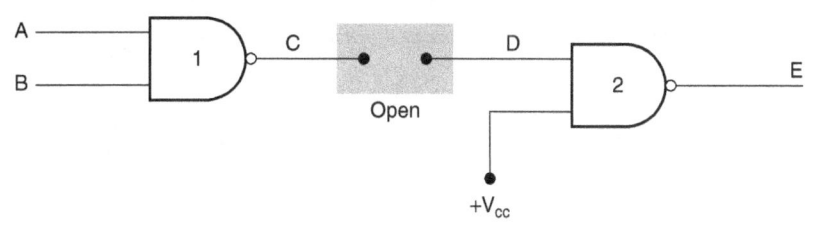

Figure 6-12

Experiment 7

Name _____

BOOLEAN THEOREMS

OBJECTIVES

1. To verify experimentally some of the Boolean theorems.
2. To introduce the student to circuit simplification.

TEXT REFERENCE

Read section 3.10.

EQUIPMENT NEEDED

Components
7404 IC;
74LS08 IC;
7432 IC;
2 toggle switches;
1 LED monitor.

Instruments
0–5 volt DC power supply;
pulse or square wave generator;
dual trace oscilloscope.

DISCUSSION

Boolean theorems are useful in simplifying expressions. You have already seen that logic circuits can be constructed from an expression, so it should follow, naturally, that simpler expressions mean simpler circuits. In this experiment, you will verify some of these theorems, discuss others, and demonstrate their impact on the physical circuit. You should also be alert to spot any theorem that might be useful in explaining a circuit fault.

PROCEDURE

a) *Boolean Theorems:* Figure 7-1 lists the univariate (single variable) Boolean theorems. Mount a 7404 IC, a 74LS08 IC, and a 7432 IC on the circuit board. Connect V_{cc} to +5 V and GND to power ground on each IC. Set the pulse generator to 1 kHz. To verify each theorem, connect the circuit for that theorem as shown in Figure 7-1, connecting the pulse generator to an "x" input, GND to an input labelled "0,"

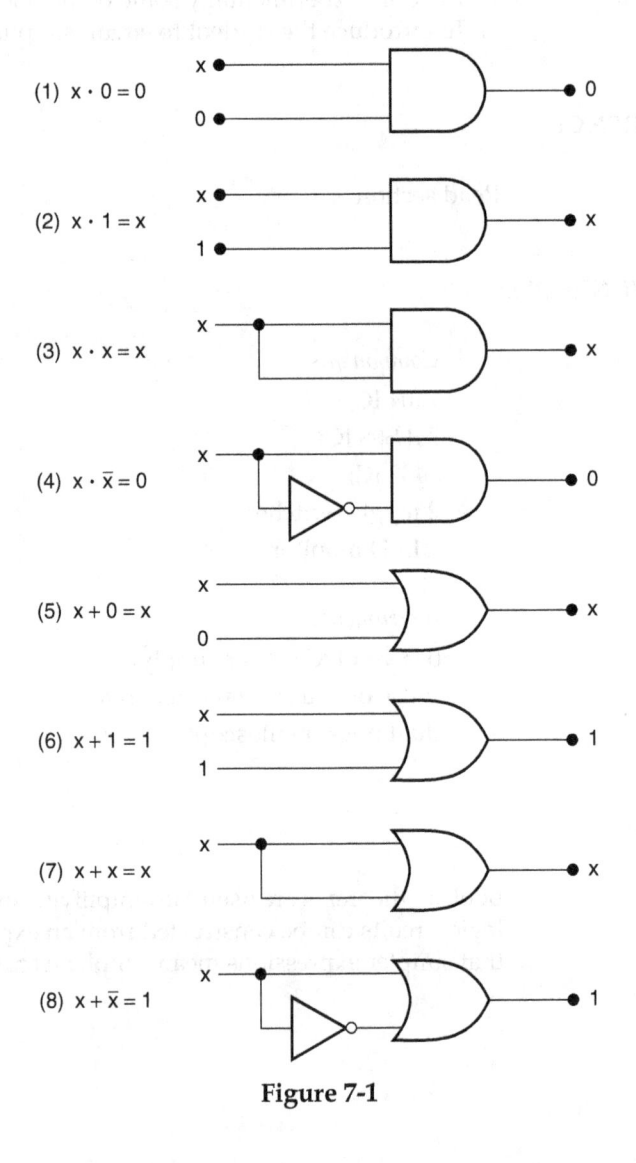

(1) $x \cdot 0 = 0$

(2) $x \cdot 1 = x$

(3) $x \cdot x = x$

(4) $x \cdot \bar{x} = 0$

(5) $x + 0 = x$

(6) $x + 1 = 1$

(7) $x + x = x$

(8) $x + \bar{x} = 1$

Figure 7-1

PROCEDURE

and V_{cc} to an input labelled "1." Monitor the output with one vertical input of the oscilloscope and the pulse generator with the other.

b) Connect the circuit for theorem 1, and observe the output displayed on the oscilloscope. The display should be constant with a level of 0 V. You probably suspect that a circuit like this one could be replaced with a single wire connected to ground, and you would be right. Theorem 1 permits this simplification.

c) Connect the circuit for theorem 2. The output should now be identical to the signal from the pulse generator. Confirm this by comparing the two waveforms on the oscilloscope. It should be plain that this circuit can be replaced by a single wire connected to signal x.

d) Connect the circuit for theorem 3. Again, the output should be identical to the signal from the pulse generator. Verify that this is so. This circuit, like that of theorem 2, can be replaced by a single wire to signal x.

e) Connect the circuit for theorem 4. Observe that the output is constant at 0 V. This circuit can be replaced by a wire to ground.

f) All of the univariate theorems involving AND gates have now been verified. You should now verify the remainder of the theorems, all of which involve OR gates.

g) *Theorem 14:* Figure 7-2 shows the circuit for testing another Boolean theorem, $x + xy = x$. Using $A = C = x$ and $B = y$, connect a toggle switch to input y and the pulse generator to x. Monitor both the pulse generator and the output of

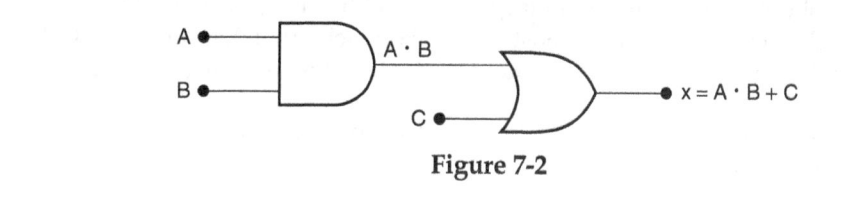

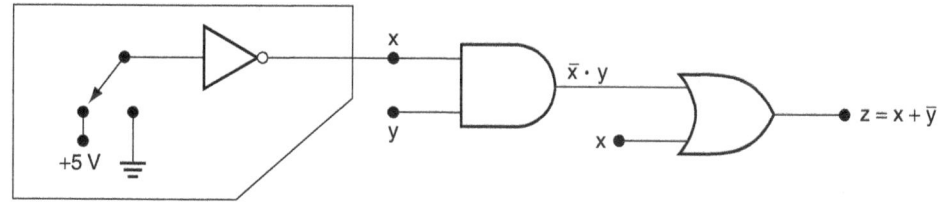

Figure 7-2

Figure 7-3

this circuit on the oscilloscope. You should observe that the two waveforms are identical.

h) Set switch y alternately to 0 and to 1. Switch y should have no effect on the output. This implies that input y serves no purpose, and since signal x is passed through the circuit unchanged, the entire circuit can be replaced with a single wire connected to x.

i) *Theorem 15:* Modify the circuit on the board by removing the pulse generator connection to the AND gate and inserting an inverter between a toggle switch and the AND gate input just disconnected. The Boolean expression for the new circuit is $x + \bar{x}y$. You are to verify that this circuit is equivalent to an OR gate with x and y as its inputs. Connect an LED monitor to the output of the circuit.

j) Set toggle switches x and y to all the input combinations listed in Table 7-1, and record the output observed in the table.

Compare Table 7-1 with the truth table for an OR gate. They should be identical, which shows that the circuit tested can be replaced by a single OR gate with inputs x and y.

Table 7-1

Input Data Switches		Output LED Monitor (On/Off)	Output Logic Level (0/1)
x	y		
0	0		
0	1		
1	0		
1	1		

k) *Review:* This concludes the experiment on Boolean theorems. To test your understanding of the results of this experiment, complete the following statements:

1. In Experiment 4, enabler and inhibitor circuits were discussed. The Boolean theorem that describes the AND inhibitor is _____.
2. The OR enabler circuit is described by theorem _____.
3. In troubleshooting a circuit containing an AND with one input shorted to LOW, one would discover [a signal, a constant HIGH (stuck-HIGH), a constant LOW (stuck-LOW)] at the output of the gate. The Boolean theorem that explains this is _____.
4. Theorem 6 guarantees a _____ output if an input to an OR gate is shorted to V_{cc}.

Experiment 8

Name _____

SIMPLIFICATION USING BOOLEAN THEOREMS

OBJECTIVE

To use Boolean theorems to simplify logic circuits.

TEXT REFERENCE

Read section 3.10.

EQUIPMENT NEEDED

Components
7404 IC (2);
74LS08 IC;
74LS11 IC;
74LS27 IC;
7432 IC;
4 toggle switches;
1 LED monitor.

Instruments
0–5 volt DC power supply.

DISCUSSION

In Experiment 7, you were introduced to the Boolean theorems and their usefulness in explaining digital circuits and their simplification. In this experiment, you will continue your investigation of the application of Boolean theorems to simplification.

PROCEDURE

a) Examine the logic circuit in Figure 8-1, and write the Boolean expression for output x: _____.

b) Make a truth table for expression x using Table 8-1.

c) Construct the circuit of Figure 8-1 on the circuit board. Connect toggle switches to inputs A, B, C, and D. Connect x to an LED monitor.

d) Verify the operation of your circuit by setting the toggle switches to each set of input values in Table 8-1 and comparing the outputs observed to the corresponding outputs in the table.

e) In the space provided below, simplify x using Boolean theorems. List the theorem used in each step of the simplification.

f) Draw the logic diagram for the simplified expression.

PROCEDURE

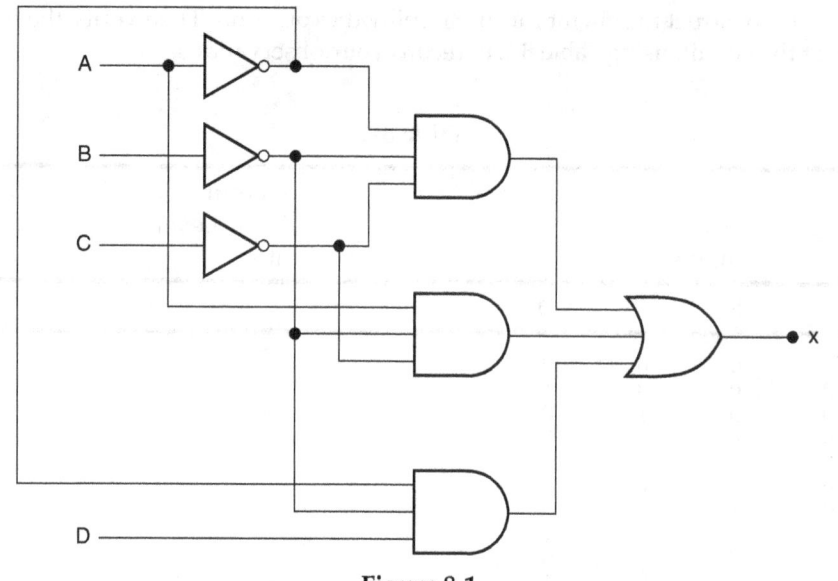

Figure 8-1

Table 8-1

Inputs				Output Logic Level at x
A	B	C	D	
0	0	0	0	
0	0	0	1	
0	0	1	0	
0	0	1	1	
0	1	0	0	
0	1	0	1	
0	1	1	0	
0	1	1	1	
1	0	0	0	
1	0	0	1	
1	0	1	0	
1	0	1	1	
1	1	0	0	
1	1	0	1	
1	1	1	0	
1	1	1	1	

g) Construct the circuit for the simplified expression. Then verify the operation of the circuit, using Table 8-2 to record your observations.

Table 8-2

Inputs				Output Logic Level at x
A	B	C	D	
0	0	0	0	
0	0	0	1	
0	0	1	0	
0	0	1	1	
0	1	0	0	
0	1	0	1	
0	1	1	0	
0	1	1	1	
1	0	0	0	
1	0	0	1	
1	0	1	0	
1	0	1	1	
1	1	0	0	
1	1	0	1	
1	1	1	0	
1	1	1	1	

h) *Review:* This completes the exercise on Boolean simplification. To test your understanding of the experiment, answer the following questions:

1. Boolean simplification can eliminate variables from an expression. For example, ABC + AB$\overline{C}$ = AB. Describe how you would verify that the two expressions are equivalent using truth tables.

2. Can all Boolean sum-of-product expressions be simplified? If so, state why; if not, give a couple of examples.

Experiment 9

Name _____

DEMORGAN'S THEOREMS

OBJECTIVES

1. To verify experimentally DeMorgan's two theorems.
2. To investigate the use of DeMorgan's theorems in circuit simplification.
3. To demonstrate the extension of DeMorgan's theorems to three variables.

TEXT REFERENCE

Read section 3.11.

EQUIPMENT NEEDED

Components
7404 IC;
74LS10 IC;
74LS11 IC;
74LS27 IC;
3 toggle switches;
1 LED monitor.

Instruments
0–5 volt DC power supply; logic probe (optional).

DISCUSSION

In Experiments 7 and 8, you investigated several rules of Boolean algebra. Now, in the current experiment, you are introduced to two more rules of Boolean algebra, known collectively as DeMorgan's theorems. The two theorems are:

1) $\overline{(x + y)} = \overline{x} \cdot \overline{y}$

2) $\overline{(x \cdot y)} = \overline{x} + \overline{y}$

You should note that each theorem permits you to simplify expressions involving inverted sums or products.

Example: Simplify $\overline{X + YZ}$. Using the first DeMorgan theorem, we can write $\overline{X + YZ} = \overline{X} \cdot \overline{YZ}$. Using the second theorem, we have $\overline{X} \cdot \overline{YZ} = \overline{X} \cdot (\overline{Y} + \overline{Z})$, and we can write the right-hand side of the equation as $\overline{X} \cdot \overline{Y} + \overline{X} \cdot \overline{Z}$. Here the inverter signs invert only single variables.

PROCEDURE

a) *Verifying DeMorgan's theorems:* Construct the circuit on the right-hand side of Figure 9-1(a) using a three-input AND gate. Tie the unused input to V_{cc}. Connect toggle switches to circuit inputs X and Y. You will monitor the output of the circuit with an LED monitor or logic probe.

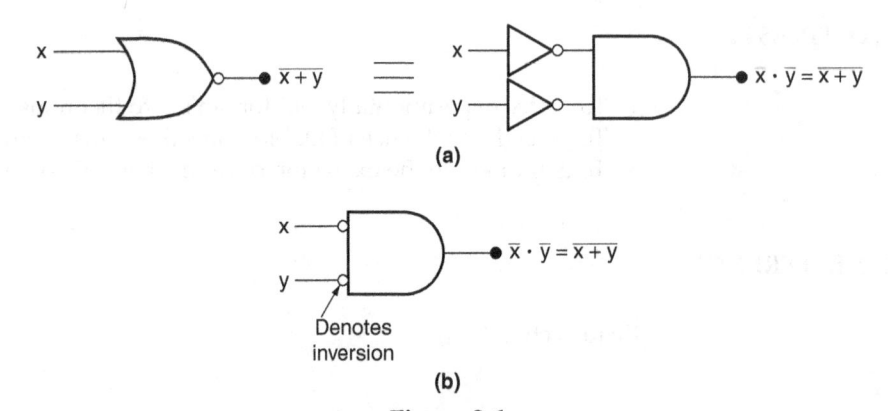

Figure 9-1

Table 9-1

Inputs		Output
X	Y	$\overline{X} \cdot \overline{Y}$
0	0	
0	1	
1	0	
1	1	

b) Set the toggle switches to each input combination in Table 9-1, and record the output you observe in the table.

Verify that Table 9-1 is identical to the truth table for a NOR gate. This part of the experiment shows that an AND with inverted inputs is equivalent to a NOR gate.

c) Disconnect the unused input from V_{cc}, and connect it to a toggle switch. Repeat step b, using Table 9-2 to record your observations.

PROCEDURE

Table 9-2

Inputs			Output
X	Y	Z	$\overline{X} \cdot \overline{Y} \cdot \overline{Z}$
0	0	0	
0	0	1	
0	1	0	
0	1	1	
1	0	0	
1	0	1	
1	1	0	
1	1	1	

Verify that Table 9-2 is identical to the truth table for a three-input NOR gate. You have demonstrated that the DeMorgan theorem being tested can be extended to three variables. Similarly, this theorem could be verified for any number of inputs.

d) Construct the circuit on the right-hand side of Figure 9-2(a) using a three-input OR gate (construct it with a three-input NOR and an inverter). Tie the unused input to ground. Connect toggle switches to circuit inputs x and y. You will monitor the output of the circuit with an LED monitor or logic probe.

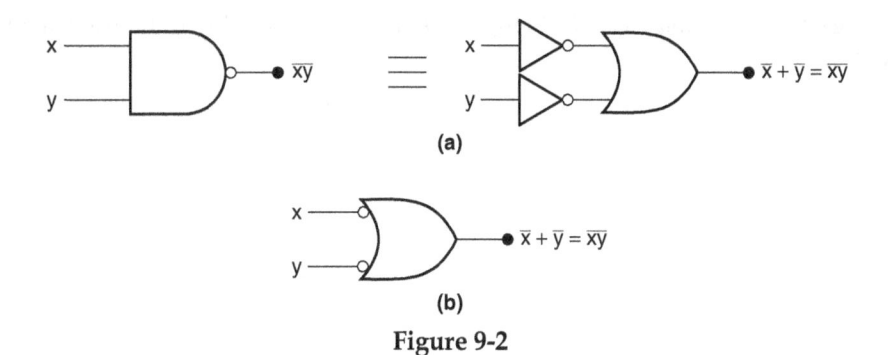

Figure 9-2

e) Set the toggle switches to each input combination in Table 9-3, and record the output you observe in the table.

Table 9-3

Inputs		Output
X	Y	$\overline{X} + \overline{Y}$
0	0	
0	1	
1	0	
1	1	

Verify that Table 9-3 is identical to the truth table for a NAND gate. This part of the experiment shows that an OR with inverted inputs is equivalent to a NAND gate.

f) Disconnect the unused input from ground and connect it to the output of an inverter. Connect the input of this inverter to a toggle switch. Repeat step b, using Table 9-4 to record your observations.

Table 9-4

Inputs			Output
X	Y	Z	$\overline{X} + \overline{Y} + \overline{Z}$
0	0	0	
0	0	1	
0	1	0	
0	1	1	
1	0	0	
1	0	1	
1	1	0	
1	1	1	

Verify that Table 9-4 is identical to the truth table for a three-input NAND gate. You have demonstrated that the DeMorgan theorem being tested can be extended to three variables. Similarly, this theorem could be verified for any number of inputs.

g) *Simplification using DeMorgan's theorems:* Draw a logic diagram for the expression

$$(\overline{A} + \overline{B} \cdot C) \cdot (\overline{A} + B \cdot \overline{C})$$

h) Construct the circuit using the diagram you drew in step g. Connect toggle switches to inputs A, B, C and an LED monitor to the circuit output. Set the toggle switches to each input combination listed in Table 9-5, and record the output value observed in the table.

Table 9-5

Inputs			Output
A	B	C	$(\overline{A} + \overline{B} \cdot C) \cdot (\overline{A} + B \cdot \overline{C})$
0	0	0	
0	0	1	
0	1	0	
0	1	1	
1	0	0	
1	0	1	
1	1	0	
1	1	1	

PROCEDURE

   i) Based on your results in Table 9-5, write an equivalent expression for the circuit: _____.

   j) Use DeMorgan's theorems to prove that the expression in step i is equivalent to the original expression in step g.

   k) *Review:* This concludes the exercises on DeMorgan's theorems. To test your understanding of the theorems, complete the following statements:

1. DeMorgan's theorems can be used to show that a NAND gate may be replaced by a(n) _____ gate with inverted inputs.
2. The theorems may also be used to show that a NOR gate may be replaced by a(n) _____ gate with inverted inputs.

Use DeMorgan's theorems to show the following:

3. An AND gate is equivalent to a NOR gate with inverted inputs.

4. An OR gate is equivalent to a NAND gate with inverted inputs.

Experiment 10

Name _____

THE UNIVERSALITY OF NAND AND NOR GATES

OBJECTIVES

1. To demonstrate the universality of the NAND gate.
2. To demonstrate the universality of the NOR gate.
3. To construct a logic circuit using all NOR gates.
4. To construct a logic circuit using all NAND gates.

TEXT REFERENCE

Read section 3.12.

EQUIPMENT NEEDED

Components
7400 IC;
7402 IC;
74LS10 IC;
74LS27 IC;
3 toggle switches;
1 LED monitor.

Instruments
0–5 volt DC power supply.

DISCUSSION

Most digital circuits consist of combinations of the basic logic gates. In this experiment, you will investigate the use of NAND and NOR gates in constructing digital circuits. You will demonstrate that NAND gates can be used to obtain all the other gates, and, similarly, NOR gates can be used to do the same thing. For example, you will show that an AND gate is simply an inverted NAND or a NOR gate with inverters in its inputs. These simple facts help a designer to design simpler circuits by reducing chip count, and they help a troubleshooter to understand and troubleshoot them.

PROCEDURE

a) Install a 7400 IC on the circuit board. Connect V_{cc} to +5 V and GND to power ground.

b) *NOT from NAND:* Connect the circuit shown on the left side of Figure 10-1(a). Connect a toggle switch to circuit input A and an LED monitor to output x. Verify that the circuit is an inverter by setting the input switch to 0 and then 1 and observing the output LED.

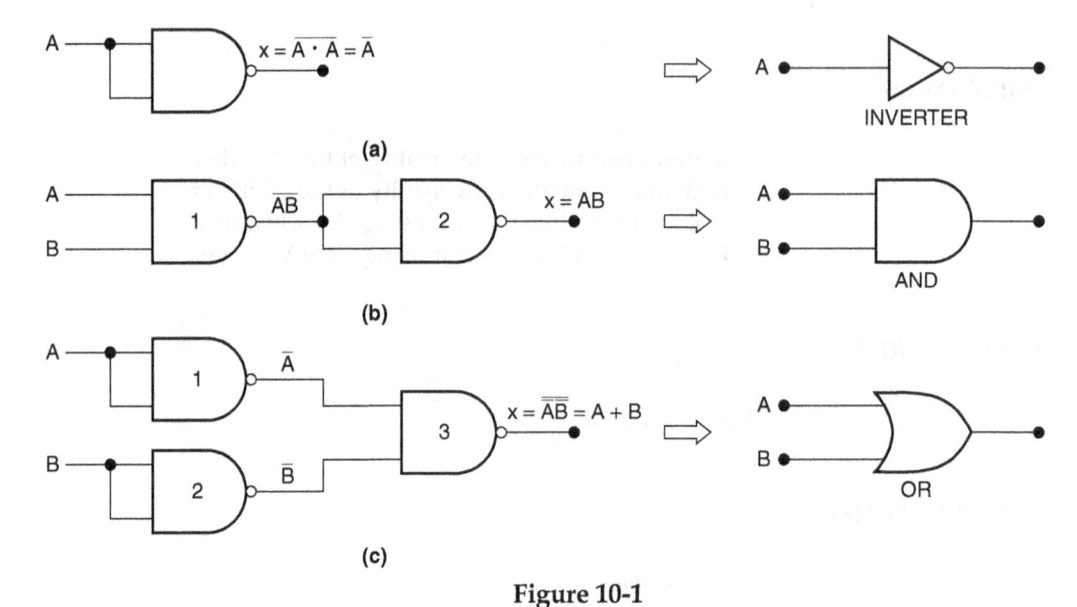

Figure 10-1

c) *AND from NAND:* Connect the circuit on the left side of Figure 10-1(b). Connect toggle switches to circuit inputs A and B. Connect an LED monitor to output x. Verify that the circuit performs the AND function by setting the toggle switches to each set of input combinations in Table 10-1 and observing the output LED monitor. Record your observations in the table.

Table 10-1

Inputs A	B	Output X
0	0	
0	1	
1	0	
1	1	

PROCEDURE

d) *OR from NAND:* Connect the circuit on the left side of Figure 10-1(c). Connect toggle switches to circuit inputs A and B. Connect an LED monitor to output x. Verify that the circuit performs the OR function by setting the toggle switches to each set of inputs in Table 10-2 and observing the output. Record your observations in the table.

Table 10-2

Inputs A B	Output X
0 0	
0 1	
1 0	
1 1	

e) Mount a 7402 IC on the circuit board. Connect V_{cc} to +5 V and GND to power ground.

f) *NOT from NOR:* Connect the circuit on the left-hand side of Figure 10-2(a). Connect a toggle switch to circuit input A and an LED monitor to output x. Verify that the circuit is an inverter by setting the input switch to 0 and then to 1 and observing the output.

g) *OR from NOR:* Connect the circuit on the left-hand side of Figure 10-2(b). Connect toggle switches to circuit inputs A and B and an LED monitor to output x. Verify that the circuit performs the OR function by setting the toggle switches to each input combination listed in Table 10-3 and observing the output LED monitor. Record your observations in the table.

Table 10-3

Inputs A B	Output X
0 0	
0 1	
1 0	
1 1	

h) *AND from NOR:* Connect the circuit on the left-hand side of Figure 10-2(c). Connect toggle switches to circuit inputs A and B and an LED monitor to output x. Verify that the circuit performs the AND function by setting the toggle switches to each input combination listed in Table 10-4 and observing the output LED monitor. Record your observations in the table.

Table 10-4

Inputs A B	Output X
0 0	
0 1	
1 0	
1 1	

(a) $x = \overline{A+A} = \overline{A}$ ⇒ A — INVERTER

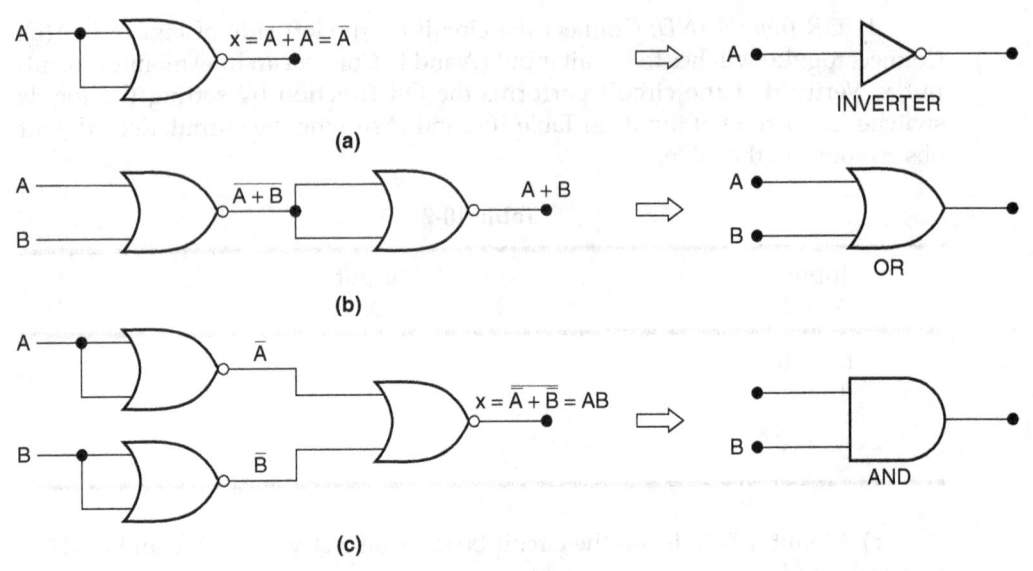

(b) $\overline{A+B}$ → $A+B$ ⇒ OR

(c) $x = \overline{\overline{A}+\overline{B}} = AB$ ⇒ AND

Figure 10-2

i) *Constructing a circuit using all NORs:* Use Table 10-5 to make a truth table for the circuit in Figure 10-3(a). Redraw and construct the circuit in Figure 10-3(a) using only NOR gates.

Table 10-5

Inputs			Output
A	B	C	X
0	0	0	
0	0	1	
0	1	0	
0	1	1	
1	0	0	
1	0	1	
1	1	0	
1	1	1	

Connect toggle switches to circuit inputs A, B, C, and D. Connect an LED monitor to output x. Verify that the circuit is equivalent to the original circuit by setting the toggle switches to each input combination in Table 10-5 and observing the output. Compare the outputs to the corresponding values for x in the truth table.

j) *Constructing a circuit using all NANDs:* Use Table 10-6 to make a truth table for the circuit in Figure 10-3(b). Redraw and construct the circuit of Figure 10-3(b) using only NAND gates.

PROCEDURE

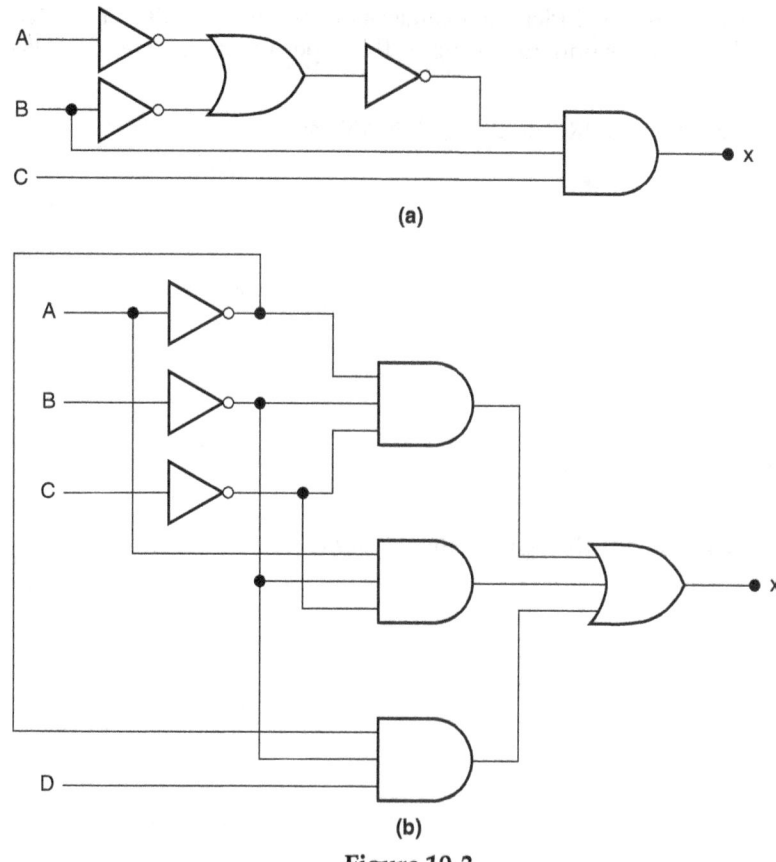

Figure 10-3

Connect toggle switches to circuit inputs A, B, C, and D. Connect an LED monitor to output x. Verify that the circuit is equivalent to the original circuit by setting the toggle switches to each input combination in Table 10-6 and observing the output. Compare the outputs to the corresponding values for x in the truth table.

Table 10-6

Inputs				Output
A	B	C	D	X
0	0	0	0	
0	0	0	1	
0	0	1	0	
0	0	1	1	
0	1	0	0	
0	1	0	1	
0	1	1	0	
0	1	1	1	
1	0	0	0	
1	0	0	1	
1	0	1	0	
1	0	1	1	
1	1	0	0	
1	1	0	1	
1	1	1	0	
1	1	1	1	

k) *Review:* This concludes the exercises on the universality of NAND and NOR gates. To test your understanding of the experiment, answer the following questions:

1. Draw a circuit for a NOR gate using all NANDs:

2. Draw a circuit for a NAND gate using all NORs:

Experiment 11

Name _____

IMPLEMENTING LOGIC CIRCUIT DESIGNS

OBJECTIVE

To design and construct logic circuits given either a truth table or a set of statements that describe the circuit's behavior.

TEXT REFERENCES

Read sections 4.1 through 4.4.

EQUIPMENT NEEDED

Components
Selected NAND and/or NOR ICs;
4 toggle switches;
1 LED monitor.

Instruments
0–5 volt DC power supply.

DISCUSSION

In this experiment, you are asked to design several combinatorial circuits. A combinatorial circuit is, as the name suggests, a circuit composed of a combination of logic gates. It acts on its inputs and gives an output based on its logic function. A combinatorial circuit can be designed directly from a truth table or from a description of its logic function. In either case, you are called upon to derive the simplest circuit possible. You should ask your instructor which circuits you will be required to build.

PROCEDURE

a) Table 11-1 is a truth table that describes the logic circuit you are to design and construct in this step. Study it carefully, and then draw a logic circuit in the space below using a *minimum* number of logic gates. Show your circuit to the instructor before constructing it. When you have completed the construction, test the circuit and compare your results with the truth table.

Table 11-1

Inputs			Output
A	B	C	X
0	0	0	1
0	0	1	0
0	1	0	1
0	1	1	1
1	0	0	1
1	0	1	0
1	1	0	0
1	1	1	1

b) Design a circuit whose output is HIGH only when a majority of inputs A, B, and C are LOW. Use Table 11-2 to make a truth table for the circuit, then draw a logic diagram that corresponds to the truth table. Show your circuit design to the instructor before constructing it. Test the circuit using the truth table you made for it.

PROCEDURE

Table 11-2

Inputs			Output
A	B	C	X
0	0	0	
0	0	1	
0	1	0	
0	1	1	
1	0	0	
1	0	1	
1	1	0	
1	1	1	

c) Design a logic circuit that has two signal inputs A_0 and A_1 and a control input S so that it functions according to the requirements given in Figure 11-1. This type of circuit is called a *multiplexer* (covered in Experiment 34). Show your circuit design to the instructor before constructing it. Test the completed circuit using the truth table given in Figure 11-1.

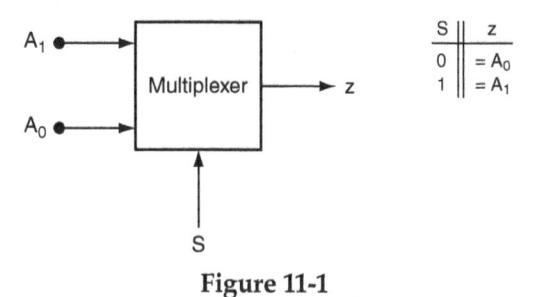

Figure 11-1

d) In step b, you designed a simple majority tester. There are many variations of this problem, one of which is the following:

A board of directors has four members—a president, vice president, a secretary, and a treasurer. In order that voting by the board members never results in a tie, the president is given two votes, and all members must vote. For a motion to carry, three "yes" ("yes" = logic 1; "no" = logic 0) votes are required. Otherwise, the motion fails to carry. Design this tester using all NAND gates.

Experiment 12

Name _____

EXCLUSIVE-OR AND EXCLUSIVE-NOR CIRCUITS

OBJECTIVES

1. To investigate the operation of the exclusive-OR circuit.
2. To investigate the operation of the exclusive-NOR circuit.
3. To investigate the operation of the 7486 quad exclusive-OR IC.

TEXT REFERENCE

Read section 4.6.

EQUIPMENT NEEDED

Components
7400 IC (2);
74LS86 IC;
2 toggle switches;
1 LED monitor.

Instrument
0–5 volt DC power supply.

DISCUSSION

Two Boolean expressions occur quite frequently in designing combinatorial circuits:

$$1)\ x = \overline{A}B + A\overline{B}$$

and

$$2)\ y = AB + \overline{A}\,\overline{B}$$

Expression 1 defines the exclusive-OR function to be one that yields an output that is HIGH whenever its inputs are different. Similarly, expression 2 defines the exclusive-NOR function to be one that yields an output that is HIGH whenever its inputs are the same. While these circuits are combinatorial circuits, they have been given their own symbols, and both have been implemented with integrated circuits. In this experiment, you will investigate the behavior of both circuits and the exclusive-OR integrated circuit implementation.

PREPARATION

Part (c) of this experiment requires the student to design a binary comparator circuit. This may take a substantial amount of time to do, and therefore the student is encouraged to do this work prior to the laboratory class for a more efficient use of the laboratory time.

PROCEDURE

a) Figure 12-1(a) shows the sum-of-products circuit for the exclusive-OR function and its associated truth table. The equivalent logic symbol for the circuit is shown in Figure 12-1(b).

b) *Sum-of-products exclusive-OR:* Implement the circuit of Figure 12-1 using all NAND gates. Draw your circuit in the space provided below. Connect toggle switches to circuit inputs A and B. You will monitor the output with an LED monitor. Using Table 12-1, set the toggle switches to each input combination listed, and observe the effect on the circuit output. Record your observations in the output column of the table. Your observations should be the same as those of the truth table in Figure 12-1.

PROCEDURE

Table 12-1

Inputs A B	Output X
0 0	
0 1	
1 0	
1 1	

A	B	x
0	0	0
0	1	1
1	0	1
1	1	0

$x = \overline{A}B + A\overline{B}$

(a)

$x = A + B = \overline{A}B + A\overline{B}$

EX-OR

(b)

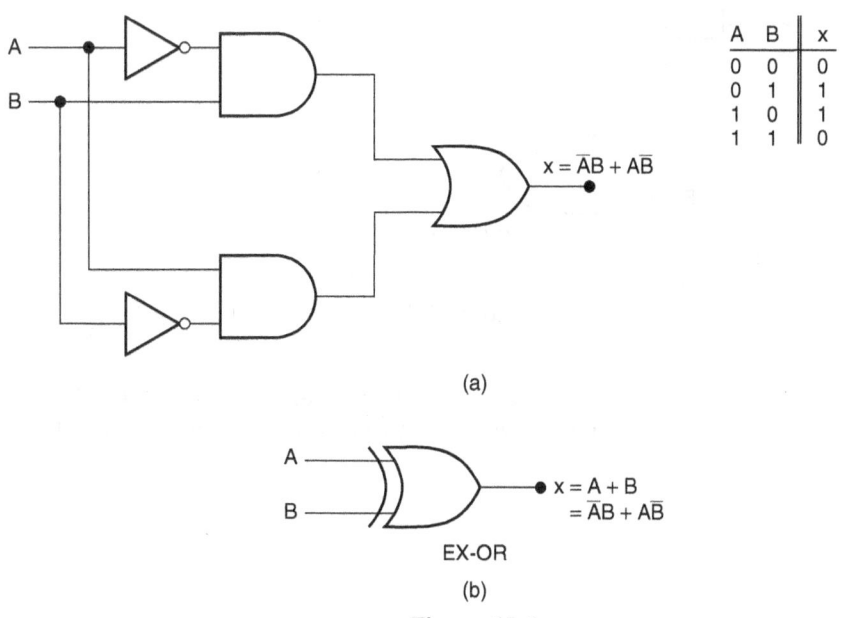

Figure 12-1

c) *Sum-of-products exclusive-NOR:* Disconnect the monitor from x and reconnect x to an inverter. Connect the monitor to the output of the inverter. The new circuit performs the exclusive-NOR function and is equivalent to the one illustrated in Figure 12-2. Using Table 12-2, verify that this is so by setting the toggle switches to each combination listed and recording the output states that you observe in the table.

Table 12-2

Inputs A B	Output X
0 0	
0 1	
1 0	
1 1	

A	B	X
0	0	1
0	1	0
1	0	0
1	1	1

$x = AB + \overline{A}\overline{B}$

(a)

$x = \overline{A \cdot B} = AB + \overline{A}\overline{B}$

EX-NOR

(b)

Figure 12-2

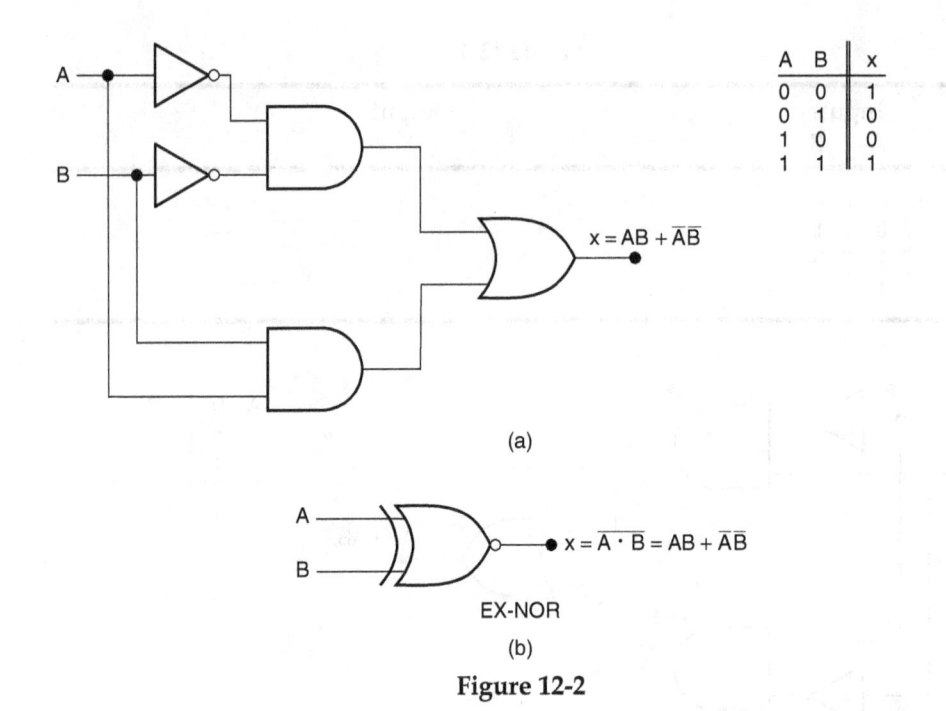

d) *74LS86 IC exclusive-OR:* Refer to the data sheet for a 74LS86 IC. This IC contains four exclusive-OR circuits. Draw the pin layout diagram for the 74LS86.

e) Mount a 74LS86 on the circuit board. Connect V_{cc} to +5 V and GND to power ground. Connect toggle switches to one of the exclusive-OR circuits of the IC, and connect an LED monitor to its output. Using Table 12-3, verify that the circuit performs the exclusive-OR function. Record your observations in the table.

Table 12-3

| Inputs | | Output |
A	B	X
0	0	
0	1	
1	0	
1	1	

PROCEDURE

f) Disconnect the LED monitor from the output of the 74LS86 circuit under test, and insert an inverter between the monitor and the circuit. Verify that the new circuit performs the exclusive-NOR function. Use Table 12-4 to record your results.

Table 12-4

Inputs A	B	Output X
0	0	
0	1	
1	0	
1	1	

g) You will now demonstrate that the placement of the inverter in steps c and f could just as well have been in one of the inputs in order to obtain the exclusive-NOR function from an exclusive-OR circuit. Remove the inverter added to the 74LS86 IC circuit in step f, and insert it between the toggle switch and input A. Reconnect the LED monitor to the output of the 74LS86 circuit under test. Now verify that the circuit still performs the exclusive-NOR function. Record your observations in Table 12-5.

Table 12-5

Inputs A	B	Output X
0	0	
0	1	
1	0	
1	1	

h) *Review:* This concludes the investigation of the exclusive-OR and exclusive-NOR circuits. To test your understanding of the circuits, answer the following questions:

1. The exclusive-OR function can be described as a circuit whose output is HIGH only when its two inputs are _____.
2. The exclusive-NOR function can be described as a circuit whose output is HIGH only when its two inputs are _____.
3. A simple application of the exclusive-OR circuit is in the addition of two 1-digit binary numbers. The output of the exclusive-OR is called the SUM bit. However, the exclusive-OR cannot always indicate the true sum since the circuit cannot indicate the carry that results in the case of 1 + 1. Design a circuit that can add any two 1-bit numbers and indicate both the sum and carry of such an addition. This circuit will have two inputs and two outputs. It is called a *half adder*.
4. Use Boolean algebra to prove the results obtained in step g.

Experiment 13

Name _____

DESIGNING WITH EXCLUSIVE-OR AND EXCLUSIVE-NOR CIRCUITS

OBJECTIVES

1. To investigate the application of an exclusive-OR circuit in a parity generator circuit.
2. To investigate the application of an exclusive-OR as a controlled inverter.
3. To investigate the application of an exclusive-NOR circuit in a digital comparator circuit.

TEXT REFERENCE

Read section 4.6.

EQUIPMENT NEEDED

Components
7404 IC (2);
7408 IC (2);
74LS11 IC;
74LS27 IC;
74LS86 IC (2);
6 toggle switches;
3 LED monitors.

Instruments
0–5 volt DC power supply;
pulse or square wave generator;
dual trace oscilloscope.

DISCUSSION

In the previous experiment, you were introduced to the exclusive-OR and exclusive-NOR circuits. You discovered that these circuits can be used to compare the level of two inputs. Indeed, the exclusive-OR and exclusive-NOR are basically digital comparators. In the current experiment, you will design several circuits using the exclusive-OR and exclusive-NOR circuits. In practice, you will probably invert an exclusive-OR to get an exclusive-NOR, since the common TTL IC implementation of the exclusive-NOR (74LS266) requires special output wiring and is used in very special applications.

PROCEDURE

a) *Parity generator application:* Design a four-bit even parity generator that uses exclusive-OR circuits. This circuit should have four inputs and an output that is HIGH only when an odd number of inputs are HIGH. Use Table 13-1 to make a truth table for the circuit. Draw your logic circuit in the space provided below. Have your instructor approve the circuit design before you construct and test it.

PROCEDURE

b) *Controlled inverter application:* Disassemble the circuit of step a except for a single 74LS86 IC. Connect the output of the square wave generator to one input of a 74LS86 exclusive-OR circuit. Connect a toggle switch to the other input. Monitor both the generator output and the output of the exclusive-OR with the dual trace oscilloscope. Trigger the oscilloscope on the generator output. Set the toggle switch to LOW, and observe the relationship between the two waveforms. They should be the same. Now set the toggle switch to HIGH, and observe the relationship between the two waveforms. Compare the two waveforms.

Table 13-1

Inputs				Output
A	B	C	D	X
0	0	0	0	
0	0	0	1	
0	0	1	0	
0	0	1	1	
0	1	0	0	
0	1	0	1	
0	1	1	0	
0	1	1	1	
1	0	0	0	
1	0	0	1	
1	0	1	0	
1	0	1	1	
1	1	0	0	
1	1	0	1	
1	1	1	0	
1	1	1	1	

c) *Binary comparator application:* Figure 13-1 represents a three-bit relative magnitude detector that determines if two numbers are equal and, if not, which number is larger. There are three inputs, defined as follows:

1) M = 1 only if the two input numbers are equal.
2) N = 1 only if $X_2X_1X_0$ is greater than $Y_2Y_1Y_0$.
3) P = 1 only if $Y_2Y_1Y_0$ is greater than $X_2X_1X_0$.

Design the circuit for the detector. Since the circuit is too complex to use a truth table, you might want to refer to Example 4.16 in the text to get an idea about how to get started. Draw your logic diagram in the space provided. Show the circuit to your instructor before constructing and testing the circuit.

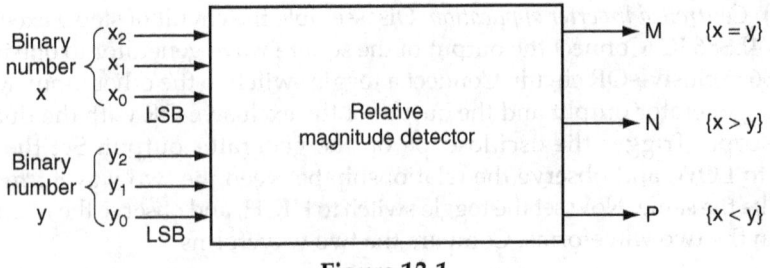

Figure 13-1

d) *Review:* This concludes the investigation of exclusive-OR and exclusive-NOR applications. To test your understanding of this experiment, answer the following questions:

1. How would you modify the even parity generator circuit you designed to produce an *odd* parity generator?

2. If a signal is applied to input A of an exclusive-OR gate and input B is controlled with a toggle switch, then whenever input B is HIGH, the output is [A, $\overline{A}$, 1, 0]. When B is LOW, the output is [A, $\overline{A}$, 1, 0].

3. The exclusive-OR circuit may be used to simplify some sum-of-products designs. Give an example of this.

Experiment 14

Name _____

TROUBLESHOOTING EXCLUSIVE-OR AND COMBINATORIAL CIRCUITS

OBJECTIVES

1. Troubleshoot 74LS86 IC faults.
2. Troubleshoot combinatorial circuits.

TEXT REFERENCES

Read sections 4.6 and 4.9 through 4.13. Also read Appendix A of this manual, and review Experiment 3.

EQUIPMENT NEEDED

Components
74LS86 IC;
2 toggle (spdt) switches.

Instruments
0–5 volt DC power supply;
logic pulser;
logic probe.

DISCUSSION

In Experiment 3, you learned how to troubleshoot the basic gates. You repeated this exercise in Experiment 6 with NAND and NOR gates. Before proceeding to troubleshoot combinations of these gates, you should spend a few minutes learning to troubleshoot the exclusive-OR, since it is found in many digital applications. Since the IC faults of the 74LS86 are isolated by using the techniques developed in the earlier troubleshooting exercises, we will concentrate on the faults that produce symptoms peculiar to the exclusive-OR circuit. You will then be asked to troubleshoot a combination circuit using the procedures found in this lab manual and the text.

PROCEDURE

a) Refer to Figure 14-1, and wire a 74LS86 EX-OR gate as shown. Recall that when the inputs to an EX-OR gate are both the same, the output is LOW. Therefore, if the output of an EX-OR appears to be stuck LOW, one possible trouble might be shorted inputs. Touch the probe tip to one of the inputs and the pulser to the other input. You should observe that the pulsing on one of the inputs is detected by the probe on the other. This indicates a short in the EX-OR's inputs.

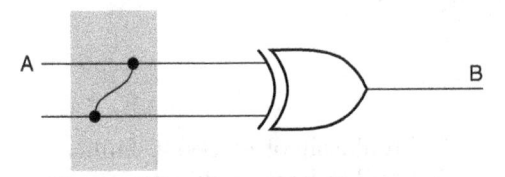

Figure 14-1

b) Refer to Figure 14-2, and wire a 74LS86 EX-OR gate as shown. Connect point A to a data switch and set the switch HIGH, but leave points B and C disconnected. Place the probe tip on point C, and observe that the output is LOW. Toggle the data switch, and observe that the output is always the inverse of the level at A. If the EX-OR is connected as a controlled inverter, this gate would be behaving properly. Now place the probe tip on point D. You should observe that the level there is floating. This tells you that the input (point D) is open. To verify that the open is external, place the pulser tip on point D and the probe tip on point C. Set the data switch at A to LOW. You should observe that the probe now indicates a pulsing condition. (If the open had been internal, you would still see a LOW at point C. You would also not detect a floating condition at the suspect input's pin.)

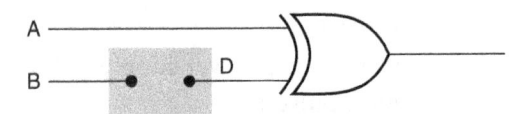

Figure 14-2

PROCEDURE

c) *Troubleshooting combinatorial circuits:*

1) Tear off the circuit diagram (Figure 14-2) and study it carefully, including the notes in the lower right-hand corner. Note the following:

- D, C, B, and A are inputs to the logic circuit.
- Each IC is identified as Z1, Z2, etc. Logic gates with the same Z number are on the same IC chip. For example, the two NAND gates labeled "Z4" are on the same 7400 NAND gate chip.
- The numbers on each logic gate input and output are pin numbers on the IC chip.
- The "balloons" TP1, TP2, etc., indicate test points that will be checked during testing or troubleshooting.

D	C	B	A	X	Y
0	0	0	0		
0	0	0	1		
0	0	1	0		
0	0	1	1		
0	1	0	0		
0	1	0	1		
0	1	1	0		
0	1	1	1		
1	0	0	0		
1	0	0	1		
1	0	1	0		
1	0	1	1		
1	1	0	0		
1	1	0	1		
1	1	1	0		
1	1	1	1		

Notes:

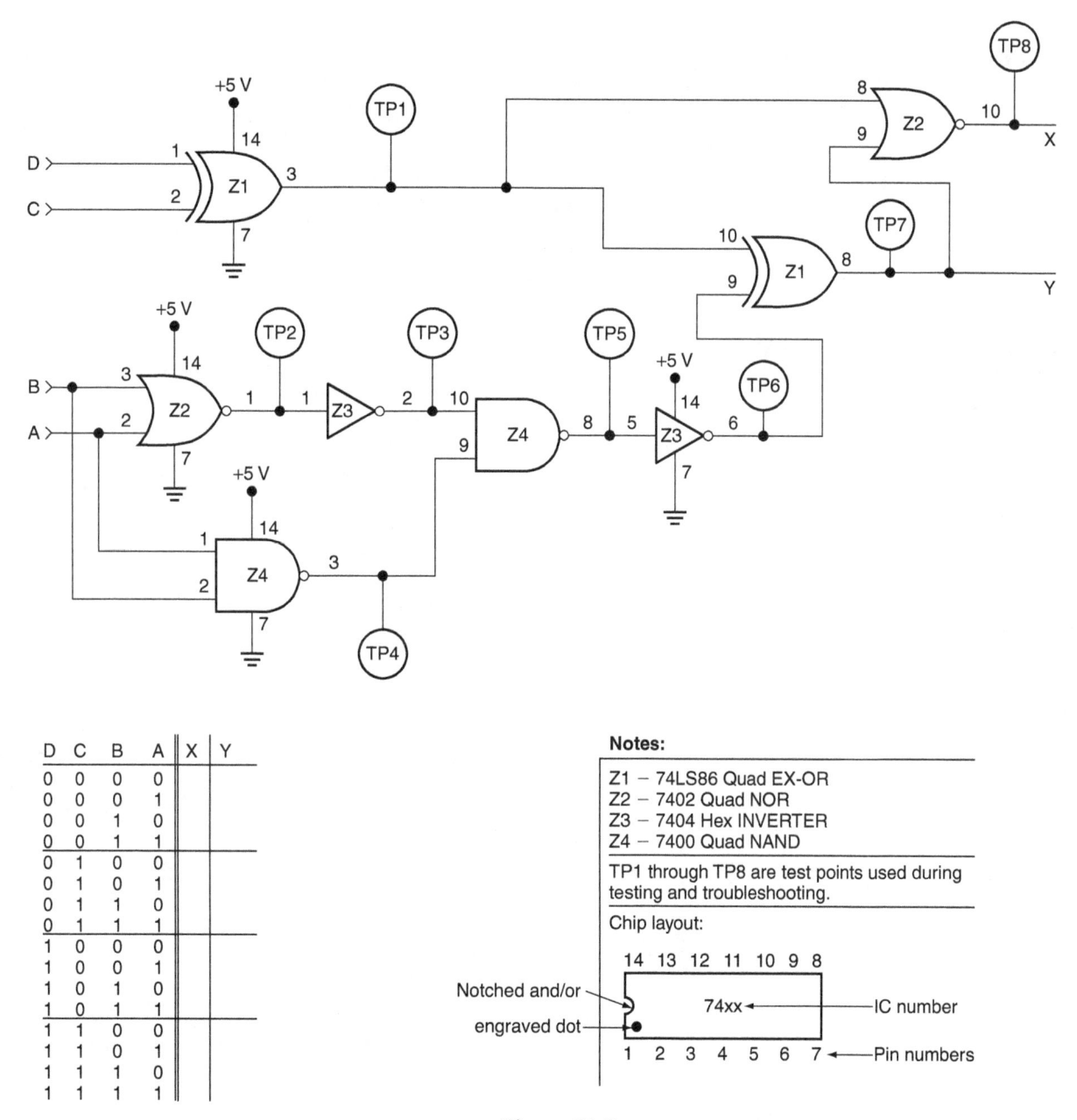

Z1 – 74LS86 Quad EX-OR
Z2 – 7402 Quad NOR
Z3 – 7404 Hex INVERTER
Z4 – 7400 Quad NAND

TP1 through TP8 are test points used during testing and troubleshooting.

Chip layout:

```
              14 13 12 11 10 9 8
Notched and/or  ┌─────────────┐
                │    74xx ◄───┼── IC number
engraved dot ──►•             │
                └─────────────┘
               1  2  3  4  5 6 7 ◄── Pin numbers
```

Figure 14-3

2) The four ICs used in this circuit should be inserted into the circuit board with a left-to-right orientation (i.e., Z1 on the left, then Z2 and Z3, with Z4 on the right). Wire the circuit according to Figure 14-3.
3) Connect inputs A, B, C, and D to toggle switches. Use your logic probe to test the circuit operation by trying each of the 16 input conditions and monitoring the outputs X and Y with the logic probe.
4) If the results do not match your predicted values, you will have to troubleshoot your circuit by following the logic levels through the circuit, starting at the inputs. Use your logic probe and pulser to locate the fault.
5) Once the circuit is operating normally, have your instructor or lab assistant introduce a fault, then use your troubleshooting procedure to locate this fault.
6) Repeat step 5 as frequently as time permits.

Experiment 15

Name _____

FLIP-FLOPS I: SET/CLEAR LATCHES AND CLOCKED FLIP-FLOPS

OBJECTIVES

1. To investigate the operation of the NOR gate SET/CLEAR latch.
2. To investigate the operation of the NAND gate SET/CLEAR latch.
3. To investigate the operation of an edge-triggered J-K flip-flop, the 74LS76 IC.
4. To investigate the operation of an edge-triggered D flip-flop, the 7474 IC.

TEXT REFERENCES

Read sections 5.1, 5.2, 5.4 through 5.7, and 5.9.

EQUIPMENT NEEDED

Components
7400 IC (2);
7402 IC (2);
74LS74A IC;
74LS76 IC;
normally HIGH pushbutton switch (2) and normally LOW pushbutton switch (2), all debounced;
2 LED monitors.

Instruments
0–5 volt DC power supply;
pulse or square wave generator;
dual trace oscilloscope;
logic probe (optional).

DISCUSSION

All of the previous experiments have been concerned with learning the fundamentals of logic gates and combinatorial circuits. Recall that an output of such a device or circuit responds to changes in its inputs and that when its inputs are removed, the output may not be sustained. In this experiment, you will be introduced to a device that can sustain a given output even when its inputs are removed. Such a device is said to possess memory. Examples of memory devices include flip-flops, which are the topic for this experiment. The following classes of flip-flops are investigated in this experiment:

- SET/CLEAR latches
- Edge-triggered J-K flip-flops
- Edge-triggered D flip-flops

SET/CLEAR Latches

The most fundamental flip-flop is the SET/CLEAR latch. Two types of SET/CLEAR latches are investigated in the current experiment:

- NAND gate SET/CLEAR latch
- NOR gate SET/CLEAR latch

The input levels to these devices determine the outputs. SET/CLEAR latches do not have a clock input, and so they are said to operate *asynchronously*.

Edge-Triggered J-K Flip-Flops

The J-K flip-flop eliminates the ambiguous condition. In place of this invalid condition, the J-K has a "toggle" condition, a characteristic of this flip-flop. Normally, a J-K flip-flop can be operated synchronously, since its J and K inputs need a separate clock to cause the flip-flop to change states. A J-K flip-flop can also be operated asynchronously and have SET and CLEAR inputs to facilitate this.

Edge-Triggered D Flip-Flops

The D flip-flop is a J-K flip-flop with an inverter between the J and the K inputs. This causes the flip-flop to SET or CLEAR with only one synchronous signal input. Like the J-K flip-flop, the D flip-flop also has an asynchronous mode.

PROCEDURE

a) *NOR gate SET/CLEAR latch:* Examine closely and then wire the NOR gate latch shown in Figure 15-1. Connect normally LOW pushbutton switches to the CLEAR and SET inputs to the circuit. You will monitor circuit outputs Q and $\overline{Q}$ with LED monitors.

PROCEDURE

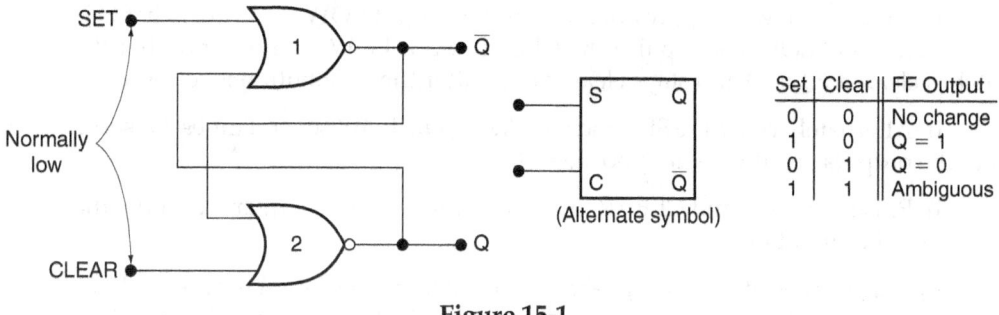

Figure 15-1

Set	Clear	FF Output
0	0	No change
1	0	Q = 1
0	1	Q = 0
1	1	Ambiguous

b) Turn the power supply on, and note the states of both LEDs: Q = _____; $\overline{Q}$ = _____.

Predicting the states of a latch when power is first applied is impossible, so the values just recorded are random.

Clear Q by momentarily pulsing the CLEAR input HIGH. If Q is already LOW, pulsing the CLEAR input will have no effect on the circuit.

c) Pulse the SET input HIGH, and observe the effects on the circuit outputs: Q = _____; $\overline{Q}$ = _____.

Note that releasing the pushbutton does not cause Q to change from its new state. Why? _____.

Now pulse the SET input HIGH again. What effect does this have on the circuit outputs? _____.

d) Pulse the CLEAR input HIGH, and observe that Q changes back to LOW and stays LOW even after the pushbutton is released.

e) Alternately pulse the SET and CLEAR inputs HIGH several times. Note that the outputs are always at opposite states.

f) Press and hold the SET and CLEAR inputs HIGH at the same time. Note that both outputs are now LOW. Release the pushbuttons, and note the states of the outputs. Are they both still LOW? _____.

Now pulse both SET and CLEAR inputs simultaneously several times, and note the effects on the outputs. If you pulse the circuit in this manner enough times, you will probably get random results. This is because the circuit response to this input condition is unpredictable.

g) *NAND gate SET/CLEAR latch:* Examine closely and then wire the NAND gate SET/CLEAR latch shown in Figure 15-2. Connect normally HIGH pushbutton switches to the SET and CLEAR inputs of the circuit.

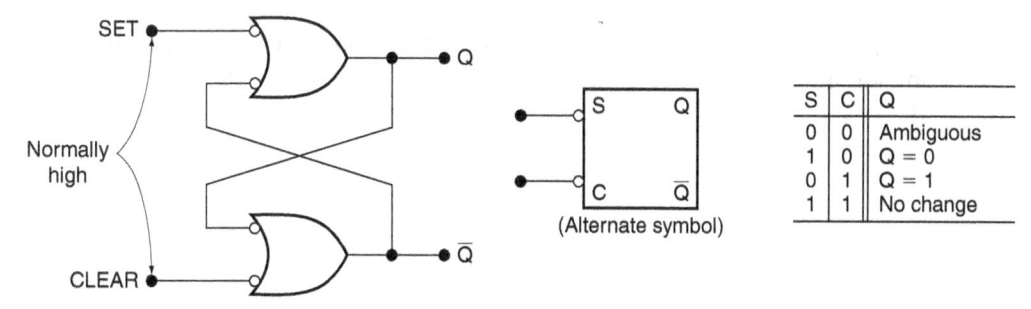

S	C	Q
0	0	Ambiguous
1	0	Q = 0
0	1	Q = 1
1	1	No change

Figure 15-2

h) Turn the power supply on. Pulse the SET input LOW, and verify that Q is HIGH and $\overline{Q}$ is LOW. Now pulse the CLEAR input LOW, and observe that the latch is cleared (Q = 0) and stays cleared even after the pushbutton is released.

i) Alternately pulse the SET and CLEAR inputs LOW several times. Observe that the outputs are always at opposite states.

j) Pulse both SET and CLEAR inputs LOW simultaneously, and observe the effects on the circuit outputs.

k) *Edge-triggered J-K flip-flop—74LS76 IC:* Refer to the data sheet for a 74LS76 IC. Notice the small arrowhead and bubble on the clock input. The arrowhead indicates that the device is sensitive only to a transition and the bubble indicates inversion. Thus, the 74LS76 is a negative-edge triggered device.

Draw the pin layout diagram for the 74LS76 IC:

Install a 74LS76 IC on the circuit board, and make the following connections:

1) Connect V_{cc} and DC SET to +5 V, GND to power ground.
2) Connect toggle switches to J and K inputs.
3) Connect a normally LOW pushbutton switch to the clock (CLK) input.
4) Connect a normally HIGH pushbutton switch to DC CLEAR.
5) Connect LED monitors to outputs Q and $\overline{Q}$ (or use a logic probe to monitor the outputs).

Turn the power supply on, and observe the states of Q and $\overline{Q}$. If Q = 1, then pulse DC CLEAR momentarily LOW. Note that this input clears the flip-flop immediately without a clock signal and that the input is active LOW.

l) *74LS76 synchronous operation:* In this step, you will observe that the J and K inputs can be used to change the output state of the flip-flop. You will also observe that in order for these inputs to effect a change, a clock pulse must be applied. For this reason, the J, K, and CLK inputs are referred to as *synchronous* inputs. Verify this by performing the following steps:

1) Change the J and K input switch settings, and observe that nothing happens to Q.
2) Set J = 1 and K = 1, and apply a positive-going transition at CLK. Do this by pressing and holding the CLK pushbutton switch. What happens to Q? _____.

PROCEDURE

3) Repeat step 2 using a negative-going transition at CLK. Do this by releasing the pushbutton switch. What happens to Q? _____. This proves that the flip-flop responds to only negative-going transitions. Apply several more pulses to the CLK input. What happens? _____.

4) If Q is LOW, pulse the CLK input so that Q is HIGH. Set J = K = 0, and note that nothing happens to Q. Pulse the CLK input momentarily, and observe that nothing happens to Q. Why?

___

5) Set J = 0 and K = 1, and note that nothing happens to Q. Pulse the CLK input momentarily. What happens to Q?

___

Apply several more pulses to the CLK input, and observe that Q remains in the LOW state.

6) Change J to 1 and then back to 0, and note that nothing happens to Q. Pulse the CLK input momentarily. You should observe that Q remains LOW. This proves that the J and K input states present *at the time of the proper clock transition* are the ones transferred to the flip-flop output.

7) Set J = 1, K = 0. Note that nothing happens to Q. Apply a clock pulse, and observe that Q will go HIGH. Apply several more clock pulses. What happens to Q?

___

m) Disconnect the pushbutton switch at the CLK input, and replace it with the output of a square wave generator set to 1 MHz (or the highest frequency obtainable). Connect the oscilloscope to observe the clock signal and output Q. Draw the waveforms displayed on the oscilloscope on Timing Diagram 15-1.

Verify that the flip-flop changes states on the negative-going transitions and does not change states on the positive-going transitions. What is the frequency of the Q waveform compared to the clock waveform? _____.

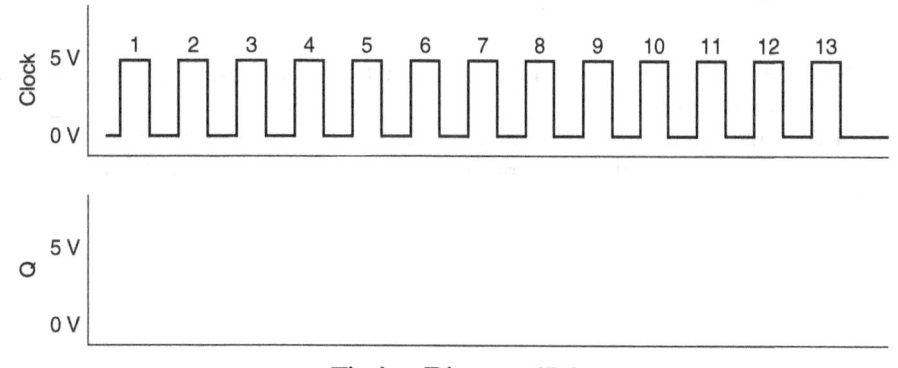

Timing Diagram 15-1

n) *74LS76 asynchronous operation:* The DC SET and DC CLEAR inputs are *asynchronous* inputs that operate independently from the synchronous inputs (J, K, CLK). These inputs are often labeled $\overline{PRE}$ and $\overline{CLR}$ (for preset and clear). The overbar indicates active LOW. The asynchronous inputs *override* the synchronous inputs when activated. Verify this by holding the DC CLEAR input LOW and observe that the flip-flop output stops toggling even though clock pulses are still being applied. Q will remain LOW, until the first clock pulse after the DC CLEAR pushbutton is released.

o) Disconnect the jumper connection from DC SET to V_{cc} at the V_{cc} end only, and touch this wire to ground. You should now observe that the flip-flop output stops toggling and remains HIGH as long as DC SET is held LOW.

p) *Edge-triggered D flip-flop—74LS74A IC:* Refer to the data sheet for a 74LS74A IC, and draw its pin layout diagram:

The 74LS74A IC has two individual positive edge-triggered D flip-flops with separate clock inputs and DC SET and DC CLEAR inputs.

Install a 74LS74A IC on the circuit board, and make the following connections to one of the D flip-flops:

1) Connect V_{cc} and DC SET to +5 V, GND to power ground.
2) Connect a toggle switch to the D input.
3) Connect a normally HIGH pushbutton switch to the CLK input.
4) Connect a normally HIGH pushbutton switch to DC CLEAR.
5) Connect LED monitors to Q and $\overline{Q}$ (or monitor the outputs with a logic probe).

q) *74LS74A synchronous operation:* Apply power and monitor the Q output. Observe that nothing happens when you toggle the D input switch back and forth. This is because the D input is a synchronous input that operates with the CLK input.

Clear Q to 0 by momentarily pulsing the DC CLEAR input LOW. Set D to 1, and apply a negative-going transition at CLK. Do this by pressing and holding the CLK pushbutton LOW. What happens to Q?

_____

Now apply a positive-going pulse at CLK by releasing the pushbutton switch. What happens? _____.
This proves that the flip-flop responds only to positive-going transitions.

Make D = 0, and pulse CLK momentarily. This should clear Q back to 0.

PROCEDURE

r) *74LS74A asynchronous operation:* For both DC SET and DC CLEAR, verify the following:

1) The inputs are active LOW and do not require a pulse at CLK to become activated.
2) The inputs override the synchronous input signals.

s) *Review:* This concludes the first set of exercises on flip-flops and latches. In Experiment 16, you will continue your investigation of latches and flip-flops with D latches and master/slave flip-flops. To test your understanding of the principles demonstrated in this experiment, answer the following questions:

1. The Q output of a NAND gate latch can change from 1 to 0 only when S = _____ and C = _____, while a NOR gate latch can change from 1 to 0 only when S = _____ and C = _____.
2. The CLEAR conditions for a NAND gate latch are S = _____ and C = _____, while the CLEAR conditions for a NOR gate latch are S = _____ and C = _____.
3. The J and K input levels of a 74LS76 J-K flip-flop are transferred to the Q and $\overline{Q}$ outputs on the _____ at CLK.
4. The DC SET and DC CLEAR inputs to a 74LS76 J-K flip-flop are active _____ inputs and [operate with, operate independently from] the CLK signal.
5. The D input level of a 74LS74A D flip-flop is transferred to the Q output of the flip-flop on the _____ at CLK.

Experiment 16

Name _____

FLIP-FLOPS II: D LATCH; MASTER/SLAVE FLIP-FLOPS

OBJECTIVES

1. To investigate the operation of a D latch, the 74LS75 IC.
2. To investigate the operation of a master/slave flip-flop constructed from basic gates.

TEXT REFERENCES

Read sections 5.8, 5.9, and 5.13.

EQUIPMENT NEEDED

Components
74LS75 IC;
74LS10 IC (2);
7400 IC;
normally HIGH pushbutton switch and normally LOW pushbutton switch, both debounced;
2 LED monitors.

Instruments
0–5 volt DC power supply;
logic probe (optional).

DISCUSSION

In this experiment, you are to investigate the behavior of the 74LS75 D Latch and a master/slave flip-flop. The D latch is level-triggered and is frequently used to interface one processing unit with another. You will be asked to compare this flip-flop to the 74LS74A D flip-flop you examined in Experiment 15. Also in Experiment 15, you investigated edge-triggered flip-flops. Recall that these flip-flops have set-up and hold times that must be satisfied if the circuits that employ them are to work properly. In this experiment, you will examine a generic J-K flip-flop constructed from basic gates. The master/slave flip-flop is no longer used in new designs, but will serve here to enable you to see how basic gates can be easily transformed into flip-flops. The wiring is a little more involved than you have done in most experiments, so it is important to be careful as you install the wiring. One way that is very helpful is to use a wire list showing the inputs on each IC. A basic wire list is provided as an example in this experiment.

PROCEDURE

a) *74LS75 IC D latch operation:* Refer to the data sheet for a 74LS75 IC, and draw its pin layout diagram:

Note that the 74LS75 has four D latches. The latch CLK inputs are tied together in pairs resulting in dual two-bit D latches. You will use only one of the D latches for this experiment, so examine Figure 16-1 closely for the proper connections to be made.

b) Install a 74LS75 IC on the circuit board, and make the connections shown in Figure 16-1. Connect a toggle switch to D_1, a normally LOW pushbutton switch to CLK, and LED monitors to Q_1 and $\overline{Q}_1$. When the circuit is completed, perform the following steps:

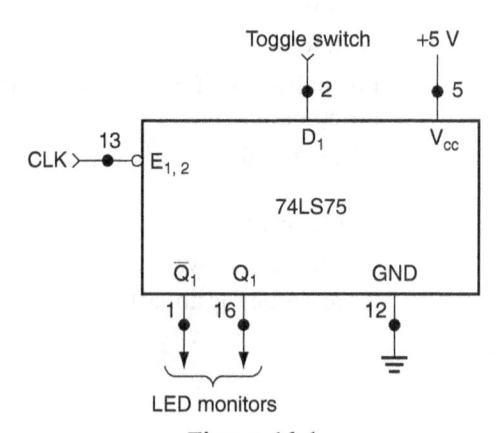

Figure 16-1

PROCEDURE

1) Turn the power supply on and monitor the outputs of the latch. Change the toggle switch back and forth a few times, and note that there is no effect on Q_1. This is because the latch is in the latch mode, and the data inputs are not enabled. Set $D_1 = 0$.
2) Press and hold the CLK input HIGH. Observe that Q_1 is LOW. Change D_1 back and forth a few times. What happens to Q_1? _____. Now set $D_1 = 1$, and release the CLK pushbutton. What happens to Q_1?

---

3) Change D_1 back and forth several times. Observe that Q_1 does not change. This proves that the data at D_1 is *latched* on the negative-going transition of the clock signal and that the output at Q_1 *follows* the data at D_1 while the clock signal is HIGH.

c) Master/slave flip-flop operation: A generic master/slave flip-flop can be formed simply from two gated R-S latches by adding cross-coupled feedback. By building the circuit using discrete gates, you can test the inner workings of the master/slave flip-flop. The circuit is shown in Figure 16-2. To add a $\overline{\text{PRE}}$ and $\overline{\text{CLR}}$ function, the last latch is a 3-input NAND gate. An additional 3-input NAND gate (U3-3) is used to gate the clock to the first flip-flop in order to inhibit the clock when either $\overline{\text{PRE}}$ or $\overline{\text{CLR}}$ is LOW. Although this gate inverts the clock, it saves chip count over adding an AND gate. In addition, one of the 3-input NAND gates is connected as an inverter (U1–3).

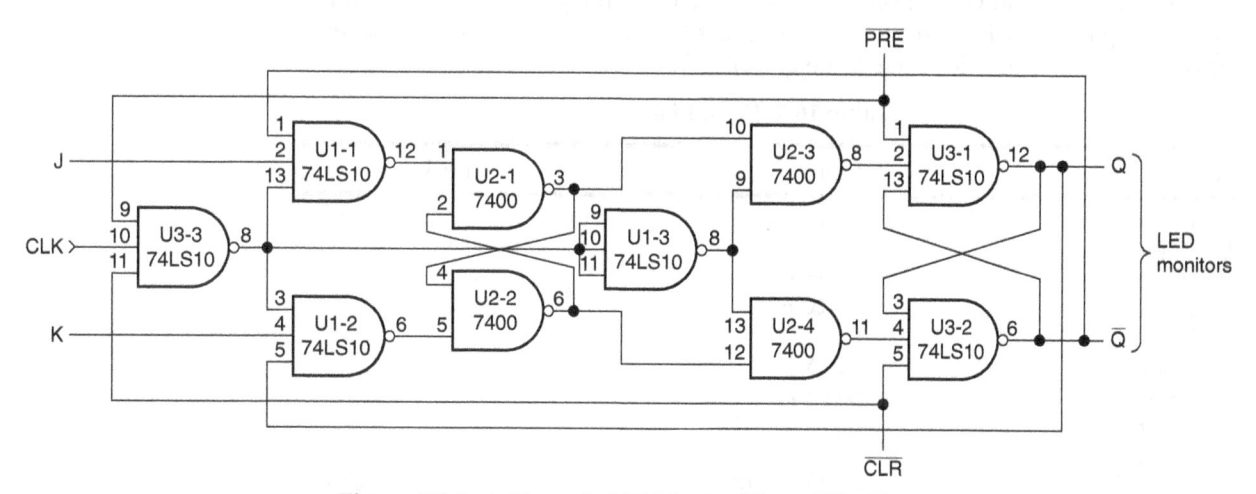

Figure 16-2 A Generic J-K Master/Slave Flip-Flop

For reference, the circuit is also shown as a Multisim simulation in Figure 16-3 with a function generator shown instead of a pushbutton for the clock. If you have Multisim, you might want to put this simulation together in Multisim as shown. By running the function generator at 1 kHz, you can observe the waveforms on the simulated oscilloscope. You can substitute a switch for the generator to examine details of the circuit and add probes to test logic within the flip-flop.

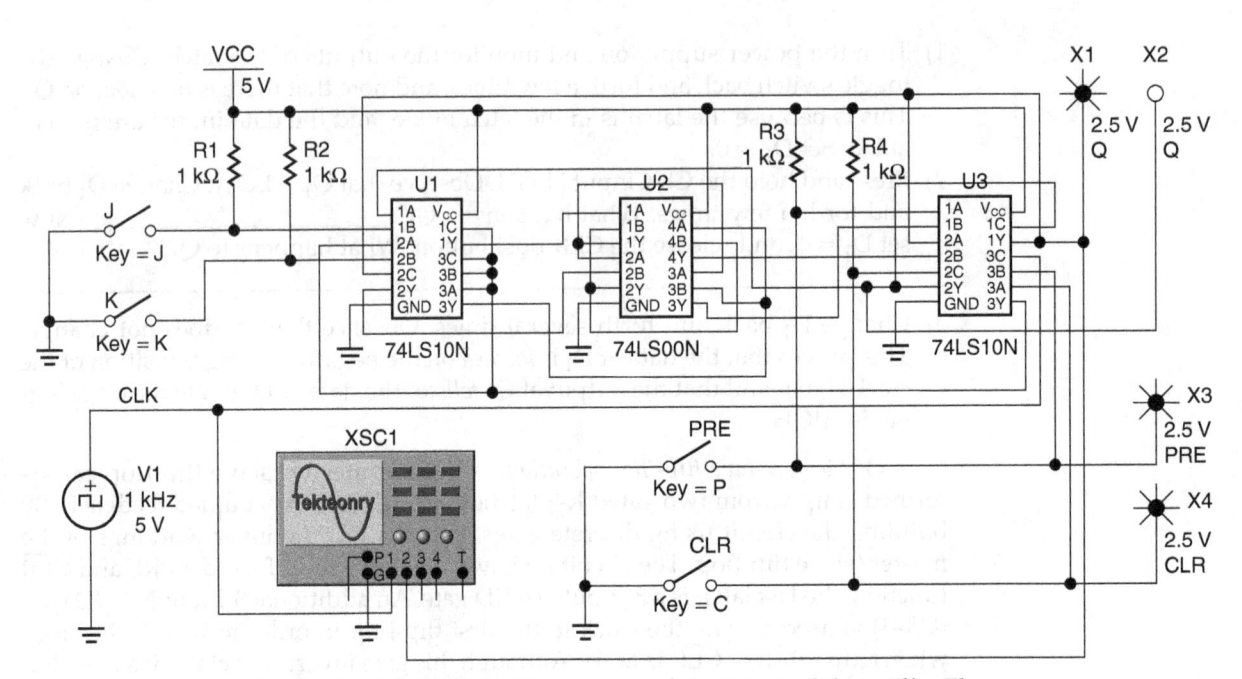

Figure 16-3 Multisim Simulation of the Master/Slave Flip-Flop

Install a 7LS10, a 7400, and a second 74LS10 IC on the circuit board, numbering them U1, U2, and U3 respectively. Wire the circuit in Figure 16-2. The circuit is fairly complex, so it is useful to use a wire list for the ICs as given in Table 16-1 (showing signal inputs for each of the gates).

Table 16-1 Wire List

To	From	Checked
U1 1A pin 1	U3 2Y pin 6	
U1 1B pin 2	J	
U1 1C pin 13	U3 3Y pin 8	
U1 2A pin 3	U1 1C pin 13	
U1 2B pin 4	K	
U1 2C pin 5	U3 1Y pin 12	
U1 3A pin 9	U1 3B pin 10	
U1 3B pin 10	U1 3C pin 11	
U1 3C pin 11	U1 1C pin 13	
U2 1A pin 1	U1 1Y pin 12	
U2 1B pin 2	U2 2Y pin 6	
U2 2A pin 4	U2 1Y pin 3	
U2 2B pin 5	U1 2Y pin 6	
U2 3A pin 10	U2 1Y pin 3	
U2 3B pin 9	U1 3Y pin 8	
U2 4A pin 13	U2 3B pin 9	
U2 4B pin 12	U2 2Y pin 6	
U3 1A pin 1	$\overline{PRE}$	
U3 1B pin 2	U2 3Y pin 8	
U3 1C pin 13	U3 2Y pin 6	
U3 2A pin 3	U3 1Y pin 12	
U3 2B pin 4	U2 4Y pin 11	
U3 2C pin 5	$\overline{CLR}$	
U3 3A pin 9	$\overline{CLR}$	
U3 3B pin 10	$\overline{PRE}$	
U3 3C pin 11	CLK (from gen)	

PROCEDURE

Make the following connections to the ICs:

1) Connect V_{cc} and DC SET to +5 V, GND to power ground on each of the three ICs.
2) Connect toggle switches to the J and K inputs.
3) Connect a normally LOW pushbutton switch to CLK.
4) Connect a normally HIGH pushbutton switch to $\overline{PRE}$.
5) Connect a normally HIGH pushbutton switch to $\overline{CLR}$.
6) Connect LED monitors to Q and $\overline{Q}$ (or monitor the outputs with a logic probe).

d) *Synchronous operation:* To test the synchronous operation of the master/slave flip-flop, do the following steps:

1) Set J = 1 and K = 0. Turn the power on and note the states of Q and $\overline{Q}$. If the flip-flop is not cleared (Q = 0), then pulse the $\overline{CLR}$ input LOW momentarily.
2) Press and hold the CLK input HIGH. You should observe that this has no effect on the outputs. Now release the pushbutton. What happens to Q?

___

Pulse the CLK input several more times, and note that this has no effect on the outputs.

3) Change J to 0. Note that this has no effect on the outputs. Pulse the CLK input several times. You should observe that this also has no effect on the outputs. Why?

___

4) Change K to 1, and note that Q does not change. Press and hold the CLK input HIGH. What happens to Q? _____ Now release the CLK pushbutton. What happens to Q now?

___

Pulse the CLK input several more times, and note that Q does not change.

5) Change J to 1. Note that Q remains LOW. Press and hold the CLK pushbutton HIGH. What happens to Q? _____

Release the pushbutton. What happens to Q now? _____

Pulse the CLK input several more times. You should observe that Q changes states on each CLK pulse.

___

e) In step d, you should have observed that the flip-flop loaded the J and K inputs only when the CLK was HIGH, and they were transferred to Q and $\overline{Q}$. on a negative-going transition at CLK. Now you will observe the chief disadvantage of the master/slave: Data at the J and K inputs can affect the flip-flop outputs any time while the CLK input is HIGH. Set J = 0 and K = 1. Clear the flip-flop by momentarily pulsing the $\overline{CLR}$ input LOW. Press and hold the CLK pushbutton HIGH. Change J to 1 and then back to 0. Now release the pushbutton. You should observe that Q changes to 1 even though J = 0 and K = 1 at the time of the negative-going transition. This demonstrates that, should an unwanted glitch or noise spike occur on J or K while the CLK input is HIGH, it may cause the flip-flop outputs to be invalid when CLK goes LOW.

f) *Asynchronous operation:* The $\overline{PRE}$ and $\overline{CLR}$ inputs are asynchronous inputs. Verify that this is so.

g) *Review:* This concludes the exercises on flip-flops and latches. To test your understanding of the principles covered in this experiment, answer the following questions:

1. The data input of a 74LS75 D latch is enabled when the CLK input is _____.
   The data present at the D input is latched to the output when _____.
2. How did the flip-flop in this experiment differ from the 74LS76 in Experiment 15?

3. How does a latch differ from a flip-flop?

4. Why was it necessary to gate the clock using the $\overline{\text{PRE}}$ and $\overline{\text{CLR}}$ inputs?

Experiment 17

Name _____

SCHMITT TRIGGERS, ONE-SHOTS, AND ASTABLE MULTIVIBRATORS

OBJECTIVES

1. To investigate the operation of a 7414 IC Schmitt trigger inverter.
2. To investigate the operation of a 74121 IC nonretriggerable one-shot (OS) multivibrator.
3. To investigate the application of the one-shot multivibrator in timing and control circuits.
4. To investigate the operation of a 555 IC timer in the astable mode.

TEXT REFERENCES

Read sections 5.20, 5.21, and 5.23.

EQUIPMENT NEEDED

Components
555 IC;
7400 IC;
7414 IC;
74121 IC (2);
1N914 diode (2);
normally LOW pushbutton switch (2);
100 µF capacitor;

0.01 µF capacitor;
680 pF capacitor;
6.8 k-ohm resistor;
10 k-ohm resistor;
33 k-ohm resistor;
student-selected resistors and capacitors;
2 LED monitors.

Instruments
0–5 volt DC power supply;
12.6 VAC power supply;
pulse or square wave generator;
dual trace oscilloscope.

DISCUSSION

In Experiments 15 and 16, you investigated the behavior of flip-flop or bistable devices. In the current experiment, you will turn your attention briefly to devices that are somewhat related to flip-flops: the Schmitt trigger, one-shot, and astable multivibrator.

Schmitt Trigger

When logic circuit inputs have slow rise or fall times, the circuits can fail to operate correctly. For this reason, devices are necessary to square up the incoming signals. One device is the Schmitt trigger, whose symbol and response is shown in Figure 17-1. Note that as the input waveform exceeds the positive-going threshold, V_{t+}, the output changes from HIGH to LOW. Once the output goes LOW, it stays there even after the input drops back below V_{t+}. The input must drop below V_{t-}, the negative-going threshold voltage, before the output can change from LOW to HIGH. In the current experiment, you will demonstrate that a 74LS14 Schmitt trigger may be used to produce a clock signal from a sine wave source.

One-Shots

A one-shot (OS) is a monostable device, that is, a device with one stable state. It may be triggered by an input signal to switch to the opposite states, at which it will remain for a fixed amount of time, usually dependent on some external RC time constant. After the fixed time has elapsed, the one-shot returns to its stable state. You will investigate the behavior of the 74121 IC one-shot.

Astable Multivibrators

Astable multivibrators have no stable state. Instead, they can be wired to oscillate and produce clock signals. One popular IC that can be wired as an astable multivibrator is the 555 timer, which you will investigate in the current experiment.

PROCEDURE

PROCEDURE

a) *7414 Schmitt trigger inverter operation:* Refer to the data sheet for a 7414 IC, and draw its pin layout diagram:

b) Construct the circuit illustrated in Figure 17-1. Make sure that the protective diodes D1 and D2 are connected properly before applying power.

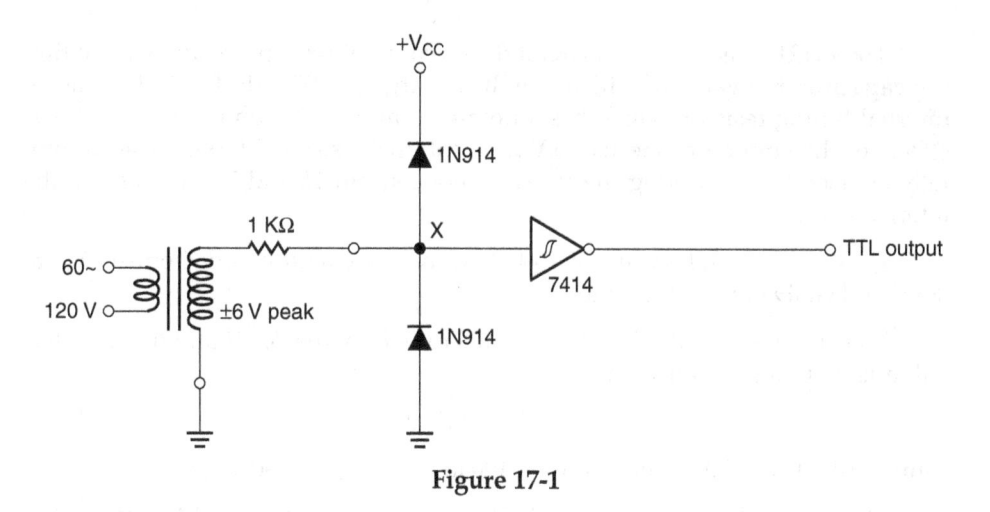

Figure 17-1

c) Connect one vertical input of the oscilloscope to point "x" of the circuit and the other vertical input to the TTL output. Apply power and adjust the oscilloscope for a few cycles' display. With the vertical position controls, superimpose the two waveforms on one another. Be sure to keep the vertical gain of the two inputs the same. Draw the resulting oscilloscope display using Timing Diagram 17-1.

Timing Diagram 17-1

d) Estimate the values of the two threshold points:

$$V_{t+} = \underline{\qquad}$$

$$V_{t-} = \underline{\qquad}$$

Look up the values for V_{t+} and V_{t-} given on the data sheet for the 7414 and compare these values to your estimated values.

e) Refer to the data sheet for a 74121 IC, and draw its pin layout diagram:

The 74121 IC is a one-shot circuit that is complete except for an external timing capacitor, the value of which may be as large as 1000 µF. The 74121 has an internal timing resistor, which has a nominal value of 2 k-ohm. This resistor is effective when pin 9 is connected to V_{cc}. An external resistor (2 k-ohm to 40 k-ohm) may be used by connecting the resistor between pin 11 and V_{cc}, bypassing the internal resistor.

f) *74121 IC operation:* Install a 74121 IC on the circuit board. Connect V_{cc} to +5 V and GND to power ground.

You are to bypass the 2 k-ohm internal resistor of the 74121 and use an external resistor. Using the formula

$$T_p = 0.7 R_t C_t$$

compute T_p if R_t = 33 k-ohm and C_t = 100 µF: _____ seconds.

g) Connect R_t and C_t to the 74121. Connect a normally LOW pushbutton switch to input B of the 74121. Connect A_1 and A_2 to LOW. Connect LED monitors to Q and $\overline{Q}$ of the 74121.

h) Turn the power supply on. Note that Q = 0 and $\overline{Q}$ = 1. Pulse input B HIGH, and observe that the outputs have changed states. In a time approximately equal to that computed in step b, the outputs should return to their original states. When Q has returned to 0, pulse input B repeatedly several times. You should observe that once Q has changed to 1, the OS does not respond to any more triggering until Q returns to 0.

i) Disconnect the pushbutton switch from input B, and replace it with the output of a square wave generator. Change R_t and C_t so that T_p = 5 milliseconds (close as possible). Remember that R_t can range from 2 to 40 k-ohm, and C_t can be as large as 1000 µF. Use a x10 probe, and connect one vertical input of the oscilloscope to output Q of the OS. Connect the output of the generator to the other vertical input. Trigger the oscilloscope on the positive-going transition of the square wave signal. Set the generator to 50 Hz. Draw the waveforms displayed on the oscilloscope using Timing Diagram 17-2.

PROCEDURE

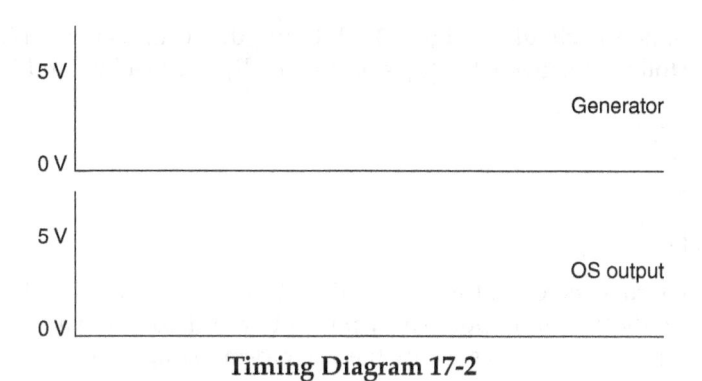

```
5 V
                                                          Generator
0 V

5 V
                                                          OS output
0 V
```
Timing Diagram 17-2

Change the generator frequency to 1 kHz. What effect does this have on the frequency of the OS output? _____. Does the pulse width of the OS output change? _____.

j) *Using one-shots to delay a pulse:* In this step, you will cascade another one-shot to the circuit of the first one-shot already on the circuit board. The purpose of this circuit is to produce a pulse that is *delayed* (with respect to the generator signal) by an amount of time that is approximately equal to the T_p of the first OS (5 m sec). The width of the delayed pulse will be approximately equal to the T_p of the second OS. For this exercise, you will use a T_p of 1 millisecond for the second OS.

Install another 74121 IC on the circuit board. Connect V_{cc} to +5 V and GND to power ground. You will cascade the second OS (OS2) to the first (OS1). Choose R_t and C_t so that $T_p = 1$ millisecond. Connect inputs A_1 and A_2 to ground and the B input to the Q output of OS1. Monitor the output of the generator, which is set to 50 Hz, and the output of OS2 with the dual trace oscilloscope. Draw the waveforms displayed, using Timing Diagram 17-3.

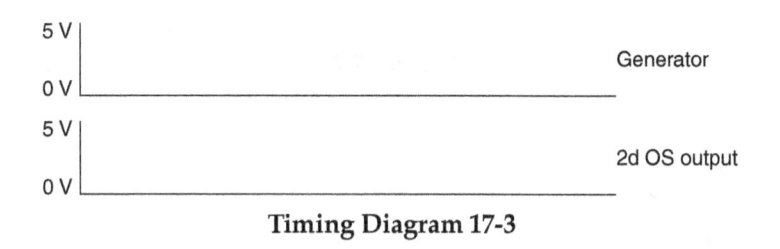

```
5 V
                                                          Generator
0 V

5 V
                                                          2d OS output
0 V
```
Timing Diagram 17-3

k) *555 IC timer operation:* Refer to the data sheet for a 555 IC timer, and draw its pin layout diagram:

l) Examine the circuit in Figure 17-2. Using the formulas given in the text, compute the following, given that $R_A = 6.8$ k-ohm, $R_B = 10$ k-ohm, and $C = 680$ pF:

1) $t_1 =$ _____
2) $t_2 =$ _____
3) $T =$ _____
4) $f =$ _____
5) duty cycle = _____

m) Construct the circuit in Figure 17-2. Connect one vertical input of the oscilloscope to the circuit output. Apply power to the circuit, and adjust the oscilloscope to display a few cycles. Draw the waveform using Timing Diagram 17-4.

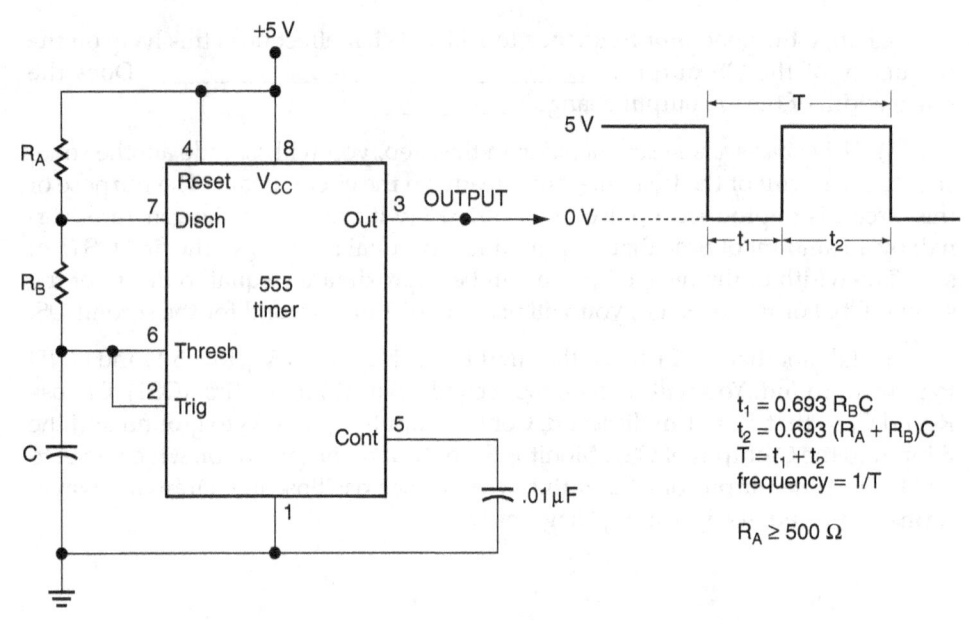

$t_1 = 0.693\, R_B C$
$t_2 = 0.693\, (R_A + R_B) C$
$T = t_1 + t_2$
frequency $= 1/T$

$R_A \geq 500\, \Omega$

Figure 17-2

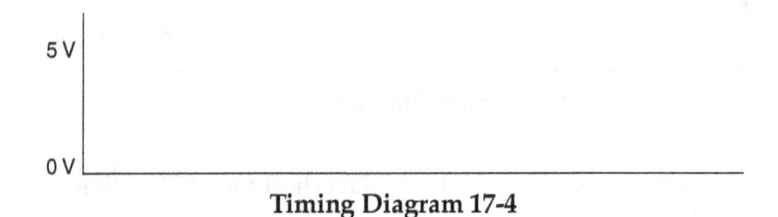

Timing Diagram 17-4

n) Estimate the period of the output waveform: _____. Is this value close to the value computed in step 1? If not, recheck the values you computed in step 1, and then, if the period estimated in this step is still appreciably different, check your circuit.

o) *Review:* This concludes the exercises on the Schmitt trigger, one-shot, and astable multivibrator. To test your understanding of the principles covered in this experiment, answer the following questions:

PROCEDURE

1. a) Why are the 1N914 diodes in Figure 17-1 necessary?

   b) What type of input signal to the circuit would make the diodes unnecessary?

2. Show how you would wire the 74121 IC inputs to respond to a negative-going transition (NGT).

3. The maximum value for R_t is _____, while the maximum value for C_t is _____.

4. If R_t = 10 k-ohm, what value must C_t be to produce a pulse having a width of 7 seconds? C_t = _____.

5. Explain the pulse delay circuit of step j.

6. An OS always returns to the Q = _____ state.

7. True or false: The output pulse width of a nonretriggerable OS is independent of the input frequency and pulse width.

8. In Figure 17-2, if R_A = 10 k-ohms, R_B = 2.2 k-ohms, and C = 680 pF, compute the following values:

   t_1 = _____

   t_2 = _____

   T = _____

   f = _____

   Duty cycle = _____

Experiment 18

Name _____

DESIGNING WITH FLIP-FLOP DEVICES

OBJECTIVES

1. To investigate the application of J-K flip-flops in counting circuits.
2. To investigate the application of D flip-flops in data registers.
3. To investigate parallel data transfer.
4. To investigate the application of J-K flip-flops in shift register circuits.
5. To design a programmable counter.

TEXT REFERENCES

Read sections 5.17 through 5.19.

EQUIPMENT NEEDED

Components
7400 IC;
7404 IC;
74LS74A IC (2);
74LS76 IC (2);
3 toggle switches;
normally HIGH pushbutton switch;
normally LOW pushbutton switch;
6 LED monitors.

Instruments
0–5 volt DC power supply;
pulse or square wave generator;
dual trace oscilloscope.

DISCUSSION

Now that you have become acquainted with the fundamentals of flip-flops, you are ready to apply your knowledge in designing and constructing circuits that use flip-flops. Two very important applications of flip-flops are counters and registers.

Counters

A counter is a digital device capable of producing an output, which represents a sequence of numbers. The output count of the counter is binary and is usually triggered by a clock signal. Each bit of the binary count is produced by a flip-flop. Among the various types of counters used in digital circuits is the so-called MOD counter. The MOD counter has a count sequence that may start at 0 and end at some number $N-1$ and repeat. The number N is called the modulus (MOD-number) of the counter and is equal to 2^M, where M is the number of flip-flops. The number N can be changed for a given MOD counter. In the current experiment, you will investigate these principles and also be asked to design a counter whose MOD-number can be changed by "programming" the counter.

Registers

While counters are a very important application of flip-flops, probably the most common application is the register. A register is simply a collection of flip-flops used for storing information (data) temporarily. In the current experiment, you will construct a simple register for storing data, and then investigate two methods of transferring data to and from such registers.

PROCEDURE

a) *Counter application:* Figure 18-1 shows the circuit for a three-bit binary counter. Examine the circuit closely, then construct it. Arrange the order of the flip-flops exactly as the diagram shows it.

b) Connect a normally HIGH pushbutton switch to the clock input of flip-flop X_0. Connect LED monitors to the Q outputs of each flip-flop. The order of the LEDs is important, since the output of the rightmost flip-flop will represent the LSB of the count and the leftmost the MSB. Connect all DC SET inputs to V_{cc} and all DC CLEAR inputs to a single normally HIGH pushbutton switch.

c) Turn the power on, and clear the counter by pulsing the DC CLEARs LOW momentarily. The number stored in the counter is indicated by the LEDs, which should all be OFF (i.e., the number should be 000_2). Test the counter circuit by pulsing the clock input and observing the count indicated by the LED monitors. Record your observations in Table 18-1.

Your results should indicate that the counter counts to a maximum of 7 and recycles to 000 on the eighth clock pulse.

PROCEDURE

Figure 18-1

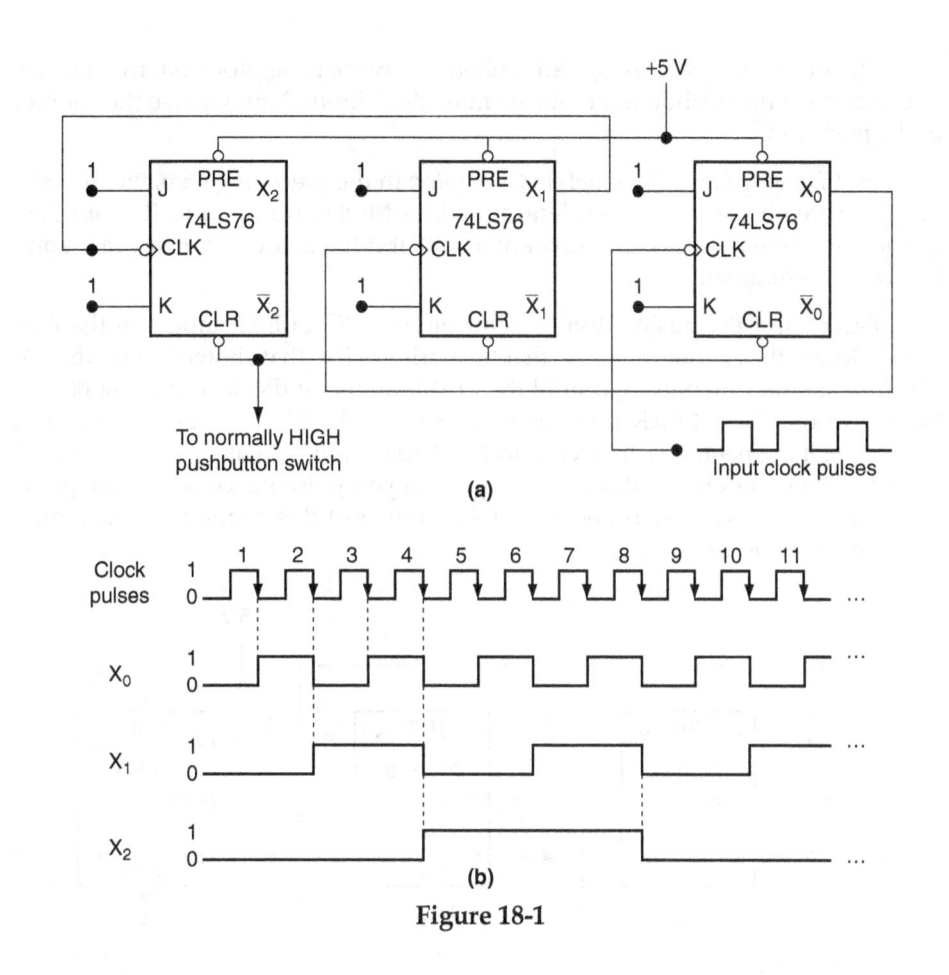

Table 18-1

Clock Pulse	Output State		
	X_2	X_1	X_0
0	0	0	0
1			
2			
3			
4			
5			
6			
7			
8			

Disconnect the pushbutton switch from the clock input of the counter. Connect the output of a square wave generator, set at 10 kHz, to the clock input of the counter and to one vertical input of a dual trace oscilloscope. Connect the other vertical input of the oscilloscope to the Q output of flip-flop X_0. Trigger on output Q of flip-flop X_0. What is the frequency of the signal at X_0? _____. Move the oscilloscope input from X_0 to X_1. Trigger on X_1. What is the frequency of this signal? _____. Finally, move the oscilloscope input from X_1 to X_2. Trigger on X_2. What is the frequency of this signal? _____. Based on your observations, what is the MOD-number of this counter? _____.

Disconnect the oscilloscope and the square wave generator from the counter, and reconnect the pushbutton to the counter clock input. You will use the counter in the next step.

d) *Changing the MOD-number of a counter:* In the preceding step, you investigated a three-bit counter. You determined its MOD-number and its count sequence. You will now modify the counter so that it has a new count sequence and a new MOD-number.

Disconnect the pushbutton switch from the DC CLEAR inputs of the flip-flops. Modify the counter circuit so that the wiring is like that shown in Figure 18-2. Clear the counter by pulsing it until the LED monitors indicate a count of 000, or by lifting the DC CLEAR line connection from the NAND gate output and grounding it momentarily, then reconnecting it to the NAND gate. Using Table 18-2, record the counter output states you observe as you pulse the counter through its new count sequence. Determine the MOD-number of this counter by examining the counter sequence: _____.

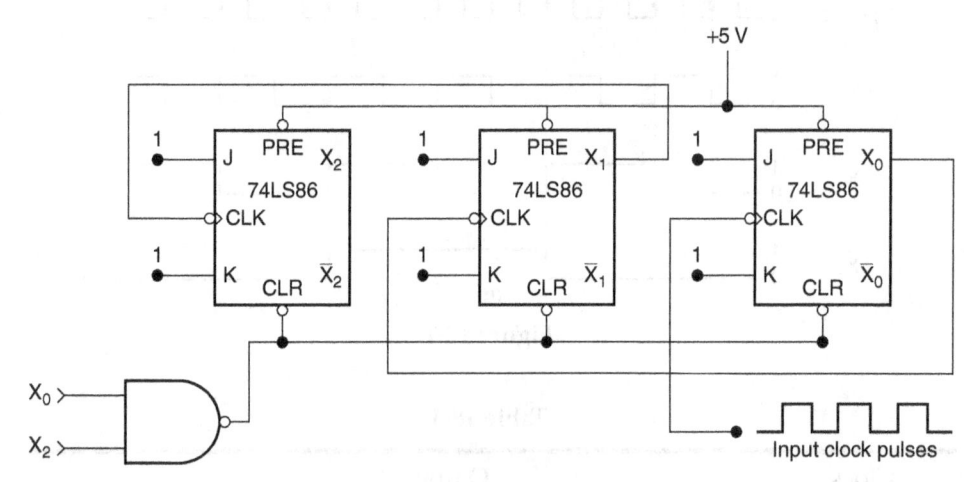

Figure 18-2

Table 18-2

Clock Pulse	Output State X_2 X_1 X_0
0	0 0 0
1	
2	
3	
4	
5	
6	
7	
8	

PROCEDURE

e) *Parallel data transfer:* Many times it is necessary to store binary data temporarily. A device that performs this function is called a *register*. A register can be a set of D flip-flops, one for each bit to be stored. Figure 18-3 shows two 3-bit registers, X and Y. While each has the same storage capacity, the data in X may be transferred to Y, where it may be stored as long as necessary, thus permitting register X to receive new data. All three bits are transferred simultaneously whenever register Y receives the transfer pulse.

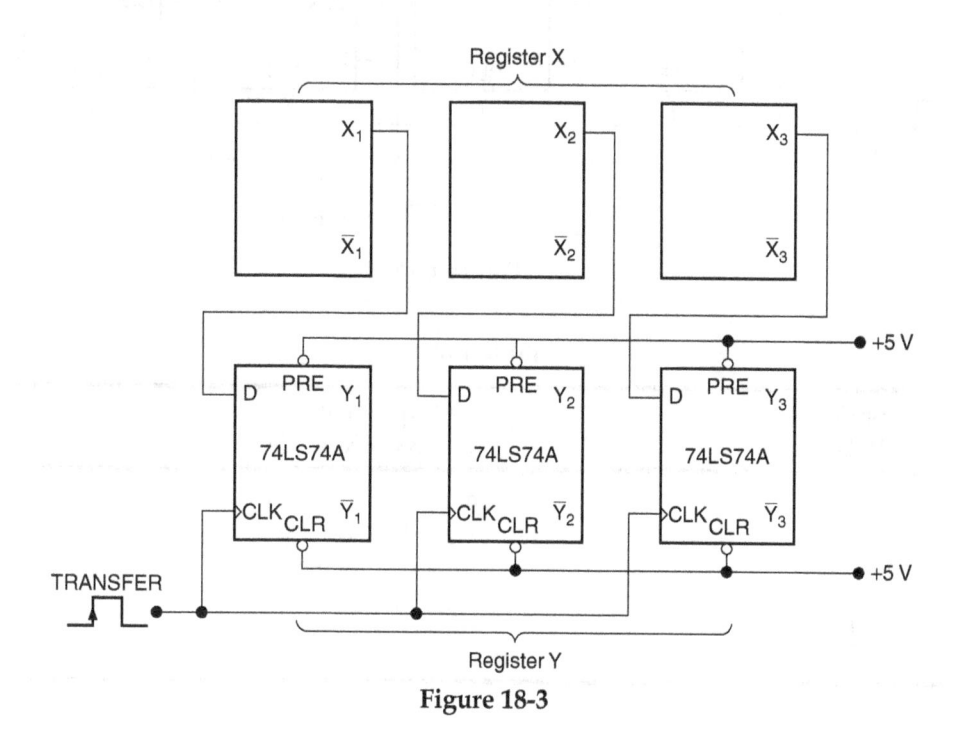

Figure 18-3

f) Build register Y using 74LS74A D flip-flops. Connect the D inputs of register Y to the outputs of the three-bit counter on the circuit board. Connect LED monitors to the outputs of register Y. Finally, connect all clock inputs of register Y together to a single normally LOW pushbutton switch.

g) Clear the counter by using the method of step d, then pulse the transfer pushbutton switch HIGH momentarily. You should observe that the LEDs of the counter and the register are OFF. Now pulse the counter once, and again pulse the transfer pushbutton HIGH momentarily. You should observe that the contents of the counter (001) are transferred to the register, all three bits simultaneously. Repeat this procedure several times, verifying that the data in the counter is transferred to the register each time.

h) *Serial shift register application:* In steps e–g you constructed and tested a parallel shift register. Actually, the proper name for this register is a parallel in/parallel out (P.I.P.O.) register. In this step, you will investigate another type of shift register, a serial shift register. Examine the circuit in Figure 18-4 closely, then modify and construct the circuit using four 74LS76 J-K flip-flops.

Connect a toggle switch to the data input and a normally HIGH pushbutton switch to the shift pulse input. Connect LED monitors to the Q outputs of each flip-flop. Finally, connect all DC SET inputs to V_{cc} and all DC CLEAR inputs to a single normally HIGH pushbutton switch.

Figure 18-4

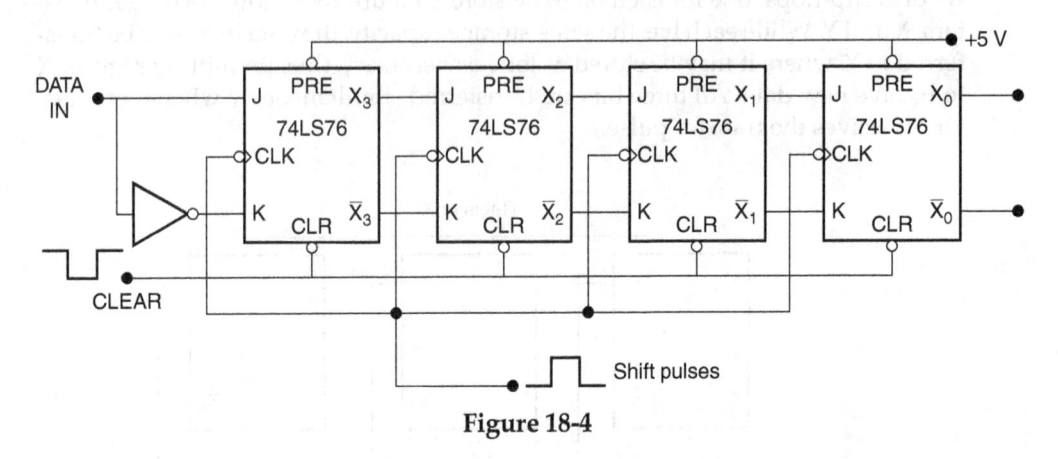

Table 18-3

Shift Pulse	Output State			
	X_3	X_2	X_1	X_0
0	0	0	0	0
1	1	0	0	0
2				
3				
4				
5				

Clear the register by pulsing the CLEAR line LOW momentarily. Set the data toggle switch HIGH and pulse the shift line once. The LEDs should indicate 1000, reading from left to right. Set the data switch LOW, and pulse the shift line four more times, each time observing the output LEDs. Record your observations in Table 18-3.

You should have observed that the 1, which was shifted into the flip-flop X_3 on shift pulse 1, was subsequently shifted to the right until it was shifted out of flip-flop X_0 on the last shift pulse applied (shift pulse 5). Each time the 1 was shifted, a 0 replaced it.

If the purpose of the serial shifting is to load a number into the register, this may be accomplished by setting or resetting the data input toggle switch after each shift pulse is applied. For example, to shift the number 0101 into the register, you would set the toggle switch to 1 before the first shift pulse is applied, then reset the switch before applying the next shift pulse. Repeating this procedure one more time will leave the desired number in the register. Try shifting this number into the register. Next, try shifting other numbers into the register, such as 1011.

Data can be taken from this register in two ways, parallel or serial. The serial output can be taken at the Q output of flip-flop X_0, while parallel data can be taken from the Q outputs of each flip-flop in a manner like that of steps e–g.

i) *A programmable counter:* Design a four-bit counter that is controlled by two control lines x and y and that behaves according to Table 18-4.

PROCEDURE

Table 18-4

Program Switch		Counter Mode
X	Y	
0	0	No Count
0	1	Mod 5
1	0	Mod 10
1	1	Mod 12

Draw the logic diagram in the space provided below.

j) *Review:* This concludes the exercises on flip-flop applications. To test your understanding of the principles investigated in this experiment, answer the following questions:

1. The counter you constructed in this experiment was called a MOD-8 counter or a divide-by-8 counter. Using the results recorded in Table 18-1, explain how the counter can be used to divide the frequency of a signal by 8.

2. Suppose you were to add another flip-flop to the circuit, making a total of four flip-flops. The maximum count this counter would indicate would be _____, and the MOD number would be _____. Its frequency division capability would be _____.

3. The text explains the use of the DC SET and CLEAR inputs to transfer data *asynchronously* into a J-K flip-flop. Explain the difference between this type of transfer and the parallel transfer method employed in steps e–g.

4. Making use of your observations recorded in Table 18-3, how many shift pulses are necessary to serially transfer data from one 8-bit register to another? _____.
5. Based on your knowledge of parallel and serial data transfer, does this experiment verify the relative speeds of the two methods? _____.
6. Explain how the modification you made to the counter in step d changed the count sequence.

7. How would you modify a binary counter to cause it to stop counting at a particular count even though clock pulses are continually applied?

Experiment 19

Name _____

TROUBLESHOOTING FLIP-FLOP CIRCUITS

OBJECTIVES

1. To practice troubleshooting counting circuits containing IC flip-flops.
2. To practice troubleshooting register circuits containing IC flip-flops.
3. To investigate a flip-flop device timing problem, clock skew.

TEXT REFERENCES

Read sections 5.3 and 5.24.

EQUIPMENT NEEDED

Components
7400 IC;
74LS74A IC (2);
74LS76 IC (2);
normally HIGH pushbutton switch (2) and normally LOW pushbutton switch (2), all debounced;
4 LED monitors.

Instruments
0–5 volt DC power supply;
pulse or square wave generator;
dual trace oscilloscope;
ohmmeter;
logic pulser;
logic probe.

DISCUSSION

Circuits containing flip-flops can develop many of the faults that occur in combinatorial circuits. For example, flip-flops can develop opens and shorts at their inputs and outputs the same as gates do. Unlike gates, however, flip-flops possess memory, and it is this characteristic that can cause flip-flop circuits to behave differently than combinatorial circuits. For example, when trying to determine the cause for a given flip-flop output condition, you must remember that the signals causing the condition may not be present at the inputs to the flip-flop.

In the current experiment, you will first construct a simple asynchronous counting circuit and verify that the counter operates properly. Remember, it is rather difficult to troubleshoot a circuit without knowing how the circuit normally works. You will then have someone put a "bug" (fault) into the circuit for you to find. As always, do not look for the bug, and even if you should suspect what the bug is, use your troubleshooting techniques to isolate the problem. Also, before coming into the lab to perform the experiments, place various bugs into the circuit on paper and try to predict the response of the circuit.

The second exercise is similar to the first. The device used here is a shift register that is synchronous. The last exercise will give you an opportunity to investigate one type of timing problem in flip-flop circuits referred to as clock skew.

PROCEDURE

a) *Troubleshooting flip-flop counters*: Examine the circuit and its timing diagram in Figure 19-1. Since you will be clocking the counter manually at first, use Table 19-1 to make a state table for the counter. Construct the counter. Use a pushbutton switch to clear the counter. Use a logic pulser, if available, to clock the counter and a logic probe to monitor the output of each flip-flop. If a pulser is not available, use a debounced pushbutton switch.

Table 19-1

Clock Pulse	Output State		
	X_2	X_1	X_0
0	0	0	0
1			
2			
3			
4			
5			
6			
7			
8			

b) To begin the exercise, have your instructor or lab assistant place a fault into your counter while your back is turned. Now test the circuit and make a state table using Table 19-2. If the two are different, your circuit is indeed not operating

PROCEDURE

correctly. In the space provided below, list as many possible faults as you can that could cause the circuit to malfunction in this manner.

Figure 19-1

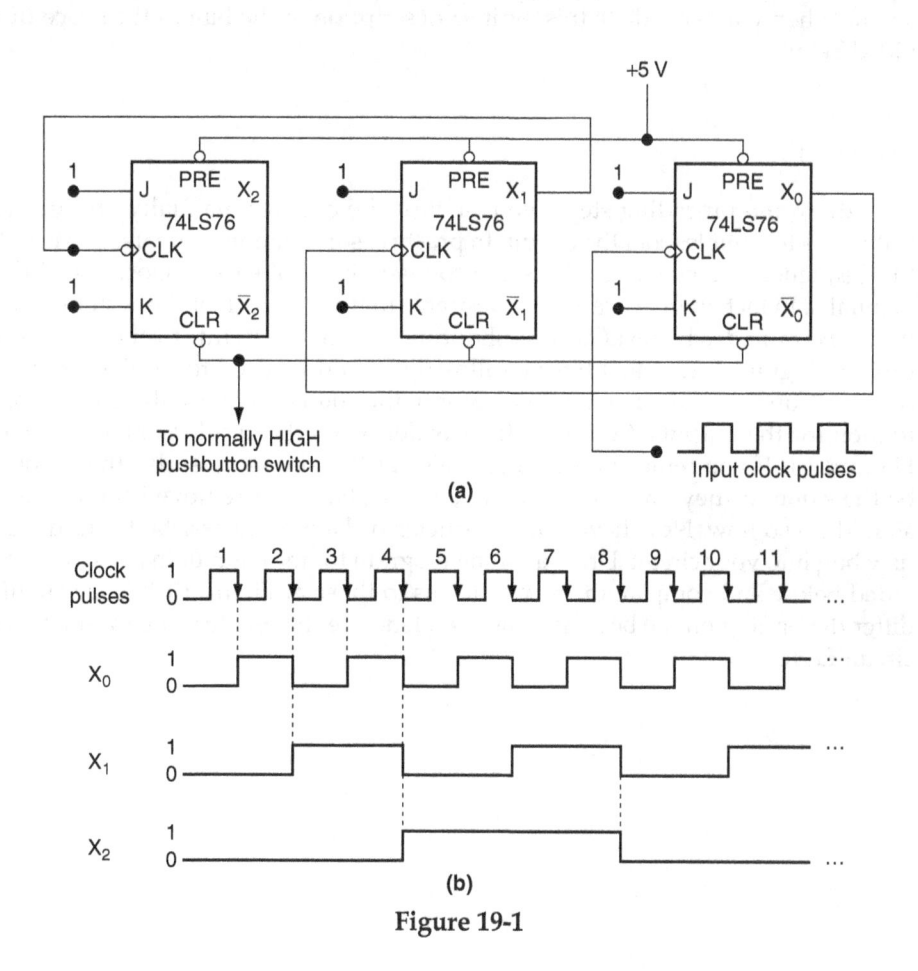

Table 19-2

Clock Pulse	Output State		
	X_2	X_1	X_0
0	0	0	0
1			
2			
3			
4			
5			
6			
7			
8			

c) Once you have listed as many faults as you can, proceed to troubleshoot the circuit. When you have isolated the bug, remove it and check the circuit once again. When you have done this, write a description of the bug in the space provided below.

d) In the preceding steps, you clocked the counter manually and used a state table to troubleshoot the circuit. In practice, some counters are used to divide the frequency of a signal, and in some counters it is not easy to clock the circuit manually. In fact, in many cases it is desirable to use the clock of the system that is being examined whenever troubleshooting a counter circuit. In these cases, a timing diagram and a dual trace oscilloscope could be used to troubleshoot the counter. Connect a square wave generator to the counter and use the oscilloscope to monitor the outputs. Compare the waveforms on the oscilloscope to those of Figure 19-1. If you removed the bug from steps b and c successfully, they should be the same. If they are not the same, you still have not removed the bug and should do so now. Now have your instructor or laboratory assistant introduce a new bug into your circuit. Draw a timing diagram for the counter in the space provided below and compare these waveforms to those of Figure 19-1. They should differ depending on the bug that was installed. Use the oscilloscope to isolate the circuit fault.

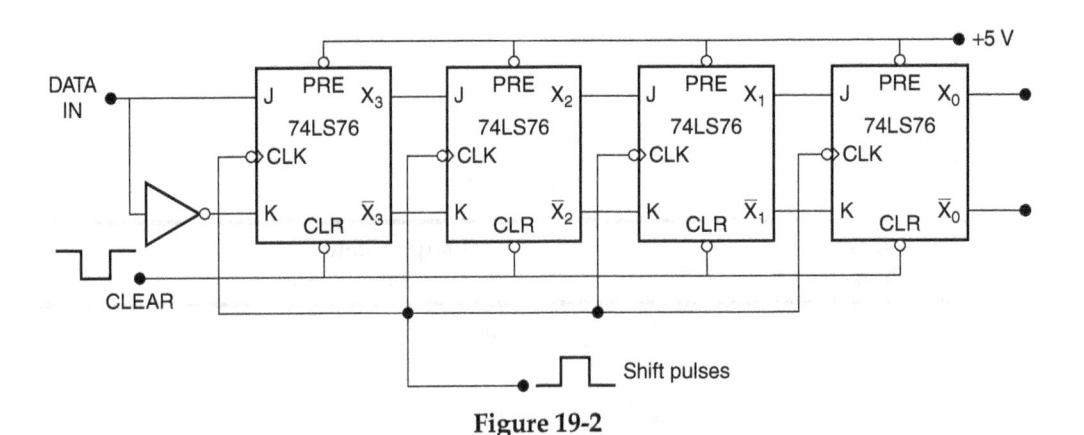

Figure 19-2

PROCEDURE

Table 19-3

Shift Pulse	Output State			
	X_3	X_2	X_1	X_0
0	0	0	0	0
1	1	0	0	0
2				
3				
4				
5				

e) *Troubleshooting flip-flop shift registers*: Examine the circuit and timing diagram of Figure 19-2. Use Table 19-3 to make a state table for the circuit. Construct the circuit. If a pulser is available, use the pulser for the clock. If a pulser is not available, connect a normally HIGH pushbutton switch (debounced) to the circuit's clock input. Connect the circuit's CLEAR input to a normally HIGH pushbutton switch. Connect a toggle switch to the data input (DATA IN).

f) Pulse the CLEAR input LOW momentarily to clear the register. Set the data input switch to HIGH. Pulse the clock input once and then set the data input switch LOW. Test the register by pulsing the clock enough times to cause the register to go through all of its remaining states, and check the output of each flip-flop after each clock pulse. Compare your observations with Table 19-3. When the register is operating correctly, have your instructor introduce a fault into the register circuit.

g) Test your circuit, and make a state table using Table 19-4. Compare your observations with Table 19-3. If they are not the same, you must troubleshoot. Begin by listing in the space provided below as many possible faults as you can that would cause the circuit to behave the way it currently does.

Table 19-4

Shift Pulse	Output State			
	X_3	X_2	X_1	X_0
0	0	0	0	0
1	1	0	0	0
2				
3				
4				
5				

h) Use the pulser (or debounced switch) and probe to isolate the fault. When you believe you have found the fault, write a brief description of the bug below.

---

i) *Troubleshooting timing problems—clock skew*: Construct the circuit in Figure 19-3. If a logic pulser is available, use the pulser to clock the circuit. Connect a normally HIGH pushbutton switch to the CLEAR circuit. You will monitor the flip-flop outputs with a logic probe.

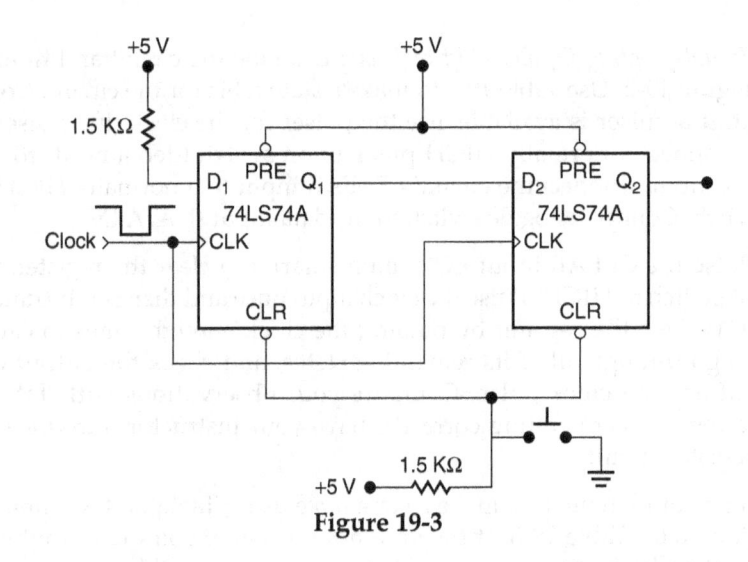

Figure 19-3

j) Test the circuit using the timing diagram provided in Figure 19-4. When the circuit is operating correctly, modify the circuit according to Figure 19-5. Now retest the circuit. You should notice that the circuit may not be functioning correctly, due to the additional propogational delay inserted in this step. If it still works correctly, add two inverters between the NAND gate and the CLK input of FF_2, and retest the circuit operation.

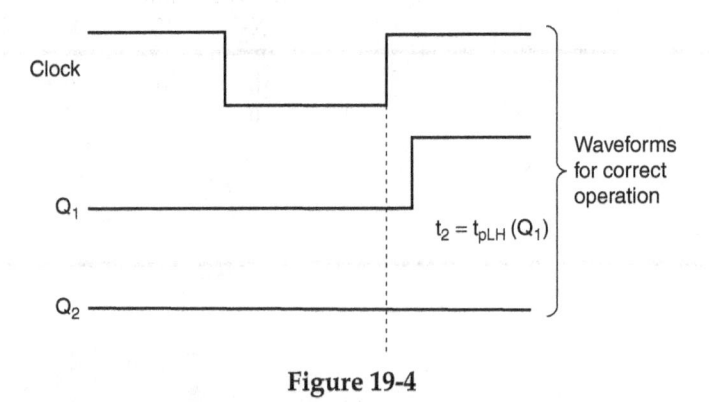

$t_2 = t_{pLH}(Q_1)$

Waveforms for correct operation

Figure 19-4

PROCEDURE

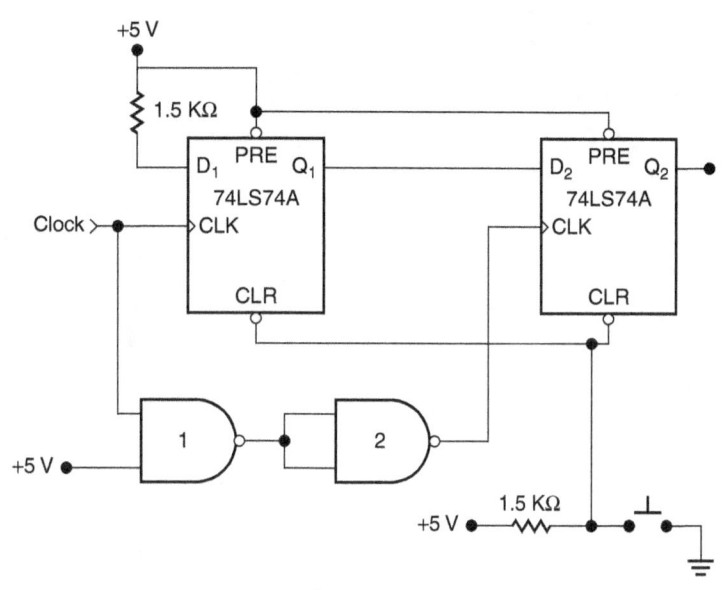

Figure 19-5

k) *Review*: This concludes the exercises on troubleshooting flip-flop devices. To test your understanding of the principles investigated in this experiment, answer the following questions:

1. What would happen to the counter in Figure 19-1 if $\overline{X}_1$ becomes shorted to ground?

2. In Figure 19-5, if FF_1 and FF_2 are both 74LS74A IC D flip-flops, and G_1 and G_2 are 74L00 IC NAND gates, would it seem likely that clock skew could be a problem? _____.

3. Answer review question 2 if FF_1 and FF_2 are both 74LS74A IC D flip-flops and G_1 and G_2 are 7400 IC NAND gates. _____

Experiment 20

Name _____

BINARY ADDERS AND 2'S COMPLEMENT SYSTEM

OBJECTIVES

1. To investigate the operation of a half adder.
2. To investigate the operation of a full adder.
3. To investigate the operation of a two-bit ripple adder.
4. To investigate the operation of a 74LS283 IC adder.
5. To investigate the operation of a 2's complement adder/subtractor circuit.

TEXT REFERENCES

Read sections 6.1 through 6.4 and sections 6.9 through 6.15.

EQUIPMENT NEEDED

Components
7400 IC;
7404 IC;
7432 IC;
74LS74A IC (4);
74LS283A IC;
74LS86 IC;
6 toggle switches;
normally HIGH pushbutton switch;
normally LOW pushbutton switch (2);
10 LED monitors.

Instrument
0–5 volt DC power supply.

DISCUSSION

At the heart of digital computers and calculators is the arithmetic unit. Depending on the system, this unit can range in complexity from the very simple, such as an adder, to a unit that possesses the capabilities of computing values for special functions, such as trigonometric and logarithmic functions. No matter how complex the unit is, its most important function is still addition.

At a minimum, arithmetic units usually consist of two flip-flop registers, one of which is called the *accumulator*, and the arithmetic circuits, which are special combinatorial circuits. The arithmetic circuit is fed data (operands) by the two registers and stores its answer into the accumulator.

In the current experiment, you will investigate one type of arithmetic circuit, the parallel binary adder. This circuit adds two binary numbers by operating on pairs of bits (a bit from one operand and the corresponding bit from the other operand) with units called full adders or half adders and generating carry bits that are fed to the next highest significant adder. Full adders have a third input to receive the carry bit output of the previous adder. Half adders do not and are thus used only to add the least significant bits of the two numbers being added. You will then investigate an IC four-bit parallel adder, the 74LS283, and use this IC to construct a 2's complement adder/subtractor.

PROCEDURE

a) *Half adder*: In review question 3, Experiment 12, you were asked to design the circuit for a half adder. Redraw your circuit in the space provided below.

PROCEDURE

b) Construct the half adder. Connect toggle switches to the two inputs and LED monitors to both outputs. When construction is completed, test the half adder, using Table 20-1 to record your observations on the sum and carry outputs.

Table 20-1

Inputs		Outputs	
A	B	S	C
0	0		
0	1		
1	0		
1	1		

c) *Full adder*: Draw the circuit for a full adder:

d) Construct the full adder. Connect toggle switches to each input and LED monitors to both outputs. When construction is completed, test the full adder, using Table 20-2 to record your observations on the sum and carry outputs.

Table 20-2

Inputs			Outputs	
A	B	C_i	S	C
0	0	0		
0	0	1		
0	1	0		
0	1	1		
1	0	0		
1	0	1		
1	1	0		
1	1	1		

e) *Two-bit ripple adder*: Connect the half adder and full adder together, according to the diagram shown in Figure 20-1 to form a two-bit ripple adder.

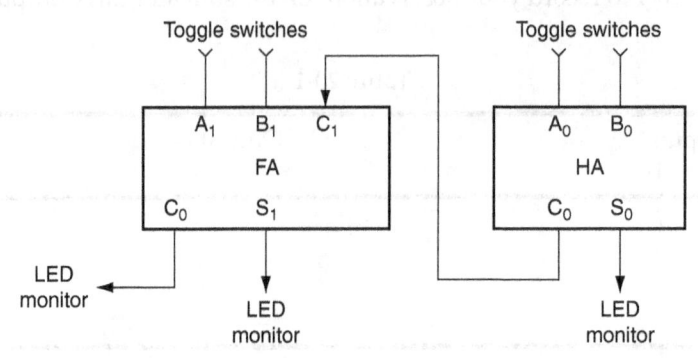

Figure 20-1

Test the operation of the adder by setting the toggle switches to several different values and observing the sum and carry indicated by the LED monitors. Demonstrate the circuit operation for your instructor.

f) *74LS283A IC adder operation*: Refer to the data sheet for a 74LS283 IC, and draw its pin layout diagram:

g) Install a 74LS283A IC on the circuit board, and make the following connections:

1) Connect V_{cc} to +5 V and GND to power ground.
2) Connect C_0 to power ground.
3) Connect toggle switches to inputs A_0 through A_3 and B_0 through B_3.
4) Connect LED monitors to sum outputs S_0 through S_3 and also to C_4. Note that S_0 through S_3 may be shown on the data sheet as Σ_0 through Σ_3.

Table 20-3

Inputs								Outputs			
A_3	A_2	A_1	A_0	B_3	B_2	B_1	B_0	C_4	S_3	S_2	S_1	S_0
0	0	1	1	0	0	0	1					
0	1	1	1	1	0	0	1					
1	0	1	1	0	1	0	1					
1	1	1	1	1	1	1	1					

PROCEDURE

h) Verify that the adder is operating correctly by entering the input values listed in Table 20-3 and recording your observations on the outputs in the table.

i) *Adder/Subtractor*: Examine the circuit of Figure 20-2. The circuit is for a parallel adder/subtractor using the 2's complement number system. Investigate the possibility of simplifying the circuit somewhat by examining the AND-OR circuits. The right simplification will permit you to replace the ADD and SUB control lines with a single control line, X. When X = 0, the circuit will function as an adder; when X = 1, it will function as a subtractor. When you believe that you have found a way to simplify the circuit, draw the simplified diagram below, and show it to your instructor or laboratory assistant. HINT: consider using exclusive-OR circuits as controlled inverters.

j) Construct the circuit you drew in step i. Connect toggle switches to each D input of register B (not shown). Also, connect a toggle switch to control line X. Connect LED monitors to the 74LS283 sum outputs (S_3 through S_0) and carry output C_4, switch X, and register A outputs (A_3 through A_0).

The clock inputs of register A flip-flops should be connected to a single normally LOW pushbutton switch (ADD/SUB pulse input). The clock inputs of register B should all be tied to a single normally LOW pushbutton switch (TRANSFER line, not shown). Tie the DC SET inputs for both registers A and B HIGH. Finally, tie the DC CLEAR inputs for register A to a normally HIGH pushbutton switch, so that the register may be cleared to begin a new sequence of operations.

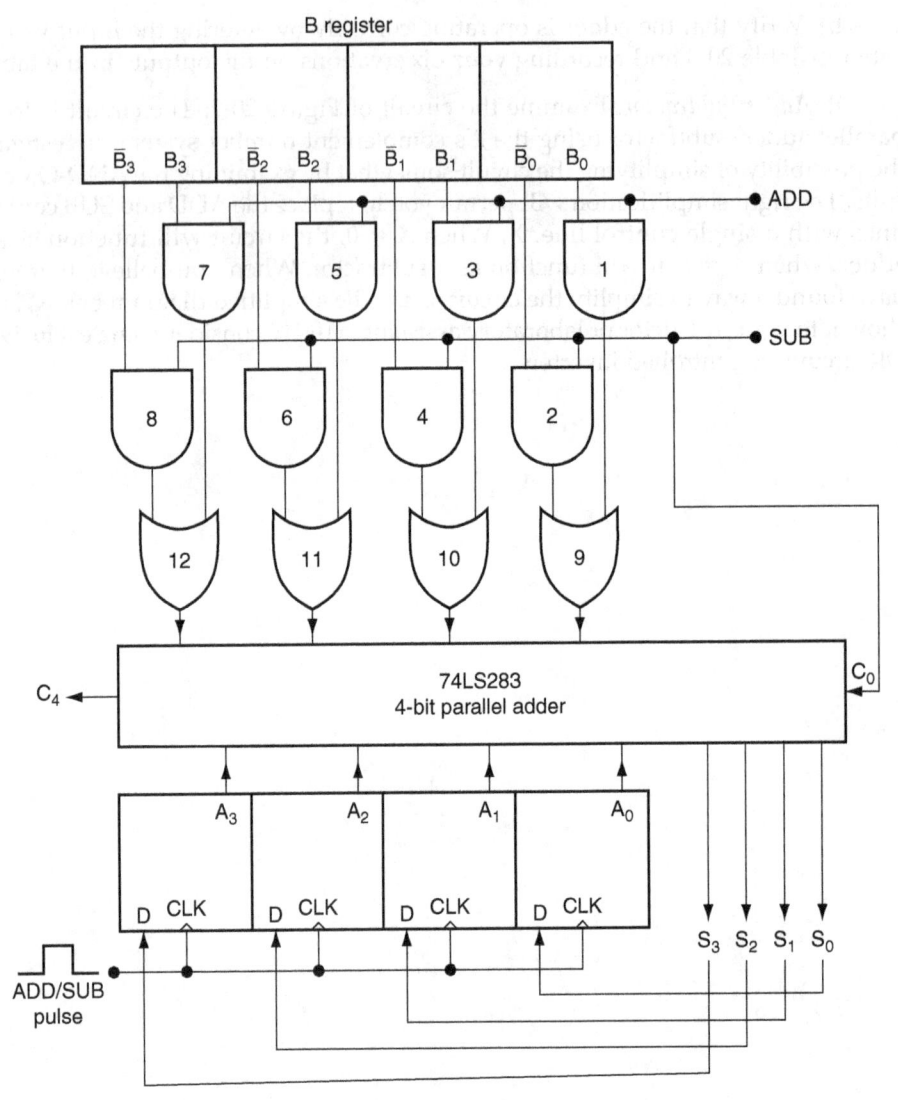

Figure 20-2

k) The circuit operates in the following manner:

1) To clear the unit, pulse the DC CLEAR pushbutton switch momentarily LOW.
2) To ADD a number to register A, set X to 0. To SUBTRACT a number from register A, set X to 1.
3) To enter a number to be added or subtracted, set toggle switches B_3 through B_0 to the value of the number, then pulse the TRANSFER line momentarily HIGH to clock the data into register B. The LED monitors at the 74LS283 sum and carry outputs should now indicate the new sum (or difference) of [A] and [B].
4) Pulse the ADD/SUB line momentarily HIGH to clock the sum outputs from the 74LS283A into register A. The LEDs at the outputs of register A should now indicate the new accumulated sum (or difference).
5) Repeat steps 2–4 until all numbers have been added (or subtracted). The final result will be indicated by the register A output monitors.

PROCEDURE

l) Perform the following operations with the adder/subtractor unit:

1) $1 + 2 + 3 + 4 = $ _____.
2) $+7 + (-1) + (-2) + (-3) = $ _____.
3) $+1 - (-3) - (-2) = $ _____.
4) $-3 + (-2) + (-3) - (+4) = $ _____.
5) $9 - 1 - 3 - 2 = $ _____.
6) $+1 - (+5) - (+4) = $ _____.
7) $5 - 4 - 6 - 7 = $ _____.

m) *Review*: This concludes the exercises on binary adders. To test your understanding of the principles investigated in this experiment, answer the following questions:

1. A half adder may be made from a full adder by _____.
2. How many outputs must an adder have if it is to add two 8-bit numbers? _____.
3. What is the range of SIGNED numbers that a 74LS283 adder can operate on? _____.
4. Explain how register A, in the 2's complement adder/subtractor, acts like an accumulator in a digital arithmetic unit.

Experiment 21

Name _____

ASYNCHRONOUS IC COUNTERS

OBJECTIVES

1. To investigate the operation of the 74LS93 IC counter.
2. To investigate a method of changing the MOD-number of the 7493 IC counter.
3. To investigate the cascading of 74LS93 IC counters.
4. To investigate propagation delays in asynchronous counters.

TEXT REFERENCES

Read sections 7.1 and 7.2.

EQUIPMENT NEEDED

Components
74LS93 IC (2);
4 LED monitors;
normally HIGH pushbutton switch (debounced).

Instruments
0–5 volt DC power supply;
pulse or square wave generator;
dual trace oscilloscope.

DISCUSSION

In Experiment 18, you investigated flip-flop asynchronous counters. Recall that in asynchronous counters, the output of one flip-flop serves as the clock for the next flip-flop. In the current experiment and the next one, you will investigate integrated circuit (IC) asynchronous counters.

The flip-flops in a typical asynchronous binary counter are connected in their toggle mode. That is, the flip-flops toggle on the arrival of the correct clock edge. The connections between flip-flop outputs and clock inputs determine the direction characteristics of the counter, that is, whether it counts up or down. The counting range is determined by the number of flip-flops and how the clear circuits are connected.

Counters are used not only for counting, but they also may be used to divide the frequency of digital signals and to time and sequence events in a control system. To meet these needs, IC manufacturers have provided counters with varying bit widths and MOD numbers. These counters can be cascaded to extend the counting range and the MOD number. Recall that the overall MOD number of cascaded counters is the product of the individual counter MOD numbers.

In the current experiment, you will investigate the 74LS93 IC counter. The 74LS93 consists internally of a single toggle flip-flop, a three-bit (MOD 8) counter, and a gated reset circuit. The flip-flop and counter may be wired together to form a four-bit counter. With appropriate external connections to the reset inputs, the counter's MOD number may be reduced. Finally, the 74LS93, as well as all asynchronous counters, produces a significant propogation delay between the input and output of the counter.

PROCEDURE

a) Refer to the data sheet for the 74LS93 IC. This IC contains four flip-flops that may be arranged as a MOD-16 ripple counter. To do this, Q_0 must be tied externally to $\overline{CP_1}$. The MSB of this counter is Q_3 and the LSB is Q_0. The counter's MOD-number may be changed by making the appropriate external connections.

Draw the pin layout diagram for this IC:

b) *74LS93 IC operation:* Connect the circuit of Figure 21-1. Connect a normally HIGH pushbutton switch to input $\overline{CP_0}$ and LED monitors to outputs Q_3 through $\overline{Q_0}$.

c) Pulse $\overline{CP_0}$ and observe the counter sequence displayed on the LEDs. It should count from 0000 to 1111 and then recycle to 0000. Note that the NAND gate inputs R0(1) and R0(2) have no effect on the counter, since they are both tied LOW.

PROCEDURE

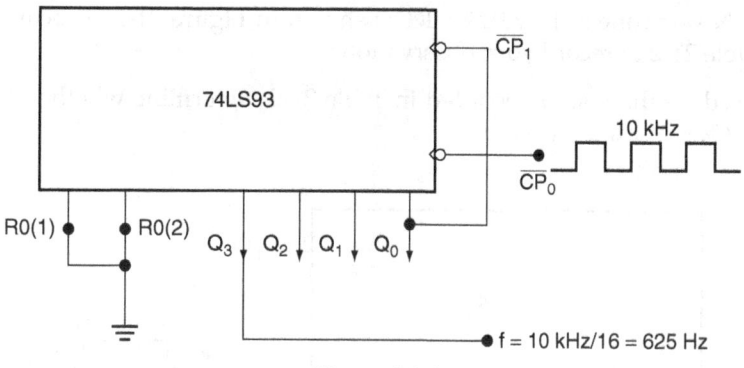

Figure 21-1

d) Disconnect the pushbutton switch at input $\overline{CP_0}$. Connect a square wave generator to this input, and set the generator to 10 kHz. Monitor the $\overline{Q_3}$ output with one vertical input of the oscilloscope and the generator output with the other input. Set the horizontal sweep so that you can verify that there is one $\overline{Q_3}$ pulse for every 16 generator pulses. Is the signal at $\overline{Q_3}$ a square wave? _____.

e) *Changing the 74LS93 MOD-number:* Disconnect the pulse generator from the counter, and reconnect the pushbutton switch in its place. Disconnect R0(1) and R0(2) from ground and connect one of them to Q_3 and the other to Q_2.

f) Pulse $\overline{CP_0}$ repeatedly, and observe the count sequence displayed on the LEDs. Record this sequence of output states in Table 21-1.

Table 21-1

Input Pulse Applied	Output States				Decimal Number
	Q_3	Q_2	Q_1	Q_0	
None	0	0	0	0	0
1	0	0	0	1	1
2	0	0	1	0	2
3					
4					
5					
6					
7					
8					
9					
10					
11					
12					
13					
14					
15					

g) What is the MOD-number of the counter in step f? _____. Verify the MOD-number you gave by disconnecting the pushbutton switch at $\overline{CP_0}$ and applying a 10 kHz square wave to this input. Then measure the frequency of the signal at Q_3. Is this output a square wave? _____.

h) Now connect the 74LS93 IC as shown in Figure 21-2. Repeat steps c-f, using Table 21-2 to record your observations.

Based on the results recorded in Table 21-2, determine whether or not the signal at Q_3 is a square wave: _____.

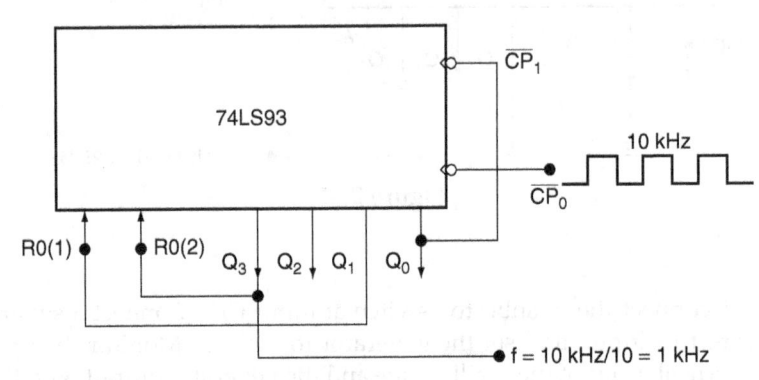

Figure 21-2

Table 21-2

Input Pulse Applied	Output States				Decimal Number
	Q_3	Q_2	Q_1	Q_0	
None	0	0	0	0	0
1	0	0	0	1	1
2	0	0	1	0	2
3					
4					
5					
6					
7					
8					
9					
10					
11					
12					
13					
14					
15					

i) *Cascading 74LS93 IC counters:* Connect the circuit shown in Figure 21-3. Apply a 6 kHz square wave to input $\overline{CP_0}$ of the MOD-10 counter. With the oscilloscope, determine the frequency of the signal at Q_3 of the MOD-6 counter: _____. What is the MOD-number of this counter arrangement? _____.

j) *Propagation delays in asynchronous counters:* Change the MOD-number of the cascaded counter of step i to 256. Adjust the frequency of the generator to 1 MHz. Monitor both the generator output and the counter output on the dual trace oscilloscope. Trigger the oscilloscope on Q_3 of the second MOD-16 counter. Measure the delay time t_d between the falling edge of the generator signal and the rising edge of the signal at Q_3. This measurement is for the TOTAL delay of the counter. Compute the *average* propagation delay (t_{pd}) for a single flip-flop using the following formula:

PROCEDURE

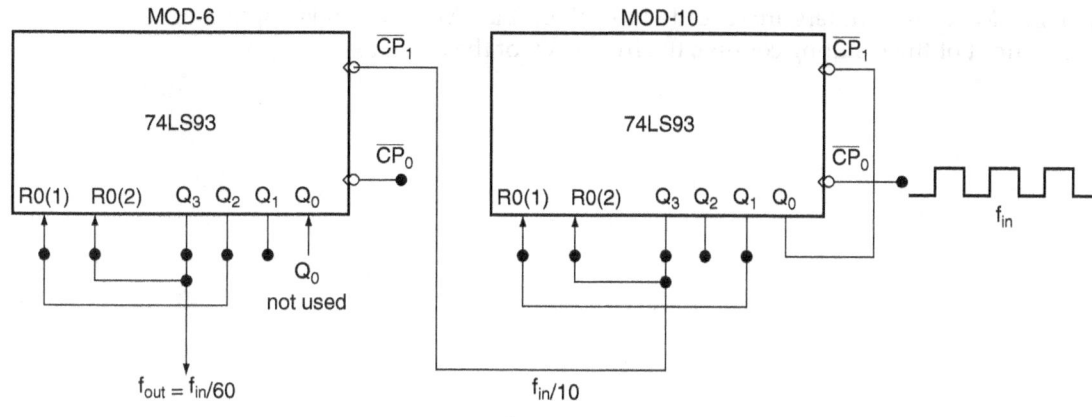

Figure 21-3

MOD-6 74LS93: R0(1), R0(2), Q3, Q2, Q1, Q0; CP1, CP0; $f_{out} = f_{in}/60$; Q0 not used

MOD-10 74LS93: R0(1), R0(2), Q3, Q2, Q1, Q0; CP1, CP0; $f_{in}/10$; f_{in}

$$t_{pd} = \frac{t_d}{N}$$

where N is the number of flip-flops in the counter. What is f_{max} for this counter? _____.

k) *Review:* This concludes the exercises on asynchronous IC counters. To test your understanding of the principles covered in this experiment, answer the following questions:

1. Draw a diagram showing how a 74LS93 IC can be wired as a MOD-12 counter.

2. If the outputs Q_3 through Q_0 of a 74LS93 MOD-16 counter are each inverted, what is the count sequence at Q_3 through Q_0?

3. Explain why the Q_3 output in steps e and f is not a square wave.

4. Three MOD-10 counters are cascaded together. If a 1 MHz signal is applied to the input of the resulting counter, the frequency of the output is _____.

Experiment 22

Name _____

A BCD COUNTER

OBJECTIVES

1. To investigate the operation of the 74LS163 IC counter.
2. To demonstrate that BCD counters may be configured from a standard binary counter.
3. To demonstrate reading the output of a counter with seven-segment display units.
4. To investigate the cascading of 74LS163 IC counters configured as BCD counters.

TEXT REFERENCES

Read sections 7.7 and 9.2.

EQUIPMENT NEEDED

Components
7447A IC (2);
74LS163 IC (2);
7400 IC;
FND507 seven-segment display units (2);
normally HIGH pushbutton switch (debounced);
4 toggle switches;
4 LED monitors.

Instruments
0–5 volt DC power supply;
pulse or square wave generator;
dual trace oscilloscope.

DISCUSSION

Many digital systems use BCD to display decimal numbers. An example where BCD is employed is the output indicator of a frequency counter (see Experiment 24). The CMOS CD4510B is a MOD-10 counter in IC form and can be used for direct implementation of a MOD-10 counter. An alternative is to use a TTL MOD-16 IC synchronous counter and modify it for MOD-10. This experiment will show how you can accomplish this. (The 74160 counter is an older TTL MOD-10 counter, but is no longer available.)

74LS163 IC Counter

The 74163 is a 4-bit synchronous counter featuring an internal carry look-ahead for application in fast counting circuits. The counter can load a binary pattern when a clock pulse is applied and LOAD is held LOW. In the current experiment, you will investigate the 74LS163 operating in both stand-alone and cascaded modes and see how to configure it with external gates as a BCD counter.

A synchronous counter reduces (but does not always eliminate) output spikes ("glitches"), which are usually associated with asynchronous counters. (Note that glitches can still occur in synchronous counters because the internal flip-flops can have very slight differences in delay times.) Glitches are normally not a problem for circuits configured for visual displays because the glitch occurs so fast, it is not seen. This is okay for visual displays, but in programs such as critical positioning controls, glitches cannot be tolerated and can be avoided by latching the output after the counter has had time to settle.

BCD Displays

It is often desirable to display the count of a BCD counter for convenience in reading base ten numbers. In many applications, a seven-segment LED display is used. The outputs of BCD counters must be converted from BCD into seven-segment codes and then applied to the seven-segment devices through current booster circuits called *drivers*. Both of these functions are found in the 74LS47A BCD-to-seven-segment decoder/driver. In this experiment, you will learn how to connect a 74LS47A and a seven-segment LED device to function as a BCD display unit.

PROCEDURE

  a) Refer to the data sheet for the 74LS163 IC, which is a 4-bit binary counter. The MSB of this counter is Q_D, and the LSB is Q_A. Draw the pin layout diagram for this IC:

  b) *74LS163 IC operation:* In this step you will investigate the basic operation of the 74LS163 IC configured as a BCD counter. Install a 74LS163 IC and a 7400 IC on the circuit board, and wire the counter so that it is like that of Figure 22-1. Use toggle switches for the Enable inputs (Enable-P and Enable-T) and LOAD and either a debounced pushbutton or a TTL compatible function generator operating at 1 Hz for the Clock input. The A, B, C, and D inputs can be connected to switches or to ground. Connect LED monitors to Q_A, Q_B, Q_C, and Q_D (Q_A = LSB; Q_D = MSB).

PROCEDURE

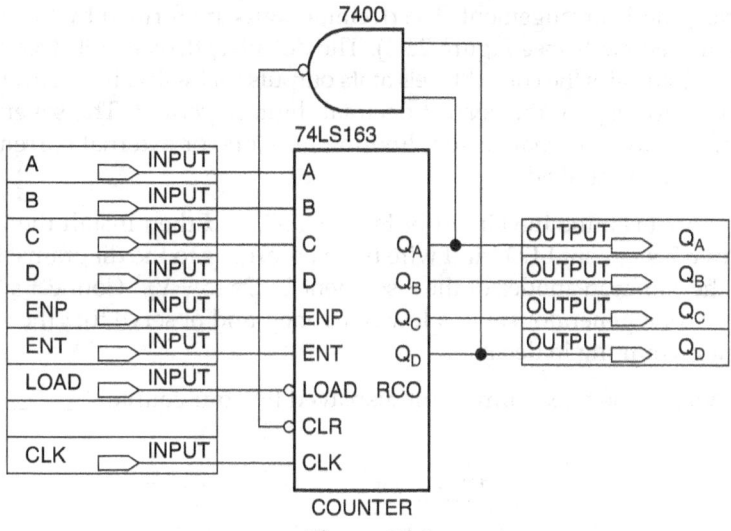

Figure 22-1

c) Clear the counter by setting inputs A through D LOW and set the LOAD switch to LOW and then toggle the clock. This loads all zeros into the counter. Note that the CLR input could do this, but we are using it to clear the count after the digit 9 is displayed, so LOAD is a better option.

d) Set the (count) Enables (ENT and ENP) HIGH to enable counting.

e) Pulse the clock repeatedly (or connect the function generator if you are using one), and observe the count sequence displayed on the LEDs. Record your observations in Table 22-1. Note that the counter does have ten different states, from 0 to 9.

Table 22-1

CLK	Q_D	Q_C	Q_B	Q_A
0				
1				
2				
3				
4				
5				
6				
7				
8				
9				
10				

f) *Cascaded 74LS163 IC operation:* In this step you will investigate the operation of two cascaded 74LS163 IC counters configured as BCD counters by decoding the output at 9 and using the decoded output to clock the next counter. Install another 74LS163 IC on the circuit board, and wire the counters so that they are like that of Figure 22-2. Notice that the RCO outputs are not used because of the BCD operation.

g) *Displaying BCD counters:* The count of the BCD counter can be more conveniently displayed in decimal. One of the most common devices used to display BCD counters is the seven-segment LED driven by a BCD-to-seven-segment decoder/driver, such as a 7447A IC. The seven segments that make up the display device each consist of one or two LEDs, and all are connected in a common cathode or common anode arrangement. The decimal digits are formed by turning on the appropriate segments (see Figure 22-3). The decoder/driver unit decodes a BCD number and supplies the correct levels at its outputs that will cause a seven-segment display unit to display the correct decimal digit. *Important*: The seven-segment display must have internal current limiting resistors, or external current limiting resistors must be installed.

h) Now examine the circuit of Figure 22-2, and then install the 7447A ICs and the two 7-segment LEDs and wire them according to the diagrams. Be sure to connect the common anodes of the 7-segment LEDs to $+V_{cc}$. Connect a TTL compatible function generator set to a low frequency, and observe that the seven-segment unit is counting in decimal.

i) What is the maximum count displayed by your counter? _____

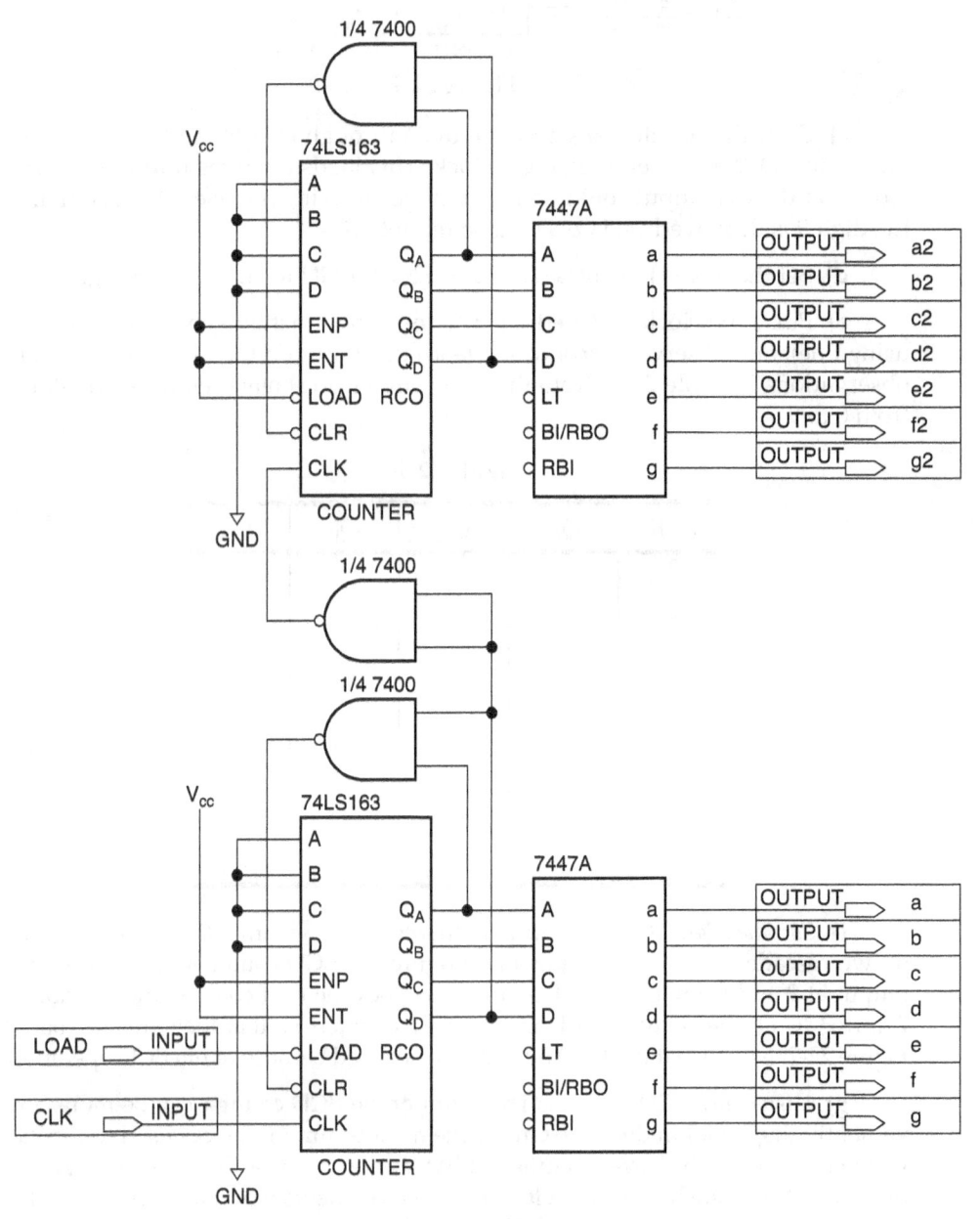

Figure 22-2

PROCEDURE

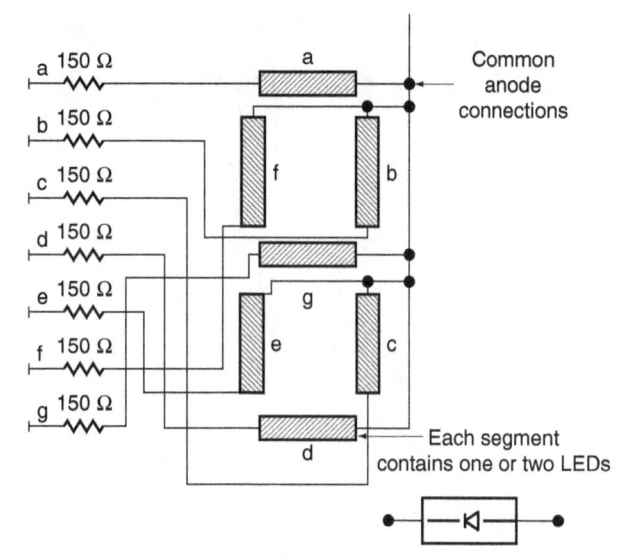

Figure 22-3

Experiment 23

Name _____

SYNCHRONOUS IC COUNTERS

OBJECTIVES

1. To investigate the operation of the 74LS193 IC counter as an UP counter.
2. To investigate the parallel load function of the 74LS193 IC.
3. To investigate the operation of the 74LS193 IC as a DOWN counter.
4. To investigate the cascading of 74LS193 IC counters.

TEXT REFERENCES

Read sections 7.6 through 7.9.

EQUIPMENT NEEDED

Components
7404 IC;
74LS93 IC; (2)
74LS193 IC;
normally HIGH pushbutton switch and two normally LOW pushbutton switches (all debounced);
4 toggle switches;
8 LED monitors.

Instruments
0–5 volt DC power supply;
pulse or square wave generator;
dual trace oscilloscope.

DISCUSSION

In Experiments 21 and 22, you investigated some of the characteristics of asynchronous counters. One undesirable characteristic is the accumulation of propagation delays from clock input to clock output. Another undesirable characteristic that is directly related to propagation delay is the production of unwanted "glitches" at the outputs of decoders that are used to decode the counter. This type of glitch will be investigated in Experiment 25.

Recall that in asynchronous counters, each flip-flop receives its clock input from the output of the previous flip-flop. In another type of counter, each flip-flop clock input is connected to a common clock. This virtually eliminates the propagation delay problem and reduces the number of glitches appearing at the output of decoders significantly. This type of counter is referred to as a synchronous counter.

Flip-Flop Synchronous Counters

Examine Figure 23-1 closely. This figure gives the basic logic diagram for a four-bit synchronous counter. Note that all flip-flops have their J and K inputs tied together as in the asynchronous counter, but, except for the LSB flip-flop, these common connections are not tied to V_{cc}. They are instead connected to the outputs of the previous flip-flop. This means that these flip-flops are in the toggle mode part of the time and in the no-change mode the rest of the time.

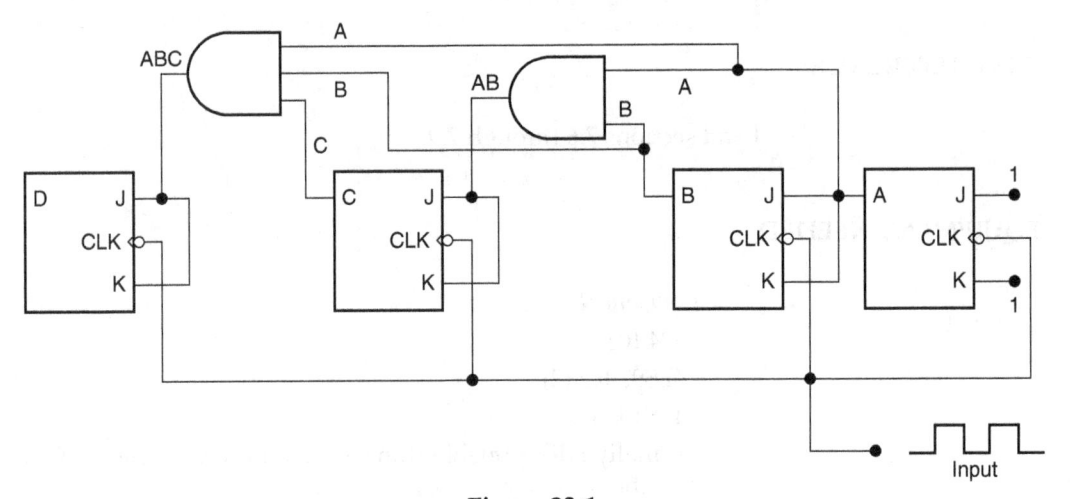

Figure 23-1

Two other differences are outstanding. One is the identifiable characteristic of synchronous counters that the flip-flop clock inputs are tied in parallel. Recall that a weakness of an asynchronous counter is that the individual flip-flop clock inputs are not connected together and that each flip-flop must wait on the output of the previous flip-flop to be clocked. If these inputs are tied together, then they are clocked simultaneously. The other difference is the presence of gates, which force the flip-flops to count in sequence. You should convince yourself that they are necessary by eliminating them. Simply connect the J and K inputs to the normal output of the previous flip-flop and derive a state table for such a counter (refer to Figure 23-2). You should observe that whenever a flip-flop's output is

PROCEDURE

HIGH, the next flip-flop will toggle on the next clock pulse. Thus the counter will "skip" certain counts, such as 3 and 4, as well as others.

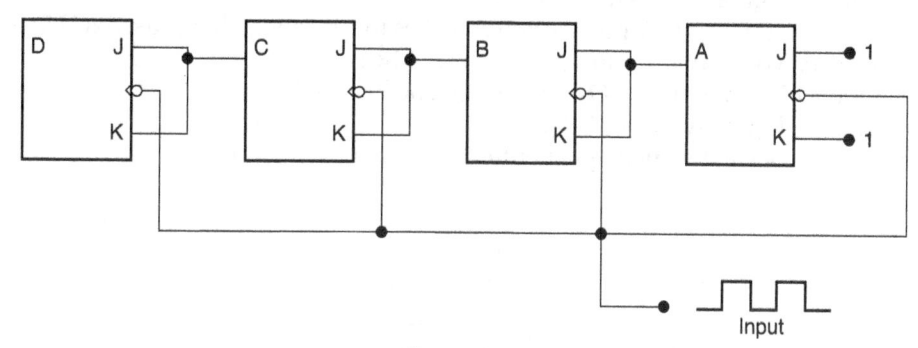

Figure 23-2

74LS193 IC Synchronous Up/Down Counter

The 74LS193 IC is representative of the trend toward versatility in the manufacture of ICs. It can count either up or down and can be preset. These features make the chip *programmable*, since they may be selected automatically, even while the circuit is operating. You will investigate other such ICs in future experiments.

The 74LS193's up/down feature is selected by applying the clock signal to the up or down clock inputs. For example, if it is desired that the counter count up, the clock is sent to the CP_U input. The preset feature is selected by setting the $\overline{PL}$ input LOW. Whatever is currently at the parallel inputs (P_0–P_3) will be loaded into the counter flip-flops.

Other features of the 74LS193 are the terminal count outputs, $\overline{TC_U}$ and $\overline{TC_D}$. These outputs go LOW when the counter reaches 1111 and 0000, respectively. Thus, they may be used in clocking another 74LS193 whenever cascading is desired.

PROCEDURE

a) Refer to the data sheet for the 74LS193 IC. This IC is a synchronous, four-bit, presettable UP/DOWN counter. The MOD-number may be changed by making the appropriate external connections. Draw the pin layout diagram for the 74LS193.

b) *74LS193 IC operation as a MOD-16 UP counter:* Construct the circuit of Figure 23-3. Make the following connections to the 74LS193 IC:

1) Connect toggle switches to P_3 through P_0.
2) Connect normally LOW pushbutton switches to CP_U and MR inputs. (NOTE: if necessary, you may use a toggle switch for MR.)
3) Connect LED monitors to Q_3 through Q_0 and also at $\overline{TC_U}$.
4) Connect a toggle switch to CP_D.
5) Connect a normally HIGH pushbutton switch to $\overline{PL}$.

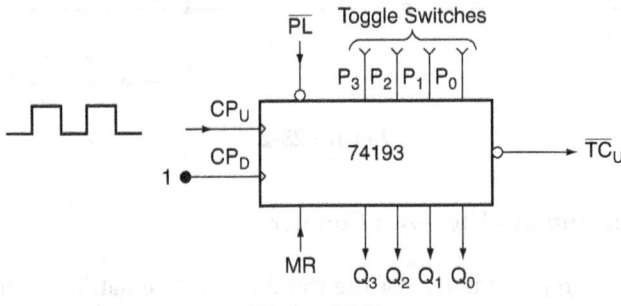

Figure 23-3

c) Set the toggle switches to the parallel inputs so that $P_0 = P_1 = P_2 = P_3 = 0$. Set CP_D HIGH. Clear the counter to 0000 by pulsing MR HIGH. Note that $\overline{TC_U}$ (terminal count-up mode) is HIGH.

d) Pulse CP_U HIGH a couple of times, and note that the counter counts UP. Pulse the counter until a count of 1111 is displayed. Record the state of $\overline{TC_U}$: ___. Now pulse the counter one more time. Now record the value of $\overline{TC_U}$: ___. Record the observations you have made in Table 23-1.

Table 23-1

Input Pulse Applied	Q_3	Q_2	Q_1	Q_0	Decimal Number	$\overline{TC_U}$
None	0	0	0	0	0	1
1						
2						
3						
4						
5						
6						
7						
8						
9						
10						
11						
12						
13						
14						
15						
16						

PROCEDURE

e) Set CP_D LOW. Pulse the CP_U input several times. What happens? _____. Return CP_D to HIGH and pulse CP_U a few more times. You should observe that the counter does not count as long as CP_D is LOW.

f) Using the parallel inputs to preset the counter, set the toggle switches at the parallel inputs so the $P_3 = P_1 = 1$ and $P_2 = P_0 = 0$ and pulse $\overline{PL}$ LOW. The LEDs should now indicate 1010. Pulse the counter until the $\overline{TC_U}$ output LED indicates 0, observing the output LEDs as you do so. What sequence of numbers does the counter count? _____.
Pulse the counter one more time. What is the count now? _____.

g) *74LS193 IC operation as a MOD-16 DOWN counter:* Disconnect the toggle switch from CP_D and exchange it for the pushbutton switch at CP_U and vice versa. Disconnect the LED from $\overline{TC_U}$ and reconnect it at $\overline{TC_D}$. Clear the counter by pulsing MR HIGH momentarily. Pulse the counter through its count sequence, and record your observations in Table 23-2.

Table 23-2

Input Pulse Applied	Output States				Decimal Number	$\overline{TC_D}$
	Q_3	Q_2	Q_1	Q_0		
None	0	0	0	0	0	0
1						
2						
3						
4						
5						
6						
7						
8						
9						
10						
11						
12						
13						
14						
15						
16						

h) Set the parallel input toggle switches to 1010 and pulse $\overline{PL}$ LOW. Note that $\overline{TC_D}$ is now at 1. Pulse the counter until $\overline{TC_D}$ indicates 0, observing the output LEDs as you do so. What sequence of numbers does the counter count? _____. Pulse the counter one more time. What is the count now? _____.

i) *Cascading 74LS193 IC counters:* Two or more MOD-16 counters may be cascaded to form a MOD-256 counter. In this step, you will cascade a 74LS93 MOD-16 counter with a 74LS193 MOD-16 (UP) counter. Use the 74LS193 to count the lower four bits of the count and the 74LS93 to count the upper four bits. Pass $\overline{TC_U}$ through an inverter, and connect the output of the inverter to the clock input of the 74LS93. Connect a square wave generator to the CP_U input of the 74LS193. Use

LED monitors to observe the count sequence of the counter. Draw a circuit diagram for the counter below.

j) Set the generator to a frequency low enough to easily observe the LEDs. Verify that the counter is a MOD-256 counter. Demonstrate the counter for your instructor.

k) *Review:* This concludes the exercises on synchronous IC counters. To test your understanding of the principles covered in this experiment, answer the following questions:

1. Draw a diagram showing how a 74LS193 IC might be used as a MOD-10 UP counter.

PROCEDURE

2. Draw a diagram showing how two 74LS193 ICs might be cascaded to form a MOD-100 DOWN counter.

3. What conditions must be satisfied in order for the 74LS193 to count DOWN?

4. If a six-bit synchronous counter is to be built, what type of gate is required to feed the J and K inputs of the counter's MSB flip-flop? _____
5. Repeat question 4 for a ten-bit synchronous counter. _____
6. The three-input AND gate in Figure 23-1 can be replaced by a two-input AND gate. How?

7. Does the circuit suggested by question 6 have an advantage over the circuit in Figure 23-1? Explain why (or why not).

Experiment 24

Name _____

IC COUNTER APPLICATION: FREQUENCY COUNTER

OBJECTIVE

To investigate the use of IC counters in a frequency counter.

TEXT REFERENCE

Read section 10.5.

EQUIPMENT NEEDED

Components
7400
74LS10 IC;
7447A IC (2);
74LS76 IC;
74LS163A IC (3) (configured for BCD; see Experiment 22);
74LS93 IC (2);
74LS273 IC;
74121 IC;
seven-segment LED display units (2);
1N457 diode;
1 k-ohm resistor;
2.2 k-ohm resistors (2);
560 pF capacitor (value not critical).

151

Instruments
0–5 volt DC power supply;
6.3 VAC, 60 Hz power source;
pulse or square wave generator;
dual trace oscilloscope.

DISCUSSION

In previous experiments, frequency measurements have been made using an oscilloscope. Although accurate enough for many types of measurements, the procedure is time-consuming. In this experiment you will construct a model for another instrument capable of measuring frequency directly, without the operator having to make time-frequency conversions. This instrument is a *frequency counter.*

The frequency counter you will construct is composed of three major units:

1) Sampling unit

2) Counter unit

3) Decoder/display unit

The counter will be able to display a frequency count of 00–99 Hz. Of course, the frequency counter can be modified to extend this range considerably. The accuracy of the counter depends on the accuracy of the sampling unit. This unit is driven by an extremely accurate 60 Hz signal. After being divided down to 0.5 Hz by the MOD-120 counter, the pulse produced by the sampling unit controls the amount of time an unknown frequency is sampled. The amount of sampling time is very close to one second. Therefore, the counter unit (MOD-100 BCD) counts the unknown frequency for one second, displays the number of pulses counted during this period, then recycles and counts another sample. The BCD display will indicate the frequency of the unknown signal.

PROCEDURE

a) Examine Figure 24-1 very carefully. Put IC pin numbers for each connection on the diagram, including those not shown. In the latter case, pencil in these connections. Show your completed diagram to your instructor for approval.

b) Wire the circuit of Figure 24-1. Since the circuit is complex, cross off each connection you make with a yellow pencil or marker.

c) *Testing the frequency counter:* Apply a 50 Hz signal to the input of the frequency counter. Observe the displays. The displays should count up to a final value of between 48 and 52. This may change by + or − 1 count with each reading, since the sample pulses and the input signal are not synchronized. The start or end of a sample interval may sometimes catch the falling edge or rising edge of an input pulse, and sometimes it will not.

If the displays do not give the expected results, turn the power off and cross-validate your circuit wiring against the logic diagram. Visually inspect for unconnected V_{cc} and GND pins on each IC. Also check the reset-to-zero pins on the BCD counters—at least one on each IC should be grounded.

Figure 24-1

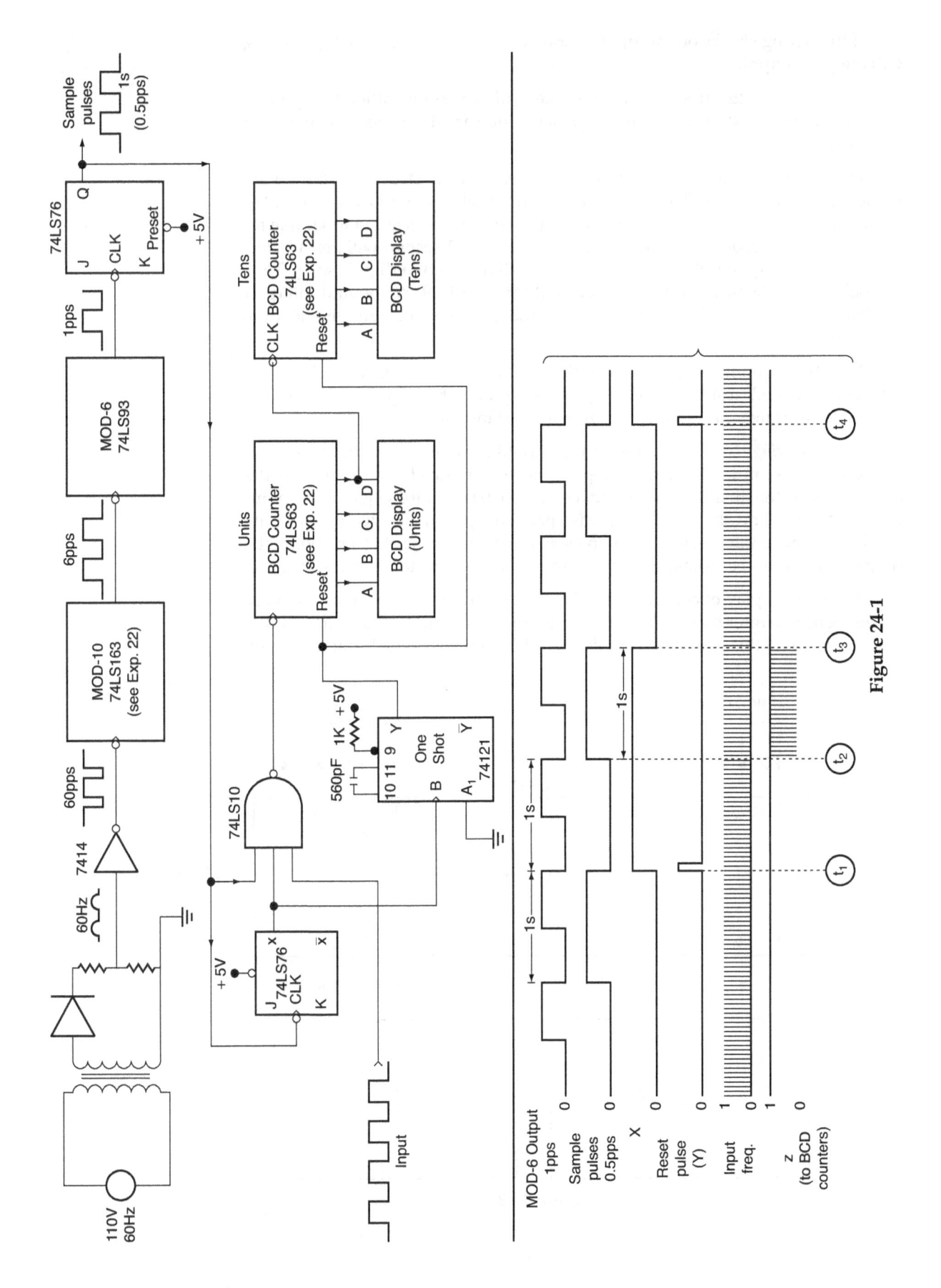

If the wiring checks out all right, you will have to continue troubleshooting with the power applied.

d) If the results in step c were successful, try some other frequencies between 0 and 100 Hz. Verify that the displays indicate the correct frequency in each case.

e) *Changing the sample time:* The counter you have constructed will measure frequencies ranging from 0 to 99 Hz. To measure higher frequencies, you could add another BCD counting unit. Another way would be to *decrease* the sample interval. For example, decreasing the sample interval to 0.1 second will permit your counter to count up to 999. However, you will still have a two-digit display. Thus the counter will display frequencies from 100–999 as 10–99. You would interpret the display as 100–990. This results in some inaccuracy since you have lost the units digit.

Modify the counter so that the sample interval is 0.1 second. When the modification is installed, test the counter by applying a 250 Hz signal at the counter input. Demonstrate your modified counter to your instructor.

f) *Eliminating the blinking display:* You should have observed that when you decreased the sample interval in step e, you experienced a reduction in the amount of time the final count was displayed. In fact, the entire sequence of counting, displaying, and then recycling was speeded up. Further decreases in the length of a sample interval will result in even less display time. At some point, the sample time would be so small that the displays will be constantly changing.

The only display of concern is the final count. In this step, you will modify the frequency counter so that it changes only whenever there is a change in the final count. To do this, you will use a 74LS273 IC, which is an eight-bit latch with

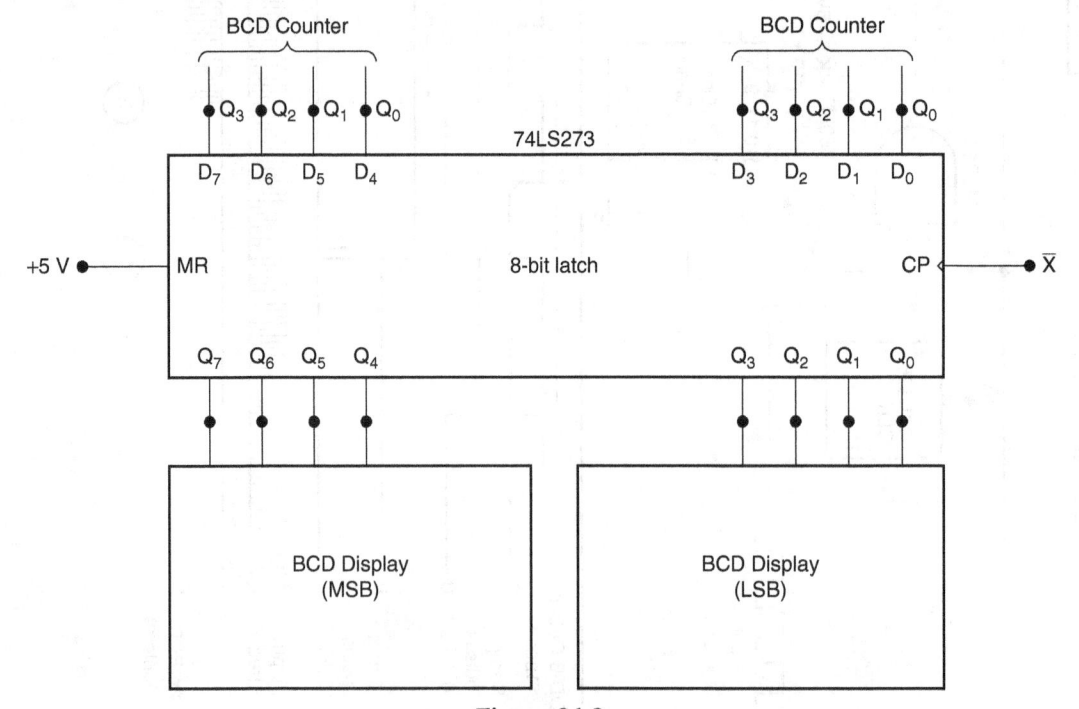

Figure 24-2

PROCEDURE

CLEAR, to latch the output of the counter when it reaches the end of a sample interval. The latched output will then be displayed. Figure 24-2 shows the circuit modification you are to make to the frequency counter.

Refer to the data sheet for a 74LS273 IC and draw its pin layout diagram:

Incorporate this modification into the frequency counter. Check the modified counter out for various input frequencies, and demonstrate it for your instructor.

g) *Review:* This concludes the exercises on applications of IC counters. To test your understanding of the principles covered in this experiment, answer the following questions:

1. Assume that the unknown frequency input to the counter is 50 Hz. Study the waveforms shown in Figure 24-1. Describe what the BCD counters and display units are doing during each of the following time intervals:

   a) $t_1 - t_2$:

   b) $t_2 - t_3$:

   c) $t_3 - t_4$:

2. Why is $\overline{X}$ used in Figure 24-2?

Experiment 25

Name _____

TROUBLESHOOTING COUNTERS: CONTROL WAVEFORM GENERATION

OBJECTIVES

1. To investigate the application of counter decoding in generating control waveforms.
2. To compare asynchronous and synchronous counters used in control circuits.
3. To practice troubleshooting counter circuits.

TEXT REFERENCES

Read sections 7.1 through 7.10.

EQUIPMENT NEEDED

Components
7400 IC;
74LS76 IC;
74LS93 IC;
74LS193 IC;
normally HIGH pushbutton switch (debounced).

Instruments
0–5 volt DC power supply;
pulse or square wave generator;
dual trace oscilloscope;
logic pulser;
logic probe;
ohmmeter or DMM.

DISCUSSION

In Experiment 22, we investigated a way to display the count of a BCD counter. The use of the seven-segment LED display unit was found to be superior to the usual single LED, because the former shows us the count in decimal. In the current experiment, you will investigate another reason for decoding counters. Often counters are used in controlling the sequence of steps in a process. For example, if a process consists of five steps, a MOD-5 counter could be used to count the steps and decoders could be used to detect the presence of the current count. The outputs of the decoders could then be used to initiate each step of the process.

We have already used this type of counter decoding in controlling the MOD-number of flip-flop counters in Experiment 18. A NAND gate was used to detect the presence of the MOD-number, and the resulting LOW output was applied to the flip-flop CLEAR inputs, thus causing the counter to recycle. The NAND gate decoder is called an active-LOW (or one-LOW) decoder, since it outputs a LOW when it detects the "right" number and outputs a HIGH otherwise (normal output). Other circuits may require an active-HIGH (one-HIGH) or AND gate decoder.

In the current experiment, you will construct a simple waveform-generating device that produces two control signals. These signals could be used to control two other devices by turning them on and off at regular intervals. You will then use the circuit to practice troubleshooting counter and counter decoder circuits.

Troubleshooting IC Counters

IC counters can have the same faults as gates and flip-flops. For example, they may have open (externally and internally) inputs and outputs, shorted (externally and internally) inputs and outputs, and other internal faults. A logic probe and pulser can be used to isolate many of the IC faults, and an ohmmeter (or DMM) can be used to pinpoint shorts.

As in previous troubleshooting exercises, you should know the operation of the circuit before attempting troubleshooting. Next, you should record your observations on how the circuit operates with a bug in it. Once this is accomplished, you should list as many possible faults as you can. You will use this list and your troubleshooting tools to isolate the fault.

A simple 74LS93 fault is open master reset inputs. Suppose the outputs of your counter do not respond to the input clock and appear to be stuck at 0000. The possible faults include an open clock input to the counter, counter outputs shorted to ground, and open master reset inputs. If the resets are supposed to be grounded to V_{ss}, place the tip of your logic probe on the master reset input pins. If it indicates a floating condition on both, you have an external open in the reset circuit. If the probe indicates ground, then it is possible that the IC has an internal open. If you have eliminated the other possible faults, replace the IC and retest the circuit.

Another counter problem is glitches. Glitches commonly occur in asynchronous counter decoder outputs. However, any binary counter whose MOD-number has been changed to a number that is not a power of two produces glitches. The current experiment will provide an opportunity for you to investigate this problem.

PROCEDURE

PROCEDURE

a) *Basic control circuit:* Figure 25-1 shows the circuit for a simple control waveform circuit that produces two control signals. It consists of a counter and two counter decoders that control the two flip-flops, from which the output waveforms are taken. Draw a diagram showing how you would implement this circuit, using a 74LS93 IC counter for the four-bit counter.

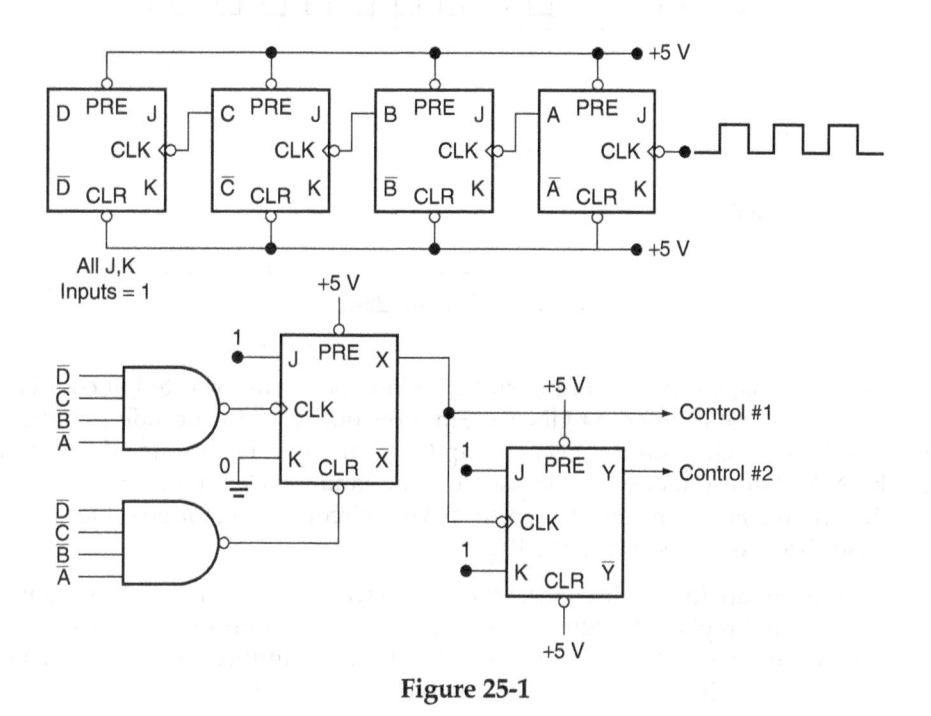

Figure 25-1

b) Draw all circuit output waveforms you expect to observe on an oscilloscope, assuming an input square wave at 10 kHz and no glitches. Use Timing Diagram 25-1.

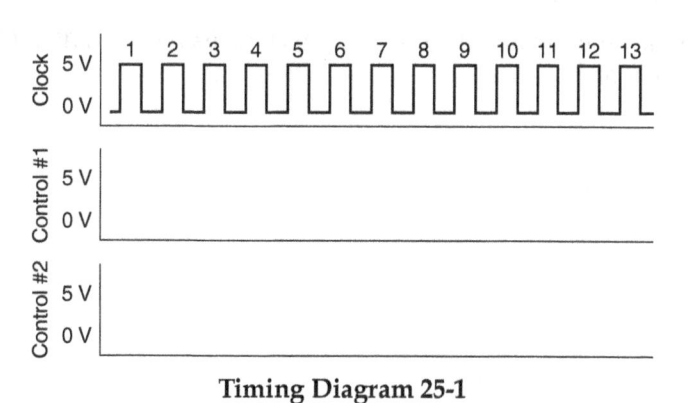

Timing Diagram 25-1

Show the waveforms and the circuit you drew in step a to your instructor for approval. Then construct the circuit on the circuit board.

c) Apply a 10 kHz square wave to the counter clock input, and monitor both control outputs on the oscilloscope. Trigger the oscilloscope on either control output signal. Draw the waveforms you *observe* on the oscilloscope. Do this even if the waveforms are not as expected. Use Timing Diagram 25-2.

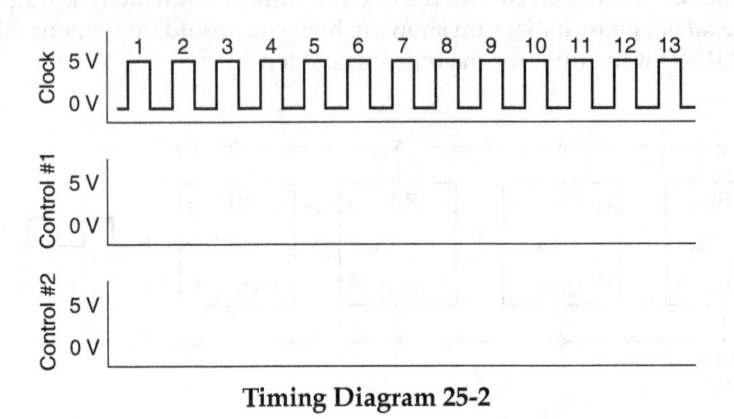

Timing Diagram 25-2

d) *Decoding glitches:* If your circuit failed to operate as expected, it could be because the circuit is decoding glitches. Since the output of the decoders controls the flip-flops, glitches could trigger the flip-flops erroneously. Examine the output of the NAND decoders with the oscilloscope, and check for the presence of glitches. If no glitches are found, then check your circuit wiring for possible errors, and also check for a possible faulty IC.

If glitches are found, disconnect the square wave generator from the counter clock input, and replace the square wave generator with a normally HIGH push-button switch. Pulse the counter clock input until a malfunction occurs. Record the count at the point where the malfunction occurs: _____.

Glitches are commonly found at the output of decoders that are driven by asynchronous counter outputs. One way to eliminate the glitches, although not foolproof, is to use synchronous counters instead of asynchronous counters. In the next step, you will reconstruct the control waveform generator using a 74LS193 IC, a synchronous counter.

e) Redraw the circuit of Figure 25-1, this time using a 74LS193 IC wired as a MOD-16 UP counter.

PROCEDURE

Show your circuit to your instructor before proceeding.

f) Repeat step c, using Timing Diagram 25-3 to record your observations.

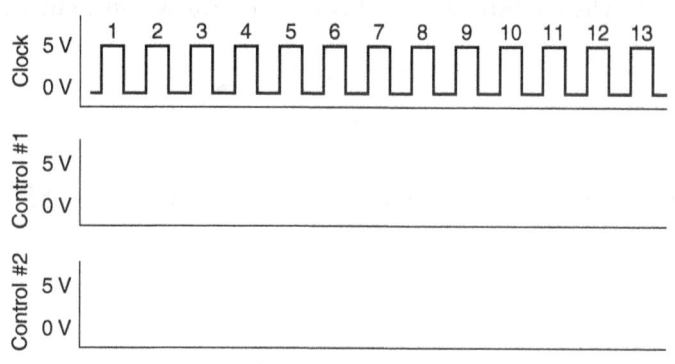

Timing Diagram 25-3

g) Check the outputs of the decoders for glitches. If the decoder outputs are free of glitches but the waveforms are not what is expected, check for faulty circuit wiring and ICs.

h) Now that you have your circuit operating normally, have your instructor or lab assistant put a bug into your circuit. Remember, if you see the bug, ignore it. Also, after you have drawn the observed waveforms (using Timing Diagram 25-4), list all the possible faults you can think of, in the space provided below, before beginning the fault isolation process.

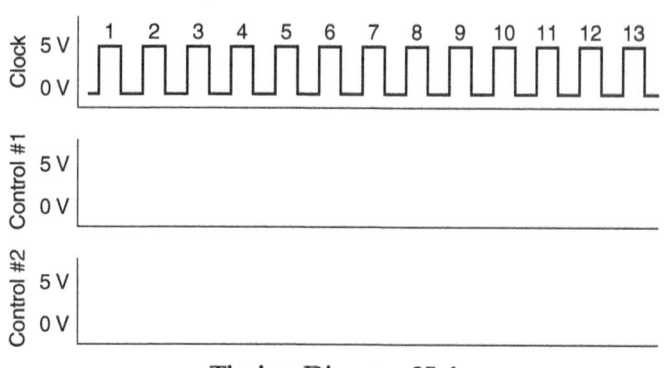

Timing Diagram 25-4

i) *Review:* This concludes the exercises on troubleshooting counters. To test your understanding of the principles covered in this experiment, answer the following questions:

1. Explain how a glitch is produced, and why it may cause faulty circuit operation.

2. Why is the use of synchronous counters to eliminate glitches not foolproof?

3. What other method of glitch elimination are you familiar with?

4. Describe how the control circuit you constructed in this experiment operates.

Experiment 26

Name _____

SHIFT REGISTER COUNTERS

OBJECTIVES

1. To investigate the operation of a ring counter.
2. To investigate the operation of a Johnson counter.

TEXT REFERENCE

Read section 7.20.

EQUIPMENT NEEDED

Components
74HC195 IC (1);
normally HIGH pushbutton switch, debounced (3);
4 LED monitors.

Instruments
0–5 volt DC power supply;
pulse or square wave generator;
dual trace oscilloscope.

DISCUSSION

Shift-register counters are flip-flop devices with feedback, in some manner, from the counter's last flip-flop output to the counter's first flip-flop input. In this experiment, you will use the 74HC195 IC. This is a CMOS circuit that is composed of four flip-flops. The inputs to the first flip-flop are labeled J and $\overline{K}$ (note inversion of the K input). Also, you will see that the clear function is labeled $\overline{MR}$ (for Master Reset) and the load function is labeled $\overline{PE}$ (for Parallel Enable).

There are two shift-register counters that are commonly found in digital systems: ring counters and Johnson counters. These two counters are covered in the current experiment. The normal output of a ring counter's last flip-flop is connected to the input of the first flip-flop, thus permitting the current contents of the counter to be continually circulated. On the other hand, the inverted output of a Johnson counter's last flip-flop is connected to the input of the first flip-flop. Thus, the Johnson counter also circulates the inverse of the counter's current contents.

In the current experiment, you will investigate both of these counters. You will study the basic characteristics of each counter. You should become familiar with the waveforms associated with each. While you will not be asked to construct decoders for the counters, you will be queried about the decoding requirements.

PROCEDURE

a) *Ring counter:* Connect the circuit of Figure 26-1. Connect the $\overline{PE}$ (Parallel Enable) to a pushbutton that can load the pattern into the counter—a single one and three zeros. Connect a debounced normally HIGH pushbutton to the clock input (CP). Connect LED monitors to the Q_A, Q_B, Q_C, and Q_D outputs.

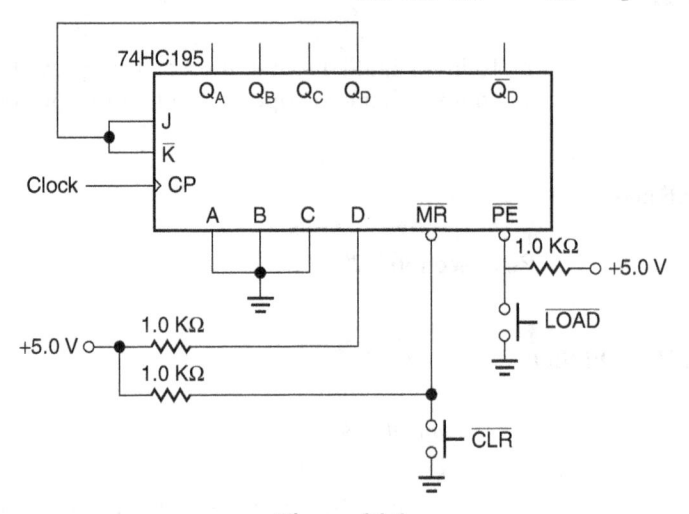

Figure 26-1

b) Initialize the counter by holding the $\overline{LOAD}$ pushbutton LOW and pulsing the clock. The LEDs should indicate a count of 1000. Pulse the clock input. The counter output is now _____. Pulse the clock two more times. Observe that the 1 now occupies the rightmost position of the counter display. Now pulse the clock once more. Where is the 1 positioned now? _____

c) Verify that it does not matter which flip-flop was initially loaded with a 1. Do this by rearranging the A, B, C, and D inputs and loading the new arrangement. Pulse the clock and notice that the pattern is the same as before, but the starting value is different.

d) Disconnect the clock input from the pushbutton switch, and replace it with the output of a square wave generator set at 1 kHz. Display the generator output and Q_3 on the oscilloscope and observe the time relationship between the two signals. Draw the waveforms on Timing Diagram 26-1. Repeat this procedure for each of the other outputs of the counter.

The MOD-number of this counter is _____. The outputs of the counter [are, are not] square waves. The frequency of each output is _____.

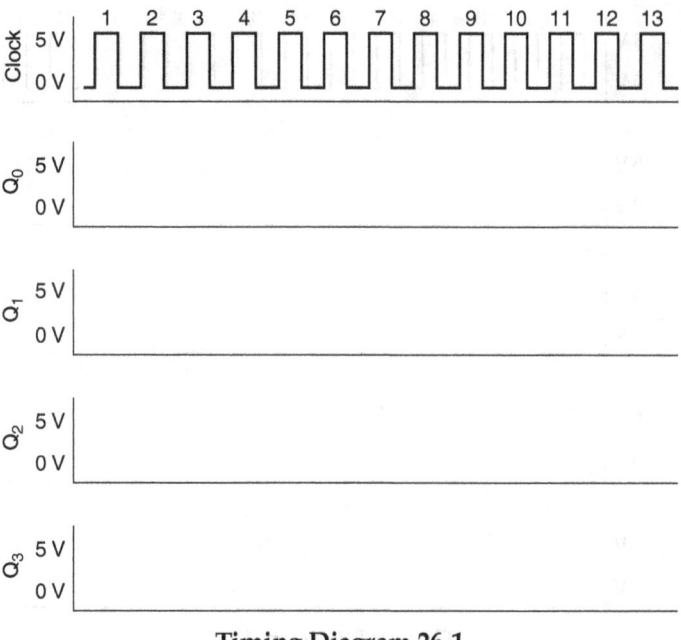

PROCEDURE

Timing Diagram 26-1

e) *Johnson counter:* Rewire the ring counter so that it is a Johnson counter. Disconnect the square wave generator, and reconnect the pushbutton switch to the clock input of the counter. Clear the counter. Verify the operation of the Johnson counter by pulsing the counter LOW eight times while observing the output. Record your observations in Table 26-1.

Table 26-1

Shift Pulse	Output State Q_3	Q_2	Q_1	Q_0
0	0	0	0	0
1				
2				
3				
4				
5				
6				
7				
8				

f) Repeat step d, using Timing Diagram 26-2.

The MOD-number of this counter is _____. The outputs of the counter [are, are not] square waves. The frequency of each output is _____.

g) *Review:* This concludes the exercises on shift register counters. To test your understanding of the principles covered in this experiment, answer the following questions:

1. To start a ring counter, _____ flip-flop must be _____.
2. The MOD-number of a ring counter is equal to _____.
3. The MOD-number of a Johnson counter is equal to _____.
4. Compare the waveforms of the ring and Johnson counters. Which counter needs decoding gates in order to decode its outputs? _____.

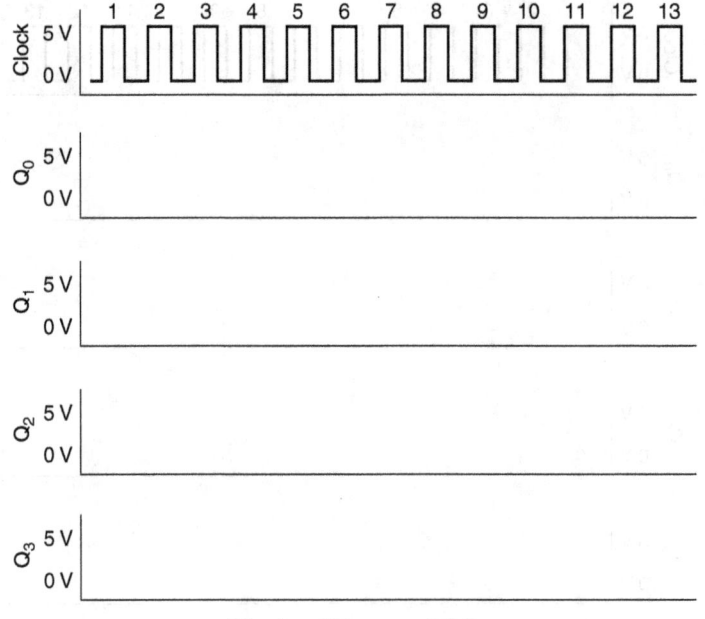

Timing Diagram 26-2

Experiment 27

Name _____

IC REGISTERS

OBJECTIVES

1. To investigate the operation of a 74194A IC four-bit shift register.
2. To investigate the operation of the 74194A IC as a ring counter.
3. To investigate parallel-to-serial data conversion using a 74194A IC.

TEXT REFERENCE

Read sections 7.15 through 7.19.

EQUIPMENT NEEDED

Components
74LS79A IC;
74LS194A IC;
2 normally HIGH debounced pushbutton switches;
6 toggle switches;
4 LED monitors.

Instruments
0–5 volt DC power supply;
pulse or square wave generator;
dual trace oscilloscope.

DISCUSSION

Much of what goes on in a digital system is associated with data transfer. Recall that there are two types of data transfer: serial and parallel. For a system to run at top speed, parallel transfer is desired within the system. However, data coming into the system from a device or going out to a device is often in serial form. To accommodate a serial input into a parallel system, some type of serial-to-parallel data conversion is necessary. To accommodate a serial output from a parallel system, a parallel-to-serial conversion would be necessary. Of course, parallel-to-parallel and serial-to-serial data transfers are also sometimes necessary.

One means of making these conversions is with registers. A typical IC register is a 74LS194A bidirectional universal shift register. This register can load data in parallel, load and shift right or left serially, and/or shift out in parallel. Like the 74LS193 IC, the 74LS194A must be programmed to make full use of its capabilities. For example, the shift right serial mode of the register may be selected by setting $S_0 = 1$ and $S_1 = 0$ before the next active clock edge arrives. Be sure to study the mode select table for the 74LS194A given in the data sheets in this manual carefully before beginning the procedure below.

PROCEDURE

a) Refer to the data sheets for a 74LS74A IC and a 74LS194A IC, and draw their pin layout diagrams:

b) *74LS194A serial operation:* Install a 74LS194A IC on the circuit board, and make the following connections:

1) Connect toggle switches to A through D, SR SER, S_0, and to S_1.
2) Connect a normally HIGH pushbutton switch to CLK and $\overline{CLR}$.
3) Connect LED monitors to Q_A through Q_D.

c) Set S_1 and S_0 to HIGH, SR SER to LOW, and A through D to LOW. Note that this has no effect on the outputs. Pulse $\overline{CLR}$ LOW momentarily to clear the register if it is not already cleared. Return S_0 and S_1 to LOW.

d) Verify that neither the serial data input (SR SER) nor the parallel data inputs (A through D) have an effect on the register as long as both S_0 and S_1 are LOW when CLK is pulsed LOW. Do this by setting SR SER = 1, D = A = 1, and C = B = 0 and momentarily pulsing CLK LOW.

e) Now set S_0 to HIGH, and pulse CLK. You should observe that the LEDs indicate 1000. Set SR SER to LOW, and pulse CLK three more times. You should observe that the 1 is shifted one position to the right on each pulse. The output now reads _____.

f) Set S_1 HIGH, and pulse CLK LOW momentarily. Observe that the register output changes to the value represented by the parallel data input switches (1001). This type of data transfer is called _____.

g) *74LS194A wired as a ring counter:* Figure 27-1 shows the 74LS194A wired as a ring counter. Examine the circuit, then make the necessary changes to the 74LS194A

PROCEDURE

so that it is the same as that shown in the figure. Verify that the circuit operates as a four-bit ring counter.

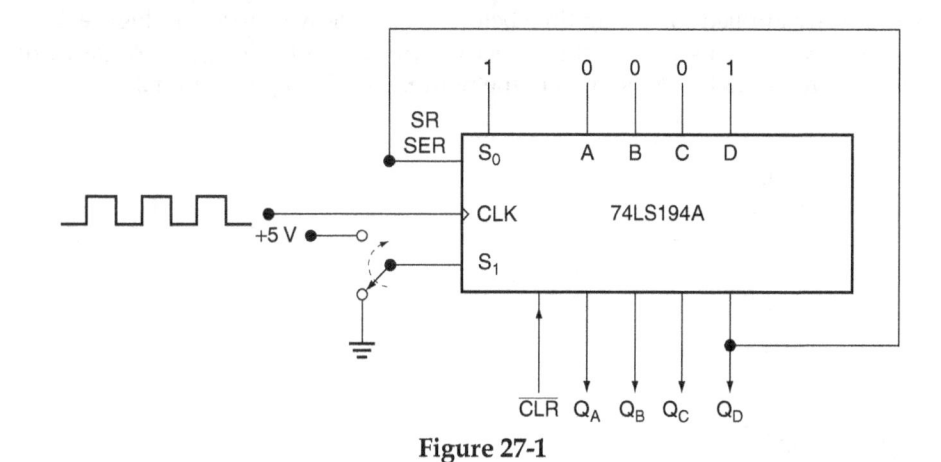

Figure 27-1

h) *Parallel-to-serial data conversion:* Examine the circuit of Figure 27-2. This circuit operates as a parallel-to-serial data converter. It first loads the data present at A through D and then shifts the data out of Q_D. The shifting of the loaded data is controlled by the occurrence of a START pulse. Since the START pulse occurs

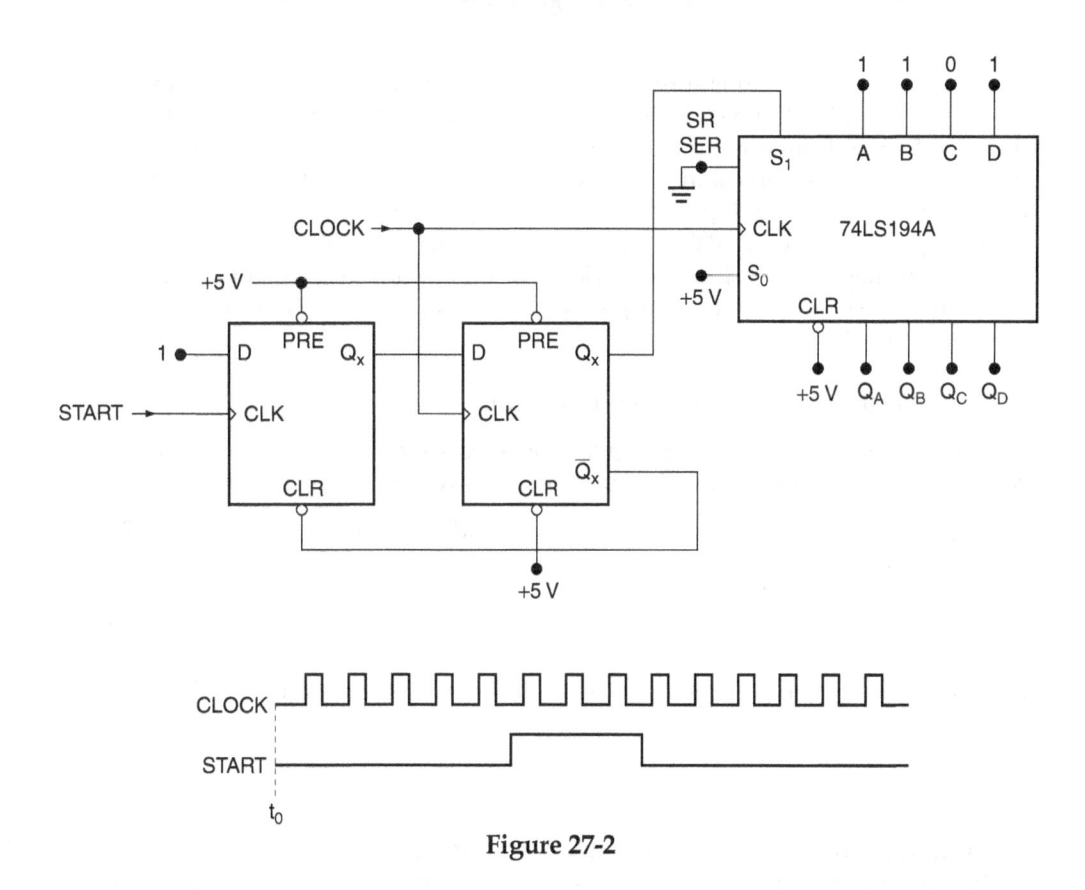

Figure 27-2

asynchronously to the clock pulses, the two flip-flops are used to synchronize the loading and shifting of the 74LS194A.

Assume that the START pulse has been inactive and that clock pulses have been continuously applied for a long time before t_0 (see the waveforms in Figure 27-2). Draw the waveforms you might expect to appear at Q_x, $\overline{Q}_x$, Q_y, and Q_D in response to the START pulse shown in the figure. Use Timing Diagram 27-1.

Timing Diagram 27-1

i) Wire the circuit in Figure 27-2. Use a normally HIGH pushbutton switch for the START pulse. Connect a square wave generator set at 100 Hz to input CLK of the 74LS194A. Connect toggle switches to A through D, and set these switches to 1101. Connect one vertical input of the oscilloscope to the output of the generator. Use the other vertical input to monitor first Q_x, $\overline{Q}_x$, Q_y, and Q_D, in that order. Do this by pulsing the START pushbutton several times while connected to each output. You may have to slow the clock down or speed it up so that the outputs are easily observed. Verify that output waveform Q_D is the serial representation for the parallel data.

Demonstrate the circuit for your instructor.

j) *Review:* This concludes the exercises on IC registers. To test your understanding of the principles covered in this experiment, answer the following questions:

1. Why is it necessary to synchronize the load and shift operations of the 74LS194A?

2. Explain how the flip-flops accomplish this synchronization in the circuit of Figure 27-2.

3. The number of clock pulses necessary to transfer data in parallel is ___, while the number of clock pulses necessary to transfer data serially depends on _____.

PROCEDURE

4. Show how you would modify the circuit of Figure 27-2 to change it into an eight-bit converter.

5. Show how you would wire the 74LS194A as a ring counter that shifts left instead of right.

Experiment 28

Name _____

DESIGNING WITH COUNTERS AND REGISTERS

OBJECTIVES

1. To practice designing systems using IC counters.
2. To practice designing systems using IC registers.

TEXT REFERENCE

Read Chapter 7.

EQUIPMENT NEEDED

Components and instruments are selected by the student according to her or his design needs.

DISCUSSION

The exercises below will give you an opportunity to practice designing devices using flip-flops, counters, and registers. Before constructing any of these projects, you should check with your instructor as to which exercises are to be completed from design through construction and testing. You should also be sure you know what documentation is required to be turned in.

For your own personal records, you should keep a notebook containing all draft schematics and text from each project as well as a copy of the final schematic and text. When a proposed solution does not work, make a notation on the back of its schematic about what is wrong and why. If you practice this technique, you will seldom repeat the same mistakes on other design projects. You will appreciate your efforts when you are assigned large projects.

PROCEDURE

a) *System clock:* Design a clock with a 100 kHz master oscillator and a 6.667 kHz symmetrical TTL compatible clock signal as its output.

b) *Variable timer:* Design a circuit that, when started, will produce a HIGH output for N milliseconds and then a LOW output for 1 second, where N is a number that can vary from 0 to 65,535 and that can be loaded into the circuit using toggle switches. Include in the design a clock oscillator to drive the circuit.

c) *Serial input with handshake:* An interface is a circuit that connects a computer to an external device. An input interface is usually capable of receiving a data word from the device—even if the computer is tied up doing something else—and store it until the computer takes it. Computers are not always ready to receive data from an interface. Similarly, interface circuits are not always ready to send data to the computer, interface units are not always ready to receive data from input devices, and finally, input devices do not always have data to send when the interface circuit is ready to receive. In order that all three units can communicate with each other, a single data line and several control lines, called *handshaking lines*, are used to interconnect them. These control lines help to minimize data loss.

A computer has a serial interface circuit that receives 10 bits of information from an input device and a (HIGH) INPUT DEVICE READY signal, and passes all bits, except the first and last bit, on to the computer in parallel when the computer is ready for it. Figure 28-1 shows the relationship among input device, interface, and computer and the format that the input device uses to send data to the interface. The following is a description of these lines:

- The input data line from the device to the interface is normally HIGH. When data is to be transmitted from the device, the device causes the input line to go LOW for one *bit time* (called the START bit), signalling that the next 10 bits are 8 bits of data followed by 2 (HIGH) STOP bits. Bit time is 100 microseconds in duration. See Figure 28-1.

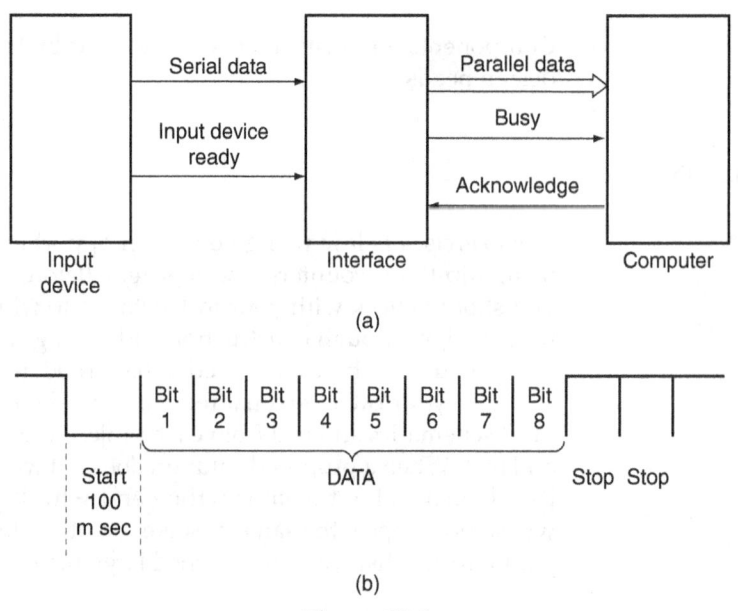

Figure 28-1

PROCEDURE

- The BUSY line is a control line from the interface to the computer. When this line is HIGH, the interface is not ready for the computer to take the data. When the line is LOW, the interface is ready.

- The ACKNOWLEDGE line is a control line from the computer to the interface. Whenever the computer has transferred the data from the interface, it drives this line LOW momentarily. The interface responds by forcing the BUSY line HIGH.

- The INPUT DEVICE READY line is a control line from the input device to the interface. As long as this line is LOW, the interface will not shift data serially. On the leading (NGT) edge of the START bit, the device drives this line HIGH, causing the interface to shift data from the input data line. At the end of the STOP bit, the input device forces the INPUT DEVICE READY line LOW, causing the interface to quit shifting data from the input data line. (You can assume that the device will not send data as long as the BUSY line is LOW.)

Design the interface unit.

Experiment 29

Name _____

TTL AND CMOS IC FAMILIES—PART I

OBJECTIVE

To investigate some of the important characteristics of TTL and CMOS IC families.

TEXT REFERENCES

Read sections 8.1 through 8.5 and sections 8.9 and 8.10.

EQUIPMENT NEEDED

Components
4023B IC;
7410 IC;
74F10 IC (optional);
74LS10 IC;
4 toggle switches;
4 LED monitors.

Instruments
0–5 volt DC power supply;
pulse or square wave generator;
dual trace oscilloscope;
logic probe;
VOM.

DISCUSSION

This experiment is the first of two in which you will investigate some family characteristics of TTL and CMOS ICs. In this experiment, you will learn how to measure average current drawn by an IC, propagation delay, and input/output characteristics of both families.

A typical small scale integration (SSI) IC has 1 to 13 logic gates (or their equivalent). Each gate on a single chip is designed to operate independently of the other gates. However, each receives its power from a single V_{CC} (or V_{DD}) connection to the IC. Measuring the current into this connection will help you to establish a relationship between the power consumption of the chip and the number of the chip's gates that have a LOW output.

Propagation delay is a measure of a gate's speed. It is necessary to have propagation delay for some digital circuits to work properly. It is important for you to understand how propagation delay "accumulates" in circuits, and how individual components like gates contribute to this accumulation, even though you will rarely make such measurements in the field. This accumulated propagation delay is often the cause of incorrect circuit operation.

Most important in the current experiment are the input/output (I/O) characteristics of gates. Knowledge of these parameters is necessary in design work and troubleshooting, for the parameters tell how gates (and other devices) behave under given conditions. For example, the maximum fan-out rating for a gate tells a designer how many loads its output can handle, while the value V_{IH} tells a technician that if a HIGH digital signal on a gate's input goes below this value, the gate's output will be unpredictable.

When you finish these exercises, you should be able to transfer this knowledge to flip-flops and other types of TTL and CMOS devices.

> NOTE 1: In this experiment and also in Experiment 30, you will be using CMOS ICs. It is important that you handle CMOS ICs carefully, since they are easily destroyed by static electricity. The ICs are shipped in a special conductive foam that keeps all pins at the same potential. Before removing the IC from the foam, be sure to discharge any potential that may exist among you, the chip, and the circuit board. Once on the board, make sure that all pins are connected (don't leave any disconnected) before applying power. When you are finished with the CMOS IC, remove it and return it to its protective foam.
>
> NOTE 2: Unless specified otherwise, $V_{DD} = V_{CC} = +5$ V and $V_{SS} = $ GND $= 0$ V.
>
> NOTE 3: Some of these exercises may not be required by the instructor.

PROCEDURE

a) Refer to the data sheets for a 4023B IC and a 7410 IC, and draw their pin layout diagrams:

PROCEDURE

b) *Measuring $I_{CC}(avg)$:* Connect each of the 4023B, 7410, 74LS10, and 74F10 ICs using the diagram of Figure 29-1. Do not connect V_{CC} (V_{DD} on 4023B) until instructed to do so. Connect GND on each IC to power ground (V_{SS} on 4023B). Connect toggle switches to one input of each 4023B gate. Tie all other 4023B gate inputs to V_{DD}. Do not connect any toggle switches to the 7410, 74LS10, and 74F10 inputs, but tie all but one input per gate to V_{CC}. Use a logic probe to check the toggle switches connected to the 4023B to make sure that they can switch between HIGH and LOW.

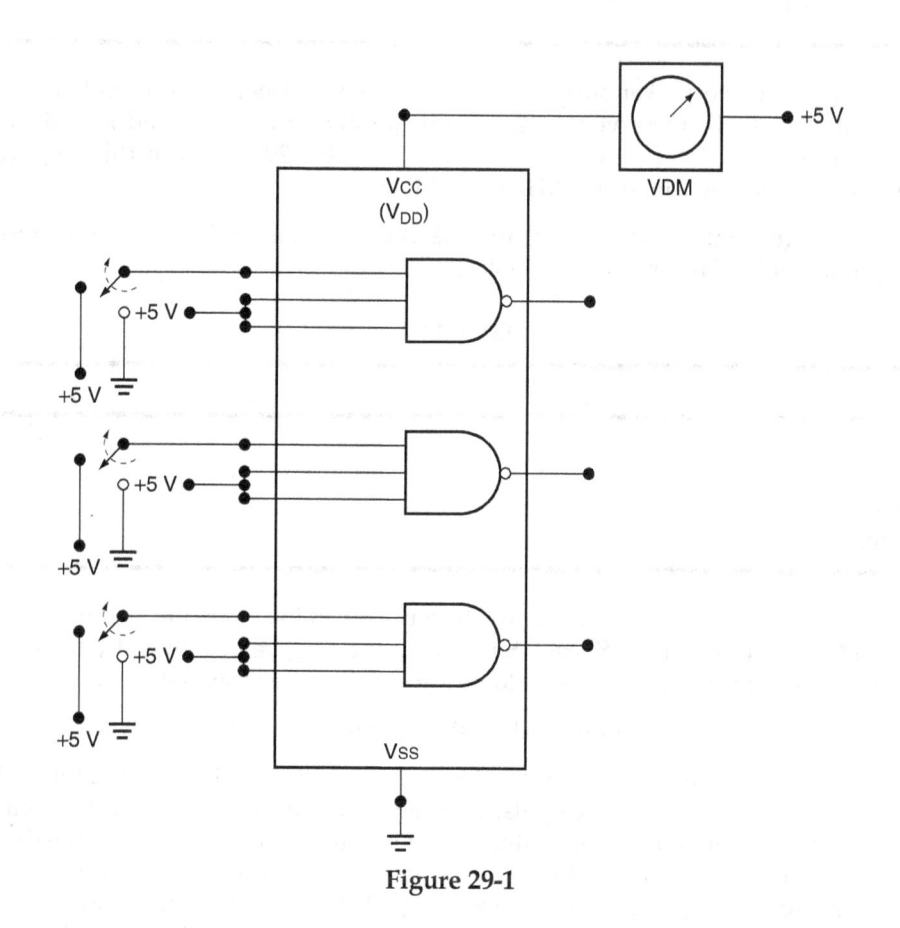

Figure 29-1

To measure the average current that an IC draws from its power supply, connect a VOM set-up to read DC milliamps between V_{CC} and +5 V. Set all three toggle switches to HIGH, and verify that each gate is operating properly with the logic probe. When all gate outputs are LOW, record the value of I_{CCL} indicated by the VOM: _____. Now set all three toggle switches to LOW. Verify that each gate output is now HIGH with the logic probe. Record the value of I_{CCH} indicated by the VOM: _____. Compute the average value for I_{CC} using the formula

$$I_{CC}(avg) = (I_{CCL} + I_{CCH})/2$$

and record this value in the "Static" column for the appropriate IC in Table 29-1.

c) Turn the power supply off. Disconnect the three toggle switches and V_{CC} connections from the inputs of the IC under test. Connect all inputs together on all three gates of the IC, and connect the square wave generator at their common

Table 29-1

Family	Static	I_{CC}(avg) 10 Hz	100 kHz	1MHz	Data Sheet
CMOS					
TTL					
LSTTL					
FTTL					

junction. Turn the power supply and square wave generator on, and set the generator to 10 Hz. Observe the value of I_{CC}(avg) on the VOM, and record this value under "10 Hz" for the appropriate IC in Table 29-1. Repeat this step for square wave frequencies of 100 kHz and 1 MHz.

d) Refer to the data sheet for the 4023B IC, and locate the values given for I_{CCL} and I_{CCH} and record them in Table 29-2.

Table 29-2

	I_{CCL}	I_{CCH}
4023B		
7410		
74LS10		
74F10		

Now compute I_{CC}(avg) using the formula given in step b. Record this value in Table 29-1 under "Data Sheet." Since I_{CCH} and I_{CCL} values given by the data sheet are maximums, I_{CC}(avg) should be greater than the observed values.

e) Repeat steps a-d for the 7410, 74LS10, and 74F10 ICs.

f) *Operating speed measurements:* Connect the circuit of Figure 29-2 for each IC that is to be tested. Connect a pulse generator set at 100 kHz signal to the circuit input and monitor both the generator output and the circuit output y on the dual trace oscilloscope. Measure the delay between the leading edge of a positive-going pulse of the generator and the leading edge of the circuit response at y as shown in the waveforms in Figure 29-2. You should note that this is the TOTAL propagation delay, t_{PD}, of the circuit and is the sum of t_{PHL} of the first gate and t_{PLH} of the second gate. An average value for one of the gates can be obtained by dividing t_{PD} by 2. Compute the average value, and record it in Table 29-3.

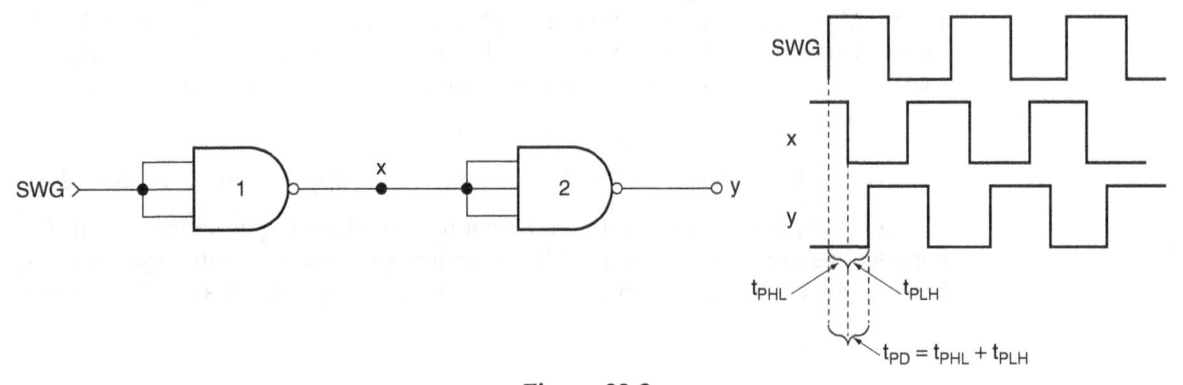

Figure 29-2

PROCEDURE

Table 29-3

Family	t_{PD} Observed	Data Sheet
CMOS		
TTL		
LSTTL		
FTTL		

g) Refer to the data sheet for the IC being tested, and compute the average delay using the maximum values that are given for t_{PHL} and t_{PLH}. Record this value in Table 29-3 also. This computed value should be greater than the observed value, since you used maximum values.

h) Repeat steps f and g for the 7410, 74LS10, and 74F10 ICs.

i) *I/O characteristics:* For one CMOS gate, remove the connection between its inputs and the inputs of the other gates. Make sure that the gate has nothing tied to its output Refer to Figure 29-3a and connect a toggle switch to the inputs of the gate you just disconnected. Set the toggle switch to HIGH and measure V_{OL}, then set the switch to LOW, and measure the value of V_{OH}. Record these values in Table 29-4, line 1.

Table 29-4

	No Load	
	V_{OL}	V_{OH}
4023B		
7410		
74LS10		
74F10		

j) Refer to Figure 29-3b and connect the output of the gate tested in step i to the inputs of one of the remaining gates. This represents three loads. Measure V_{OL} and V_{OH} of the driver gate, and record your measurements in Table 29-5, line 1.

Table 29-5

	Under Load	
	V_{OL}	V_{OH}
4023B		
7410		
74LS10		
74F10		

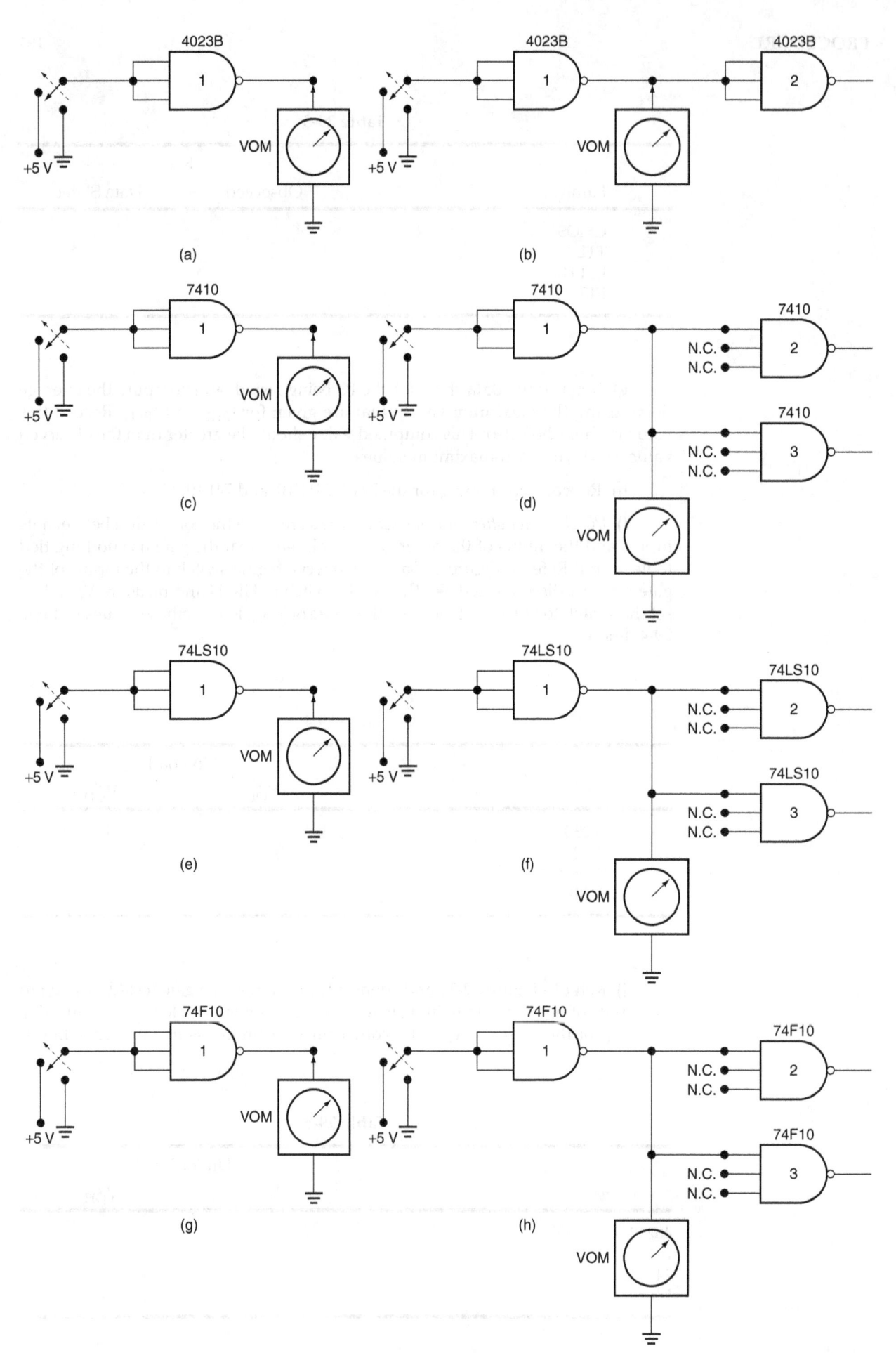

Figure 29-3

PROCEDURE

These readings should not differ appreciably from those recorded in step i. This is due to the extremely high input resistance at a CMOS input.

k) Wire the circuit of Figure 29-3c. Make sure that you do not connect anything to the output of gate 1 except the VOM. Set the toggle switch to HIGH and measure V_{OL}. Record this value on line 2 of Table 29-4 under V_{OL}. Now set the toggle switch to LOW and measure V_{OH}. Record this value on line 2 of Table 29-4 under V_{OH}.

l) Wire the circuit of Figure 29-3d. Note that only one input to each of gates 2 and 3 is connected to the output of gate 1. These two inputs combine for a total of two TTL loads. The remaining two inputs to gates 2 and 3 are left disconnected. Also, make sure that nothing is tied to the outputs of gates 2 and 3. Set the toggle switch to HIGH and measure V_{OL} of gate 1. Record this value on line 2 of Table 29-5 under V_{OL}. Now set the toggle switch to LOW and measure V_{OH} of gate 1. Record this value on line 2 of Table 29-5 under V_{OH}.

m) Wire the circuit of Figure 29-3e. Make sure that you do not connect anything to the output of gate 1 except the VOM. Set the toggle switch to HIGH and measure V_{OL}. Record this value on line 3 of Table 29-4 under V_{OL}. Now set the toggle switch to LOW and measure V_{OH}. Record this value on line 3 of Table 29-4 under V_{OH}.

n) Wire the circuit of Figure 29-3f. Note that only one input to each of gates 2 and 3 is connected to the output of gate 1. These two inputs combine for a total of two LS-TTL loads. The remaining two inputs to gates 2 and 3 are left disconnected. Also, make sure that nothing is tied to the outputs of gates 2 and 3. Set the toggle switch to HIGH and measure V_{OL} of gate 1. Record this value on line 3 of Table 29-5 under V_{OL}. Now set the toggle switch to LOW and measure V_{OH} of gate 1. Record this value on line 3 of Table 29-5 under V_{OH}.

o) Wire the circuit of Figure 29-3g. Make sure that you do not connect anything to the output of gate 1 except the VOM. Set the toggle switch to HIGH and measure V_{OL}. Record this value on line 4 of Table 29-4 under V_{OL}. Now set the toggle switch to LOW and measure V_{OH}. Record this value on line 4 of Table 29-4 under V_{OH}.

p) Wire the circuit of Figure 29-3h. Note that only one input to each of gates 2 and 3 is connected to the output of gate 1. These two inputs combine for a total of two F-TTL loads. The remaining two inputs to gates 2 and 3 are left disconnected. Also, make sure that nothing is tied to the outputs of gates 2 and 3. Set the toggle switch to HIGH and measure V_{OL} of gate 1. Record this value on line 4 of Table 29-5 under V_{OL}. Now set the toggle switch to LOW and measure V_{OH} of gate 1. Record this value on line 4 of Table 29-5 under V_{OH}.

q) For the 7410, compare the no-load measurements (line 2 of Table 29-4) with the measurements made under two loads (line 2 of Table 29-5). You should have observed an appreciable difference. This is due to the fact that TTL input resistances are not extremely high. You should have observed a larger variation between the two V_{OL} readings. This is because of the current that is sunk from a 7410 gate input back through the driver output transistor. Increasing the number of loads will increase this sink current and therefore cause V_{OL} to increase, even though the output resistance is relatively low (< 50 ohms). This is true for all TTL devices.

r) Compare the 74LS10 no-load measurements (line 3 of Table 29-4) with the measurements made under two loads (line 3 of Table 29-5). As in step q above, you should have noted an appreciable difference between the two measurements.

s) Connect the circuit of Figure 29-4a. This circuit connects one unit load to the output of gate 1, a 7410 gate. Set the toggle switch to HIGH, then measure and record I_{OL}. Compare this reading to the one-unit load factor of 1.6 mA.

$I_{OL} = $ _____ mA (1 TTL load)

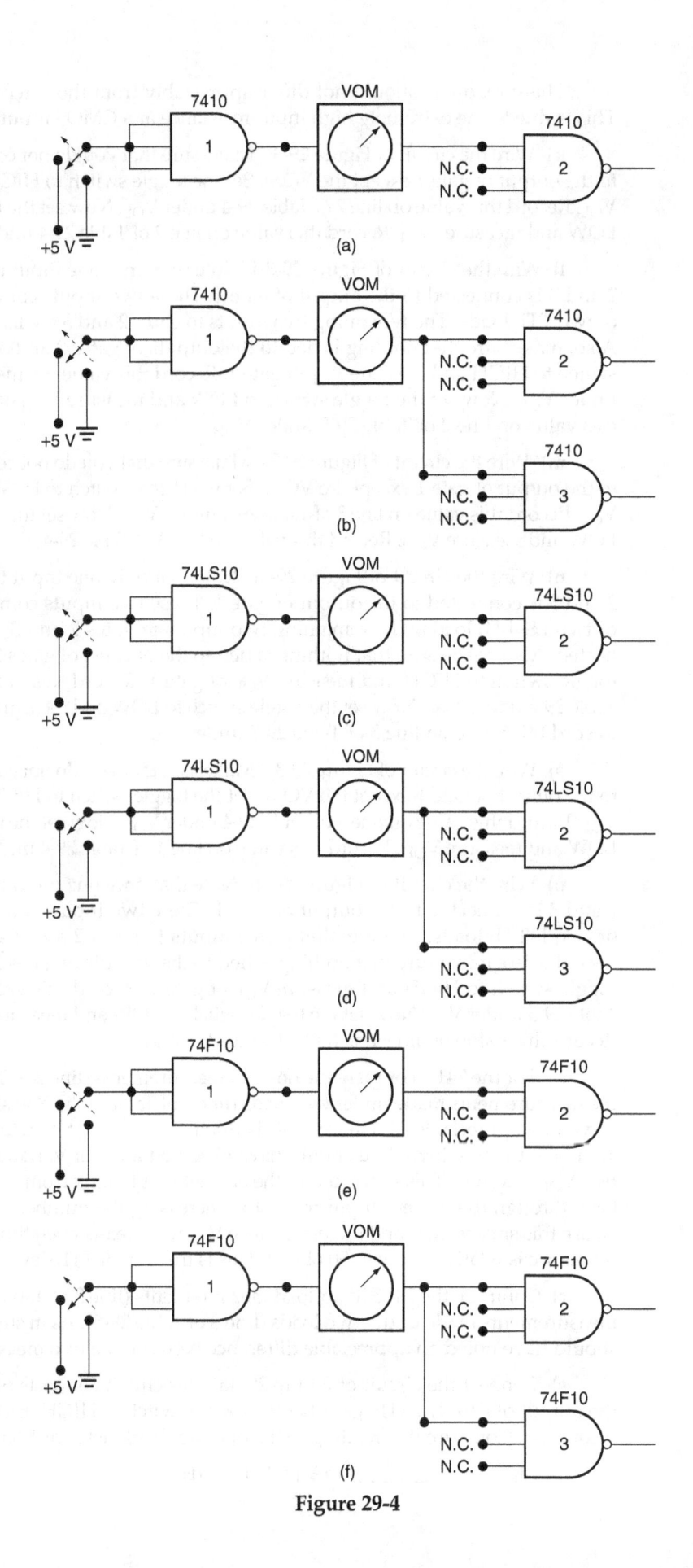

Figure 29-4

PROCEDURE

t) Connect the circuit of Figure 29-4b. This circuit connects two unit loads to gate 1. Set the toggle switch to HIGH, then measure and record I_{OL}.

I_{OL} = _____ mA (2 TTL loads)

u) Connect the circuit of Figure 29-4c. This circuit connects one LS-TTL load to the output of gate 1, a 74LS10 gate. Set the toggle switch to HIGH, then measure and record I_{OL}. Compare this reading to the one-unit load factor of 1.6 mA.

I_{OL} = _____ mA (1 LS-TTL load)

v) Connect the circuit of Figure 29-4d. This circuit connects two LS-TTL loads to gate 1. Set the toggle switch to HIGH, then measure and record I_{OL}.

I_{OL} = _____ mA (2 LS-TTL loads)

w) (Optional) Connect the circuit of Figure 29-4e. This circuit connects one F-TTL load to the output of gate 1, a 74F10 gate. Set the toggle switch to HIGH, then measure and record I_{OL}. Compare this reading to the one-unit load factor of 1.6 mA.

I_{OL} = _____ mA (1 F-TTL load)

x) (Optional) Connect the circuit of Figure 29-4f. This circuit connects two F-TTL loads to the output of gate 1. Set the toggle switch to HIGH, then measure and record I_{OL}.

I_{OL} = _____ mA (2 F-TTL loads)

y) *Review:* This concludes the first part of the investigation of TTL and CMOS families. To test your understanding of the principles covered in this experiment, answer the following questions:

1. The total propagation delay, t_{PD}, in a circuit like that of Figure 29-2 is computed by summing _____ of gate 1 and _____ of gate 2. The average value of one of the gates is _____.
2. The reason why V_{OL} increased with increasing loads for the TTL gate was _____.
3. One unit load for standard TTL when output is low is _____.
4. The fan-out for LS-TTL when the output is LOW is _____ unit load(s).
5. Compare standard TTL with CMOS in terms of the measurements you made on both in this experiment.
6. The fan-out for F-TTL when the output is LOW is _____ unit loads.

Experiment 30

Name _____

TTL AND CMOS IC FAMILIES—PART II

OBJECTIVES

1. To investigate CMOS and TTL interfacing.
2. To investigate the operation of a TTL open collector IC, the 7405.
3. To investigate the operation of a TTL tristate IC, the 74LS125A.
4. To investigate the operation of a CMOS transmission gate, the 4016B.

TEXT REFERENCES

Read sections 8.12 through 8.13 and sections 8.16 through 8.19.

EQUIPMENT NEEDED

Components
4016B IC;
4023B IC;
7405 IC;
7410 IC;
74LS10 IC;
74LS125A IC;
4 toggle switches;
1 LED monitor.

Instruments
0–5 volt DC power supply;
pulse or square wave generator;
dual trace oscilloscope;
logic probe;
VOM.

DISCUSSION

TTL-CMOS Interfacing

Interfacing two gates or circuits is not always straightforward. For example, you cannot drive the inputs of some CMOS gates (4000 and 74HC series) with a TTL output the way you drove TTL inputs with TTL outputs. One reason for this, as you discovered in Experiment 29, is that the lower limit for a HIGH (V_{IH}) at a CMOS input is 70% of V_{DD}, or about 3.5 volts if V_{DD} = 5 volts. Since V_{OH} for TTL outputs can be as low as 2.4 volts, the CMOS gate may not respond properly all the time. A solution to this problem is to use 74HCT series ICs, which are directly compatible with TTL. In the current experiment, you will investigate TTL-CMOS interfacing.

Open Collector TTL

Other interface devices are also investigated in this experiment: the open collector TTL and tristate TTL. The open collector is a common device for connecting TTL outputs to a device that operates at a slightly higher voltage or current than TTL can handle directly. Open collector devices can also be used to connect two or more TTL outputs together directly.

Tristate TTL

Tristate TTL devices have an enable input that can switch the output of the gate from a TTL HIGH or LOW to a third logic state, which is called high impedance (Hi-Z) or floating. These devices act like their standard TTL counterparts when enabled but like an open output when disabled. In fact, when the device is disabled, an ohmmeter will indicate an extremely high resistance between output and ground and output and V_{CC}. Tristate devices are primarily used to switch TTL outputs onto a bus when enabled and disconnect the outputs when disabled. Registers and other large ICs often have tristate devices built into their output circuits so that the devices may be used in bus systems.

CMOS Transmission Gates

The transmission gate is called by various other names, such as bilateral switch and analog switch. It is a circuit that switches analog or digital signals (as long as the signals are within the limits of 0 to V_{DD}), much like an electromechanical relay. The switching function is controlled by digital input logic. The gate's input and output are interchangeable.

> NOTE: See the note in Experiment 29 concerning the handling of CMOS ICs. Also, unless specified otherwise, $V_{DD} = V_{CC} = +5$ V and V_{SS} = GND = 0 V.

PROCEDURE

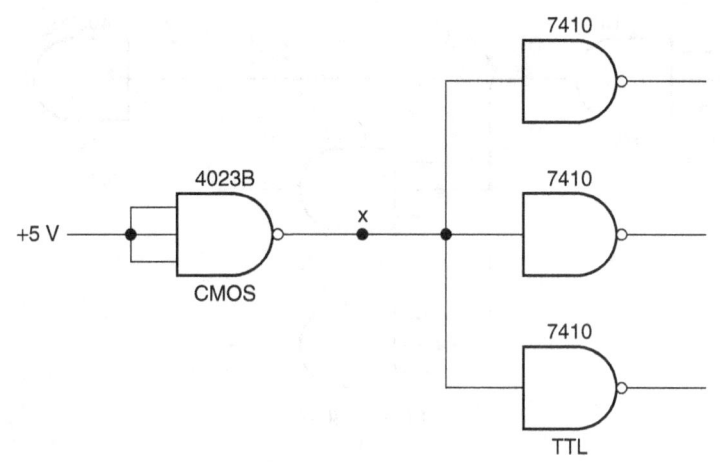

Figure 30-1

PROCEDURE

a) *CMOS driving TTL:* Connect the circuit of Figure 30-1.

Measure and record V_{OL} at point x.

V_{OL} = _____ (3 TTL loads)

You should observe that $V_{OL} > 0.8$ V, which is V_{IL} (max) for TTL. When V_{IL} exceeds 0.8 V, a TTL gate cannot be guaranteed to respond as if its input is LOW.

b) Disconnect two of the 7410 inputs from the 4023B output and observe that V_{OL} of the 4023B decreases, probably well below 0.8 V. Record this value.

V_{OL} = _____ (1 TTL load)

You should deduce that the 4023B can reliably drive one 7410 input.

c) *CMOS driving LSTTL:* Repeat step a, this time using three 74LS10 loads. You should observe that V_{OL} is well below 0.8 V, even though the 4023B is driving three loads. Why? _____.

d) *TTL driving CMOS:* Since CMOS gates draw virtually no current, you should expect no current problems when TTL gates drive CMOS gates. However, CMOS gate inputs require V_{IH} to be at least 3.5 V when $V_{DD} = 5$ V, while TTL HIGH outputs can go as low as 2.4 V under extreme conditions.

Connect the circuit of Figure 30-2, measure the voltage at x, and record it.

V_x = _____ (no pull-up)

While V_x is probably greater than 3.5 V and still sufficient for driving a CMOS input, the *safety noise margin* for the CMOS gate is greatly reduced. For this reason, a pull-up resistor would be advisable to raise the voltage at x. Use a 1 k-ohm pull-up resistor at x and measure V_x again. Record this value.

V_x = _____ (1 k-ohm pull-up)

Figure 30-2

[Circuit diagram showing a 7410 CMOS NAND gate with output x connected to a 4023B gate and two 7410 gates]

Figure 30-2

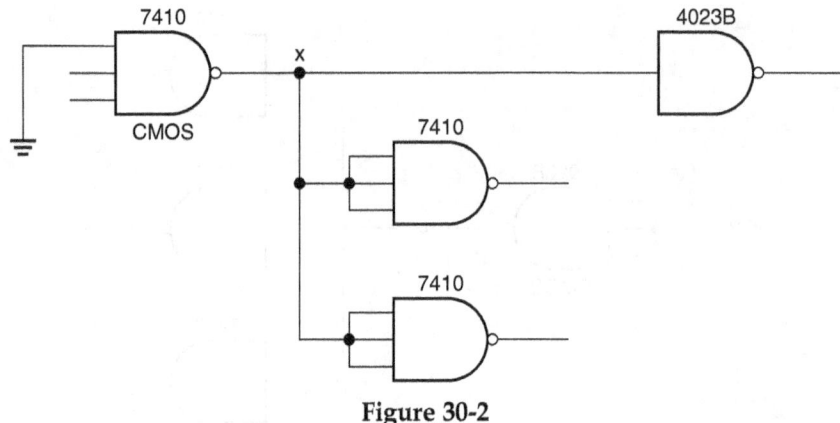

e) *TTL open collector—7405 IC operation:* Refer to the data sheet for a 7405 IC. Note that the pin layout is like the 7404 totem-pole hex-inverter. However, the 7405 inverter requires an external pull-up resistor to operate. Install a 7405 IC on the circuit board. Connect V_{CC} to +5 V and GND to power ground. Connect a toggle switch to each inverter input. Do not connect the pull-up resistor yet.

f) Toggle the input switch to one of the inverters, and use a VOM to measure V_{OL} and V_{OH} at the output of this gate. Record these values.

$V_{OL} = $ _____ (no pull-up)

$V_{OH} = $ _____ (no pull-up)

You should observe that there is no voltage in the HIGH state because there is no pull-up resistor.

g) Add a 1 k-ohm pull-up resistor, and repeat step f.

$V_{OL} = $ _____ (1 k-ohm pull-up)

$V_{OH} = $ _____ (1 k-ohm pull-up)

h) The value of the pull-up resistor used above was arbitrary. When determining the pull-up value for a particular load situation, you would compute the value using the outline given in the text. Assume three TTL loads, and compute R_C for a 7405 gate. Record your value below.

$R_C = $ _____ ohms

i) *Open collector application—WIRED-AND:* Connect the circuit of Figure 30-3. Use a standard resistor slightly larger than the one computed for R_C in step h.

This circuit configuration is called the WIRED-AND. Connect one input from each gate of the 7410 IC to act as the three TTL loads. Connect inputs D, E, and F to ground and toggle switches to inputs A, B, and C. Connect an LED monitor to output x of the WIRED-AND circuit. The circuit is now a three-input logic gate. Test the operation of the circuit by setting toggle switches A, B, and C to each input combination listed in Table 30-1 and observing the effect of these inputs on output x. Record your observations in the table.

PROCEDURE

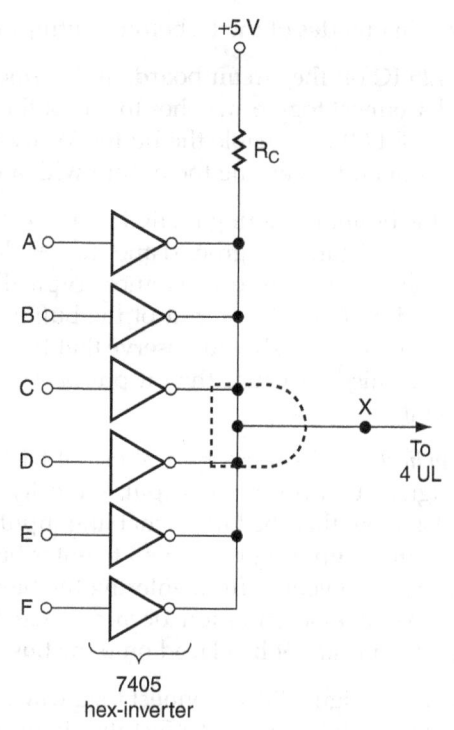

Figure 30-3

Table 30-1

Inputs			Output
A	B	C	x
0	0	0	
0	0	1	
0	1	0	
0	1	1	
1	0	0	
1	0	1	
1	1	0	
1	1	1	

j) Based on your observations in step i, the logic operation this circuit performs is the _____ operation.

k) *TTL tristate—74LS125 IC operation*: Refer to the data sheet for a 74LS125 IC, and draw its pin layout diagram:

Study the various operating modes of the IC before continuing.

l) Install a 74LS125 IC on the circuit board, and connect V_{CC} to +5 V and GND to power ground. Connect toggle switches to one of the buffer inputs and to its enabling input, $\overline{E}$. Set $\overline{E}$ LOW to enable the buffer. Verify that the buffer is enabled by toggling the input and observing the output with a logic probe.

m) Now disable the buffer by setting $\overline{E}$ HIGH. Connect the common lead of a VOM, set up to measure resistance, to ground and the "+" lead to the buffer output. You should observe a very large resistance, typically several M-ohms. Disconnect the VOM, and monitor the output of the buffer with the logic probe while toggling the input switch. You should observe that the input has no effect on the output of the buffer. This is because the output is essentially disconnected from the rest of the circuit.

n) *TTL tristate application—busses*: A typical use of tristate buffers is in connecting two or more signals to a common output, typically a data bus. Examine the circuit of Figure 30-4. Note that the buffer enabling inputs are always at complementary levels. Data from input A_0 is allowed to enter bus B when buffer 1 is enabled. Data at input A_1 is prevented from entering the bus at this time because buffer 2 is disabled. The reverse situation occurs when buffer 2 is enabled. Therefore the two sets of data cannot be placed onto the bus at the same time.

Construct the circuit of Figure 30-4. Connect toggle switches to the buffer inputs A_0 and A_1 and to the enabling input, E. Test the circuit by setting E to LOW, and observe the effect of toggling the data input switches on bus B with a logic probe. Repeat this for E set to HIGH. Record your observations in Table 30-2. Under "Bus," put A_0 or A_1, depending on which one has an effect on the bus.

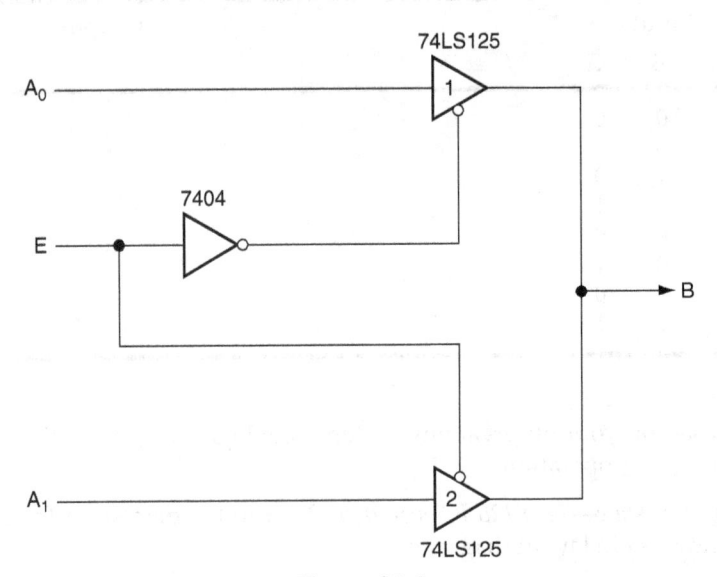

Figure 30-4

Table 30-2

E	Bus
0	
1	

PROCEDURE

o) *4016B IC operation*: Refer to the data sheet for 4016B IC, and draw its pin layout diagram:

p) Install a 4016B IC on the circuit board, and make the connections as shown in Figure 30-5. Connect a toggle switch to pin 13 and a pulse generator set to 10 kHz and with an output that ranges from 0 V to 4 V to pin 1. Connect an oscilloscope set up to measure DC to pin 2. Set the control switch to 0, and turn the power on. What do you observe on the oscilloscope display? _____.

Now set the control switch to 1. What do you observe? _____.

q) Turn the power off, and reverse the connections at pins 2 and 1. Turn the power on. What do you observe on the oscilloscope display?

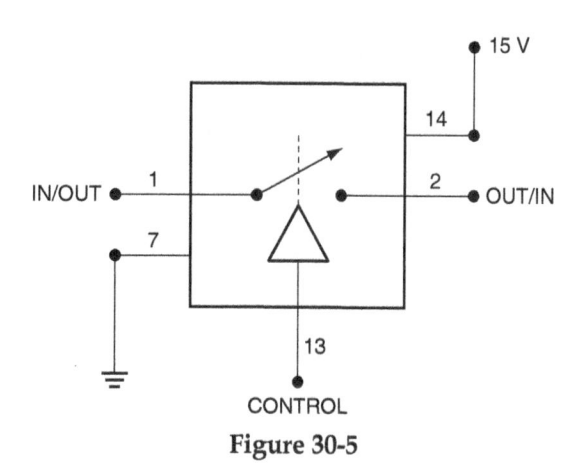

_____

Figure 30-5

r) In step p, you should have observed that the output of the gate was about 0 V when the control switch was set to 0 and was essentially the output of the generator when the control switch was set to 1. In step q, you demonstrated that the switch is bidirectional.

s) *Review*: This concludes the second set of exercises on TTL and CMOS families. To test your understanding of the principles covered in this exercise, answer the following questions:

1. When a TTL gate is used to drive a CMOS input, a pull-up resistor is used at the CMOS input because

_____.

2. When a CMOS gate is used to drive several TTL inputs, the TTL gates may not respond correctly if _____.

3. Why are buffers necessary when tying LEDs to TTL outputs?

4. TTL totem-pole outputs cannot be connected together because _____.

5. Use Boolean algebra to prove that the WIRED-AND circuit you constructed performs the function you specified in step j.

6. Explain why it is necessary to use tristate buffers when connecting two or more data sources to a bus.

7. Compare a CMOS transmission gate to a TTL tristate buffer.

Experiment 31

Name _____

USING A LOGIC ANALYZER

OBJECTIVES

1. To learn how to display and interpret digital waveforms using a 7D01 logic analyzer.
2. To practice using the various trigger modes and sources available on the 7D01.

TEXT REFERENCE

Read Appendix B in the back of this manual.

EQUIPMENT NEEDED

Components
74LS93 IC (2)

Instruments
7D01 logic analyzer;
0–5 volt DC power supply;
pulse or square wave generator;
logic probe.

DISCUSSION

Before attempting this experiment, read and be sure you understand Appendix B in the back of this manual. If your lab uses a logic analyzer other than the one this experiment features, your instructor will explain the differences and/or modify the experiment to fit the lab model.

PROCEDURE

PART 1—INITIAL CHECKOUT

a) 1. Do not connect the data probes to any signals.
2. Set the 7D01 panel controls as follows:

Vertical

```
POS. (position) . . . . . . . . . . . . . . . . . . . . . . MIDRANGE
MAG. (magnifier). . . . . . . . . . . . . . . . . . . . X1
RECORD DISPLAY TIME. . . . . . . . . . . . ∞ (fully clockwise detent)
CLOCK QUALIFIER . . . . . . . . . . . . . . . . OFF
```

Horizontal

```
POS. . . . . . . . . . . . . . . . . . . . . . . . . . . . . . . MIDRANGE
MAG. . . . . . . . . . . . . . . . . . . . . . . . . . . . . . X1
THRESHOLD VOLTAGE . . . . . . . . . . . . . TTL (+1.4 V)
EXT CLOCK POLARITY. . . . . . . . . . . . . ↑
SAMPLE INTERVAL. . . . . . . . . . . . . . . . 1 ms
DATA CHANNELS . . . . . . . . . . . . . . . . . 0–3
DATA POSITION . . . . . . . . . . . . . . . . . . POST-TRIGGER
TRIG SOURCE . . . . . . . . . . . . . . . . . . . . CH. 0 (channel 0)
EXT TRIGGER POLARITY. . . . . . . . . . . ↑
```

Word Recognizer (W.R.)

```
W.R. MODE. . . . . . . . . . . . . . . . . . . . . . . ASYNCH
FILTER . . . . . . . . . . . . . . . . . . . . . . . . . . MIN
CH. 0 through 15 . . . . . . . . . . . . . . . . . . X (center)
EXTERNAL QUALIFIER . . . . . . . . . . . . X (center)
PROBE QUALIFIER. . . . . . . . . . . . . . . . X (center)
```

b) Turn on the power to the 7D01.

c) Press the MANUAL RESET button (right above the SAMPLE INTERVAL selector). This initiates a new STORE mode. The analyzer is now sampling the various channel inputs every 1 μs.

d) To terminate the STORE mode and begin the DISPLAY mode, a TRIGGER EVENT must occur. With the TRIGGER SOURCE set to CH. 0, the logic analyzer is waiting for a positive transition on the channel 0 probe. Since this will not occur (with the probes not connected) we can use the MANUAL TRIGGER to generate a TRIGGER EVENT. Depress the MANUAL TRIGGER pushbutton.

PROCEDURE

e) Four waveform traces should appear on the CRT. They will have random logic levels, because the probes are not connected.

f) Verify that the VERT POS and VERT MAG controls can be used to vary the vertical position and spacing of the four traces. The HORIZ POS control should vary the horizontal position of the traces.

g) Change the DATA CHANNELS switch to 0–7. Eight traces should now appear. You may have to vary the VERT POS and MAG controls to get a reasonable display of the eight traces.

h) Change the DATA CHANNELS switch to 0–15 and repeat step g for 16 traces.

i) Set the DATA CHANNELS switch back to 0–3.

j) The four traces should have an intensified zone on the left end, which indicates the TRIGGER EVENT point. If you can't see this, gradually reduce the CRT intensity.

k) You should also see another intensified zone, which indicates the CURSOR point. It will be at some random point on the screen. Verify that it can be horizontally positioned on the screen using the CURSOR COARSE and FINE controls. The readout at the top of the screen will continually indicate the CURSOR position relative to the TRIGGER points in the units of one SAMPLE INTERVAL.

l) Set the CURSOR position to TRIG +45, and answer the following.

1) What is the time duration between the CURSOR and the TRIGGER point?
   _____.

2) What are the logic levels on the four channels at the CURSOR point?
   _____.

3) What is the total time duration from the beginning to the end of the trace?
   _____.

m) Change the DATA POSITION switch to CENTER, and note that the intensified trigger point has moved to the approximate center of the display. Verify that the CURSOR can be positioned before or after the TRIGGER point.

n) Change to PRE-TRIGGER and verify that the trigger point is moved to the right end of the screen.

o) Set the DATA POSITION switch back to POST-TRIGGER.

p) The current display will stay on the screen indefinitely because the DISPLAY TIME has been set to ∞. A new STORE mode can be initiated by depressing the MANUAL RESET pushbutton. Then, a new DISPLAY mode can be initiated by pressing MANUAL TRIGGER. Do this several times.

q) Repeat step q for a shorter display time.

r) Set the DISPLAY TIME back to ∞. Hit MANUAL TRIGGER to obtain the four channel traces.

s) The SAMPLE INTERVAL that is used during the STORE mode can be varied between 10 n sec and 5 m sec. However, there are certain limitations when 8 or 16 channels are being used. When 8 channels are being used, the selected SAMPLE INTERVAL must be 20 ns or greater. For 16-channel operation, the

SAMPLE INTERVAL must be 50 ns or greater. If you try to use a smaller SAMPLE INTERVAL than allowed, a light will blink on the SAMPLE INTERVAL dial. Verify this by changing the DATA CHANNELS switch and the SAMPLE INTERVAL selector switch.

t) If the SAMPLE INTERVAL is set to EXT, the SAMPLE INTERVAL is determined by an external clock signal connected to the appropriate probe.

NOTE: In this experiment, we will not be using the external SAMPLE CLOCK. Instead, we will use the 7D01's internally-generated, switch-selected SAMPLE INTERVAL.

u) Set the SAMPLE INTERVAL back to 1.0 ms.

PART 2—DISPLAYING COUNTER WAVEFORMS

a) We will now use the logic analyzer to display the waveforms from a 74LS93 IC counter. Connect a 74LS93 IC as a MOD-16 counter. Be sure that the master reset inputs are grounded. Connect the power supply to the 74LS93 IC. Now connect the clock A input of the counter to a pulse generator that is set to 10 kHz. Turn on the pulse generator, and use your logic probe to verify that all of the counter outputs are pulsing.

b) Connect the 7D01's upper data probes to the counter outputs as follows:

Ch. 0 (black wire) . . . . Output Q_A
Ch. 1 (brown wire) . . . . Output Q_B
Ch. 2 (red wire) . . . . Output Q_C
Ch. 3 (orange wire) . . . . Output Q_D

Also, connect the ground probe (white wire) to the ground of your circuit.

c) The logic analyzer can now be commanded to sample and store these waveforms by pressing the MANUAL RESET button. Do this, and note that a display of the counter waveforms appears almost immediately. This is because the signal connected to channel 0 has produced a positive transition for the TRIGGER EVENT (remember, our TRIG SOURCE is selected as Ch. 0).

NOTE: If the counter waveforms do not appear, notify your instructor immediately.

d) Check to see that the intensified TRIGGER point occurs when the channel 0 waveform has gone from LOW to HIGH. Also, note the logic levels on the other waveforms at the TRIGGER point. Record them as a four-bit number in the following order:

Ch. 3 _____, Ch. 2 _____, Ch. 1 _____, Ch. 0 _____.

e) Now position the CURSOR so that it coincides with the TRIGGER point. The 4-bit number on the bottom of the display will now indicate the waveform levels at the TRIGGER point. The results should be the same as you recorded in step d.

f) As you move the CURSOR to various other points on the waveforms, the 4-bit number should change to reflect the waveform levels at the CURSOR point. Verify this.

PROCEDURE

g) The CURSOR can be used to measure approximate time intervals between two points on a waveform. To demonstrate this, position the CURSOR just before one of the positive transitions of the Ch. 0 waveform. Note the CURSOR position relative to the TRIGGER point. Then move the CURSOR to the next Ch. 0's positive transition and note how many SAMPLE INTERVALS the CURSOR has moved. Record this number:

Number of SAMPLE INTERVALS = _____.

Record the measured value of one period of the Ch. 0 waveform. _____.

Compare this to the expected value (remember that the counter is being driven by a 10 kHz clock signal).

h) Use the same method to measure the pulse duration of the waveform on Ch. 2.

Pulse duration of Ch. 2 = _____.

i) The HORIZ MAG control can be used to horizontally magnify the waveforms by a factor of up to 10. Verify its operation.

j) When the waveforms are magnified horizontally, the HORIZ POS control can be used to find any part of the original unmagnified waveforms. Verify this.

k) Return the HORIZ MAG to ×1, and position the CURSOR at the TRIGGER point (TRIG + 0).

l) Press the MANUAL RESET button to initiate a new STORE mode and DISPLAY mode. Find the TRIGGER point on the waveforms; the CURSOR should still be at the TRIGGER point.

m) What are the waveform levels at the TRIGGER point this time? Are they the same as in steps d and e? If they are, then repeat step 1 until they change.

n) Why has the TRIGGER point changed? (Hint: What is our TRIGGER source?) _____.

o) With four channels, the duration of the waveform display is 1016 SAMPLE INTERVALS, or 1016 μs, since our SAMPLE INTERVAL is 1 μs. If we chose to use eight channels, the duration would be 508 SAMPLE INTERVALS, and the display would not show as many cycles of the counter waveforms. Check this out by switching the DATA CHANNELS to 0–7. Then press MANUAL RESET. You may have to use the VERT POS and VERT MAG controls to get all eight traces on the screen at the same time.

Note that the Ch. 0–Ch. 3 waveform displays are now spread out by a factor of 2 from what they were before, since the total time duration is 508 ms. This is similar to changing the time scale on a conventional scope.

p) Repeat step o for 16 channels, and observe what happens. What is the total time duration of the waveform now? _____.

q) Return to the 4-channel display mode.

r) We can also change the total time duration of the display by changing the SAMPLE INTERVAL. Change it to 2 μs and press MANUAL RESET. The resultant display now represents 1016 × 2.0 μs = 2032 μs time duration. Thus, there should be twice as many cycles of the various counter waveforms.

s) Change the SAMPLE INTERVAL to 0.5 μs and press MANUAL RESET. The display now represents 1016 × 0.5 μs = 508 μs. Thus, there are fewer cycles of the waveform being displayed.

t) The SAMPLE INTERVAL is usually made very small whenever we want to take a close look at a specific portion of the waveforms. For example, suppose we want to take a close look at the positive transition of the Ch. 0 waveform. The TRIG SOURCE is already set to Ch. 0, and we are using POST-TRIGGER so that the start of the display will show the positive-going edge of Ch. 0. Set SAMPLE INTERVAL to 10 n sec, and press MANUAL RESET. The display should now show the waveforms spread way out. The rising edge of Ch. 0 can now be examined with a resolution of 10 n sec. The total time duration of the display is only 1016 × 10 n sec = 10.16 μs. The other channel waveforms are not changing during this time.

u) A large SAMPLE INTERVAL is usually used when we want to see a larger portion of the waveforms. Set SAMPLE INTERVAL to 20 μs, and press MANUAL RESET. The display now represents 1016 × 20 μs = 20.32 m sec. Thus, there are many cycles of the waveforms crowded into the display. The HORIZ MAG can be used to spread out the display to make it easier to examine the various waveforms.

v) There are two important considerations that limit the maximum size of the SAMPLE INTERVAL:

1) If the interval chosen is too large, glitches or narrow pulses on the waveforms may be missed by the logic analyzer if they happen to occur in the interval between edges of the SAMPLE CLOCK. See Figure 31-1.

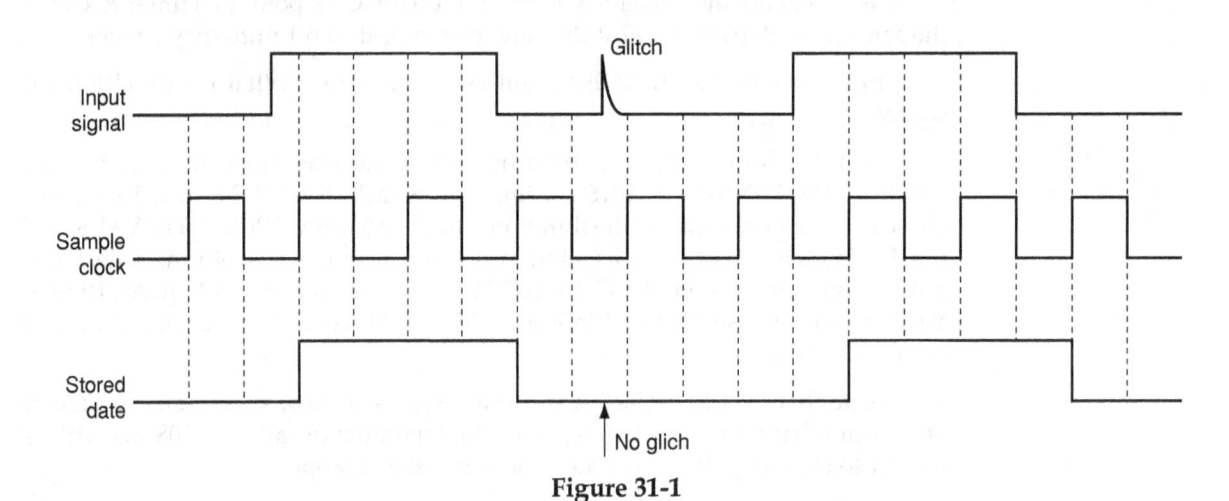

Figure 31-1

2) If the chosen SAMPLE INTERVAL is greater than the narrowest pulse width on the data waveform, the logic analyzer will not produce an accurate display of the waveform. For our waveforms, the narrowest pulse is about 100 μs on Ch. 0, 200 μs on Ch. 1, 400 μs on Ch. 2, and 800 μs on Ch. 3. Try various SAMPLE

INTERVALs greater than 100 µs and observe what happens to the displayed waveforms. Use the HORIZ MAG to spread the waveforms out. *In general, you should never use a SAMPLE INTERVAL greater than 1/2 the width of the narrowest data pulse.*

w) Set SAMPLE INTERVAL back to 1 µs and get a new display.

PART 3—USING AN EXTERNAL TRIGGER SOURCE

a) We have used MANUAL TRIGGER and Ch. 0 as the source for the TRIGGER EVENT. Another possible trigger source is EXT (external). Switch the TRIGGER SOURCE selector to EXT.

b) Press MANUAL RESET to initiate a new STORE mode. Note that the display does not appear. This is because no TRIGGER EVENT has occurred, since there is no signal connected to the EXT TRIG input jack.

c) Connect a 1x or BNC-to-alligator probe (*do NOT use a 10× probe*) from the EXT TRIG jack to a normally LOW pushbutton.

d) Press the pushbutton switch. The logic analyzer should now display the various waveforms, because the positive transition from the pushbutton output has produced the TRIGGER EVENT.

e) Change the EXT TRIG POLARITY switch to negative-going transition, and then press MANUAL RESET. Again, there is no display, because a TRIGGER EVENT hasn't occurred yet.

f) Press the pushbutton and hold it down. Why doesn't the display appear? Release the pushbutton. The display should now appear.

PART 4—USING THE WORD RECOGNIZER AS THE TRIGGER SOURCE

a) The WORD RECOGNIZER is used to generate a TRIGGER EVENT when a specific word (pattern of 0s and 1s) is present on the input data channels. For example, you might want to generate a TRIGGER when the 16 data inputs are at 1001101100101111. This binary pattern is called the TRIGGER WORD.

There are 16 three-position switches (one for each channel) that select the state of each data channel for the desired TRIGGER WORD. For example, if the desired TRIGGER WORD is 1010111000010110, the switches for channels 15, 13, 11, 10, 9, 4, 2, and 1 should be set HIGH, and the switches for the other channels should be set LOW.

Whenever a data channel is not being used, the WORD RECOGNIZER switch should be set to the middle position (X) so that it is not part of the TRIGGER WORD. Even when a data channel is being used, you can set its WR switch to X if you don't want that particular data channel to be part of the TRIGGER WORD.

b) Set the 16 WR channel switches to XXXXXXXXXXXX0101. The TRIGGER WORD is thus 0101 for the four channels we are using.

c) Set the TRIG SOURCE to WR. Then press the MANUAL RESET button.

d) Position the CURSOR at the TRIGGER point (TRIG + 0). The TRIGGER should occur when the waveform levels are 0101.

e) Display the four waveforms with a TRIGGER point at 1011. Demonstrate this to your instructor or lab assistant.

f) The WR MODE switch is currently in the ASYNC position. In this position, the WR generates a TRIGGER EVENT as soon as the conditions on the WR channel switches are met. *It does not wait for a clock signal.*

If this switch is placed in the SYNC position, the WR will generate a TRIGGER EVENT when the WR conditions are met *and* the active edge of the EXTERNAL SAMPLE CLOCK occurs. This position is used *only* when the SAMPLE INTERVAL switch is set to EXT and the external system clock is connected to the "C" probe on the Ch. 0–7 connector.

Since we are using the logic analyzer's internal SAMPLE CLOCK, the WR MODE has to be set to ASYNC or the WR will not produce a trigger. Verify this by switching to SYNC and hitting MANUAL RESET. Note that there is no display.

Return to the ASYNC mode. The display should appear.

g) There are two other switches that are part of the WR operation. One is the EXTERNAL QUALIFIER switch. It can be used to add one more condition to those needed to produce a WR TRIGGER. The other switch is the PROBE QUALIFIER. Let's examine the EXT QUALIFIER switch first.

1) Set the EXT QUALIFIER switch to HIGH.
2) Connect a toggle switch to the EXT TRIG/QUALIFIER INPUT jack (this is a dual purpose jack). Set the toggle switch to LOW.
3) Press MANUAL RESET. Note that there is no display, even though the data inputs satisfy the WR channel switch settings, because the level at the QUALIFIER INPUT does not match the EXTERNAL QUALIFIER switch setting.
4) Set the toggle switch HIGH, and note that the display appears, since the EXTERNAL QUALIFIER condition is now satisfied. The TRIGGER point on the waveform should still be 1011 (the settings on the WR channel switches).
5) Set EXT QUALIFIER back to X, and disconnect the QUALIFIER INPUT.

h) The PROBE QUALIFIER switch operates the same as the EXT QUALIFIER switch, except that it specifies the logic level required at the PROBE QUALIFIER input that is connected to the "Q" probe on the Ch. 8–15 connector.

1) Set the PROBE QUALIFIER switch to HIGH, and connect the Q probe to a toggle switch that is in the LOW state.
2) Press MANUAL RESET, and note that no display appears because the Q input is not HIGH.
3) Set the toggle switch connected to the Q probe to HIGH, and the display should appear. The TRIGGER point should still be at 1011.
4) Set the PROBE QUALIFIER switch back to X, and disconnect the Q probe.

PROCEDURE

i) The WR FILTER control (lower right-hand corner) is used to control the amount of time that the WR conditions have to be satisfied for the WR to generate a TRIGGER. This time requirement can be varied from 10 n sec to around 300 n sec. For example, if the WR FILTER setting is 100 n sec, then the WR will not generate a TRIGGER EVENT unless the WR conditions (WR channel switches and qualifiers) are met for 100 n sec or longer.

The WR FILTER can be used to prevent erratic triggering of the WR in situations where glitches or "race" conditions cause the WR conditions to be satisfied. This is demonstrated in the next step.

j) Make sure that the CURSOR is still positioned at the TRIGGER point.

k) Set the WR for a TRIGGER WORD of 0000.

l) 1) Set the DISPLAY TIME for a few seconds. The logic analyzer will now automatically perform a new STORE and DISPLAY mode every few seconds.

    2) Examine each new display, and note the readout of the waveform levels at the TRIGGER point. Since the WR is set for a TRIGGER WORD of 0000, you expect the levels at the TRIGGER point to be 0000. However, you will see the levels at the TRIGGER point varying randomly (e.g., 0000, 1000, 0100, 0010). How can this happen if we selected 0000 for the WR TRIGGER WORD?

The answer is that the 0000 condition can occur momentarily at various points on the counter waveforms as the flip-flops are in transition from one state to another. An example is shown in Figure 31-2. Examine the waveforms where the transition is being made from 0111 to 1000. Because of its propagation delay, flip-flop D doesn't go HIGH until after C has gone LOW. Thus, for a few nanoseconds, the 0000 condition is present. This can be recognized by the WR, which will then generate a TRIGGER.

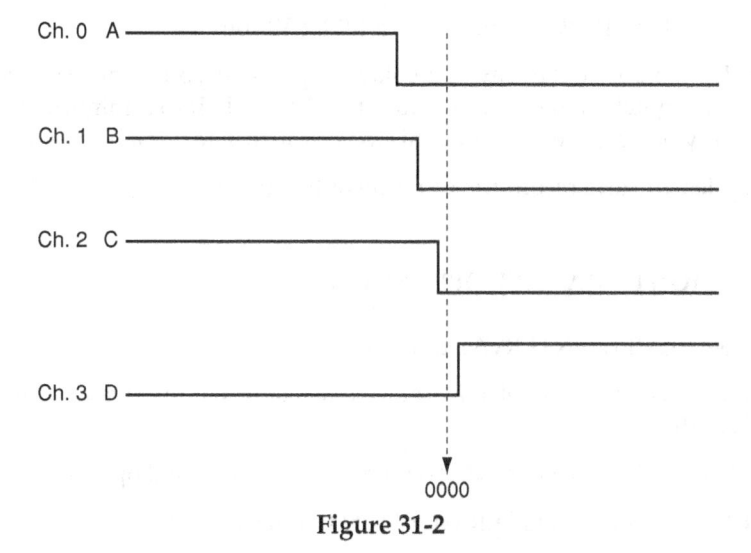

Figure 31-2

m) We can prevent this erratic operation by setting the WR FILTER high enough so that the momentary occurrences of the 0000 TRIGGER WORD will be ignored by the WR, and it will respond only to the stable 0000 condition that occurs on the waveform.

Gradually increase the WR FILTER setting until the display always shows 0000 at the TRIGGER point.

Demonstrate this to the instructor or lab assistant.

n) We can use the logic analyzer to display the race condition depicted in Figure 31-2. To do so, however, we will have to use a much smaller SAMPLE INTERVAL, since the propagation delays between flip-flops is very small.

1) Set the WR for a TRIGGER WORD of 1000.
2) Change SAMPLE INTERVAL to 10 n sec.
3) Set DISPLAY TIME to ∞ and examine the waveforms. They should appear as in Figure 31-2. You can spread them out with the HORIZ MAG control.
4) Move the CURSOR to the point on the waveforms where the temporary 0000 condition occurs. Measure and record the duration of this 0000 condition.

PART 5—DETECTING GLITCHES WITH THE LOGIC ANALYZER

a) Very narrow glitches can be detected and displayed by using a very small SAMPLE INTERVAL. To demonstrate this, we have to somehow generate a glitch. One way to do this is to change our MOD-16 counter to a MOD-10 counter by connecting the outputs of flip-flops D and B back to the counter RESET inputs. This will produce a glitch at the B output as the counter reaches 1010 and immediately gets cleared back to 0000. See the waveform in Figure 31-3.

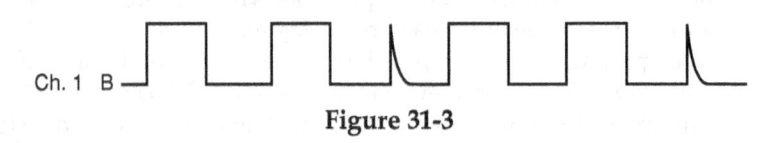

Figure 31-3

b) Wire the 74LS93 IC counter as a MOD-10 counter.

c) Try to figure out a way to display the glitch. If you want to use the WR to trigger on the glitch, remember to reduce the WR FILTER to minimum because the glitch is very narrow. Demonstrate this to your instructor or lab assistant.

d) Measure and record the duration of the glitch: _____.

PART 6—EIGHT-CHANNEL OPERATION

a) Set SAMPLE INTERVAL to 1 ms.

b) Connect the channel 4–7 probes to the same counter waveforms as channels 0–3, in the same order.

c) Set the DATA CHANNELS switch for eight-channel operation.

d) Set WR for a TRIGGER WORD of 00010001.

e) Press MANUAL RESET.

PROCEDURE

f) Find the TRIGGER point on the waveforms. It should occur at 00010001.

g) Repeat for a TRIGGER WORD of 01010101.

h) Repeat for a TRIGGER WORD of 01010001. Why is there no display?

Experiment 32

Name _____

IC DECODERS

OBJECTIVES

1. To investigate the operation of a 1-of-8 decoder IC, the 74LS138.
2. To investigate the operation of a BCD-to-decimal decoder IC, the 74LS42.
3. To examine the outputs of a decoder with a logic analyzer.
4. To investigate a method of eliminating glitches from decoder outputs.

TEXT REFERENCES

Read sections 9.1 and 9.2.

EQUIPMENT NEEDED

Components
7402 IC;
7408 IC;
74LS42 IC;
74LS93 IC;
74121 IC;
74LS138 IC;
4 toggle switches;
normally HIGH pushbutton switch, debounced;
4 LED monitors;
1 k-ohm potentiometer;
270 and 330 pF capacitors.

Instruments
0–5 volt DC power supply;
pulse or square wave generator;
dual trace oscilloscope;
logic probe
logic analyzer (optional).

DISCUSSION

You are already familiar with decoders, having investigated counter decoding in previous experiments. You are now ready to investigate representative examples of IC decoders. In this experiment you will study the behavior of two popular IC decoders, the 74LS138 and 74LS42.

74LS138 Decoder

The 74LS138 is an octal (1-of-8) decoder. It is frequently used to decode special binary codes called addresses in small computer systems. Its eight outputs are active LOW and it has three enable inputs and three data inputs. For a given octal code, one—and only one—of the eight outputs will go LOW if the chip is enabled. If the chip is not enabled, all of the outputs will remain HIGH no matter what data is present at the inputs.

74LS42 BCD-to-Decimal Decoder

In an earlier experiment, you investigated a special IC called a BCD-to-seven-segment decoder/driver. You found that it decoded a BCD counter and provided the necessary translation and power to drive a seven-segment LED display unit. the 74LS42 is a BCD-to-decimal (1-of-10) decoder, but, like the 74LS138, only one of its outputs will go LOW when a *valid* BCD code is at the data inputs. Unlike the 74LS138, the 74LS42 has no enables.

PROCEDURE

a) Refer to the data sheet for a 74LS138 IC and draw its pin layout diagram:

The 74LS138 can operate as a 1-of-8 decoder or a 3-line-to-8-line demultiplexer. In this experiment you will investigate the decoder function. The demultiplexer function will be covered in Experiment 34. Study the data sheet and familiarize yourself with the IC's functions.

PROCEDURE

b) *74LS138 decoder operation:* Install a 74LS138 on the circuit board. Connect the IC as follows:

1) Connect V_{cc} to +5 V, GND to power ground.
2) Connect toggle switches to select inputs A_2, A_1, and A_0, and enable inputs $\overline{E}_1$, $\overline{E}_2$, and E_3. The select inputs will be used to input data.

You will use a logic probe to monitor the outputs, $\overline{O}_7$ through $\overline{O}_0$.

c) Set E_3 to 1 and $\overline{E}_1 = \overline{E}_2 = 0$. Verify the decoder operating mode by setting inputs A_2, A_1, and A_0 to each input combination listed in Table 32-1 and checking the decoder outputs with the logic probe.

d) Now set $\overline{E}_1$ to 1. You should observe that the decoder is now disabled (all outputs are HIGH) and that the select inputs have no effect. Repeat this for $\overline{E}_2 = 1$ and then $E_3 = 0$.

Table 32-1

Select Inputs			Outputs							
A_2	A_1	A_0	$\overline{O}_0$	$\overline{O}_1$	$\overline{O}_2$	$\overline{O}_3$	$\overline{O}_4$	$\overline{O}_5$	$\overline{O}_6$	$\overline{O}_7$
0	0	0								
0	0	1								
0	1	0								
0	1	1								
1	0	0								
1	0	1								
1	1	0								
1	1	1								

e) *Decoder used as a device enabler:* Decoders are often used to select or enable other devices such as memory ICs and peripheral interface adapters. Each device is assigned a device number such as 0, 1, or 2. The device enable input is connected to the corresponding output of a decoder, and whenever the decoder receives the binary code for the device at its inputs, the decoder will activate the device. Let the four NOR gates of a 7402 IC simulate four devices numbered 0, 1, 2, and 3. One input of each NOR gate is connected to a toggle switch to represent data. Draw a diagram showing how a 74LS138 can be used to selectively enable the four gates.

Show your diagram to your instructor for approval. Then construct the circuit and test it, using a logic probe to monitor the outputs of the NOR gates.

f) *BCD decoder operation—74LS42:* Refer to the data sheet for a 74LS42 IC and draw its pin layout diagram:

You should note that the 74LS42 has four inputs, A_3 through A_0, and ten outputs, $\overline{O}_9$ through $\overline{O}_0$. You should also note that this decoder has no enable inputs.

g) Install a 74LS42 IC on the circuit board and also a 74LS93 IC counter. You will use the counter to supply the binary data to the 74LS42. Connect V_{cc} to +5 V and GND to power ground on each IC. Wire the 74LS93 as a MOD-16 counter and connect its outputs to the corresponding inputs of the 74LS42 (A_0 to Q_0, A_1 to Q_1, A_2 to Q_2, and A_3 to Q_3). Use a pushbutton switch to clock the counter. You will monitor the output of the decoder with a logic probe.

Clear the counter outputs. Verify that $\overline{O}_0$ is now LOW and the rest of the outputs are HIGH. Single-step the counter through its count sequence, and verify that the appropriate decoder output is LOW for each count, and that the rest of the outputs are HIGH. You should observe that any count greater than 1001 is not decoded. In other words, all outputs should be HIGH for any invalid BCD code.

h) *Observing decoder outputs with a logic analyzer:* If a logic analyzer is not available, go on to step i. Otherwise, connect the decoder outputs $\overline{O}_7$ through $\overline{O}_0$ to logic analyzer channels 0–7, respectively. Set the analyzer controls as follows (consult with your instructor to learn the exact settings for your particular model):

```
MODE........................Timing Diagram
CLOCK QUALIFIER ................Off
THRESHOLD ....................TTL
EXTERNAL CLOCK POLARITY ........Positive Edge
SAMPLE INTERVAL ................20 n sec
DATA POSITION ..................Post-Trigger
TRIGGER SOURCE .................Word Recognizer (WR)
WR FILTER ......................MAX
WR DATA ........................11111110
```

Apply a 1 MHz square wave to the counter. Start the logic analyzer so that it begins a new store mode. You should observe that the analyzer displays the decoder output waveforms almost immediately. If the screen remains blank, check the WR data switch setting (on some models this is displayed at the top of the screen) to make sure you set it correctly. If you persist in having difficulty getting a display, consult with your instructor.

PROCEDURE

Observe the waveforms, and note that they go LOW one at a time as the counter goes through its count sequence. You may see glitches on some of the waveforms.

Repeat this procedure several times. You may note that any glitches that appear may vary as to position on the screen. This is because they are very narrow, and even with a 20 n sec sample interval, the logic analyzer does not always pick them up.

Use the cursor to measure the approximate duration of one of the decoder output pulses. Record your value: _____.

The WR FILTER was set to MAX so that the WR would not be triggered by a glitch on the $\overline{O}_0$ waveform. If possible, set the WR FILTER to MIN, and repeat the measurement several times. You should see that the logic analyzer occasionally triggers on one of these glitches.

i) *Eliminating decoder glitches:* If you did not have access to a logic analyzer to do the last step, you should first apply a 1 MHz square wave to the counter input (or highest frequency obtainable) and observe decoder output $\overline{O}_0$ for glitches. Depending on the quality of your oscilloscope, you should be able to spot a few that occur randomly.

Glitches can cause serious problems, as you are already aware. You have seen one way to possibly eliminate glitches by using synchronous counters instead of asynchronous. Since that method is not always satisfactory, you will now employ another method, strobing the decoder. Strobing requires the decoder to have an enable input, like the 74LS138. The 74LS42 does not have an enable input. However, you can convert the 74LS42 into a 1-of-8 decoder WITH an enable. To do this, you simply use data inputs A_0, A_1, and A_2 as usual, and use the A_3 input as an enable. Since outputs $\overline{O}_8$ and $\overline{O}_9$ will not be needed, they will not be connected. When A_3 is set LOW, the 74LS42 outputs $\overline{O}_0 - \overline{O}_7$ will respond to the correct data input. With A_3 HIGH, they cannot respond.

Disconnect input A_3 from the counter, and connect it to a toggle switch. Set the toggle switch HIGH. Set the square wave generator to a low frequency and verify with the logic probe that none of the decoder outputs $\overline{O}_0 - \overline{O}_7$ are activated at any time. Set the toggle switch LOW, and verify that the decoder now acts like a 1-of-8 decoder.

j) Set up the circuitry shown in Figure 32-1. Note that the enable input (A_3) of the 74LS42 is to be driven by a pulse delay circuit consisting of a 74121 IC one shot. Note also the values of R_t and C_t. R_t is selected to be variable, since you will have to adjust t_p slightly in order to achieve a delay of slightly more than 0.5 microseconds.

Assuming that the count is 111 prior to the first clock pulse, draw the predicted signals at the enable input and decoder output $\overline{O}_0$.

k) With the dual trace oscilloscope, display the OS output and the output of the square wave generator. Adjust R_t so that t_p is slightly more than 0.5 microsecond. If this is not possible, change the value of C_t to 330 pF.

l) If a logic analyzer is available, get a display of the decoder waveforms $\overline{O}_7$ through $\overline{O}_0$ on the analyzer. You should observe no more glitches. If there are still some, try increasing t_p slightly. If this doesn't work, check your OS connections.

Figure 32-1

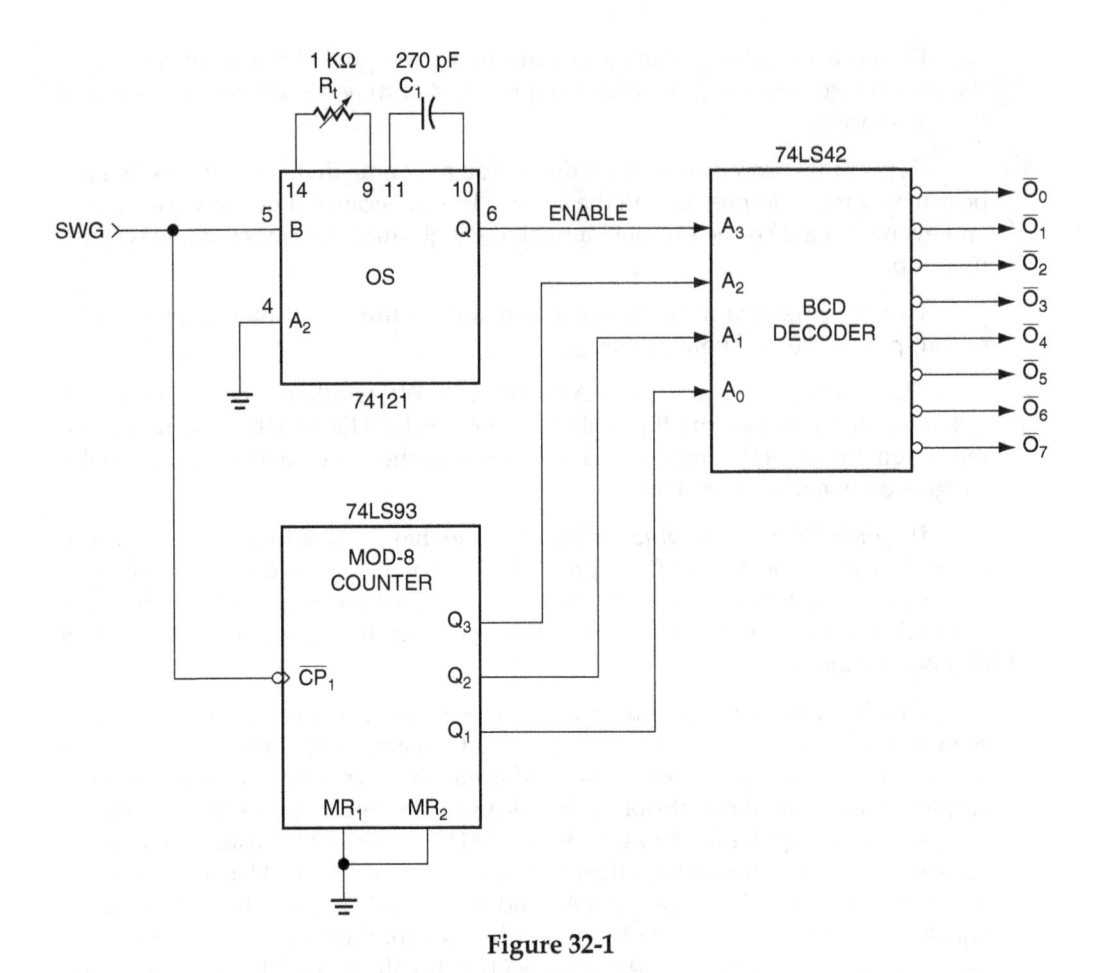

m) Once you have no glitches, reduce t_p gradually while observing the outputs of the decoder. You should, at some point, see the glitches return. This is because t_p is too narrow and the decoder is enabled during counter flip-flop transitions.

Demonstrate your circuit to your instructor.

n) *Review:* This concludes the exercises on decoders. To test your understanding of the principles covered in this experiment, answer the following questions:

1. Give a few examples of decoder applications.

2. Give the conditions necessary for the 74LS138 to decode input data.

3. In using the logic analyzer, if the WR data word is set to 11101111 when observing the 74LS42 outputs $\overline{O}_7 - \overline{O}_0$, on what output will the logic analyzer be triggered?

PROCEDURE

4. Can you think of other devices you have tested where the logic analyzer might have been useful?

5. If the clock signal used in steps j–m had been 2 MHz, what value of OS t_p would have been necessary?_____.

Experiment 33

Name _____

IC ENCODERS

OBJECTIVES

1. To investigate the operation of a decimal-to-BCD encoder, the 74HCT147 IC.
2. To investigate the application of the 74HCT147 IC in key encoding.

TEXT REFERENCE

Read section 9.4.

EQUIPMENT NEEDED

Components
7404 IC;
74LS08 IC;
74LS76 IC;
74121 IC;
74HCT147 IC;
74HC192 IC;
4 toggle switches;
normally HIGH pushbutton switch, debounced;
4 LED monitors;
33 k-ohm resistor;
1 µF capacitor;
decimal or hex keyboard with normally open contacts (recommended).

Instruments

0–5 volt DC power supply;

accurate 1 Hz square wave source;

storage oscilloscope (recommended), time interval counter, or nonstorage oscilloscope set at very low sweep speed.

DISCUSSION

In Experiment 32, you investigated IC decoders. Recall that for a given N-bit code received by the decoder, one and only one output became active. In this experiment, you will investigate the opposite of decoding: encoding. An encoder takes a single input and produces an N-bit code. For example, an octal encoder gives a three-bit code for a given octal input (digits 0–7) and a BCD encoder gives a four-bit code for a given decimal input (digits 0–9).

In the current experiment, you will investigate a 74HCT147 IC decimal-to-BCD encoder. You were introduced to this IC in Experiment 1. You will use this encoder to interface a keypad with a programmable timer circuit, so that you can input (preset) the amount of time desired by pushing a single key. Keep in mind that this is a CMOS IC, so you should observe static handling precautions.

PROCEDURE

a) Refer to the data sheet for the 74HCT147. Note that the inputs are active LOW. Grounding one (and only one) input will result in a binary code being produced at

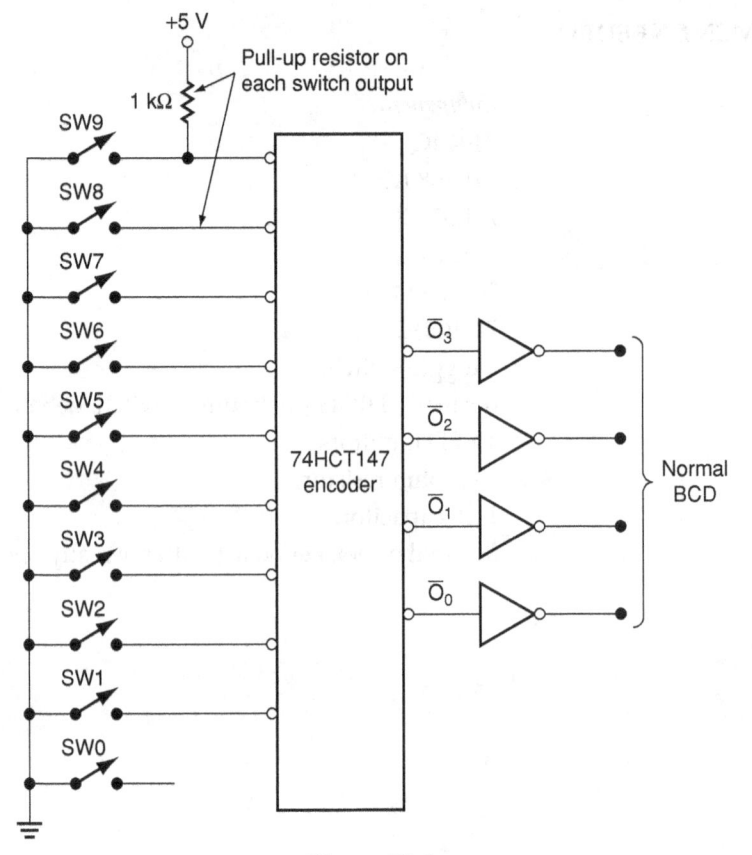

Figure 33-1

PROCEDURE

the encoder outputs, which represent the number of the input activated. Note also that the outputs are active LOW. Therefore the output code will be *inverted* BCD.

b) Install 74HCT147 and 7404 ICs on the circuit board. Connect them as shown in Figure 33-1. Connect V_{CC} to +5 V and GND to power ground for both ICs. Connect LED monitors to the output of the inverters. If a keyboard is available, connect it to the inputs of the 74HCT147. If not, simply touch a small length of wire to ground, and use this wire to activate the inputs. Verify that the 74HCT147 operates as a decimal-to-BCD encoder.

c) Study the programmable timer circuit in Figure 33-2. You will construct and test this circuit using toggle switches to provide the binary inputs to the counter. Once the circuit operates satisfactorily, you will then replace the toggle switches with the encoder unit you currently have on the board.

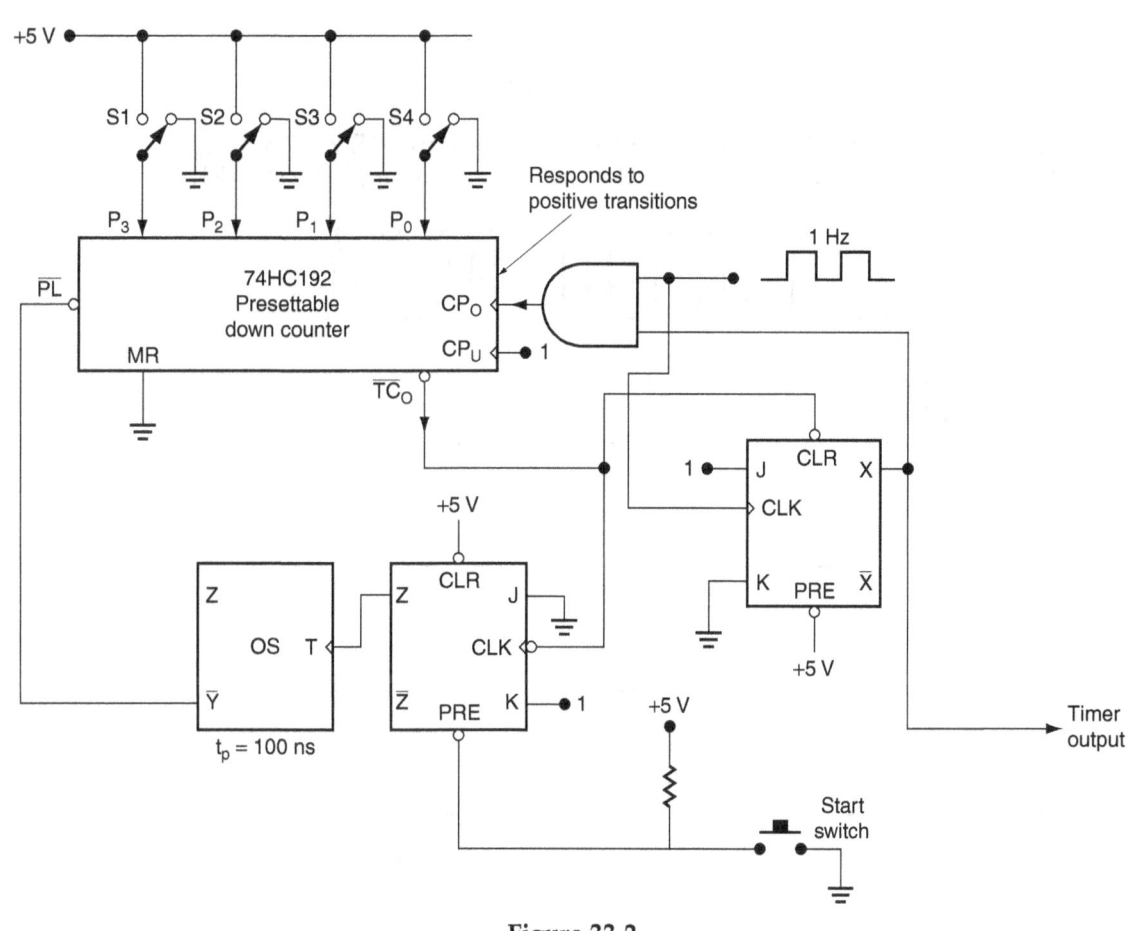

Figure 33-2

The timer output is normally LOW, and then it goes HIGH for 1–9 seconds, depending on the counter's preset conditions. The counter will be a 74HC192 BCD presettable UP/DOWN counter operated in the DOWN mode. The accuracy of the circuit depends on the accuracy of the 1 Hz square wave.

Refer to the data sheet for the 74HC192 IC counter. Note that its pin layout is the same as the 74LS193 and that it is functionally equivalent to a 74LS193 that is wired to operate as a MOD-10 counter.

Now complete the following circuit description:

1) Assume $S_1 = S_2 = 0$ and $S_3 = S_4 = 1$. Initially the counter is at 000 and $\overline{TC}_D = 0$.
2) Pressing the START switch produces a [positive-, negative-] going transition at Z.
3) This causes the OS, whose $\overline{Y}$ output goes _____, to activate the 74HC192 PL input. This loads the counter with _____, and as a result, $\overline{TC}_D$ goes _____.
4) Flip-flop X now goes to _____ on the next [positive-, negative-] going transition of the 1 Hz timing signal, causing the AND gate to allow the _____ to pass through into the counter.
5) The counter begins counting at _____. When the count reaches zero and CP_D is LOW, $\overline{TC}_D$ goes _____ and clears _____ to 0. This disables (inhibits) the _____ from getting through the AND gate to the counter.
6) The counter stays at 0000 until _____.

d) Draw the expected waveforms for the 1 Hz clock, START, Z, Y, X, CP_D, and $\overline{TC}_D$. Assume that switches S_1–S_4 are set as above. Use Timing Diagram 33-1.

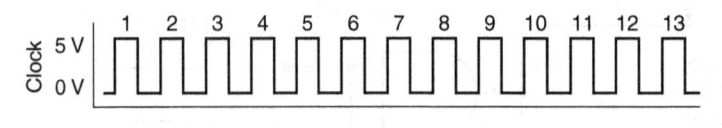

Timing Diagram 33-1

PROCEDURE

e) Show the results of steps c and d to the instructor before continuing.

f) Construct the programmable timer. Check its operation for several settings of switches S_1–S_4. Use either the storage oscilloscope or the time interval counter to measure the timer output. A nonstorage oscilloscope set at a very slow sweep can be used if necessary.

g) When the circuit is working as expected, press the START switch, and keep it depressed until after the output returns LOW. What happens when the START switch is pressed again? _____.

Draw a circuit that, when added to the programmable timer, will prevent this from happening. Obtain approval from your instructor or laboratory assistant and then install the modification before continuing.

h) Now remove switches S_1–S_4 and replace them with the outputs of the encoder circuit. Test the operation of the circuit by pressing and holding a key (or activating a 74HCT147 input) and pressing the START switch. The timer output should be a pulse whose duration, in seconds, corresponds to the key pressed (or input of the 74HCT147 activated).

i) Demonstrate the programmable timer for your instructor.

j) *Review:* This concludes the exercises on encoders. To test your understanding of the principles covered in this experiment, answer the following questions:

1. Why can't the START switch be connected directly to the OS?

2. Why can't the output of the timer be taken directly from $\overline{TC_D}$?

3. Explain the results you obtained in step g.

Experiment 34

Name _____

IC MULTIPLEXERS AND DEMULTIPLEXERS

OBJECTIVES

1. To investigate the operation of a 1-of-8 multiplexer, the 74HC151 IC.
2. To investigate the operation of a frequency selector.
3. To investigate the operation of a 3-line-to-8-line demultiplexer, the 74LS138 IC.
4. To investigate the application of multiplexers and demultiplexers in a synchronous data transmission system.

TEXT REFERENCES

Read sections 9.6 through 9.8.

EQUIPMENT NEEDED

Components
7404 IC;
7408 IC;
74LS74A IC (2);
74LS76 IC (4);
74121 IC (2);
74LS138 IC;
74HC151 IC;
74194A IC (4);
8 toggle switches;
normally HIGH pushbutton switch, debounced;
student selected capacitors and resistors.

Instruments
0–5 volt DC power supply;
pulse or square wave generator;
dual trace oscilloscope;
logic analyzer (optional).

DISCUSSION

Switches used to select data from several input sources are common in electronic systems. Digital systems use electronic circuits to simulate data selector switches called multiplexers. The multiplexer consists of several inputs, one output, and a number of SELECT inputs. When a binary code is applied to the SELECT inputs, the data with the input number represented by the code will be routed to the output. In the current experiment, you will investigate the operation of a 74HC151 IC, an 8-line-to-1-line multiplexer with a complementary output, and an enable. You will then use the multiplexer in a frequency selector.

The opposite of multiplexing is demultiplexing. A demultiplexer receives a single data line and distributes it over several outputs. Each output is selected by SELECT inputs, and each gets a "slice" of the data present on the input line. In this experiment, you will discover that the 74LS138 IC, whose decoder function was investigated in Experiment 32, can also be used as 1-line-to-8-line demultiplexer.

Finally, an elaborate, but not too complex, synchronous data transmission system is investigated. This system makes use of both the multiplexer and demultiplexer and is representative of serial communication on a small scale. You will use this system in Experiment 35, which is a troubleshooting exercise.

Note: The completed transmission system may be used in Experiment 35. Verify this with your instructor before disassembling the circuit.

PROCEDURE

a) Refer to the data sheet for 74HC151 IC and draw its pin layout diagram:

Note that the 74HC151 has two complementary outputs and an enable input.

b) *74151 IC operation:* Install a 74HC151 IC and a 74LS93 IC. Connect V_{CC} to +5 V and GND to ground for each IC. Wire the 74LS93 as a MOD-8 counter. Wire the remaining pins on the 74HC151 as follows:

PROCEDURE

1) Connect toggle switches to each of its eight inputs.
2) Connect select inputs S_2 through S_0 to outputs Q_2 through Q_0, respectively, of the counter.
3) Connect the enable input, $\overline{E}$, to ground.

Set the toggle switches so that $I_0 = I_2 = I_4 = I_6 = I_7 = 0$ and $I_1 = I_3 = I_5 = 1$. Set the clock counter to 1 kHz, and observe the output at Z and the counter clock with the dual trace oscilloscope. Trigger the oscilloscope sweep on Q_2 to get a stable display of the Z waveform. Draw the waveforms displayed on the oscilloscope. Use Timing Diagram 34-1.

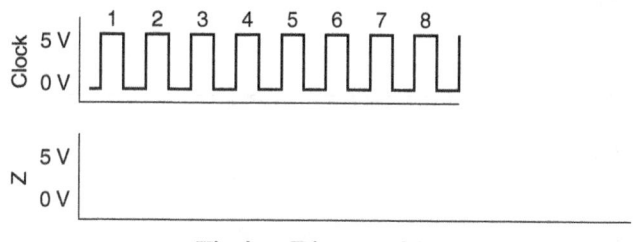

Timing Diagram 34-1

You should observe that the output at Z is the serial representation of the data at the multiplexer input. Toggle the I_2 data switch and observe the effect on the output. You should be able to locate the position of the input channel's data by observing the change in the output.

c) Lift the enable input momentarily. What happens?

d) *Frequency selector:* Figure 34-1 shows a circuit for a frequency selector. The 74HC151 multiplexer is fed various square wave frequencies with I_3 the highest and I_0 the lowest. Inputs I_4–I_7 are kept LOW. The data select inputs of the 74HC151, S_1–S_0, are used to select any of the square waves to be output at Z.

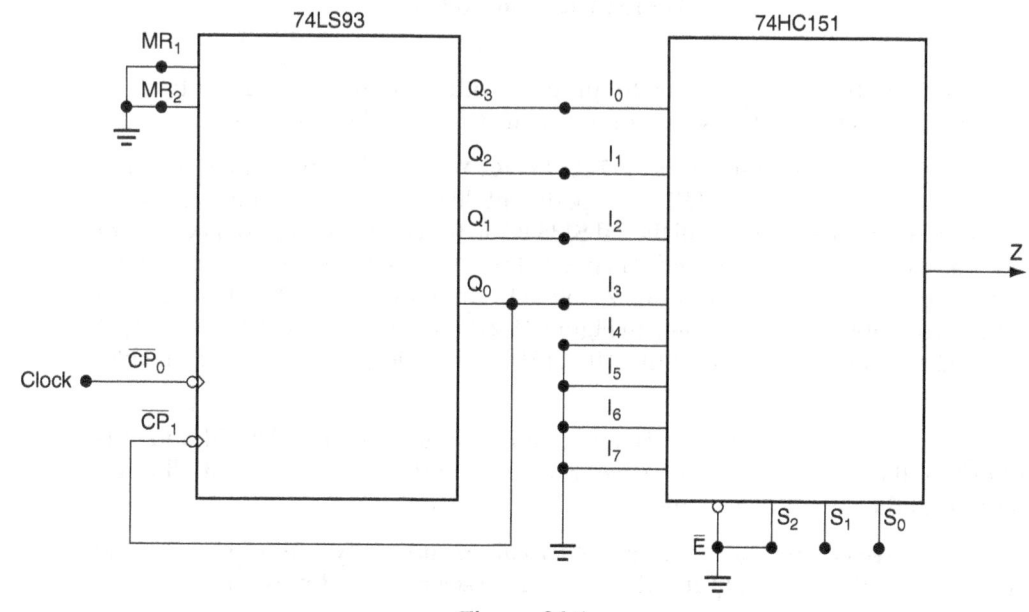

Figure 34-1

e) Wire the circuit of Figure 34-1. Use the SWG set at 5 kHz for the counter clock. Connect toggle switches to select inputs S_1 and S_0, and set them both to HIGH. If you are using an oscilloscope to monitor the circuit waveforms, use one vertical input to monitor Z and the other input to monitor the 74HC151 data inputs I_0–I_3. If you are using a logic analyzer, connect the logic analyzer to display I_1 through I_4 on channels 0–3 and the multiplexer output, Z, on channel 4. Use a sample interval that allows you to observe several cycles of the lowest frequency waveform to be displayed. Draw the waveforms that you observe on the logic analyzer, using Timing Diagram 34-2.

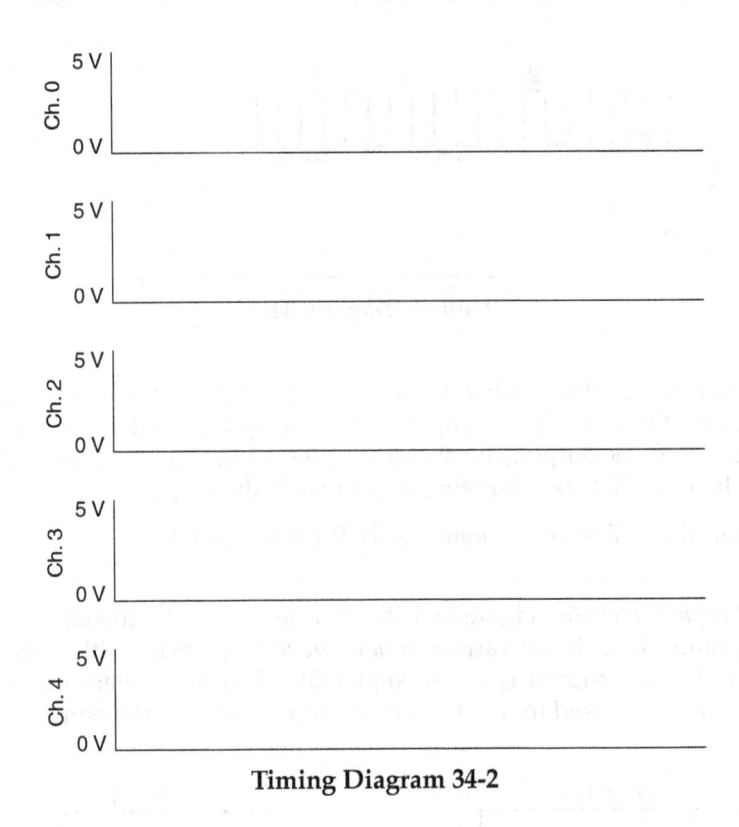

Timing Diagram 34-2

f) Vary the select inputs to produce other frequencies at output Z. Demonstrate the frequency selector for your instructor or laboratory assistant.

g) *Demultiplexer operation of the 74LS138 IC:* You are already familiar with the decoder operation of the 74LS138 (Experiment 32). You will now investigate its demultiplexer operation. Install the 74LS138 IC on the circuit board. Connect V_{CC} to +5 V and GND to power ground. Then wire the circuit of Figure 34-2. Connect the output of the 74HC151 multiplexer, Z, to the E_1 input of the 74LS138. Connect the other two enable inputs as shown in Figure 34-2. You will monitor outputs O_0–O_3 of the 74LS138 and inputs I_0–I_3 of the 74HC151 with the logic analyzer, if available, or oscilloscope.

h) Set the select inputs to 011, and monitor input I_3 of the 74HC151 and output $\overline{O}_3$ of the 74LS138. Are the waveforms the same? _____. Verify that all other outputs of the demultiplexer are inactive.

Set the select switches to several other values and verify that the signal at the selected demultiplexer output is the same as the selected multiplexer input.

PROCEDURE

Figure 34-2

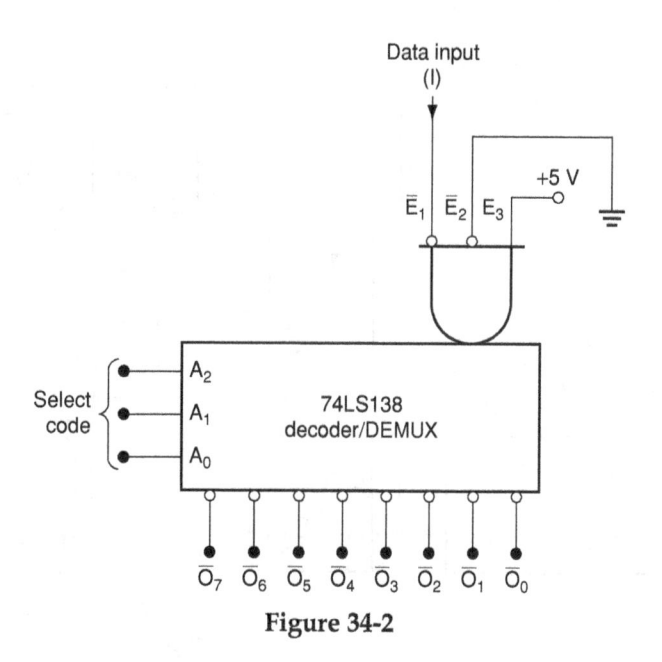

i) Application of multiplexers and demultiplexers—synchronous data transmission system: Now that you have verified the demultiplexer operation of the 74LS138 and have observed multiplexers and demultiplexers in the same circuit, you will build a synchronous data transmission system that makes use of these two operations. The operation of the data transmission system is described in the text, section 9.9. Review the operation of the system carefully, and then fill in the following statements:

Transmitter section operation:

Each register A, B, C, and D is configured as a _____ shift register. Each register will shift on the [PGT, NGT] of the shift pulse from AND gate 2. The *word counter* _____ the register data that will appear at Z. This counter counts from _____ to _____. The *bit counter* makes sure that _____ data bits from each register are transmitted through the _____ before advancing to the next register. This counter advances one count per _____ so that after _____ pulses, it recycles to 0. The word counter is incremented by _____. The Z signal contains _____ bits from each register for a total of _____ bits. The transmission system is controlled by _____.

Receiver section operation:

The receiver _____ the Z signal into _____ sets of data and _____ them to their respective outputs.

j) Construct the circuit of Figure 34-3. Use the following ICs (or their equivalent):

1) Registers A, B, C, and D - 74LS194A (4).
2) Multiplexer - 74HC151.
3) Demultiplexer - 74LS138.
4) Flip-flops W, X, and Y - 74LS74A (2).
5) MOD-4 counters - 74LS76 (4).
6) One-shots ($t_p = 1$ μS) - 74121 (2).

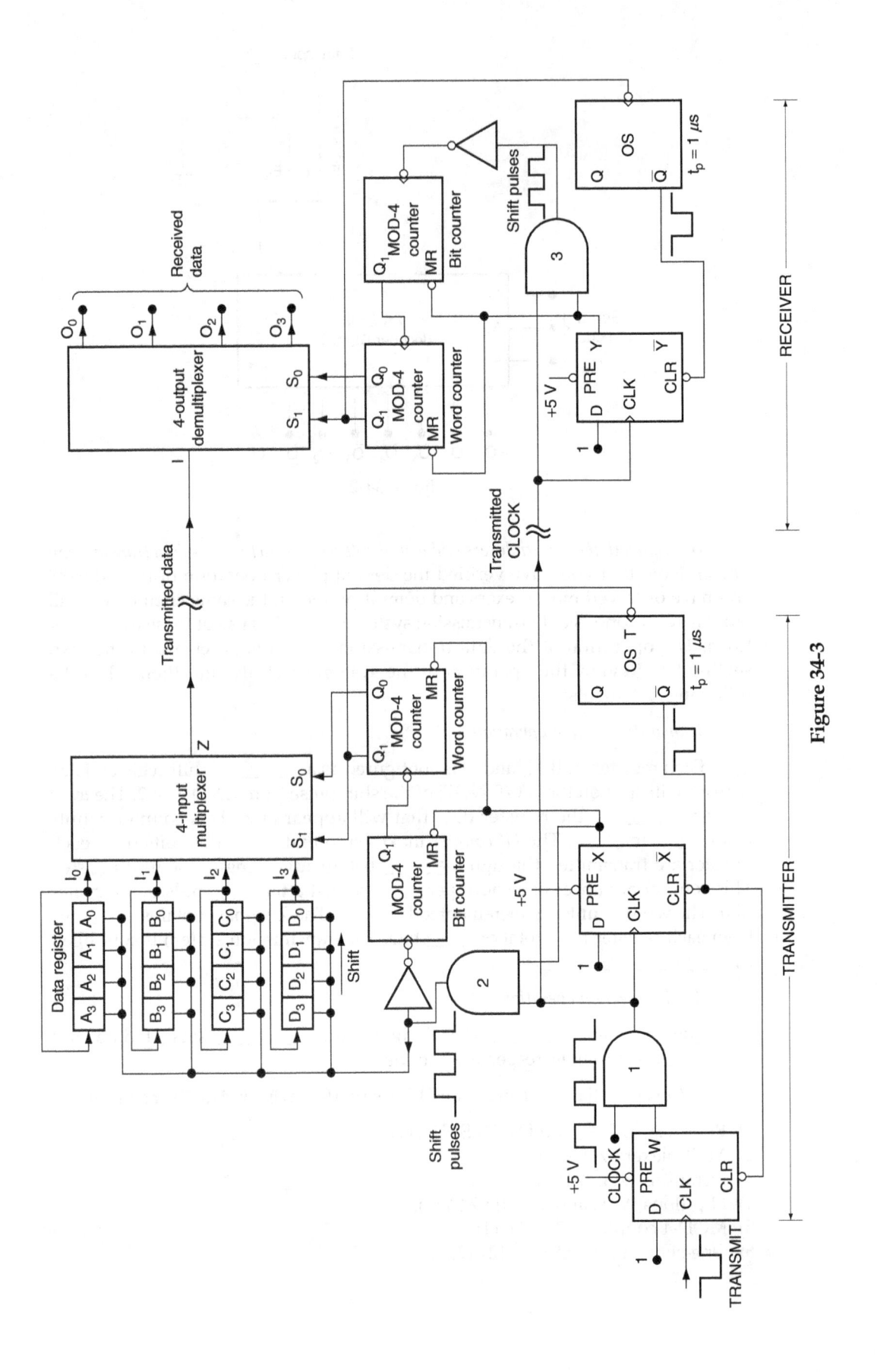

Figure 34-3

PROCEDURE

k) The 74LS194A ICs are to be wired as ring counters with parallel load capability. Connect a single toggle switch to all the register S_1 inputs. Connect all S_0 inputs to V_{CC}. The parallel data inputs will be permanently wired as follows:

Register A - [0110]
Register B - [1001]
Register C - [1011]
Register D - [0100]

The registers must be loaded manually before applying a TRANSMIT pulse. To accomplish this, set S_1 HIGH, and pulse the clock inputs LOW using a normally HIGH pushbutton switch. Then set S_1 LOW, and remove the pushbutton switch from the clock inputs.

Connect the output of AND gate 2 to the clock inputs of the registers. Output Q_D of each register should be connected to the appropriate multiplexer input and also to the right serial input SR SER. Connect the $\overline{Q}$ output of the flip-flop W to S_1 of each register.

Review Experiment 27 if necessary for the correct wiring of the 74194A as a ring counter.

l) Wire a normally HIGH pushbutton to the clock input of flip-flop W to provide the TRANSMIT pulse, and complete the rest of the circuit wiring according to the figure. Monitor the demultiplexer outputs with an oscilloscope or logic analyzer if one is available.

m) Test the circuit, and verify that the outputs of the demultiplexer are serial representations of the data stored in the corresponding register. Refer to Timing Diagram 34-3 for the waveforms that you should obtain.

If the system does not work properly, use a pushbutton switch for the clock input, and check out the operation of the circuit step-by-step.

n) Once the circuit is operating successfully, temporarily lift the connection at $\overline{MR}$ of the receiver word counter. What happens?

Reconnect the wire you just lifted. Now temporarily lift the connections at S_0 and S_1 of the 74LS151 IC. What happens?

Reconnect the select inputs, and demonstrate the circuit for your instructor.

Timing Diagram 34-3

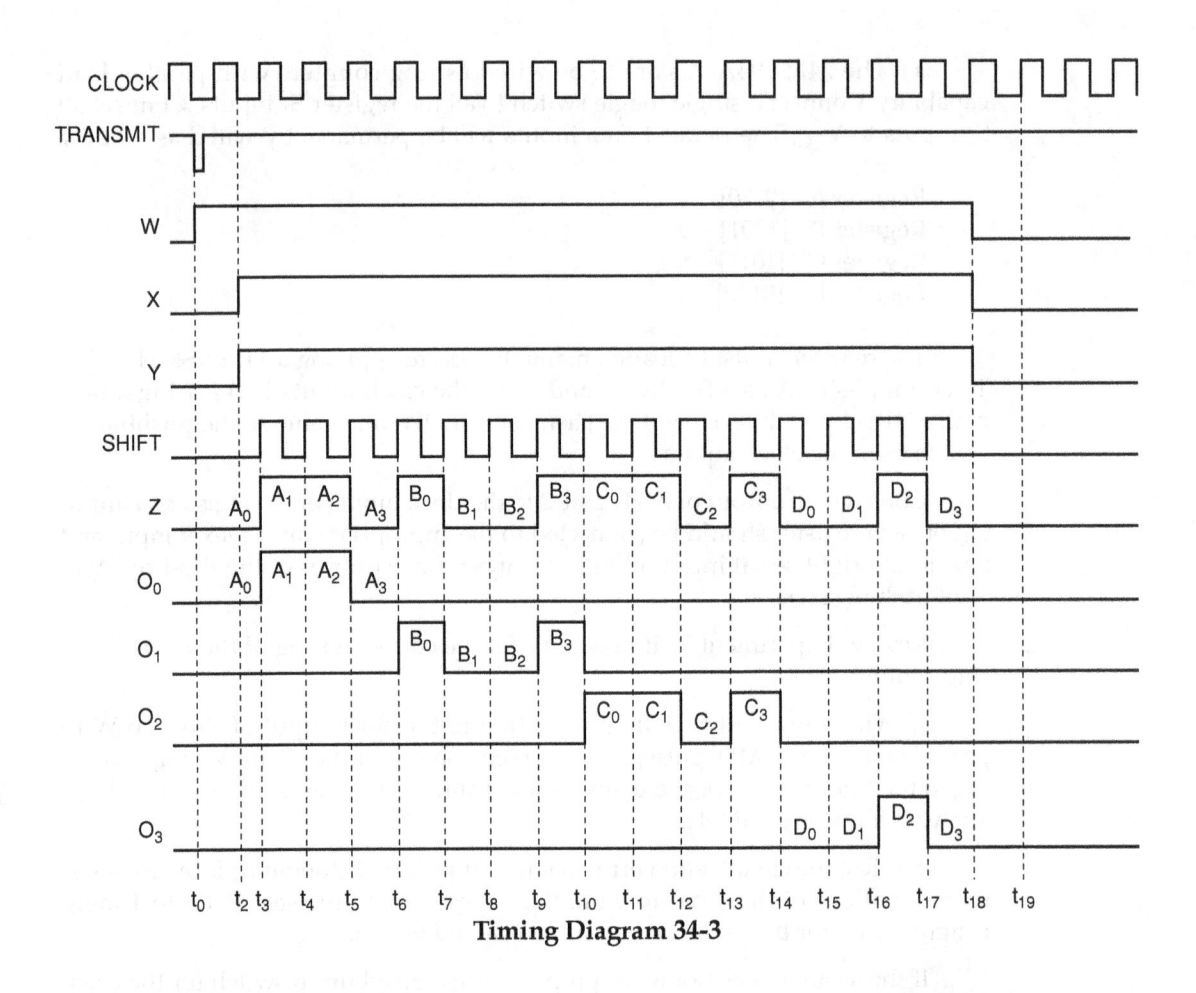

o) *Review:* This concludes the exercises on multiplexers and demultiplexers. To test your understanding of the principles covered in this experiment, answer the following questions:

1. In the data transmission system, what circuit modifications would be necessary to increase the number of channels to eight?

2. What would happen if register A's SR SER input became shorted to V_{CC}?

PROCEDURE

3. Draw a circuit diagram showing how three 74HC151 multiplexers can be arranged to form a 1-of-24 multiplexer.

Experiment 35

Name _____

TROUBLESHOOTING SYSTEMS CONTAINING MSI LOGIC CIRCUITS

OBJECTIVE

To practice troubleshooting systems containing MSI logic circuits.

TEXT REFERENCES

Read sections 9.5 and 9.9.

EQUIPMENT NEEDED

Components
74HC151 IC;
74LS76 IC (2);
74LS138 IC;
74LS93 IC;
toggle switches (8);
LED monitors (11);
OR
functioning circuit from Experiment 34.

Instruments
0–5 volt DC power supply;
logic probe;
logic analyzer (optional);
oscilloscope;
pulse generator or SWG.

DISCUSSION

Before beginning this exercise, check with your instructor to determine which exercise you are to do. Also, reread the text assignment, and work through the examples in each section. This experiment will help you to gain confidence in your ability to reason out a troubleshooting problem using what the author of the text refers to as observation/analysis.

PROCEDURE

a) *Troubleshooting a parallel-to-serial converter*: Examine Figure 35-1 closely. It is the circuit for a parallel-to-serial converter using a multiplexer and a counter. Construct the circuit, and test it until you have it operational.

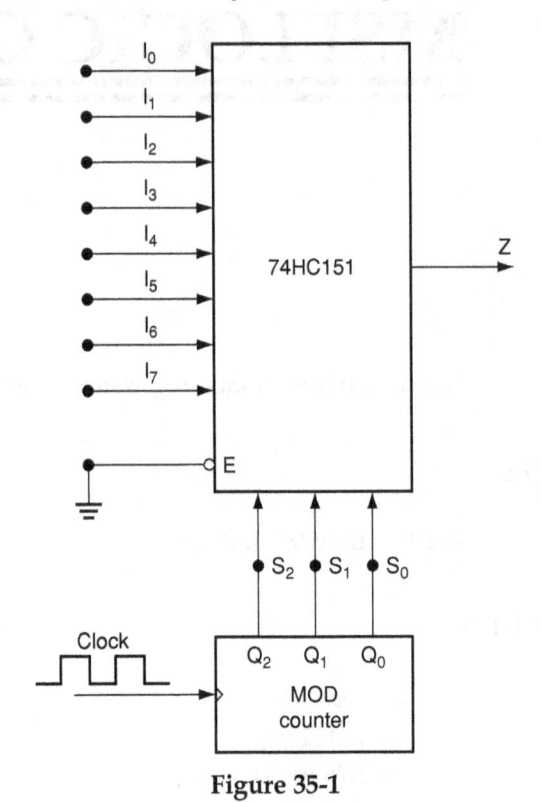

Figure 35-1

b) Have your lab partner or another student insert a fault into the circuit while you are not looking. Do not look for the bug yet. Now examine the circuit with a scope. Connect one vertical input to the clock and the other to the output of the converter. On a separate sheet of paper, draw the waveforms that you observe. Using your observations, narrow the location of the bug to a few possible faults. Then use this list and your test equipment to find the fault. Repeat this step as many times as possible.

c) Have your instructor place a bug in your circuit, and then repeat step b. Place your observations, list of possible faults, and your final solution on a separate sheet of paper.

PROCEDURE

d) *Troubleshooting a security monitoring system*: Examine Figure 35-2 closely. It is the circuit for a security monitoring system. Construct the circuit, using toggle switches for the door switches. Use a pulse generator set to a frequency sufficiently low (so that the LEDs can be monitored) to clock the counter. Test the system until you have it operational. Now review Example 9.6 in the text.

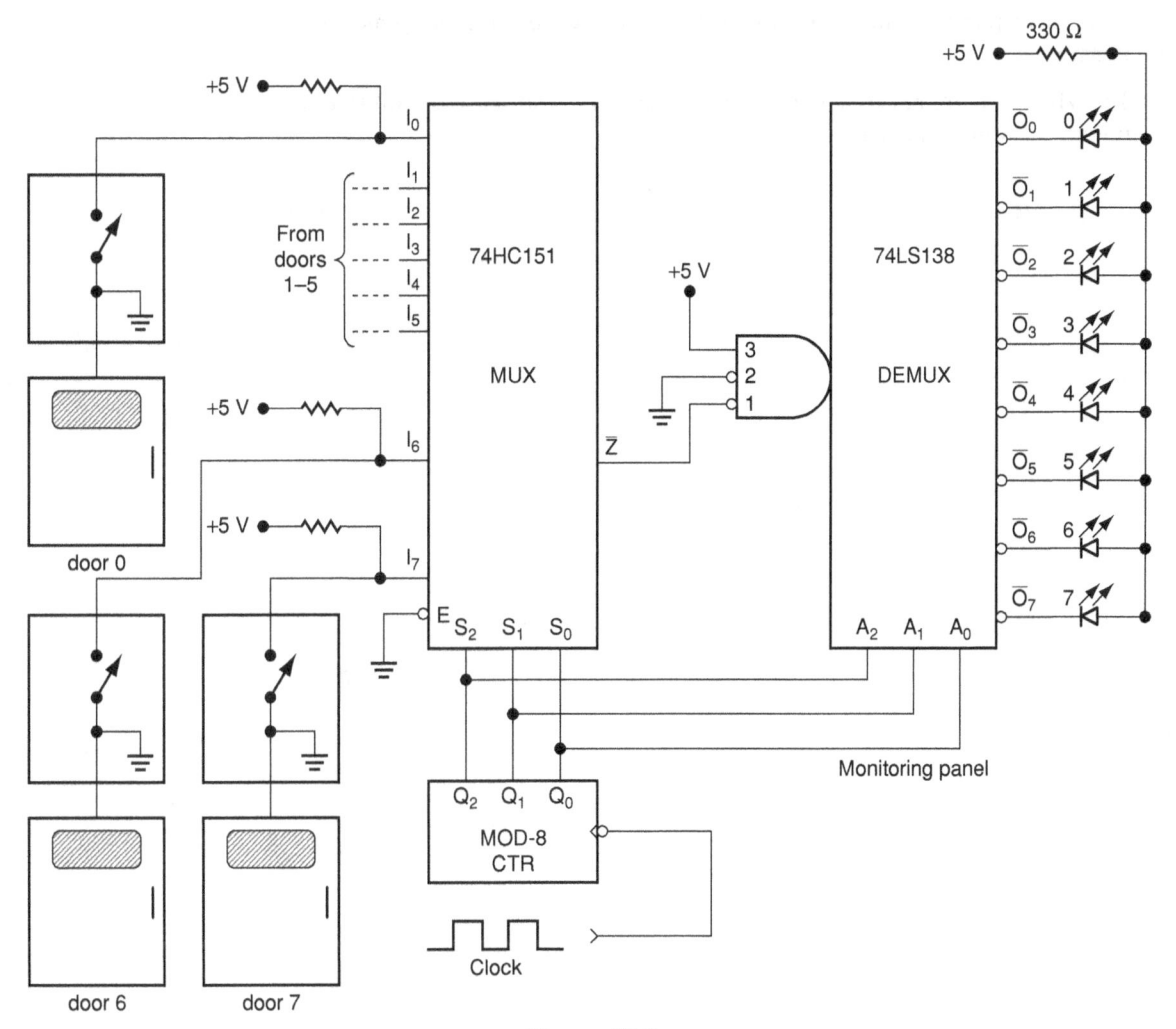

Figure 35-2

e) Have your lab partner or another student insert a fault into the circuit while you are not looking. Do not look for the bug yet. On a separate sheet of paper, draw a table that shows which LEDs are flashing (if any) for any door that opens. Also note the state of the LEDs when all doors are closed. Using your observations, narrow the location of the bug to a few possible faults. Use this list and your test equipment to find the fault. Repeat this step as many times as possible.

f) Have your instructor or laboratory assistant place a bug in your circuit, and then repeat step e. Place your observations, list of possible faults, and your final solution on a separate sheet of paper.

g) If you are to troubleshoot the synchronous data transmission system, make sure that it is functional before beginning the exercise; otherwise, you may be chasing down several bugs instead of just one. Also review Example 9.14 in the text. When you are ready, have your lab partner insert a bug into the circuit while you are not looking. Monitor the MUX and DEMUX outputs with the oscilloscope (or logic analyzer, if available) during one transmission cycle, and draw the waveforms on a separate sheet of paper. Now study the observations, and try to narrow the location of the fault to a small area in the system. List what you think are the possible faults on the same sheet of paper as your observations, then proceed to look for the fault with your troubleshooting tools. Write a description of the fault on the sheet of paper containing your other data. Repeat this step as often as time permits.

h) When you are ready, ask your instructor or laboratory assistant to place a bug in your system. Repeat step g.

Experiment 36

Name _____

IC MAGNITUDE COMPARATORS

OBJECTIVES

1. To investigate the operation of 74LS85 IC four-bit magnitude comparator.
2. To investigate the cascading of 74LS85 ICs.
3. To investigate a word recognizer application of the 74LS85 IC.

TEXT REFERENCE

Read section 9.10.

EQUIPMENT NEEDED

Components
74LS85 IC (2);
student-selected ICs and other components;
8 toggle switches;
3 LED monitors.

Instruments
0–5 volt DC power supply;
logic probe;
pulse generator or SWG.

DISCUSSION

The magnitude of any pair of real numbers A and B can be compared. A law of mathematics (called the law of trichotomy) tells us that A can be less than, equal to, or greater than B. Many systems call for this type of comparison, especially programmed systems like computers.

The 74LS85 IC can compare two 4-bit numbers and give an indication of how the two numbers compare. The ICs can be cascaded so that larger numbers may be compared. The 74LS85 IC has four pairs of data inputs, three outputs (A = B_{out}, A > B_{out}, A < B_{out}), and three cascading inputs (A = B_{in}, A > B_{in}, A < B_{in}). When cascading 74LS85 ICs, A = B_{out} is connected to A = B_{in} of the next stage, handling more significant bits, and, similarly, A > B_{out} is connected to A > B_{in}, and A < B_{out} is connected to A < B_{in}. The comparator handling the least significant nybble (group of four bits) must have A = B_{in} tied HIGH and the other cascading inputs tied LOW for the cascaded comparators to work properly.

PROCEDURE

a) *74LS85 IC operation*: Refer to the data sheet for a 74LS85 IC, and draw its pin layout diagram:

b) Install a 74LS85 IC on a circuit board. It will be referred to as IC1. Connect V_{cc} to +5 V and GND to power ground. Connect toggle switches to inputs A_0–A_3 and B_0–B_3. Connect the cascading inputs as described above. Connect LED monitors to the outputs. You will not have to make all possible comparisons, so just use the number pairs in Table 36-1 and record which output is lighted for each pair.

Table 36-1

			Inputs					LED Monitors
A_3	A_2	A_1	A_0	B_3	B_2	B_1	B_0	A > B_{out} A < B_{out} A = B_{out}
0	0	0	0	1	1	1	1	
0	0	1	0	0	1	1	1	
0	1	0	0	0	0	1	1	
1	0	0	0	0	0	0	1	
0	0	0	0	0	0	0	0	
1	1	1	1	0	0	0	0	

c) Now, connect another 7LS85 IC as shown in Figure 36-1. The new IC (IC2) will handle the least significant nybble. Connect all of its data inputs to V_{cc} with jumper wires. Now connect the outputs of this IC to the cascading inputs of IC1. Be sure to remove the jumpers to ground and V_{cc} from the cascading inputs that were connected in step a.

PROCEDURE

Figure 36-1

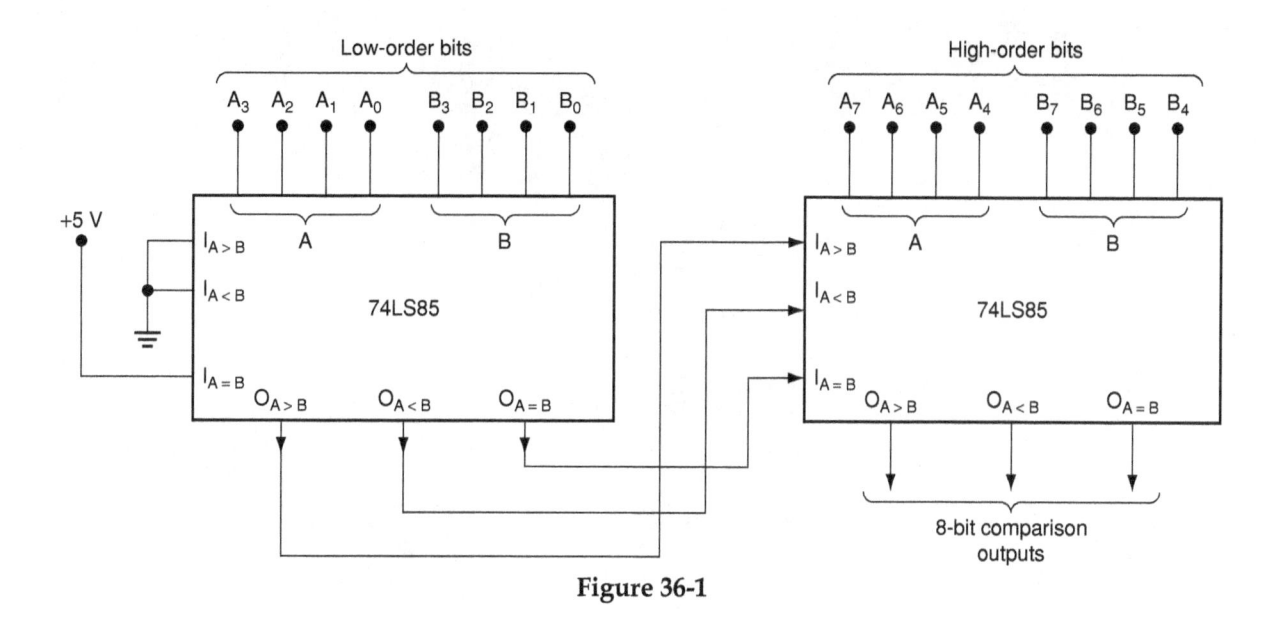

d) Set all toggle switches connected to IC1 to HIGH. Which LED is lighted? _____. Now disconnect B_3 on IC2 from V_{cc} and connect to ground. Now which LED is lighted? _____. Reconnect B_3 to V_{cc}. Remove A_3 on IC2 from V_{cc} and connect to ground. Which LED is lighted? _____. Set the toggle switch connected to A_3 on IC1 to LOW. Which LED is lighted? _____. Return the switch to HIGH, and set the toggle switch connected to B_3 to LOW. Which LED is lighted? _____.

e) *Word recognizer application*: Design a circuit that compares the output of an eight-bit counter (MOD-256) with an eight-bit word supplied by eight toggle switches. Whenever the two eight-bit words are equal, the circuit produces a LOW pulse with a time duration of 10 μs.

f) *Review*: This concludes the exercises on data buses. To test your understanding of the principles covered in this experiment, answer the following questions:

1. Design a circuit for an eight-bit comparator that will give a HIGH output only when A > B, where A and B are any two eight-bit numbers.

Experiment 37

Name _____

DATA BUSING

OBJECTIVES

1. To investigate the operation of a tristate register, the 74HC173 IC.
2. To demonstrate the application of tristate devices in bus systems.
3. To investigate register-to-register data transfer in bus systems.

TEXT REFERENCES

Read sections 9.11 and 9.12.

EQUIPMENT NEEDED

Components
7406 IC;
74LS125A IC;
74LS139A IC;
74HC173 IC (3);
74LS174 IC;
10 toggle switches;
normally LOW pushbutton switch (2), and normally HIGH pushbutton switch, all debounced;
4 LED monitors;
SK-10 circuit board.

Instruments
0–5 volt DC power supply;
logic probe.

DISCUSSION

Most modern digital computers use data buses for data transfer. Data sources come from the outputs of different devices and cannot normally be tied together. We have already seen how tristate buffers can be employed to isolate the data source outputs from other outputs. Most devices that are connected together contain registers to hold their data, and since register outputs cannot be tied together either, they will need tristate buffers at their outputs so that the registers may be tied to the bus safely. Of course, an added benefit of tristate devices is that data on a bus may flow in either direction. This minimizes the number of bus lines in the system.

74HC173 IC Tristate Register

The 74HC173 IC is a four-bit register capable of parallel I/O. The inputs are fed to the D inputs of D flip-flops through an enable circuit. Two active-LOW input enables control the input modes, of which there are two: the LOAD and HOLD modes. If both input enables are LOW, the data from the bus reaches the register's flip-flop inputs. If one or both enables are HIGH, the output of each flip-flop is fed back to the flip-flop's input. Also, there are two output enables that control the two output modes. If both output enables are LOW, all of the internal tristate buffers are enabled, connecting the flip-flop outputs to the data bus. If one or both enables are HIGH, the tristate buffers are all disabled, disconnecting the flip-flop outputs from the data bus.

The register requires clocking in the LOAD mode in order that new data may be transferred to the flip-flops. The clock does not affect the flip-flop in the HOLD mode. Normally, if a register is outputting, its inputs should be in the HOLD mode.

Data Bus System

This exercise is one of the most important in the lab manual. That is because the exercise shows how data is transferred in a data bus environment and how the transfer is controlled. As remarked in an earlier exercise, much of what goes on in a digital computer is associated with data transfer.

The basic data bus, as shown in Figures 37-1, 37-3, and 37-4, consists of three tristate registers, an input unit (switches and tristate buffers), an output unit (LEDs and latches), and a control unit (the dual decoder). The decoder receives its instruction from its inputs and connects two devices together (either register to register, or input switches to register) via the bus. The system clock then causes the data transfer to take place. This data bus system could be expanded with a little design work, but it is highly instructional as it is.

PROCEDURE

a) *74HC173 IC operation*: Refer to the data sheet for a 74HC173 IC, and draw its pin layout diagram:

The 74HC173 is a four-bit tristate register with a pair of input enables, $\overline{IE_1}$ and $\overline{IE_2}$; a pair of output enables, $\overline{OE_1}$ and $\overline{OE_2}$; and a master reset input, MR. Since you will tie the input enables together in this experiment, they will be referred to, collectively, as $\overline{IE}$. The output enables will be referred to as $\overline{OE}$. When $\overline{IE}$ and $\overline{OE}$ are both HIGH, the inputs D_3 through D_0 and the outputs O_3 through O_0 are tristated. In other words, data can neither get in nor get out. When $\overline{IE}$ is LOW and $\overline{OE}$ is HIGH, data can be clocked in and latched on a positive-going transition at CP. The output is tristated during this time. When both input AND output enables are LOW, both are enabled. There is no problem with this when both inputs and outputs are connected to the same bus, since the levels will be the same.

b) Install a 74HC173 on the circuit board. Connect V_{cc} to +5 V and GND to power ground. Connect normally LOW pushbutton switches to $\overline{CP}$ and MR. Connect toggle switches to $\overline{IE}$, $\overline{OE}$, and the data inputs. You will monitor the outputs with a logic probe. Set the data switches to LOW, the enable switches both HIGH, and pulse MR to clear the register if it is not cleared already. Now set the data switches to 1001. Verify that changing the data switches has no effect on the register by toggling $\overline{OE}$ LOW and monitoring the outputs. Now return $\overline{OE}$ to HIGH, and set $\overline{IE}$ LOW. Pulse $\overline{CP}$ momentarily HIGH, then enable the output and verify that the register is now storing the data present at the inputs. Toggle $\overline{IE}$ HIGH, and verify that the data is still present at the outputs. Change the data switches to 0110. Pulse $\overline{CP}$ HIGH. Enable the output, and verify that the data is still 1001. Now set $\overline{OE}$ HIGH and $\overline{IE}$ LOW, and pulse $\overline{CP}$ once again. Verify that the data at the output is now 0110.

c) *Operation of the data bus:* Connect the circuit of Figure 37-1 carefully. Note that all three registers have their inputs and outputs tied to the bus. You should be aware that it is undesirable to have no more than one register's *outputs* enabled at the same time. When this occurs, a condition known as *bus contention* exists. If each register is outputting different data, there is a tug-of-war between the two, which may result in chip damage. Having more than one register's inputs enabled at the same time does not present a problem. In fact, it will be desirable to do so, because it may be necessary to transfer data from one register to the other two, which is a common situation in digital systems.

Construct the circuit of Figure 37-1 on a separate SK-10 circuit board. Use its four common horizontal rows for the four bus lines. You will mount all circuit components on another board and connect them to the bus. Connect a normally LOW pushbutton switch to the common clock line. Connect toggle switches to all

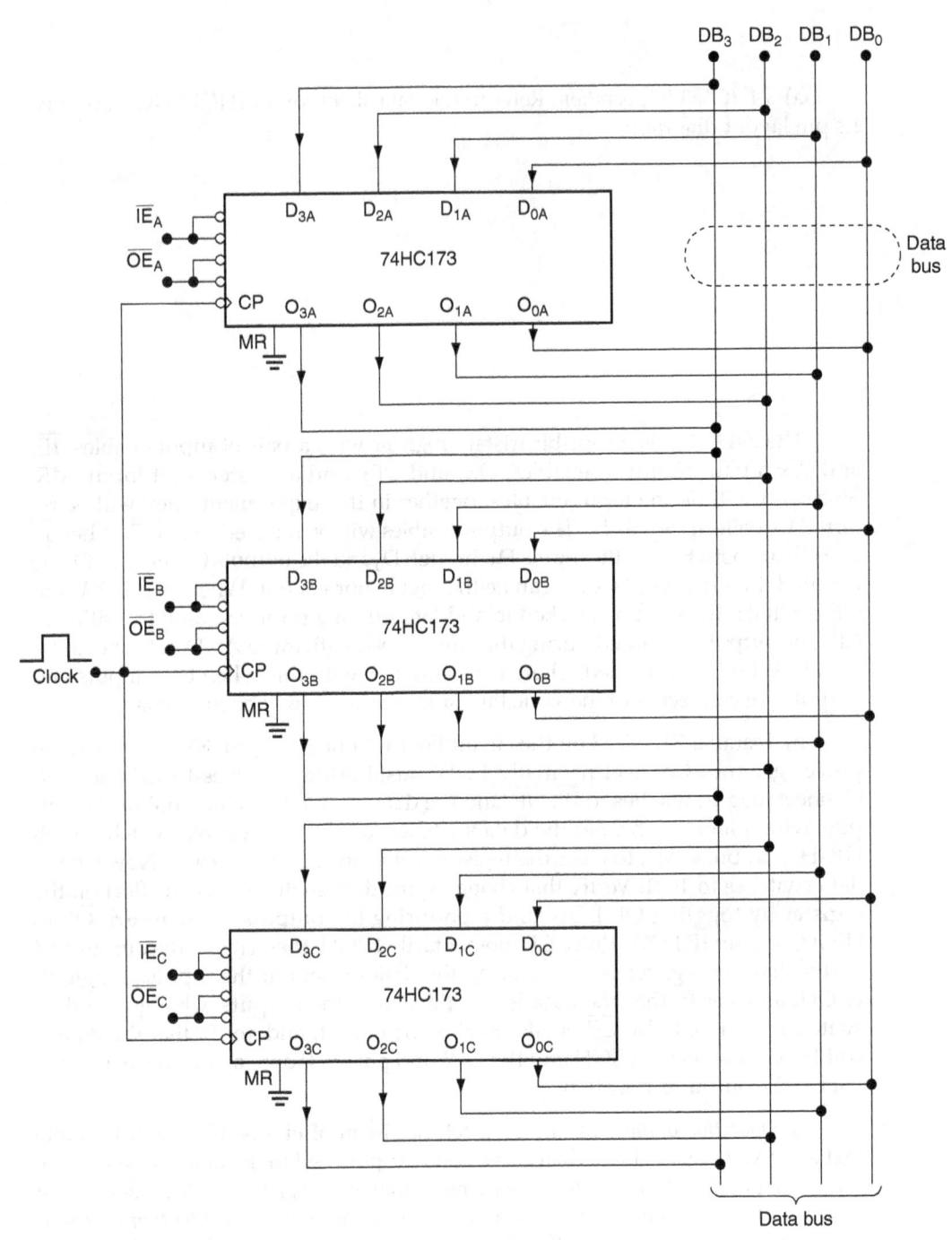

Figure 37-1

register input and output enables. Make sure that you wire all MR inputs to the normally LOW pushbutton switch connected to the register tested in step b. You will use this switch to clear all registers simultaneously.

PROCEDURE

Install 74LS125A IC on the component board. Connect V_{cc} to +5 V and GND to power ground. Connect toggle switches to the inputs of the 74LS125A tristate buffers. Connect the outputs of the buffers to the data bus. Connect all buffer enables together to a single toggle switch. Figure 37-2 illustrates these connections. The purpose of the tristate buffers is to permit data to be loaded into the registers manually without having to connect and disconnect the toggle switches prior to and following data entry. To operate the buffered data switches, you will first toggle the data switches to the desired settings and then hold the enable switch LOW. After data is loaded into the registers, release the pushbutton.

Set the data switches on the bus to 1010, and enable the buffers. Set $\overline{IE}_A$ LOW. Then pulse $\overline{CP}$ HIGH momentarily. Return $\overline{IE}_A$ to HIGH, and disable the switch buffers. Verify that only register A contains the data from the data switches by enabling first $\overline{OE}_A$ and checking the bus. Repeat this for register C.

d) After disabling all register outputs and inputs, set the switches to 0110, and enable the buffers. Enable $\overline{IE}_B$ and $\overline{IE}_C$, and pulse $\overline{CP}$ HIGH momentarily. Disable the inputs of registers B and C. Verify that register A still contains 1010 and that B and C contain 0110. Remember, enable $\overline{OE}_B$ and $\overline{OE}_C$ one at a time, and check the data bus.

e) *Register-to-register data transfer*: Now that you are familiar with entering data into a register from an external source, you will observe that the registers themselves are sources of data. Therefore, it is likely that on occasion, data from one register might be moved to another. You should note that there is a particular sequence of steps involved in manipulating the registers and that each sequence can be repeated.

Enable $\overline{IE}_A$ and $\overline{OE}_C$. Make sure that all other enables are inactive. Pulse $\overline{CP}$ HIGH momentarily. Now disable $\overline{IE}_A$ and $\overline{OE}_C$. Enable $\overline{OE}_A$, and verify that the new contents of register A are the same as what is stored in register C.

From this point on, the convention [Z] = xxxx, where Z is any register and xxxx is a binary number, will be used to refer to the contents of register Z. Thus, [B] = 1001 means that register B contains 1001.

Assume that [A] = 1011, [B] = 1000, and [C] = 0111. Study the waveforms of Figure 37-3. Write a procedure for applying these waveforms to the bus system *manually*.

Figure 37-2

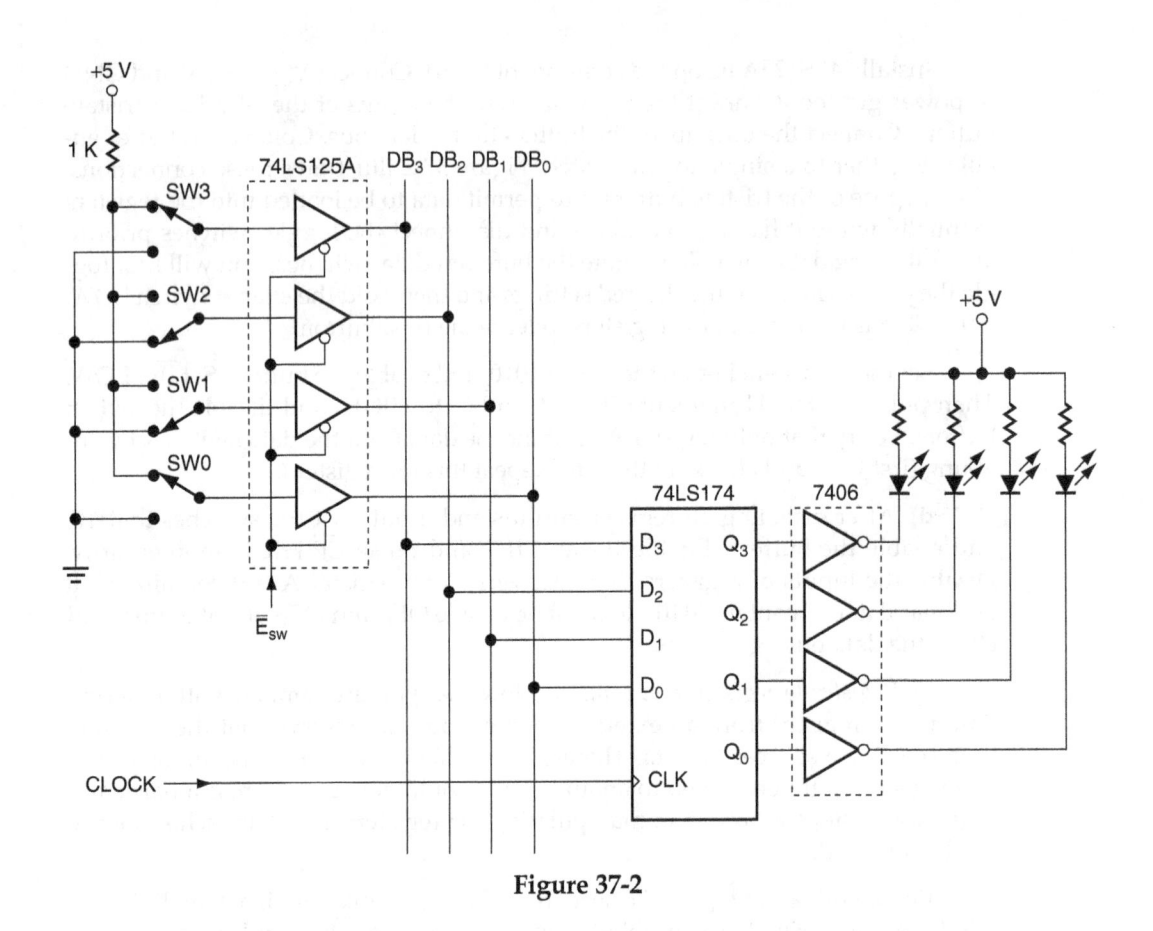

Predict [A], [B], and [C] at times t_1, t_2, and t_3, and record them in Table 37-1.

Table 37-1

Time	[A]	[B]	[C]
t_1			
t_2			
t_3			

Show the procedure and Table 37-1 to your instructor.

Now perform the procedure you wrote for this set of waveforms, and verify that Table 37-1 is correct.

f) *Latching data on the bus*: Examine Figure 37-2. You have already installed the buffered switches in step c. Now you will install the other IC, a 74LS174 register that will be used for latching any data that is on the bus during a data transfer, and display it on a set of LEDs.

Refer to the data sheet for a 74LS174 IC, and draw its pin layout diagram:

PROCEDURE

Figure 37-3

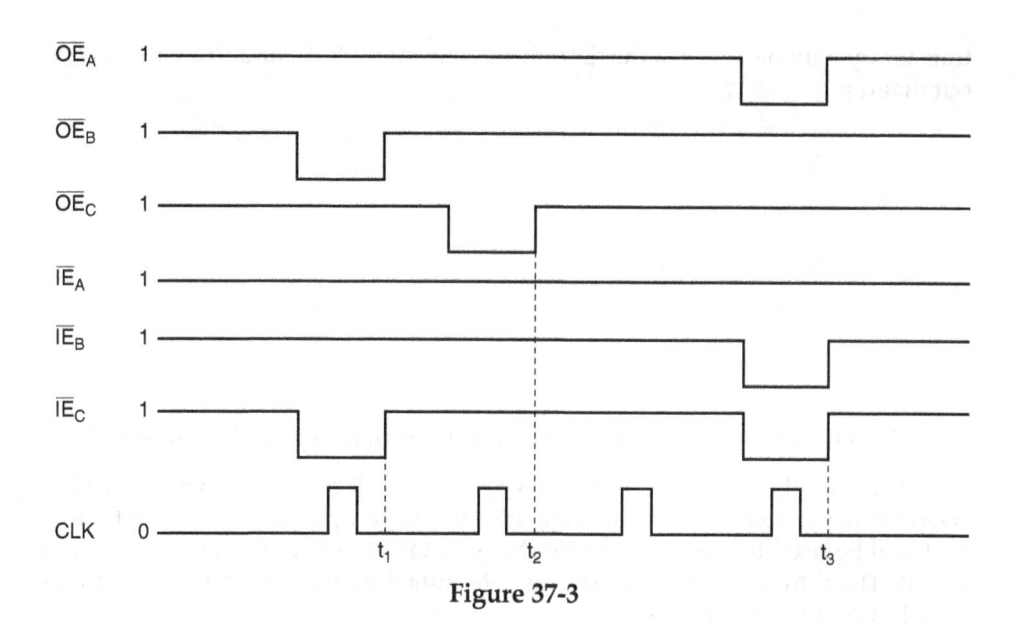

g) Add the new components to the data bus system. Make a connection between the clock pushbutton and the clock input on the 74LS174.

Write a procedure for loading the registers with the following data from the switches:

[A] = 1101, [B] = 1110, [C] = 0101.

What will be the state of the LEDs at the end of this sequence? _____.

Perform the procedure you wrote above, and verify that the state of the LEDs is as predicted.

h) *Enabling registers with decoders*: The last item to be added to the data bus is a 74LS139 decoder, which will generate the necessary enable pulses during data transfer operations. Refer to the data sheet for a 74LS139 IC, and draw its pin layout diagram:

The 74LS139 is a dual 1-of-4 decoder IC with an active LOW enable, $\overline{E}$.

Examine the circuit of Figure 37-4(a). The top decoder unit will be used to select the device that will put data onto the data bus (output select), and the bottom unit will be used to select the device that will take the data from the bus (input select). The actual devices selected are determined by the decoders' select input signals, OS_1, OS_0, IS_1, and IS_0.

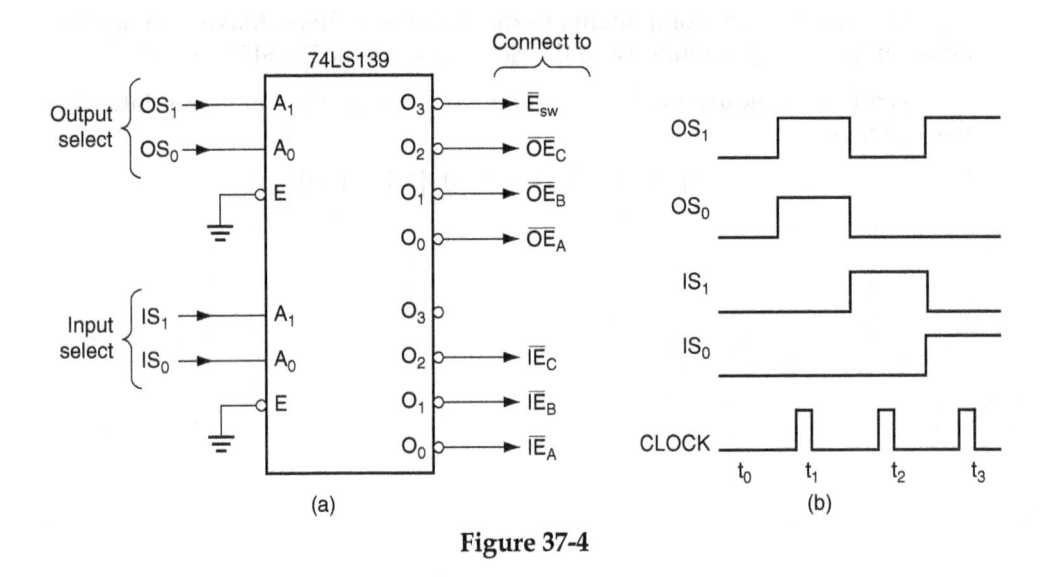

Figure 37-4

For each combination of select inputs, determine the transfer that is to take place, and record the registers affected by each transfer in Table 37-2.

Show Table 37-2 to your instructor or laboratory assistant.

Add the 74LS139 to the data bus system, and make the connections in the diagram and also the connections necessary to make the 74LS139 operate, which are not shown. Connect toggle switches to the decoder select inputs.

Clear all registers by pulsing the switch connected to the register MR inputs momentarily. Set the data switches to 1001. Set $OS_1 = OS_0 = 1$ and $IS_1 = IS_0 = 0$, and pulse the clock momentarily HIGH. What data transfer is taking place on the bus, and what is the value of the data being transferred?

PROCEDURE

Table 37-2

Selects				Data Transfer	
IS_1	IS_0	OS_1	OS_0	To Register	From Register
0	0	0	0	A	A
0	0	0	1	A	B
0	0	1	0		
0	0	1	1		
0	1	0	0		
0	1	0	1		
0	1	1	0		
0	1	1	1		
1	0	0	0		
1	0	0	1		
1	0	1	0		
1	0	1	1		
1	1	0	0		
1	1	0	1		
1	1	1	0		
1	1	1	1		

Now set $OS_1 = OS_0 = IS_0 = 0$ and $IS_1 = 1$, and pulse the clock again. What data transfer is taking place and what is the value of the data being transferred?

___

Set $OS_1 = OS_0 = IS_1 = 0$ and $IS_0 = 1$. Pulse the clock momentarily HIGH. What data transfer is taking place, and what is the value of the data being transferred?

___

Finally, set $OS_1 = OS_0 = 1$ and $OS_0 = IS_1 = 0$. Pulse the clock HIGH momentarily. What data transfer is taking place, and what is the value of the data being transferred?

___

What are the final conditions of the registers?

___

___

___

Verify this by enabling the output of each register one at a time, pulsing the clock, and observing the state of the LEDs.

Try other sequences of data transfer. Predict the final condition for each register before executing any sequence. When you are familiar enough with this type of data transfer, ask your instructor or laboratory assistant for a sequence to execute. Execute the sequence for the instructor.

i) *Review*: This concludes the exercises on data buses. To test your understanding of the principles covered in this experiment, answer the following questions:

1. Explain the term *bus contention*.

2. Why was it necessary to remove the unbuffered data switches whenever a register output was enabled?

3. Explain how tristate devices can help solve the bus contention problem.

4. Explain how the addition of the 74LS139 decoder to the data bus prevents bus contention, at least in theory.

Experiment 38

Name _____

DIGITAL-TO-ANALOG CONVERTERS

OBJECTIVE

To investigate the operation of a two-digit BCD-to-analog-converter.

TEXT REFERENCES

Read sections 11.1 through 11.5.

EQUIPMENT NEEDED

Components
MC1408 (or DAC0808) (2);
LM324 op-amp;
7447A IC (2);
74LS90 IC (2);
seven-segment display unit (2);
1 k-ohm resistor (2);
1 k-ohm potentiometer (two 10-turn potentiometers are recommended);
5 k-ohm potentiometers (two 10-turn potentiometers are recommended);
18 k-ohm resistor;
180 k-ohm resistor (2);
0.1 µF ceramic disc capacitor (2);
normally HIGH pushbutton switch, debounced.

Instruments
0–5 volt DC power supply;
+12 volt DC power supply;
−12 volt DC power supply;
DVM;
pulse or square wave generator.

DISCUSSION

In today's modern processing systems, one will rarely find a completely analog or completely digital system. Hybrid systems, which combine digital with analog, are predominant. We have already discovered the basic difference between analog and digital signals, and we know that they are incompatible, as are the devices that produce them. However, we also know that there are more analog than digital variables in the real world. If a digital system requires an input from an analog variable, we must find a way to convert the variable from analog to digital. Of course, the reverse is also true. In the current experiment and the next, you will investigate devices that perform these conversions.

Digital-to-Analog Conversion

A very simple digital-to-analog converter (DAC) is the summing amplifier circuit you most likely encountered in an electronic devices course (see Figure 38-1). The inputs of the summing amplifier are applied to an op-amp through the resistor network, and the output of the op-amp will be proportional to the sum of the weighted input currents. In practice, such a circuit is used only when accuracy is not important. IC digital-to-analog converters provide very good accuracy and are found in many applications. In this experiment, you will investigate an IC digital-to-analog convertor, the MC1408. The MC1408 is an 8-bit DAC with current output. You will also construct two digit BCD-to-analog converters, which will be used in Experiment 39 to complete a simple digital voltmeter (DVM).

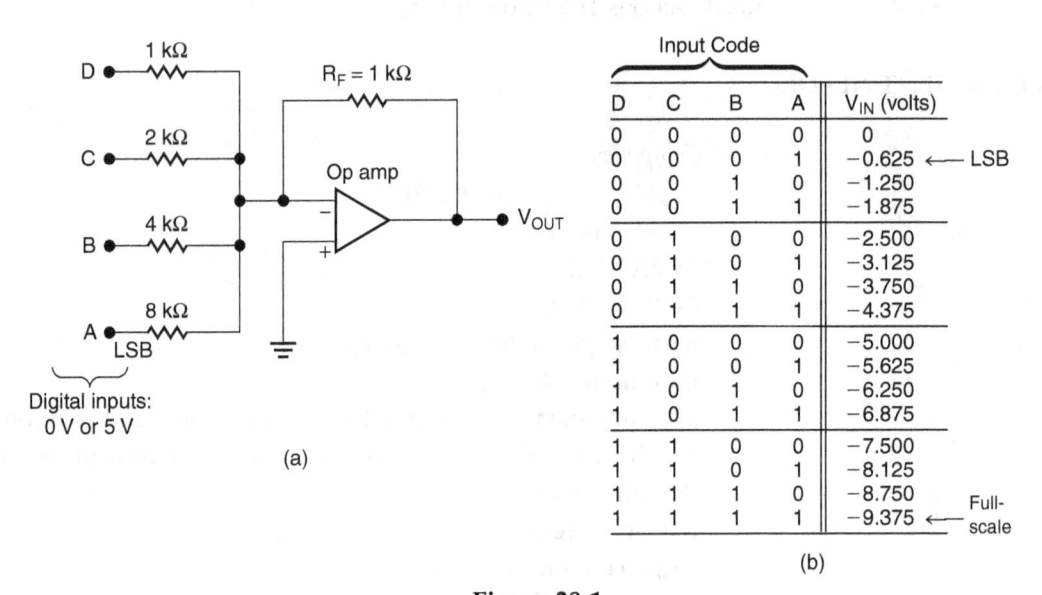

D	C	B	A	V_{IN} (volts)	
0	0	0	0	0	
0	0	0	1	−0.625	← LSB
0	0	1	0	−1.250	
0	0	1	1	−1.875	
0	1	0	0	−2.500	
0	1	0	1	−3.125	
0	1	1	0	−3.750	
0	1	1	1	−4.375	
1	0	0	0	−5.000	
1	0	0	1	−5.625	
1	0	1	0	−6.250	
1	0	1	1	−6.875	
1	1	0	0	−7.500	
1	1	0	1	−8.125	
1	1	1	0	−8.750	Full-scale
1	1	1	1	−9.375	←

Figure 38-1

PROCEDURE

a) Refer to the data sheet for an MC1408 IC, and draw its pin layout diagram:

The MC1408 accepts an 8 bit input (at D_7 through D_0) and produces an output *current* that is proportional to the binary input value. Its full-scale (F.S.) rating is 2 mA.

b) Examine and then construct the circuit in Figure 38-2. You will use this circuit later in this experiment and Experiment 39, so do not disconnect the circuit until Experiment 39 has been completed.

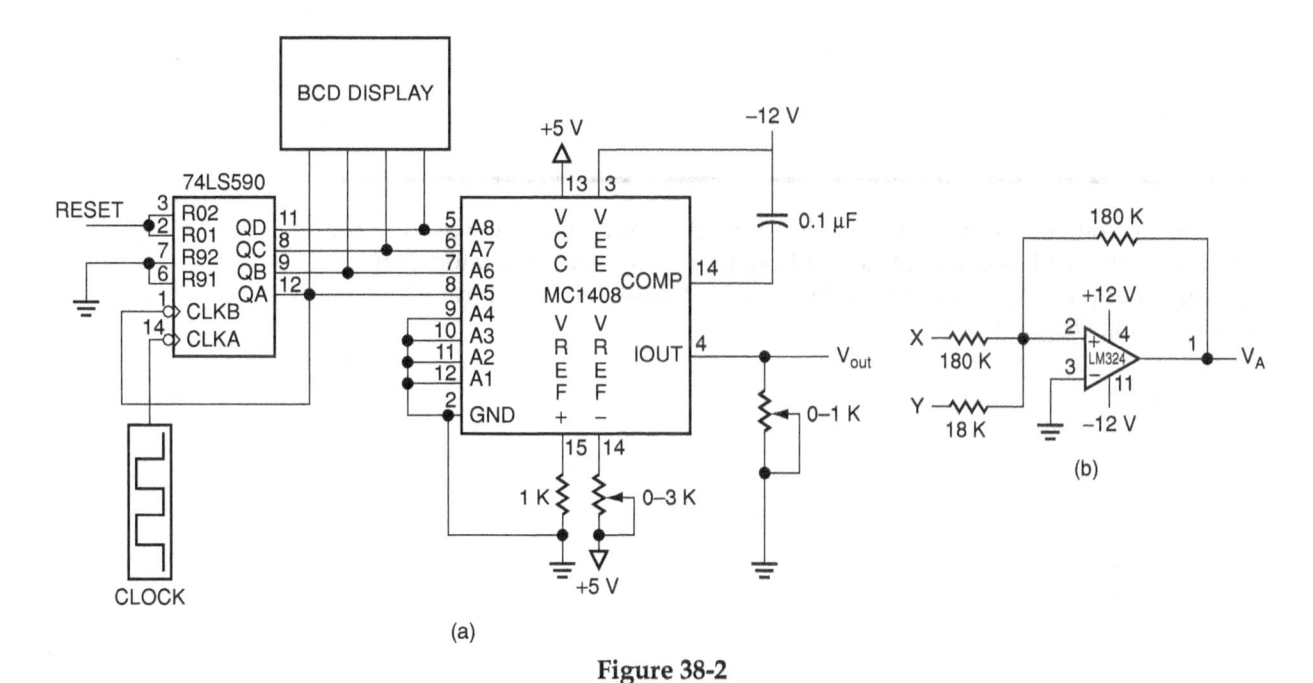

Figure 38-2

c) For each BCD value in Table 38-1, compute the corresponding value at point V_{out} and place the values in the table under V_{out} (Expected).

d) Calibrate the unit by using the following procedure:

1. Insert a milliameter between the 0–5 K potentiometer and VREF + of the DAC.

2. Apply power to the circuit and adjust the 0–5 K potentiometer until the milliameter reads 2 mA.

3. Remove the power and disconnect the milliameter. Reconnect the potentiometer to the DAC and connect the DVM to point V_{out}. Connect a normally-LOW pushbutton switch to the clock input of the 74LS90 counter.

4. Set the counter to 9 and adjust the 0–1 K potentiometer until the voltage at point V_{out} is -0.9 V $\pm$ 0.001 V, as measured with the DVM.

5. Pulse the counter through its counting range (0–9) and record the output at point V_{out} in Table 38-1 under First Unit V_{out} (Observed).

6. Compare the results in steps c and 5. How close in agreement are they?

Table 38-1

BCD Input	V_{out} Expected	First Unit V_{out} Observed	Second Unit V_{out} Observed
0			
1			
2			
3			
4			
5			
6			
7			
8			
9			

e) Disconnect the pushbutton switch from the clock input to the counter and replace it with a TTL-compatible signal from the squarewave generator. Examine the signal at point V_{out} with an oscilloscope and draw the waveform observed using Timing Diagram 38-1.

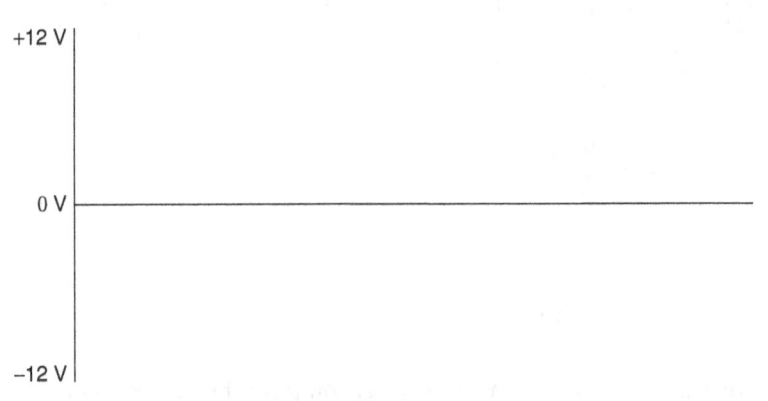

Timing Diagram 38-1

f) Construct and calibrate a second unit as in steps b–e. Use Second Unit V_{out} of Table 38-1 to record your calibration results. This unit will also be used later in this experiment and in Experiment 39, so do not disconnect it until then.

PROCEDURE

g) Construct the summing amplifier of Figure 38-2(b) and connect each of the DAC units to one of its inputs.

h) Pulse the counter so that the BCD input is 09. Measure V_A' with the DVM. It should indicate a value close to +0.9 V. Adjust the 0–1 k-ohm potentiometer at the LSD output, if necessary, until V_A' is as close to +0.9 V as possible.

i) Now set the BCD count to 90. The voltage at V_A' should now be close to +9.0 V. If necessary, adjust the 1 k-ohm potentiometer at the MSD output to bring this value as close as possible to +9.0 V. The BCD-to-analog converter is now calibrated.

j) Remove the toggle switches and install a MOD-100 BCD counter in their place as shown in Figure 38-2. Also connect BCD display units to display the counter. Clock the MOD-100 counter with a pushbutton.

k) Fill in the expected values of V_A' for each BCD number listed in Table 38-2.

Table 38-2

BCD Input	V_A' Expected	V_A' Observed
00		
05		
10		
15		
20		
25		
30		
35		
40		
45		
50		
55		
60		
65		
70		
75		
80		
85		
90		
95		
99		

l) Step the counter through its count sequence, stopping every five counts and reading V_A'. Record these values in the table. If any reading is off by more than 0.05 V from the expected value, notify your instructor or laboratory assistant.

m) Remove the pushbutton switch connected to the counter clock input, and replace it with the output of the square wave generator. Set the square wave generator to 1 kHz. Connect the oscilloscope to V_A', and draw the waveform displayed using Timing Diagram 38-2.

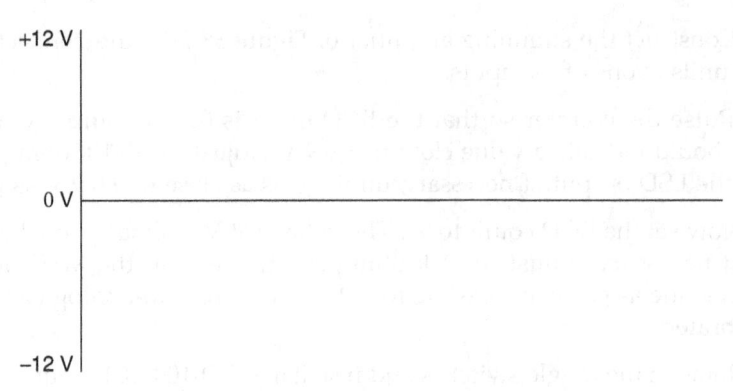

```
+12 V |
      |
      |
  0 V |_____
      |
      |
 -12 V|
```
Timing Diagram 38-2

n) *Review*: This concludes the exercises on digital-to-analog converters. Do not disassemble the converter you have on the circuit board at this time. It will be used in Experiment 39.

To test your understanding of the principles covered in this experiment, answer the following questions:

1. What is the percent resolution of the BCD-to-analog converter you constructed?
   _____%.

2. What problems would you have encountered had you not calibrated the DAC outputs before using them in the two-digit converter?

Experiment 39

Name _____

ANALOG-TO-DIGITAL CONVERTERS

OBJECTIVES

1. To investigate the operation of a TTL compatible comparator circuit.
2. To investigate the operation of a digital-ramp analog-to-digital converter in a two-digit DVM circuit.

TEXT REFERENCES

Read sections 11.8 and 11.9.

EQUIPMENT NEEDED

Components
MC1408 (or DAC0808) (2);
LM324 op-amp;
7447A IC (2);
74LS90 IC (2);
seven-segment display unit (2);
4.7 V zener diode;
1.5 k-ohm potentiometer (two 10-turn potentiometers recommended);
5 k-ohm potentiometer;
18 k-ohm resistor;
180 k-ohm resistor (2);
student-selected resistors and capacitors;
8 toggle switches;
normally HIGH pushbutton switch, debounced.

Instruments

0–5 volt DC power supply;
+15 volt DC power supply;
−15 volt DC power supply;
DVM;
pulse or square wave generator.

DISCUSSION

In Experiment 38, you investigated the DAC. You used DACs to construct a two-digit BCD-to-analog converter. Analog-to-digital converters (ADCs) and digital computers work together to measure real world variables. Thus the ADCs extend a digital computer by acting as an interface to sensors that convert analog quantities into voltages or current. In this way, whenever the computer needs information on a variable, it acts like a digital voltmeter, measures the quantity, and then stores the measurement in memory for future use.

In the current experiment, you will first investigate a comparator circuit that is TTL compatible. This type of circuit is used quite frequently in hybrid systems. It is also a component in an analog-to-digital converter. You will then complete your two-digit DVM circuit by constructing a digital-ramp ADC.

PROCEDURE

a) *TTL compatible comparator*: To construct a counter-ramp ADC, you need to add only a comparator circuit to the output of a counter-driven DAC, like the one you constructed in Experiment 38. Since the comparator output will be used to control a TTL gate, it is desirable to have the output of this comparator TTL compatible.

Examine the circuit of Figure 39-1. The circuit has two inputs. One is V_A', the input from the DAC output, and the other is V_A, the analog input. Note the zener diode at the comparator output. Since the digital output is negative or 0, the analog input must also be negative. The purpose of this diode is to limit the output excursions to a range of about −0.7 V to 4.7 V, making the output of the comparator TTL compatible.

Suppose the analog input is −2 V. If $V_A' = -1.5$ V, then the op-amp input difference will be positive, causing the output to swing positive toward +15 V. The zener diode limits this swing to +4.7 V. On the other hand, if $V_A' = -2.5$ V, the input difference will be negative, causing the output to swing negative toward −15 V. The zener diode this time will be forward-biased, so the voltage at the output will be −0.7 V, the voltage drop across the zener.

Construct the comparator. Connect one end of the 5 k-ohm potentiometer to −15 volts and the other to ground. Connect a wire to the wiper of the potentiometer. This wire will be the analog source, V_A.

Connect V_A' from the DAC to the + input of the comparator and the analog source to the − input. Adjust V_A for approximately −5 V. Pulse the MOD-100 counter until a count of around 55 is reached. Measure V_A' and the comparator output voltage, and record the measurement in Table 39-1.

PROCEDURE

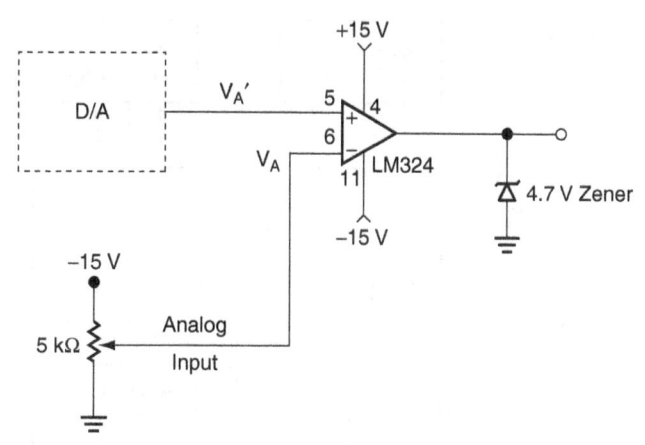

Figure 39-1

Table 39-1

V_A	Count	V_A'	Output
−5 V	55		
−5 V	45		

Now reset the counter, and pulse it until a count of around 45 is reached. Measure V_A' and the comparator output voltage, and record them in Table 39-1.

If the voltages measured are correct, proceed to the next step. If not, check the calibration of the DACs and the connections between V_A' and the comparator.

b) *Analog-to-digital converter application—digital voltmeter*: Now that the comparator is operating satisfactorily, you are ready to finish the DVM circuit. Figure 39-2 shows the complete wiring of the DVM. Make the necessary additions to the BCD-to-analog converter and the comparator, which are already on the board. One-shot OS1 controls the display time of the DVM. Its t_p should be made adjustable from about 1 second to 5 seconds or greater. One-shot OS2 provides a clear pulse for the BCD counters, and its t_p should be 10 microseconds. The AND gate is a transmission gate for the counter clock. As long as $V_A < V_A'$, the comparator output is HIGH and the AND gate is enabled, allowing the clock through to the counter. Whenever $V_A > V_A'$, the comparator switches to LOW, inhibiting the AND gate and freezing the counter at its current count.

When you have completed the circuit, disconnect the pushbutton switch from the BCD MOD-100 counter, and apply a 1 kHz square wave as clock input. Set the analog source to −5 V or slightly over. The DVM display should display 51, which represents −5.1 V. It could, however, be 50 or 52, depending on the accuracy of the calibration of the BCD-to-analog converters. Try several values of V_A between 0 and −10 V and record the results in the space provided below.

c) *Resolution of analog-to-digital converter*: With the lab DVM monitoring V_A, determine what change in V_A is needed to cause the counter display to change by one step. What is the resolution of the ADC? _____.

Figure 39-2

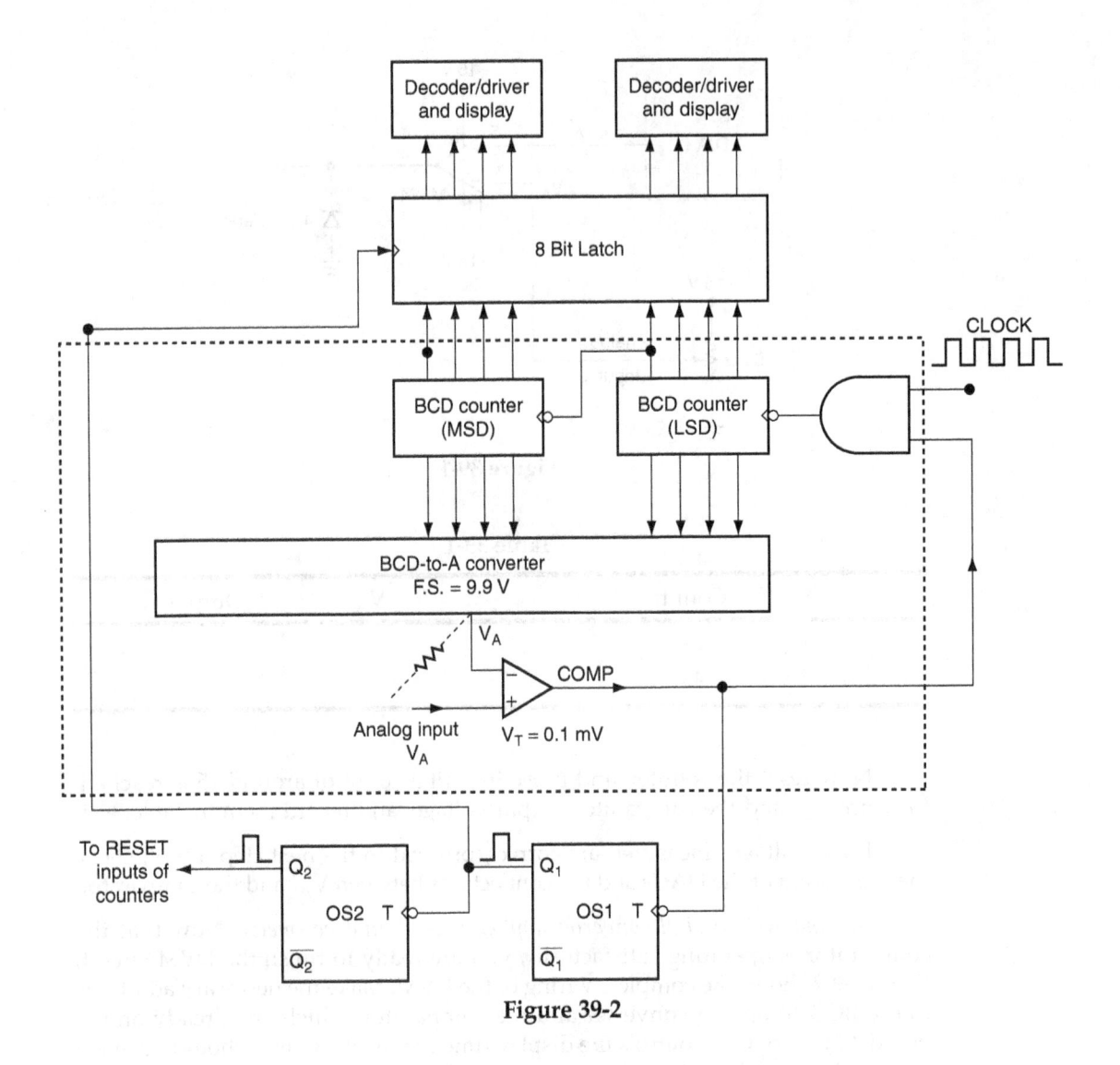

d) *Conversion time*: Change the display OS1 circuit so that t_p is 1 millisecond. Connect the dual trace oscilloscope to V_A' and the output of the comparator. Set V_A to −5 V. Observe the waveforms, and draw them, using Timing Diagram 39-1. Be sure to show proper levels. Label the conversion time and display time on the waveforms. Measure and record the conversion times for V_A = −1 V, −5 V, −9 V, and −11 V in Table 39-2.

Table 39-2

V_A	t_C
−1 V	
−5 V	
−9 V	
−11 V	

PROCEDURE

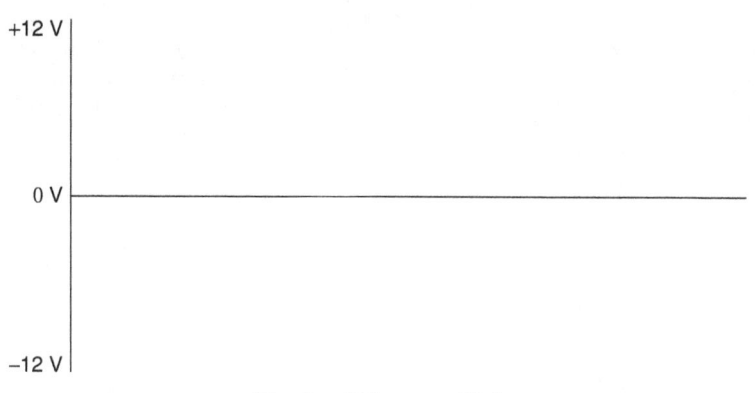

```
+12 V |
      |
      |
      |
   0 V |_____
      |
      |
      |
  -12 V|
```

Timing Diagram 39-1

e) In step d, you observed the effect of applying an analog input that is over-range (−11 V). How would you modify the DVM circuit to indicate that V_A is overrange? Draw a circuit showing the modification:

Show the circuit to your instructor for approval, and then add the modification to the DVM. Test the modified DVM. When the DVM is working properly, demonstrate the DVM to your instructor.

f) *Review*: This concludes the exercises on analog-to-digital converters. To test your understanding of the principles covered in this experiment, answer the following questions:

1. Why were the (constructed) DVM readings always higher than the actual value of V_A?

2. Explain why V_A has to be a negative value.

Experiment 40

Name _____

SEMICONDUCTOR RANDOM ACCESS MEMORY (RAM)

OBJECTIVES

1. To investigate the operation of an N-MOS static RAM memory IC.
2. To demonstrate the use of a logic analyzer in examining memory.
3. To demonstrate memory-to-register data transfer.
4. To demonstrate register-to-memory data transfer.

TEXT REFERENCES

Read sections 12.1, 12.2, 12.4, 12.11, 12.12, and 12.13.

EQUIPMENT NEEDED

Components
2114A IC;
74LS86 IC;
74LS93 IC;
74LS125 IC;
74LS139 IC;
74LS174 IC;
SK-10 circuit board;
10 toggle switches;
normally HIGH pushbutton switch (2), normally LOW pushbutton switch, debounced;
4 LED monitors.

Instruments
0–5 volt DC power supply;
logic probe;
logic analyzer (optional).

DISCUSSION

We know that memory is a characteristic of all useful digital systems. Up to now, we have been using flip-flops and registers for storing data. In Experiment 37, you investigated the tristate register and connected three of them to a data bus system. You discovered that in order to use (access) a particular register, the register has to be enabled. If data is to be stored in (written to) a register, its input must be enabled. If data is to be transferred from (read from) it, its output must be enabled. Recall that a decoder was used to do the enabling (or controlling).

Digital systems normally need hundreds of registers, and the word size needed is usually more than four bits. Memory ICs are available in various sizes, ranging from a few registers to several thousand. Large decoders are needed to access any particular register. For example, selecting one register from a 1024-word (1K) memory requires a 1-of-1024 decoder. How would you like to wire one of those? Rest at ease, for these large devices are included on the memory chip. For the same 1024-word memory, there are 10 special enable lines called address lines, which are used to externally select any register inside the chip.

A memory system that can be written to and read from is generally called read/write memory. However, semiconductor read/write memories are usually referred to as random access memory (RAM). In the current experiment, you will investigate an older 1K RAM chip, the 2114A. The word size of the 2114A is four bits. The number of data lines is four, and the chip is classified as common I/O. The chip must be enabled at the $\overline{CS}$ (chip select) input for the memory system to function. A normally HIGH $\overline{WE}$ (write enable) input keeps the chip in the read mode most of the time. Reading does not destroy the data in the memory. Whenever a WRITE operation is necessary, the $\overline{WE}$ input is pulsed LOW long enough for storage to take place.

PROCEDURE

a) In this experiment, you are to connect a 2114A memory chip to a data bus. Refer to the data sheet for a 2114A IC, and draw its pin layout diagram:

The 2114A is an N-MOS static RAM chip organized as 1024 four-bit words (1024 × 4). Its ten address inputs permit access to any of the 1024 words by applying the correct binary number representing the address to these pins. The addresses range from 0000000000 to 1111111111. Since it is more efficient to express

PROCEDURE

such large binary numbers in *hexadecimal*, we will do so from now on. The range of addresses in hexadecimal is 000_{16} to $3FF_{16}$.

The $\overline{WE}$ input is used to select the READ or WRITE operation. When the input is HIGH, the READ operation is selected; when it is LOW, the WRITE operation is selected.

The $\overline{CS}$ (chip select) input enables the chip, when LOW, so that either the READ or the WRITE can take place. When HIGH, $\overline{CS}$ disables the chip so that neither operation can take place.

The 2114A has four tristate I/O lines, I/O_3 through I/O_0. During a READ operation, the memory word selected by an address will be made available at these pins. During a WRITE operation, the word to be written must be on the lines connected to these pins.

Summarize the memory chip's operations using Table 40-1.

Table 40-1

WE	$\overline{CS}$	Operation
0	1	
1	1	
0	0	
1	0	

b) Examine the circuit of Figure 40-1. The circuit shows how the memory chip will be wired for initial testing. Later you will add other components to the bus. Read through the following steps, and make each connection as directed.

1) Mount a 2114A IC, a 74LS93 IC, and a 74LS125A IC on a circuit board. This board will be separate from the board to be used for the data bus. Connect V_{CC} to +5 V and GND to power ground for the three ICs.
2) Wire the 74LS93 as a MOD-16 counter. Connect the 74LS93 outputs, Q_3–Q_0 to address inputs A_3–A_0, respectively, of the 2114A memory chip. Make sure address inputs A_4–A_9 are grounded.
3) Connect a toggle switch to the write enable toggle input.
4) Connect a normally HIGH pushbutton switch to the device select clock input ($\overline{CS}$).
5) Connect a normally HIGH pushbutton switch to the address select clock input ($\overline{CP_0}$ input of the 74LS93).
6) Connect I/O outputs I/O_3–I/O_0 to data bus lines DB_3–DB_0, respectively.
7) Connect the outputs of the 74LS125A buffers to the data bus.
8) Connect a normally LOW pushbutton switch to the reset input (MR of the 74LS93). You will use this switch to reset the counter.

You are now ready to test the 2114A. Using the logic probe to monitor the counter outputs, clear the counter by pulsing it until its output is 0000. Set $\overline{WE}$ to HIGH and the data switches to 1001. Use the logic probe to check the data bus levels. The probe should indicate an indeterminate level. Why?

c) *2114A READ operation:* To read data from memory, $\overline{CS}$ must be made LOW while $\overline{WE}$ is HIGH. Press and hold the pushbutton connected to $\overline{CS}$, and check the data bus levels. Since the counter is at 00_{16}, the data stored at that address should

Figure 40-1

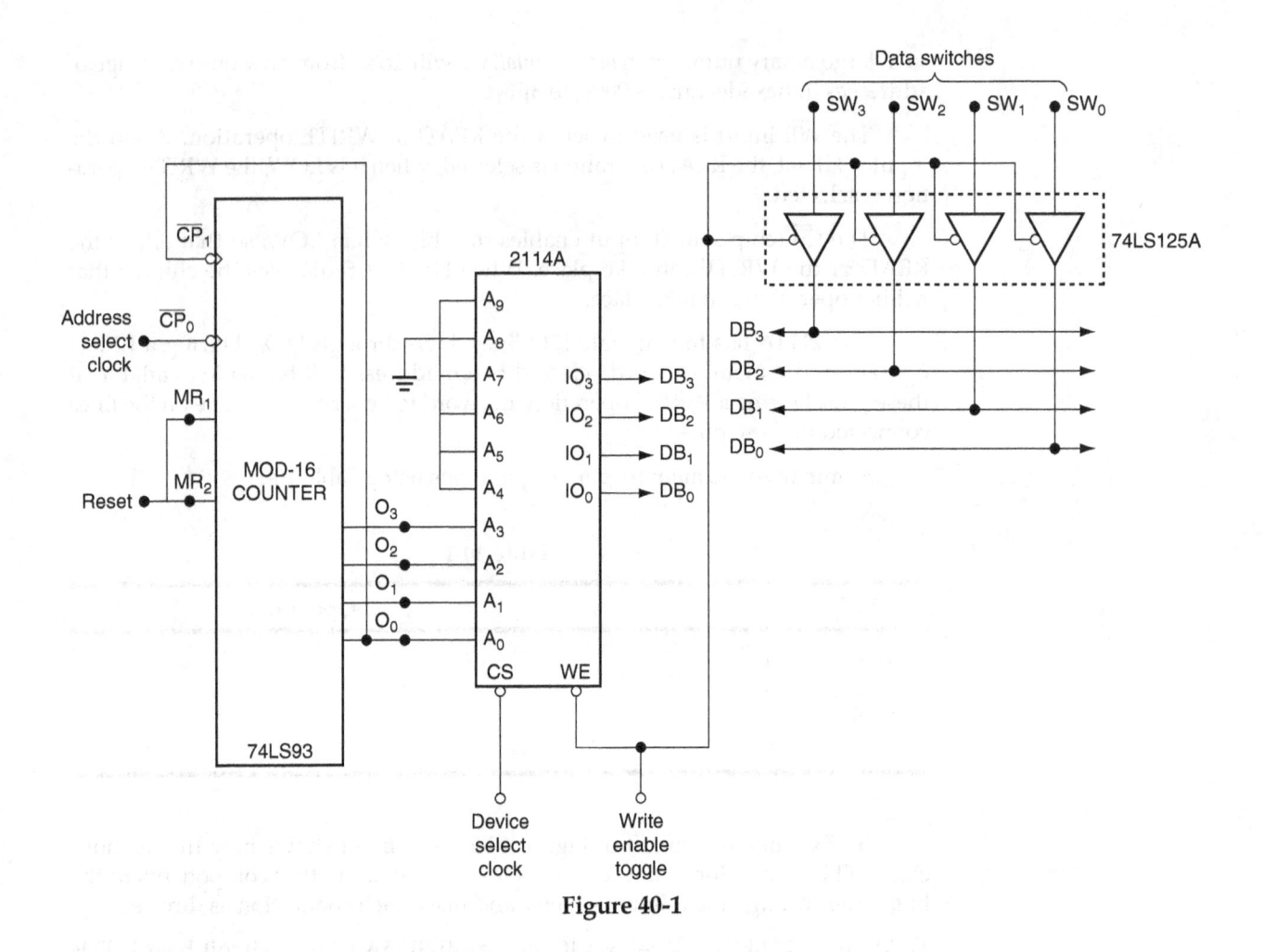

be on the data bus. Since nothing has been written into memory, whatever is on the data bus is random. Release the $\overline{CS}$ switch, and pulse the counter to 0001 (01_{16}). Press and hold the $\overline{CS}$ pushbutton switch, and check the bus levels. Again, the levels on the data bus are random. Read the data words in other memory locations in the same manner, and verify that the levels on the data bus are random.

d) *2114A WRITE operation*: To write data to memory, $\overline{CS}$ must be made LOW while $\overline{WE}$ is LOW. Set the address counter to 00_{16}. Set $\overline{WE}$ to LOW. Check the levels on the data bus now. Since the data switch buffers are enabled, the data word on the bus should be 1001. However, the data has not been written yet, since $\overline{CS}$ = 1. Momentarily pulse $\overline{CS}$ LOW. Verify that 1001 has been written to memory location 00_{16} by performing a READ as done in step c.

e) Momentarily disconnect V_{CC} from the 2114A. Then reconnect it, and READ the data word at memory location 00_{16}. It should be lost, since RAM memory is volatile.

f) Store all of the words at the memory locations shown in Table 40-2. Have your instructor check the data.

g) *Using a logic analyzer to examine memory*: If a logic analyzer is not available, go on to step h.

The contents of your memory chip can be displayed in "state table format." In order to sample and display the data, the memory must be cycled continuously

PROCEDURE

Table 40-2

Address	Data Binary	Hex
00	1111	F
01	1110	E
02	1101	D
03	1100	C
04	1011	B
05	1010	A
06	1001	9
07	1000	8
08	0111	7
09	0110	6
0A	0101	5
0B	0100	4
0C	0011	3
0D	0010	2
0E	0001	1
0F	0000	0

through each of the 16 addresses so that the contents of each memory location are placed on the data bus. Use the following procedure to set up and display memory on the logic analyzer. The instructions are very general, so you may have to consult your instructor for specific details on how to set up your particular logic analyzer.

1) Connect a 10 kHz square wave to the counter clock input.
2) Connect the logic analyzer probes as follows:

   Channels 0–3 to DB_0–DB_3
   Channels 4–7 to address lines A_0–A_3
   "C" probe to the 10 kHz clock

3) Set $\overline{WE}$ to 1 (READ mode).
4) Set the logic analyzer sample interval to EXT (positive edge).
5) Set the logic analyzer to trigger on address 0000.
6) Set the logic analyzer to state table mode. Display should be binary.
7) Have the logic analyzer start a sample cycle. Observe the tabular data displayed. Then position the cursor so that the trigger word is at the top of the table. This line should be either highlighted, blinking, or both. The table should show addresses 0000–1111 and their respective data contents. Since $\overline{CS}$ was HIGH, the data will not be correct.
8) Now hold $\overline{CS}$ LOW, and repeat step 7. The table should now display the correct memory data. This time store the sample.
9) Disconnect one of the memory's I/O lines from the data bus. Have the logic analyzer take another sample. Next, use the COMPARE mode to compare the new sample with the sample that you stored. You should observe that the missing I/O line's data position in the display is highlighted in some manner to indicate differences between old and new data.
10) Reconnect the I/O line you disconnected in step 9.

11) Disconnect the square wave generator from the counter, and replace the pushbutton switch there.

h) *Memory-to-register data transfer*: This type of transfer requires close attention to timing. During a READ operation, $\overline{WE}$ is HIGH and the $\overline{CS}$ line is pulsed LOW. After internal delays, the memory produces the data word at the I/O pins. This data can be clocked into the register at this time. For proper operation, the $\overline{CS}$ pulse has to stay LOW long enough to give the data a chance to stabilize on the data bus. If not, the register will latch erroneous data.

Refer to the data sheet for the 2114A, and determine how much time should be allowed for the memory outputs to stabilize before latching them into a register. Record this value: _____.

i) Examine the circuit in Figure 40-2. You are to add a 74LS139 IC and a 74LS174 IC to the data bus system. Install the 74LS174 IC onto the component board, and make the connections shown in Figure 40-2. Keep the 74LS174 $\overline{MR}$ input unconnected; it can be momentarily grounded to clear the register. Also make all other connections not shown that are required for the proper operation of the IC.

j) Make a table of transfer operations like you did in Experiment 37, this time for the new data bus system. Show the table to your instructor or laboratory assistant for approval.

k) Test the operation of the completed data bus system by setting the decoder select inputs to all possible combinations and verifying that the correct input and output devices are selected. What happens when the decoder select

Table 40-3

Selects				Data Transfer	
IS_1	IS_0	OS_1	OS_0	To Register	From Register
0	0	0	0		
0	0	0	1	memory	data switches
0	1	0	0		
0	1	0	1		

inputs are all LOW? _____. Check the memory READ/WRITE operations. Enter the data in Table 40-3 and then verify that the data was written.

l) *Register-to-memory data transfer*: The way the data bus is currently configured, you cannot transfer data from the 74LS174 to memory. Come up with a circuit modification that will permit this type of data transfer. Keep in mind the following considerations:

1) You cannot tie the register outputs directly to the data bus. Why?

Figure 40-2

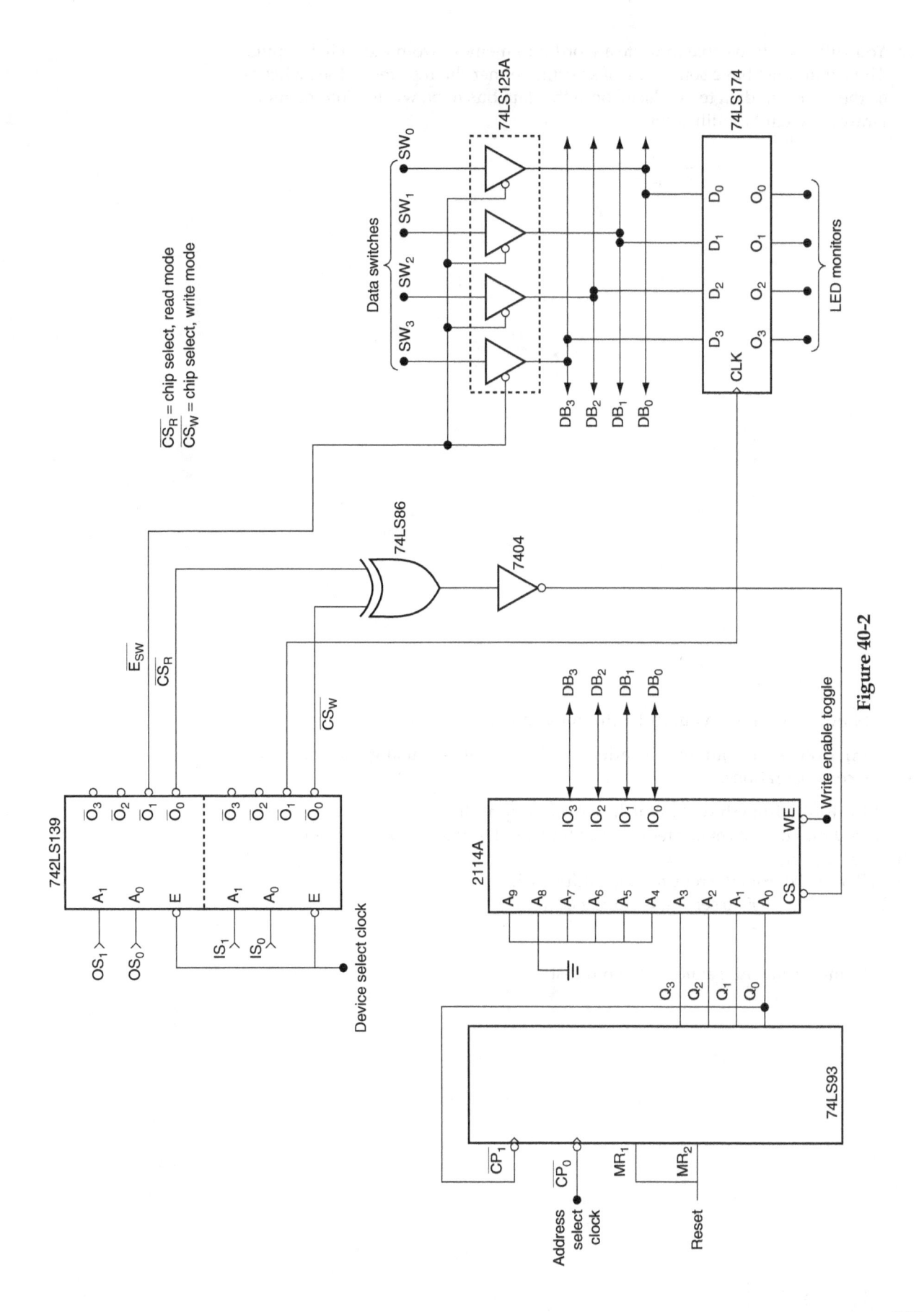

2) You still want to be able to write a word into memory from the switch inputs. Thus, you must have some way of selecting either the register or the switches as the source of data to be placed onto the data bus to be written into memory. Draw the circuit modification:

Show the circuit to your instructor for approval.

m) Construct your modification, and check it out using the following sequence of operations:

1) WRITE 1011 from the data switches to memory address 05.
2) READ the contents of address 05, and transfer it to the 74LS174 register.
3) Clear the register.
4) WRITE 0000 from the register into address 05.
5) READ address 05 to see if it is storing 0000.

Demonstrate this sequence to your instructor.

PROCEDURE

n) *Review*: This concludes the exercises on RAM memory. To test your understanding of the principles covered in this experiment, answer the following questions.

1. Explain the sequence of events illustrating how data is transferred from memory to the 74LS174 register.

2. Explain what might happen to memory if the enables to the 74LS139 were NOT pulsed LOW to enable the decoder operation, but rather tied permanently LOW.

3. Explain why timing during memory-to-register data transfer was not a critical factor in performing this experiment.

Experiment 41
(Optional)

Name _____

SYNCHRONOUS COUNTER DESIGN

OBJECTIVE

1. To learn how to design synchronous counters.

TEXT REFERENCE

Read section 7.14.

EQUIPMENT NEEDED

Components
7400 IC (1);
7402 IC (1);
7404 IC (3);
74LS10 IC (1);
74LS74A IC (2);
74LS86 IC (1);
4 LED monitors.

Instruments
0–5 volt DC power supply;
pulse or square wave generator;
dual trace oscilloscope;
logic probe.

DISCUSSION

In this experiment, you will use the step-by-step procedure described in the text in order to design and build synchronous counters of any Mod number, and compare their operation and advantages over asynchronous counters.

PROBLEM

Design and build a synchronous counter that counts as follows:
0, 2, 4, 6, 8, 10, 11, 12, 14, 15, 0, 2, 4, . . . (recycles) . . . . The *undesired* states are 1, 3, 5, 7, 9, and 13. If any of the undesired states occur upon power-up or due to noise, on the next clock pulse the counter should go to count 15.

PROCEDURE

1) Complete the table of Figure 41-1 showing the desired counting sequence for the counter.

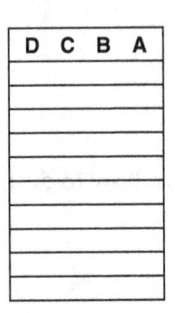

D	C	B	A

Figure 41-1

2) Complete the state transition diagram of Figure 41-2 showing all possible states, including those that are not part of the desired counting sequence.
3) Use the state transition diagram of Figure 41-2 to set up the table of Figure 41-3, which lists all **PRESENT** states and their **NEXT** states.
4) Complete the column for the **J K** *Levels*. For each **PRESENT** state, indicate the levels required at each J and K input in order to produce the transition to the **NEXT** state.
5) Use the Karnaugh maps in Figure 41-4 to derive the simplified Boolean expressions for the variables Ja, Ka, Jb, Kb, Jc, Kc, Jd, and Kd. (**Hint:** *Use DON'T CARE conditions whenever possible.*)
6) Implement the circuits for the variables of step 5 and, in the space provided on page 275, draw the complete schematic for the counter described in the original problem. *Wire it and verify its operation.* **Show it to the instructor.**

PROCEDURE

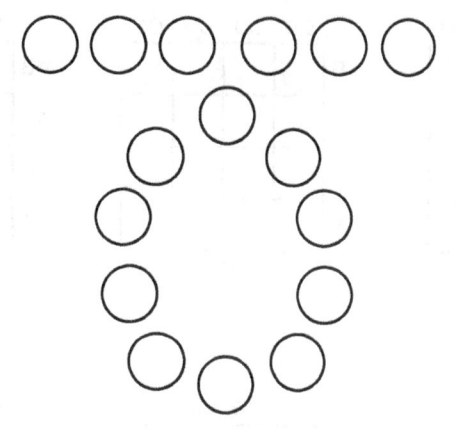

Figure 41-2

	PRESENT States D C B A	NEXT States D C B A	J K Levels			
			Jd Kd	Jc Kc	Jb Kb	Ja Ka
0						
1						
2						
3						
4						
5						
6						
7						
8						
9						
10						
11						
12						
13						
14						
15						

Figure 41-3

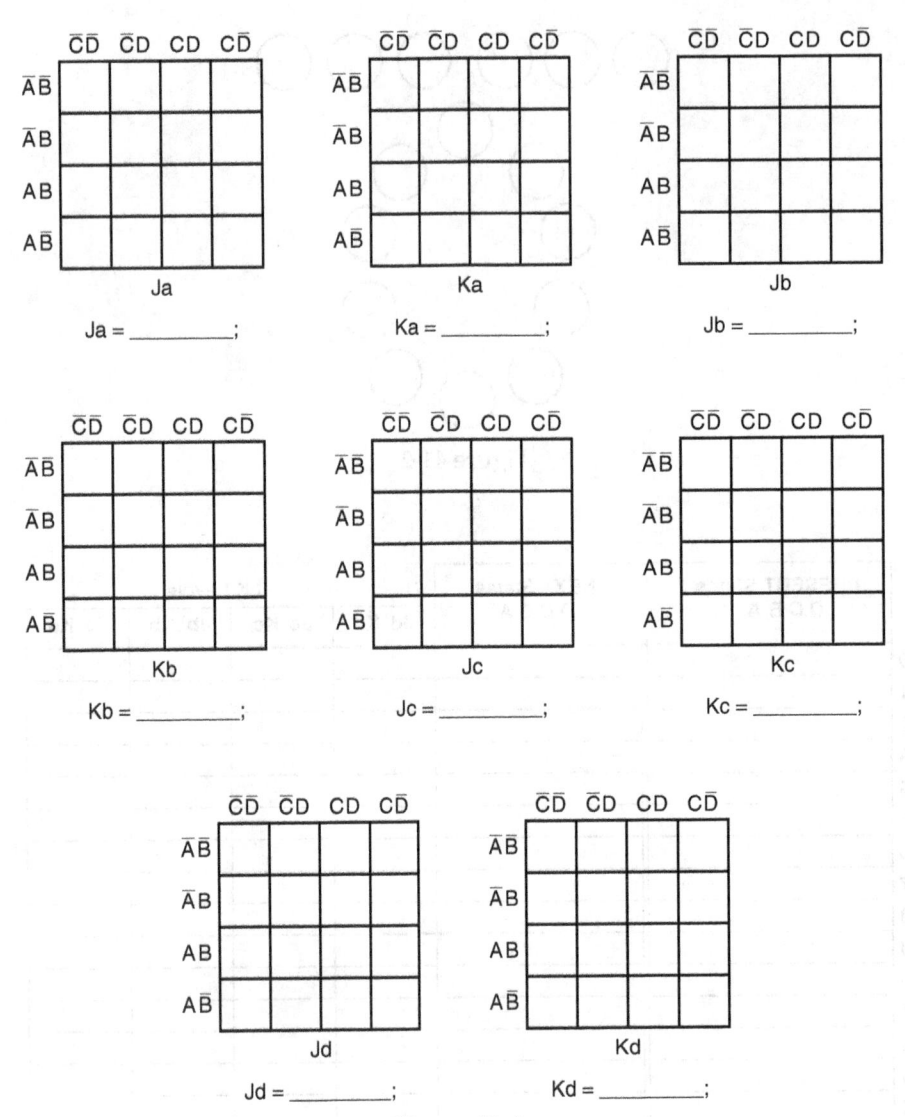

Figure 41-4

	$\bar{C}\bar{D}$	$\bar{C}D$	CD	$C\bar{D}$
$\bar{A}\bar{B}$				
$\bar{A}B$				
AB				
$A\bar{B}$				

Ja

Ja = _____ ;

	$\bar{C}\bar{D}$	$\bar{C}D$	CD	$C\bar{D}$
$\bar{A}\bar{B}$				
$\bar{A}B$				
AB				
$A\bar{B}$				

Ka

Ka = _____ ;

	$\bar{C}\bar{D}$	$\bar{C}D$	CD	$C\bar{D}$
$\bar{A}\bar{B}$				
$\bar{A}B$				
AB				
$A\bar{B}$				

Jb

Jb = _____ ;

	$\bar{C}\bar{D}$	$\bar{C}D$	CD	$C\bar{D}$
$\bar{A}\bar{B}$				
$\bar{A}B$				
AB				
$A\bar{B}$				

Kb

Kb = _____ ;

	$\bar{C}\bar{D}$	$\bar{C}D$	CD	$C\bar{D}$
$\bar{A}\bar{B}$				
$\bar{A}B$				
AB				
$A\bar{B}$				

Jc

Jc = _____ ;

	$\bar{C}\bar{D}$	$\bar{C}D$	CD	$C\bar{D}$
$\bar{A}\bar{B}$				
$\bar{A}B$				
AB				
$A\bar{B}$				

Kc

Kc = _____ ;

	$\bar{C}\bar{D}$	$\bar{C}D$	CD	$C\bar{D}$
$\bar{A}\bar{B}$				
$\bar{A}B$				
AB				
$A\bar{B}$				

Jd

Jd = _____ ;

	$\bar{C}\bar{D}$	$\bar{C}D$	CD	$C\bar{D}$
$\bar{A}\bar{B}$				
$\bar{A}B$				
AB				
$A\bar{B}$				

Kd

Kd = _____ ;

Figure 41-4

PROCEDURE

Schematic for the Synchronous Counter

QUESTIONS

1. State one advantage (a) and one disadvantage (b) that synchronous counters have over asynchronous counters.

    (a)

    (b)

2. What is the MOD number of the counter that you designed and built in this lab?
3. Verify the answer to question 2 by applying a 10 kHz TTL square wave (F_{in}) to the clock input of the counter and then by measuring the output frequency (F_{out}) at the output of the most significant flip-flop of your counter. The MOD number is then obtained by solving

$$\text{MOD} = \frac{F_{in}}{F_{out}}$$

4. Assume that the average propagation delay of each flip-flop is 15 ns and that the average propagation delay of each gate and inverter is 10 ns. Determine the approximate f_{max} for the counter.

Experiment 42
(Optional)

Name _____

PROGRAMMABLE FUNCTION SEQUENCER

OBJECTIVES

This experiment utilizes an EPROM to sequence through the hexadecimal number system, an up/down count as well as two different chase light sequence patterns. The results of each operation will be verified by a seven-segment LED display. Tasks such as this can also be accomplished by using RAMs, ROMs, or PLAs.

TEXT REFERENCES

Read section 11.9.

EQUIPMENT NEEDED

Components
2732 eprom (1);
7405 (2);
74LS93 (1);
7-segment LED, common anode (1);
SPST switches (4);
1 KΩ resistors (4);
220 Ω resistors (7).

Instruments
0–5 volt DC power supply;
pulse or square wave generator;
dual trace oscilloscope;
logic probe;
PROM programmer;
UV PROM eraser.

DISCUSSION

Refer to the circuit of Figure 42-1. IC-1 is being used as a MOD-16 counter. As the clock input is applied, it will cycle through all 16 possible states. Each one of these states will address a certain memory location of the EPROM (IC-2). The contents of each memory location (pattern of 1s and 0s) will vary according to which switch SW0-SW3 is selected (grounded). The results will be displayed on a common anode LED such as that of Figure 42-2(b).

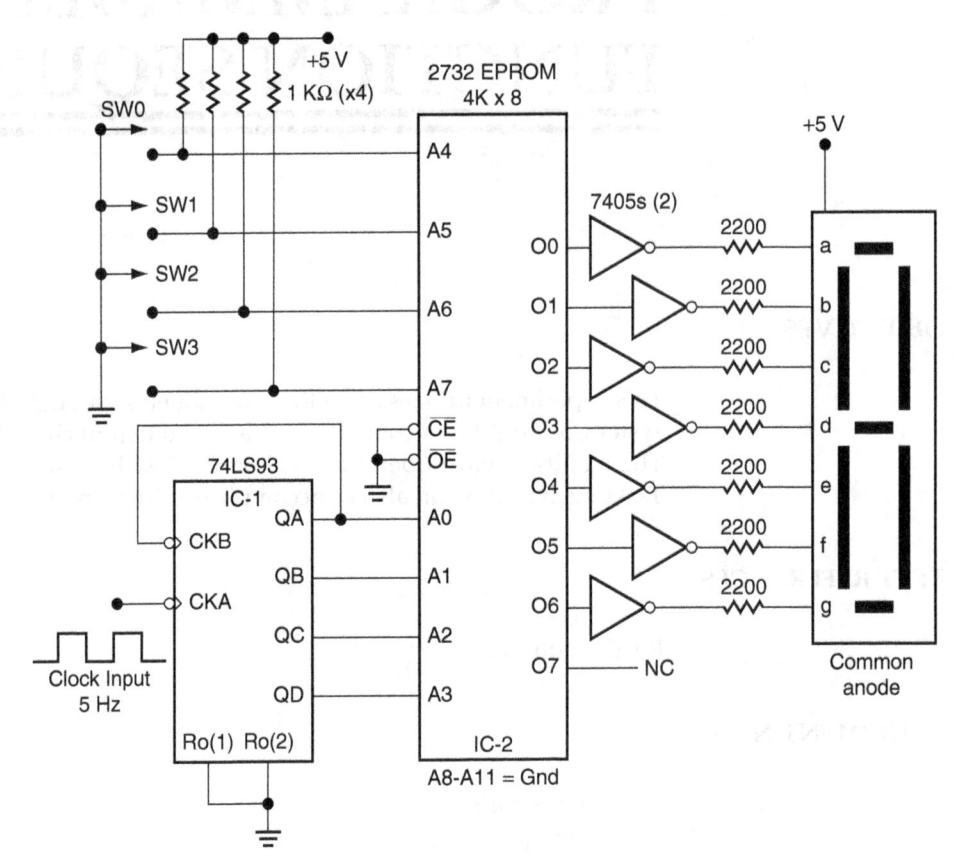

Figure 42-1

OPERATION

When SW0 is selected, the contents of the addressable locations of the EPROM should be the hexadecimal codes (0-F). Thus, the LED display should cycle through the hexadecimal number system. When SW1 is selected, the display should cycle through the up/down count sequence 1-8, 8-1 ... (up/down count sequence). SW2 selects the chase LED segment pattern sequence a,b,g,e,d,c,g,f,a ... (figure-8 LED pattern). SW3 selects the LED chase segment pattern sequence a,b,g,f,g,e,d,c,a ... (alternating circles LED pattern).

PROCEDURE

PROCEDURE

I. The first step is to make sure that you have a clean EPROM (unprogrammed). This can be done by using a W PROM eraser. In this experiment the EPROM must be programmed before being used. There are a couple of ways of programming the 2732 EPROM (either the 2716 or 2732 EPROM may be used). One way is to use one of the many commercially available programmers. Another way is to program the EPROM manually. (Refer to the EPROM's data sheet for programming instructions.)

II. Refer to the memory map of Figure 42-2(a). Determine which four blocks of memory (16 bytes each) will be used to store the data necessary for the proper operation of the circuit of Figure 42-1, as it was described above.

III. Fill out the EPROM worksheets of Figure 42-3 with the hex codes necessary to execute the program in each of the four blocks of memory determined in step (II).

IV. Use the specification data sheets in appendix B and transfer the pin numbers to all ICs used in the circuit of Figure 42-1.

V. Construct the circuit of Figure 42-1 and verify its operation. Demonstrate for the instructor.

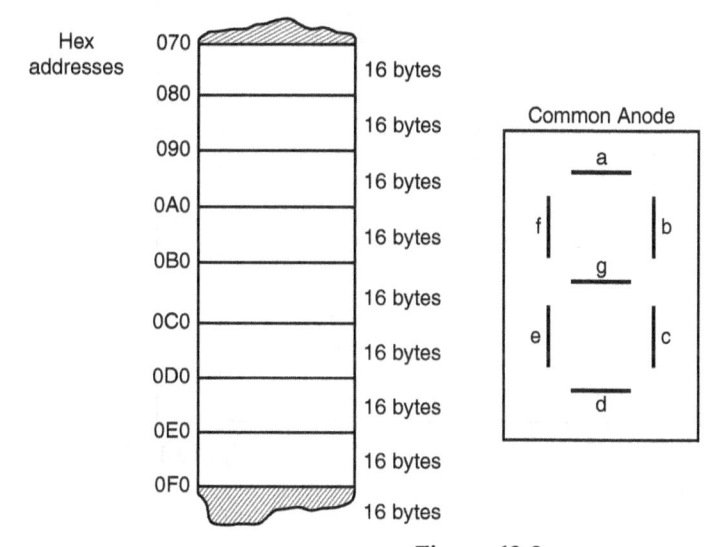

Figure 42-2

Exper. 42

Figure-8 LED pattern	
Hex Address	EPROM contents (hex)

Alternating circles LED pattern	
Hex Address	EPROM contents (hex)

Hexadecimal number system	
Hex Address	EPROM contents (hex)

Up/Down count sequences	
Hex Address	EPROM contents (hex)

Figure 42-3

PROCEDURE

QUESTIONS

a) Switches SW0–SW3 select four different sets of programming instructions. By using three switches (SW0–SW2) and one more IC, show how you would modify the circuit of Figure 42-1 so that you could have access to 8 different sets of programming instructions?

b) In the circuit of Figure 42-1 SW0–SW3 are normally HIGH and become active when they go LOW. If they were to be normally LOW and became HIGH when active, what would be the range of selectable addresses?

c) How would you modify the circuit of Figure 42-1 if a common cathode seven-segment LED was used instead?

SUPPLEMENTAL EXPERIMENTS

Experiment S1

Name _____

LOGIC GATES: OR, AND, AND NOT

OBJECTIVES

1. To become familiar with QUARTUS® II software.
2. To learn how to use QUARTUS® II software in creating schematics and waveforms.
3. To learn to use QUARTUS® II software to compile and simulate digital circuits.
4. To investigate the truth tables and associated waveforms of the basic logic gates, including the AND, OR, and NOT gates.
5. *Optional:* To program the EPM7128SLC84 CPLD.

TEXT REFERENCES

Read sections 3.1 through 3.5. Review Appendix D of this manual.

EQUIPMENT NEEDED

Components

Blank floppy, zip disk, or memory stick for storing projects;
QUARTUS® II software (Altera Corporation);
Optional: DeVry University Board eSOC with EPM7128SLC84 CPLD (or Altera University Board with EPM7128SLC84 CPLD or any other equivalent board);
Desktop Computer with minimum of INTEL Pentium PC @ 400MHz CPU running Microsoft Windows NT4SP1, 2000, or XP or better; Pentium III or 4 PC @ 400MHz running Red Hat Linux v7.3 or 8.0 or Red Hat Linux Enterprise 3; Sun Ultra running Solaris v8 or 9; HP9000 Series 700/800 running HP-UX v11.0 with ACE dated 11/1999 or later; or equivalent.

DISCUSSION

Read the Discussion in Experiment 2 of this lab manual. (If this experiment has already been completed, then a quick review of the Discussion is in order.) Next, read the software familiarization found in Appendix D of this manual. It is suggested that the first reading be light followed by a more thorough reading with the computer running the Altera software at your hand, doing the exercises as you go. The exercises in Appendix D give the model procedures for all of the Altera supplemental experiments in this lab manual. You will probably refer to Appendix D often at first.

PROCEDURE

PART 1 – The OR Gate

a) The OR Gate is covered in the QUARTUS® II tutorial (Appendix D) as Exercise 1 on page 590. Do this exercise to familiarize yourself with QUARTUS® II and the OR gate function.

b) When you have completed the exercise in Appendix D, go on to Part 2.

PART 2 – The AND Gate

a) Start the QUARTUS® II program. Set up a new project and use **basicgates2** for the name of the project. The diagram of the circuit you are to construct and its input waveforms are given in Figures S1-1 and S1-2.

b) Create a new graphic file for the schematic, *basicgates2.bdf*.

c) Place the AND gate symbol on the schematic.

d) Place the input and output connectors on the schematic. Label the inputs A and B; label the output X.

e) Wire the input and output connectors to the AND gate symbol.

f) Select **Project | Top-Level Entity**.

g) Compile the project. If there are errors, fix them and save the file before recompiling the project. If there are no errors (or perhaps a warning concerning timing at most), then go on to the next step.

h) Create a new waveform file *basicgates2.vwf* and define the inputs for simulation.

i) Select **Assignments | Settings** to access the Settings Window.

j) Click on *Simulator* at the left.

k) Choose Functional from the *Simulator Mode* pull-down menu.

l) Find and use for the *Simulator Input basicgates2.vwf* you just created above.

m) Select **Processing | Generate Functional Simulation Netlist.**

n) Simulate the project. If there are any errors and you change your circuit, remember to save and compile the circuit before re-simulating.

o) Record the output waveform X using Figure S1-2 or print out the Simulation Waveforms report from QUARTUS® II and attach it to the data collection sheet at the end of this experiment.

p) You should complete this part of the experiment by converting the timing diagram information in Figure S1-2 to Table S1-1.

PROCEDURE

q) Record and analyze results:
   1) Write the Boolean expression for X:

      X = _____

   2) Transfer the results of the timing diagram, truth table, and Boolean expression to the data collection sheets at the end of this experiment.
r) *(Optional) Program the CPLD:* If your instructor wants you to program your CPLD, refer to Appendix D, Exercise 3. A complete set of instructions will be found there. Use the same pin assignments as the OR gate in Appendix D, Exercise 3.
s) Save and close the project.

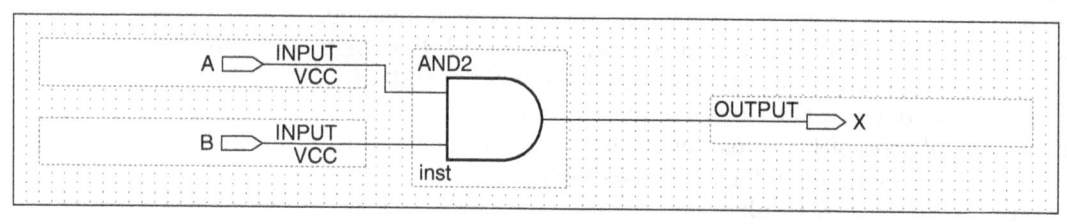

Figure S1-1

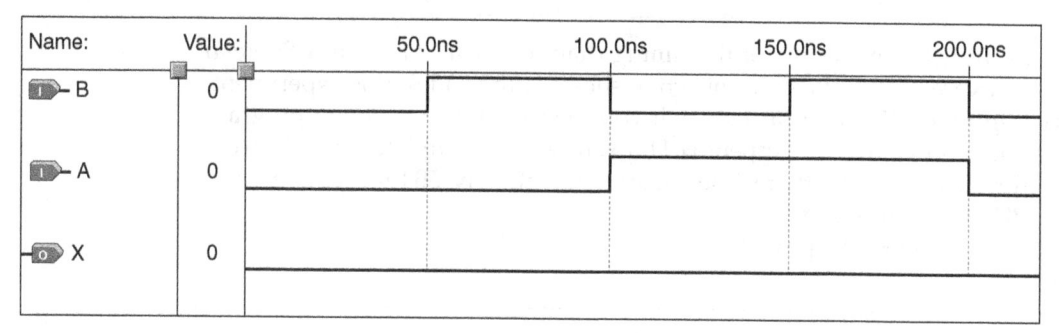

Figure S1-2

A	B	X
0	0	
0	1	
1	0	
1	1	

Table S1-1

PART 3 – The NOT Gate

a) Start the QUARTUS® II program. Set up a new project and use **basicgates3** for the name of the project. The diagram of the circuit you are to construct and its input waveforms are given in Figures S1-3 and S1-4.
b) Create a new graphic file for the schematic, *basicgates3.bdf*.
c) Place the NOT gate symbol on the schematic.
d) Place the input and output connectors on the schematic. Label the input A; label the output X.
e) Wire the input and output connectors to the NOT gate symbol.
f) Select **Project | Top-Level Entity**.

g) Compile the project. If there are errors, fix them and save the file before recompiling the project. If there are no errors (or perhaps a warning concerning timing at most), then go on to the next step.
h) Create a new waveform file *basicgates3.vwf* and define the inputs for simulation.
i) Select **Assignments | Settings** to access the Settings Window.
j) Click on *Simulator* at the left.
k) Choose Functional from the *Simulator Mode* pull-down menu.
l) Find and use for the *Simulator Input basicgates3.vwf* you just created above.
m) Select **Processing | Generate Functional Simulation Netlist.**
n) Simulate the project. If there are any errors and you change your circuit, remember to save and compile the circuit before re-simulating.
o) Record the output waveform X using Figure S1-4 or print out the Simulation Waveforms report from QUARTUS® II and attach it to the data collection sheet at the end of this experiment.
p) You should complete this part of the experiment by converting the timing diagram information in Figure S1-4 to Table S1-2.
q) Record and analyze results:
   1) Write the Boolean expression for X:

   $X = $ _____

   2) Transfer the results of the timing diagram, truth table, and Boolean expression to the data collection sheets at the end of this experiment.
r) *(Optional) Program the CPLD:* If your instructor wants you to program your CPLD, refer to Appendix D, Exercise 3. A complete set of instructions will be found there. Use pin assignments **PIN_50** for input "A" and **PIN_4** for output "X."
s) Save and close the project.

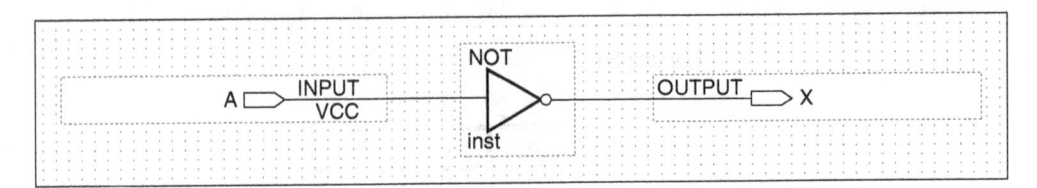

Figure S1-3

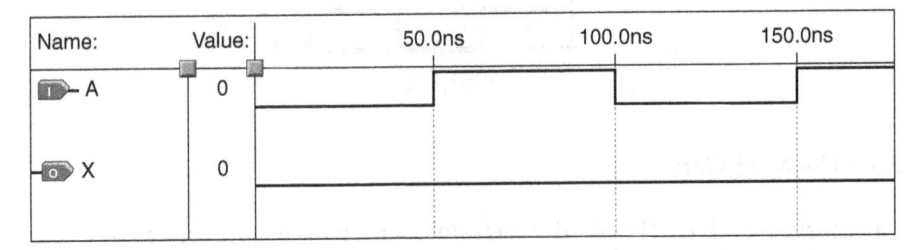

Figure S1-4

A	X
0	
1	

Table S1-2

DATA COLLECTION SHEET

Data Collection Sheets for Lab S1   Name _____

Date _____

Data collection for Part 2:

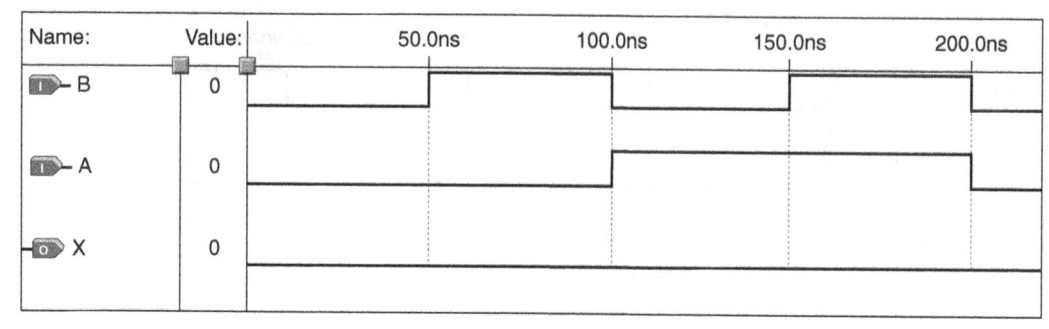

Name:	Value:	50.0ns	100.0ns	150.0ns	200.0ns
B	0				
A	0				
X	0				

Figure S1-2

A	B	X
0	0	
0	1	
1	0	
1	1	

Table S1-1

q)
1) Write the Boolean expression for X:

X = _____

Demonstrated To
Instructor/FA _____ Date _____

Part 1 [ ]   Part 2 [ ]   Part 3 [ ]

Data Collection Sheets for Lab S1   Name _____

                                        Date _____

Data collection for Part 3:

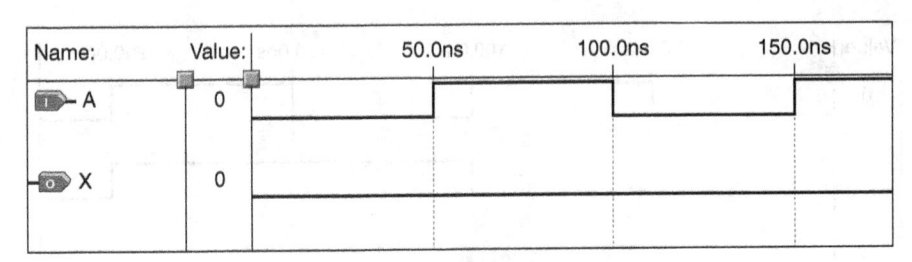

Name:	Value:	50.0ns	100.0ns	150.0ns
A	0			
X	0			

Figure S1-4

A	X
0	
1	

Table S1-2

q) Record and analyze results:
   1) Write the Boolean expression for X:

   X = _____

Demonstrated To
Instructor/FA _____ Date _____

Part 1 [ ]   Part 2 [ ]   Part 3 [ ]

Experiment S2

Name _____

BASIC COMBINATORIAL CIRCUITS

OBJECTIVES

1. To practice using QUARTUS® II software to create schematics and waveforms.
2. To practice using QUARTUS® II software to compile and simulate digital circuits.
3. To investigate the use of parentheses in Boolean expressions.
4. To implement a logic circuit from a Boolean expression.
5. *Optional:* To program the EPM7128SLC84 CPLD.

TEXT REFERENCES

Read sections 3.6 through 3.8. Review Appendix D of this manual.

EQUIPMENT NEEDED

Components

Blank floppy, zip disk, or memory stick for storing projects;

QUARTUS® II software (Altera Corporation);

Optional: DeVry University Board eSOC with EPM7128SLC84 CPLD (or Altera University Board with EPM7128SLC84 CPLD or any other equivalent board);

Desktop computer with minimum of INTEL Pentium PC @ 400MHz CPU running Microsoft Windows NT4SP1, 2000, or XP or better; Pentium III or 4 PC @ 400MHz running Red Hat Linux v7.3 or 8.0 or Red Hat Linux Enterprise 3; Sun Ultra running Solaris v8 or 9; HP9000 Series 700/800 running HP-UX v11.0 with ACE dated 11/1999 or later; or equivalent.

DISCUSSION

In Supplemental Experiment 1, you learned the characteristics of three of the fundamental logic gates: the AND, OR, and NOT. In this experiment, you will combine these gates into logic circuits and investigate their behavior, including analyzing their output waveforms and creating their function tables.

PROCEDURE

PART 1 – AND/OR Combination 1

a) Start the QUARTUS® II program. Set up a new project and use **combcircuit1** for the name of the project. The diagram of the circuit you are to construct and its input waveforms are given in Figures S2-1 and S2-2.
b) Create a new graphic file for the schematic, *combcircuit1.bdf*.
c) Place the gate symbols on the schematic according to Figure S2-1.
d) Place the input and output connectors on the schematic. Label the inputs A, B, and C; label the output X.
e) Wire the input and output connectors to the circuit as shown in Figure S2-1.
f) Select **Project | Top-Level Entity**.
g) Compile the project. If there are errors, fix them and save the file before recompiling the project. If there are no errors (or perhaps a warning concerning timing at most), then go on to the next step.
h) Create a new waveform *combcircuit1.vwf* file and define the inputs for simulation. The input waveforms should look similar to those in Figure S2-2. You should note that every combination of inputs A, B, and C from 000 to 111 is created.
i) Select **Assignments | Settings** to access the Settings Window.
j) Click on *Simulator* at the left.
k) Choose Functional from the *Simulator Mod* pull-down menu.
l) Find and use for the *Simulator Input combcircuit1.vwf* you just created above.
m) Select **Processing | Generate Functional Simulation Netlist.**
n) Simulate the project. If there are any errors and you change your circuit, remember to save and compile the circuit before re-simulating.
o) Record the output waveform X using Figure S2-2 or print out the Simulation Waveforms report from QUARTUS® II and attach it to the data collection sheet at the end of this experiment.
p) You should complete this part of the experiment by converting the waveform information in Figure S2-2 to Table S2-1.
q) Record and analyze results:
  1) Write the Boolean expression for X:

      X = _____

  2) Transfer the results of the timing diagram, truth table, and Boolean expression to the data collection sheets at the end of this experiment.

PROCEDURE

r) *(Optional) Program the CPLD:* It is suggested that you use **PIN_50** for "A," **PIN_51** for "B," **PIN_52** for "C," and **PIN_4** for "X." To program your CPLD, refer to Appendix D, Exercise 3. A complete set of instructions will be found there. Briefly,

1) *Recompile the project*: After making the pin assignments, compile the project again. When the compilation is completed, the software will display an information box showing any warnings and any detected errors. If there are errors, fix them and recompile the project.
2) *Download the project to the board:*
   a. Set all eSOC DIP switches to OFF.
   b. Connect the board to the PC's parallel printer port and apply power.
   c. Select **Tools | Programmer**.
   d. Click on **Auto Detect** in the Chain Description File.
   e. When the eSOC board is recognized, press the **Add File...** button in the Chain Description File, then locate and Add the *combcircuit1.pof* file.
   f. Select the file and device line and scroll (horizontally) to the available programming option. Check the **Program/Configure** box.
   g. Press the **Start** button. Observe the progress bar.
   h. If the program downloads successfully, test the program. For each eSOC switch used by the program, turn its corresponding DIP switch on.

s) Save and close the project.

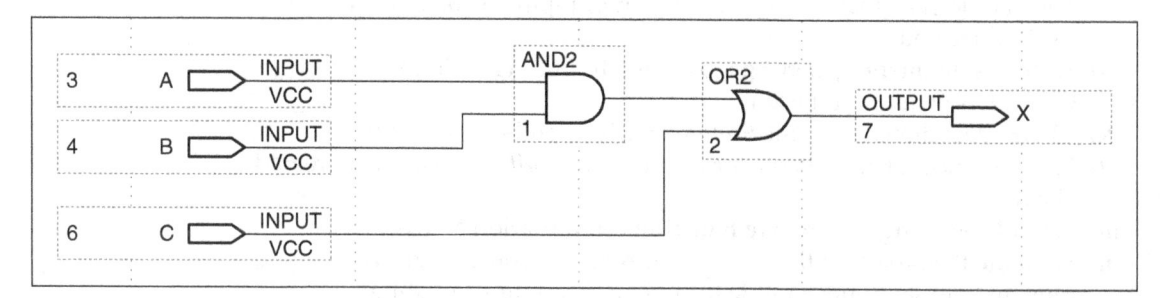

Figure S2-1

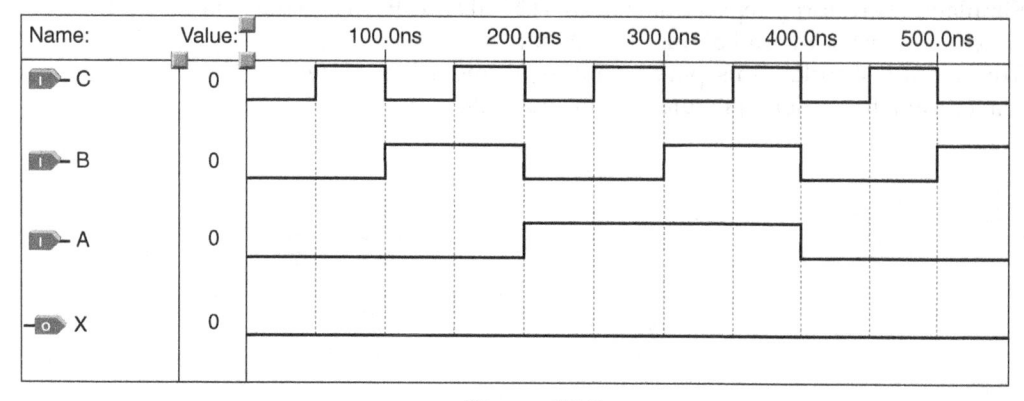

Figure S2-2

A	B	C	X
0	0	0	
0	0	1	
0	1	0	
0	1	1	
1	0	0	
1	0	1	
1	1	0	
1	1	1	

Table S2-1

PART 2 – AND/OR Combination 2

a) Start the QUARTUS® II program. Set up a new project and use **combcircuit2** for the name of the project. The diagram of the circuit you are to construct and its input waveforms are given in Figures S2-3 and S2-4.
b) Create a new graphic file for the schematic, *combcircuit2.bdf*.
c) Place the gate symbols on the schematic according to Figure S2-3.
d) Place the input and output connectors on the schematic. Label the inputs A, B, and C; label the output X.
e) Wire the input and output connectors to the circuit as shown in Figure S2-3.
f) Select **Project | Top-Level Entity**.
g) Compile the project. If there are errors, fix them and save the file before recompiling the project. If there are no errors (or perhaps a warning concerning timing at most), then go on to the next step.
h) Create a new waveform file *combcircuit2.vwf* and define the inputs for simulation. The waveforms should look similar to those in Figure S2-4. You should note that every combination of inputs A, B, and C from 000 to 111 is created.
i) Select **Assignments | Settings** to access the Settings Window.
j) Click on *Simulator* at the left.
k) Choose Functional from the *Simulator Mode* pull-down menu.
l) Find and use for the *Simulator Input combcircuit2.vwf* you just created above.
m) Select **Processing | Generate Functional Simulation Netlist.**
n) Simulate the project. If there are any errors and you change your circuit, remember to save and compile the circuit before re-simulating.
o) Record the output waveform X using Figure S2-4 or print out the Simulation Waveforms report from QUARTUS® II and attach it to the data collection sheet at the end of this experiment.
p) You should complete this part of the experiment by converting the waveform information in Figure S2-4 to Table S2-2.

PROCEDURE

q) Record and analyze results:
  1) Write the Boolean expression for X:

     X = _____

  2) Transfer the results of the timing diagram, truth table, and Boolean expression to the data collection sheets at the end of this experiment.

r) *(Optional) Program the CPLD:* It is suggested that you use **PIN_50** for "**A**," **PIN_51** for "**B**," **PIN_52** for "**C**," and **PIN_4** for "**X**." To program your CPLD, refer to Appendix D, Exercise 3. A complete set of instructions will be found there. Briefly,

  1) *Recompile the project*: After making the pin assignments, compile the project again. When the compilation is completed, the software will display an information box showing any warnings and any detected errors. If there are errors, fix them and recompile the project.
  2) *Download the project to the board:*
     a. Set all eSOC DIP switches to OFF.
     b. Connect the board to the PC's parallel printer port and apply power.
     c. Select **Tools | Programmer**.
     d. Click on [Auto Detect] in the Chain Description File.
     e. When the eSOC board is recognized, press the [Add File...] button in the Chain Description File, then locate and Add the *combcircuit2.pof* file.
     f. Select the file and device line and scroll (horizontally) to the available programming option. Check the **Program/Configure** box.
     g. Press the [Start] button. Observe the progress bar.
     h. If the program downloads successfully, test the program. For each eSOC switch used by the program, turn its corresponding DIP switch on.

s) Save and close the project.

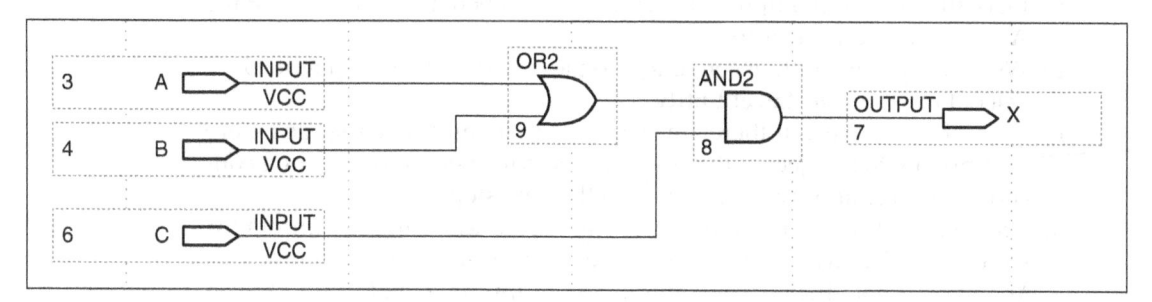

Figure S2-3

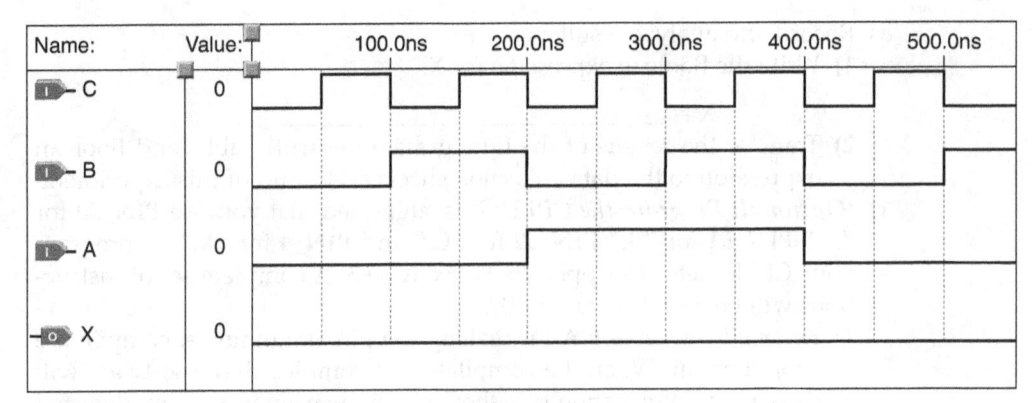

Name:	Value:	100.0ns	200.0ns	300.0ns	400.0ns	500.0ns
C	0					
B	0					
A	0					
X	0					

Figure S2-4

A	B	C	X
0	0	0	
0	0	1	
0	1	0	
0	1	1	
1	0	0	
1	0	1	
1	1	0	
1	1	1	

Table S2-2

PART 3 – Circuits with Inverters

a) Start the MAX+PLUS® II program. Set up a new project and use **combcircuit3** for the name of the project. The diagram of the circuit you are to construct and its input waveforms are given in Figures S2-5 and S2-6.

b) Create a new graphic file for the schematic, *combcircuit3.gdf*.

c) Place the gate symbols on the schematic according to Figure S2-5.

d) Place the input and output connectors on the schematic. Label the inputs A, B, and C; label the output X.

e) Wire the input and output connectors to the circuit as shown in Figure S2-5.

f) Select **Project | Top-Level Entity**.

g) Compile the project. If there are errors, fix them and save the file before recompiling the project. If there are no errors (or perhaps a warning concerning timing at most), then go on to the next step.

h) Create a new waveform file *combcircuit3.vwf* and define the inputs for simulation. The waveforms should look similar to those in Figure S2-6. You should note that every combination of inputs A, B, C, and D from 0000 to 1111 is created.

i) Select **Assignments | Settings** to access the Settings Window.

j) Click on *Simulator* at the left.

k) Choose Functional from the *Simulator Mode* pull-down menu.

l) Find and use for the *Simulator Input combcircuit3.vwf* you just created above.

m) Select **Processing | Generate Functional Simulation Netlist**.

PROCEDURE

n) Simulate the project. If there are any errors and you change your circuit, remember to save and compile the circuit before re-simulating.
o) Record the output waveform X using Figure S2-6 or print out the Simulation Waveforms report from QUARTUS® II and attach it to the data collection sheet at the end of this experiment.
p) You should complete this part of the experiment by converting the waveform information in Figure S2-6 to Table S2-3.
q) Record and analyze results:
   1) Write the Boolean expression for X:

      $X = $ _____

   2) Transfer the results of the timing diagram, truth table, and Boolean expression to the data collection sheets at the end of this experiment.
r) *(Optional) Program the CPLD:* It is suggested that you use **PIN_50** for "A," **PIN_51** for "B," **PIN_52** for "C," **PIN_54** for "D," and **PIN_4** for "X." To program your CPLD, refer to Appendix D, Exercise 3. A complete set of instructions will be found there. Briefly,
   1) *Recompile the project*: After making the pin assignments, compile the project again. When the compilation is completed, the software will display an information box showing any warnings and any detected errors. If there are errors, fix them and recompile the project.
   2) *Download the project to the board:*
      a. Set all eSOC DIP switches to OFF.
      b. Connect the board to the PC's parallel printer port and apply power.
      c. Select **Tools | Programmer**.
      d. Click on [Auto Detect] in the Chain Description File.
      e. When the eSOC board is recognized, press the [Add File...] button in the Chain Description File, then locate and Add the *combcircuit3.pof* file.
      f. Select the file and device line and scroll (horizontally) to the available programming option. Check the **Program/Configure** box.
      g. Press the [Start] button. Observe the progress bar.
      h. If the program downloads successfully, test the program. For each eSOC switch used by the program, turn its corresponding DIP switch on.
s) Save and close the project.

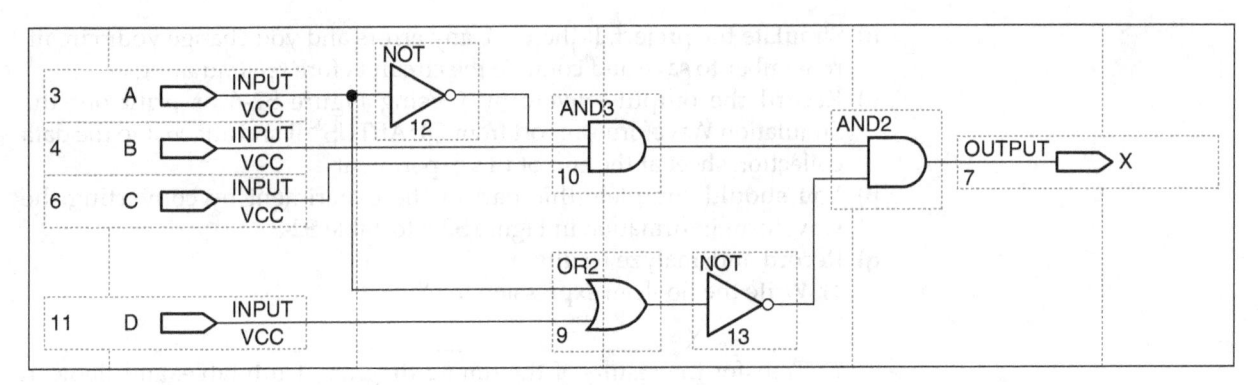

Figure S2-5

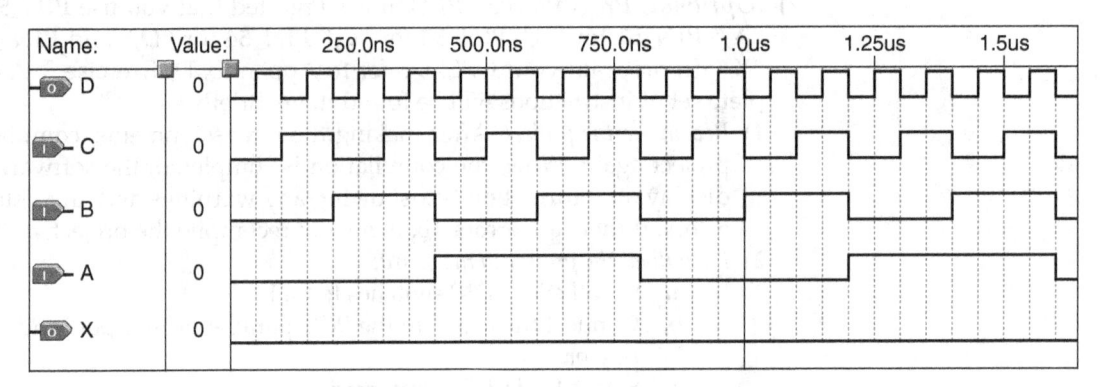

Figure S2-6

A	B	C	D	X
0	0	0	0	
0	0	0	1	
0	0	1	0	
0	0	1	1	
0	1	0	0	
0	1	0	1	
0	1	1	0	
0	1	1	1	
1	0	0	0	
1	0	0	1	
1	0	1	0	
1	0	1	1	
1	1	0	0	
1	1	0	1	
1	1	1	0	
1	1	1	1	

Table S2-3

DATA COLLECTION SHEET

Data Collection Sheets for Lab S2   Name _____

Date _____

Data collection for Part 1:

q)

1) Write the Boolean expression for X:

X = _____

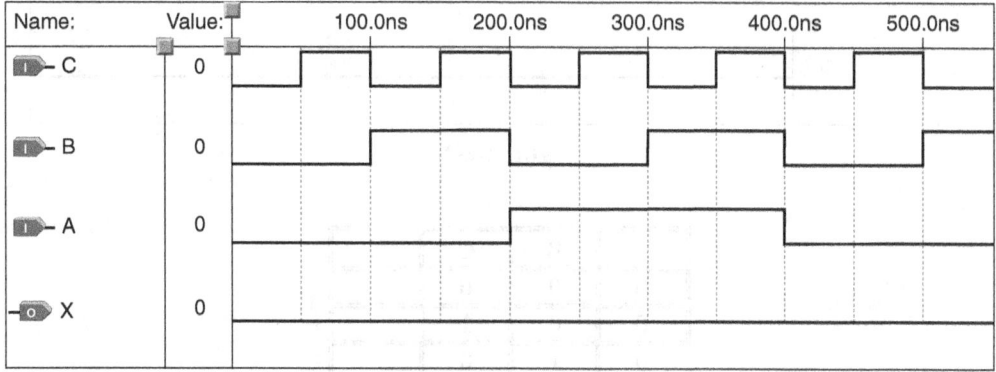

Name:	Value:	100.0ns	200.0ns	300.0ns	400.0ns	500.0ns
C	0					
B	0					
A	0					
X	0					

Figure S2-2

A	B	C	X
0	0	0	
0	0	1	
0	1	0	
0	1	1	
1	0	0	
1	0	1	
1	1	0	
1	1	1	

Table S2-1

Data collection for Part 2:

q)

1) Write the Boolean expression for X:

X = _____

Demonstrated To
Instructor/FA _____ Date _____

Part 1 [ ]   Part 2 [ ]   Part 3 [ ]

Data Collection Sheets for Lab S2 Name _____

Date _____

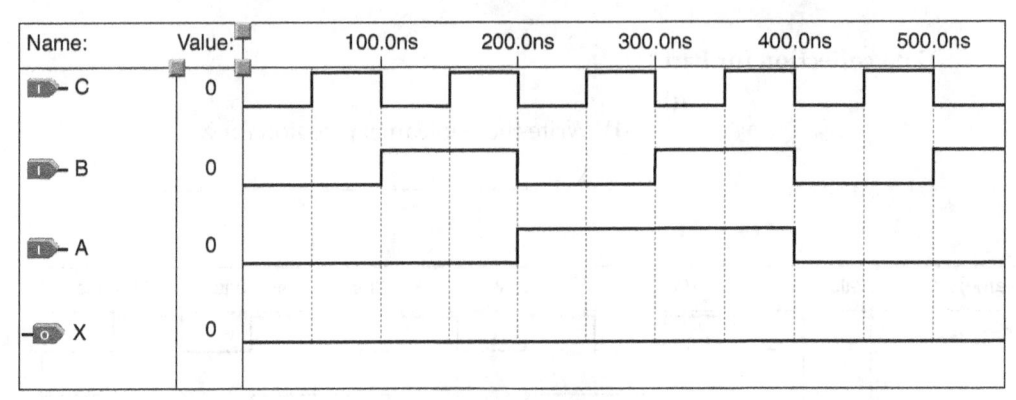

Name:	Value:	100.0ns	200.0ns	300.0ns	400.0ns	500.0ns
C	0					
B	0					
A	0					
X	0					

Figure S2-4

A	B	C	X
0	0	0	
0	0	1	
0	1	0	
0	1	1	
1	0	0	
1	0	1	
1	1	0	
1	1	1	

Table S2-2

Data collection for Part 3:
q)
  1) Write the Boolean expression for X:

  X = _____

| Demonstrated To |
| Instructor/FA _____ Date _____ |
| |
| Part 1 [ ]   Part 2 [ ]   Part 3 [ ] |

DATA COLLECTION SHEET

Data Collection Sheets for Lab S2   Name _____

Date _____

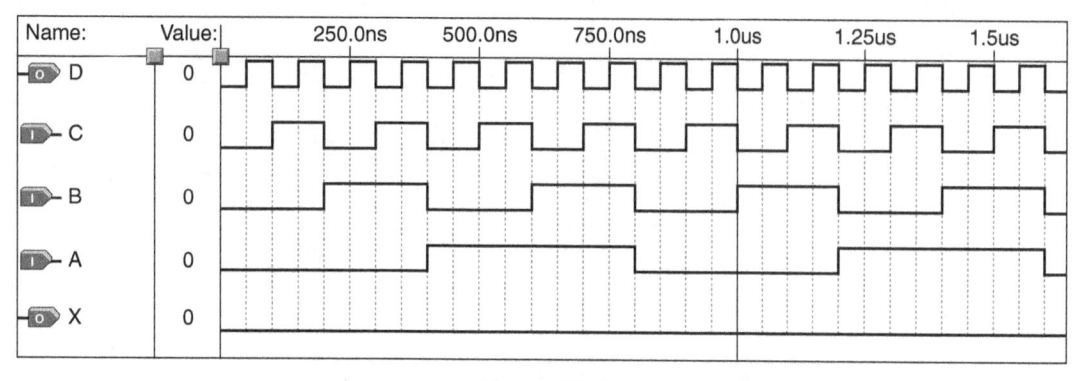

Figure S2-6

A	B	C	D	X
0	0	0	0	
0	0	0	1	
0	0	1	0	
0	0	1	1	
0	1	0	0	
0	1	0	1	
0	1	1	0	
0	1	1	1	
1	0	0	0	
1	0	0	1	
1	0	1	0	
1	0	1	1	
1	1	0	0	
1	1	0	1	
1	1	1	0	
1	1	1	1	

Table S2-3

Demonstrated To
Instructor/FA _____ Date _____

Part 1 [ ]   Part 2 [ ]   Part 3 [ ]

Experiment S3

Name _____

LOGIC GATES: NOR AND NAND

OBJECTIVES

1. To practice using QUARTUS® II software.
2. To practice using QUARTUS® II software to create schematics and waveforms.
3. To practice using QUARTUS® II software to compile and simulate digital circuits.
4. To investigate the truth tables and associated waveforms of the basic logic gates, including the NAND and NOR gates.
5. *Optional:* To program the EPM7128SLC84 CPLD.

TEXT REFERENCES

Read section 3.9. Review Appendix D of this manual.

EQUIPMENT NEEDED

Components

Blank floppy, zip disk, or memory stick for storing projects;

QUARTUS® II software (Altera Corporation);

Optional: DeVry University Board eSOC with EPM7128SLC84 CPLD (or Altera University Board with EPM7128SLC84 CPLD or any other equivalent board);

Desktop computer with minimum of INTEL Pentium PC @ 400MHz CPU running Microsoft Windows NT4SP1, 2000, or XP or better; Pentium III or 4 PC @ 400MHz running Red Hat Linux v7.3 or 8.0 or Red Hat Linux Enterprise 3; Sun Ultra running Solaris v8 or 9; HP9000 Series 700/800 running HP-UX v11.0 with ACE dated 11/1999 or later; or equivalent.

DISCUSSION

In Experiment S1, you learned the characteristics of three of the fundamental logic gates: the AND, OR, and NOT. You will now be introduced to two of the remaining logic gates: the NAND and NOR. The NAND and NOR gates are nothing more than inverted AND and OR gates, respectively. That is important, but not the most important thing. The fact that a NAND or a NOR can be used to create all other gates is most important, because this fact has made them more popular in use than the others.

PROCEDURE

PART 1 – The NOR Gate

a) Start the QUARTUS® II program. Set up a new project and use **basicgates4** for the name of the project. The diagram of the circuit you are to construct and its input waveforms are given in Figures S3-1 and S3-2.
b) Create a new graphic file for the schematic, *basicgates4.bdf*.
c) Place the NOR gate symbol on the schematic.
d) Place the input and output connectors on the schematic. Label the inputs A and B; label the output X.
e) Wire the input and output connectors to the NOR gate symbol.
f) Select **Project | Top-Level Entity**.
g) Compile the project. If there are errors, fix them and save the file before recompiling the project. If there are no errors (or perhaps a warning concerning timing at most), then go on to the next step.
h) Create a new waveform file *basicgates4.vwf* and define the inputs for simulation. The input waveforms should look similar to those in Figure S3-2.
i) Select **Assignments | Settings** to access the Settings Window.
j) Click on *Simulator* at the left.
k) Choose Functional from the *Simulator Mode* pull-down menu.
l) Find and use for the *Simulator Input* *basicgates4.vwf* you just created above.
m) Select **Processing | Generate Functional Simulation Netlist**.
n) Simulate the project. If there are any errors and you change your circuit, remember to save and compile the circuit before re-simulating.
o) Record the output waveform X on Figure S3-2 or print the Simulation Waveforms report from QUARTUS® II and attach it to the data collection sheet at the end of this experiment.
p) You should complete this part of the experiment by converting the timing diagram information in Figure S3-2 to Table 3-1.
q) Record and analyze results:
   1) Write the Boolean expression for X:

   X = _____

   2) Transfer the results of the timing diagram, truth table, and Boolean expression to the data collection sheets at the end of this experiment.

PROCEDURE

r) *(Optional) Program the CPLD:* It is suggested that you use **PIN_50** for "**A**," **PIN_51** for "**B**," and **PIN_4** for "**X**." To program your CPLD, refer to Appendix D, Exercise 3. A complete set of instructions will be found there. Briefly,

1) *Recompile the project*: After making the pin assignments, compile the project again. When the compilation is completed, the software will display an information box showing any warnings and any detected errors. If there are errors, fix them and recompile the project.
2) *Download the project to the board:*
    a. Set all eSOC DIP switches to OFF.
    b. Connect the board to the PC's parallel printer port and apply power.
    c. Select **Tools | Programmer**.
    d. Click on [Auto Detect] in the Chain Description File.
    e. When the eSOC board is recognized, press the [Add File...] button in the Chain Description File, then locate and Add the *basicgates4.pof* file.
    f. Select the file and device line and scroll (horizontally) to the available programming option. Check the **Program/Configure** box.
    g. Press the [Start] button. Observe the progress bar.
    h. If the program downloads successfully, test the program. For each eSOC switch used by the program, turn its corresponding DIP switch on.

s) Save and close the project.

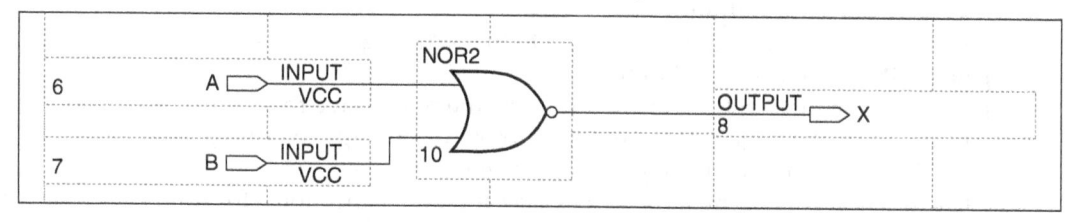

Figure S3-1

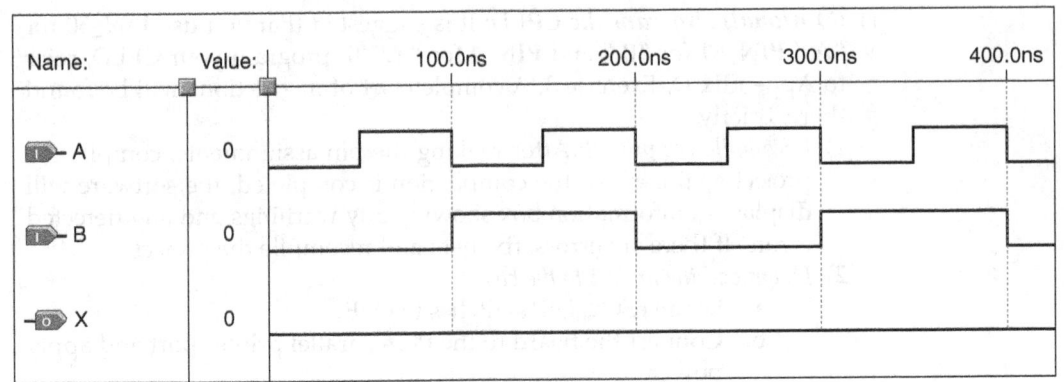

Figure S3-2

A	B	X
0	0	
0	1	
1	0	
1	1	

Table S3-1

PART 2 – The NAND Gate

a) Start the QUARTUS® II program. Set up a new project and use **basicgates5** for the name of the project. The diagram of the circuit you are to construct and its input waveforms are given in Figures S3-3 and S3-4.
b) Create a new graphic file for the schematic, *basicgates5.bdf*.
c) Place the NAND gate symbol on the schematic.
d) Place the input and output connectors on the schematic. Label the inputs A and B; label the output X.
e) Wire the input and output connectors to the NAND gate symbol.
f) Select **Project | Top-Level Entity**.
g) Compile the project. If there are errors, fix them and save the file before recompiling the project. If there are no errors (or perhaps a warning concerning timing at most), then go on to the next step.
h) Create a new waveform file *basicgates5.vwf* and define the inputs for simulation. The input waveforms should look similar to those in Figure S3-4.
i) Select **Assignments | Settings** to access the Settings Window.
j) Click on *Simulator* at the left.
k) Choose Functional from the *Simulator Mode* pull-down menu.
l) Find and use for the *Simulator Input* *basicgates5.vwf* you just created above.
m) Select **Processing | Generate Functional Simulation Netlist.**
n) Simulate the project. If there are any errors and you change your circuit, remember to save and compile the circuit before re-simulating.

PROCEDURE

o) Record the output waveform X on Figure S3-4 or print the Simulation Waveforms report from QUARTUS® II and attach it to the data collection sheet at the end of this experiment.

p) You should complete this part of the experiment by converting the timing diagram information in Figure S3-4 to Table 3-2.

q) Record and analyze results:
   1) Write the Boolean expression for X:

      X = _____

   2) Transfer the results of the timing diagram, truth table, and Boolean expression to the data collection sheets at the end of this experiment.

r) *(Optional) Program the CPLD:* It is suggested that you use **PIN_50** for "A," **PIN_51** for "B," and **PIN_4** for "X." To program your CPLD, refer to Appendix D, Exercise 3. A complete set of instructions will be found there. Briefly,
   1) *Recompile the project*: After making the pin assignments, compile the project again. When the compilation is completed, the software will display an information box showing any warnings and any detected errors. If there are errors, fix them and recompile the project.
   2) *Download the project to the board:*
      a. Set all eSOC DIP switches to OFF.
      b. Connect the board to the PC's parallel printer port and apply power.
      c. Select **Tools | Programmer**.
      d. Click on [Auto Detect] in the Chain Description File
      e. When the eSOC board is recognized, press the [Add File...] button in the Chain Description File, then locate and Add the *basicgates5.pof* file.
      f. Select the file and device line and scroll (horizontally) to the available programming option. Check the **Program/Configure** box.
      g. Press the [Start] button. Observe the progress bar.
      h. If the program downloads successfully, test the program. For each eSOC switch used by the program, turn its corresponding DIP switch on.

s) Save and close the project.

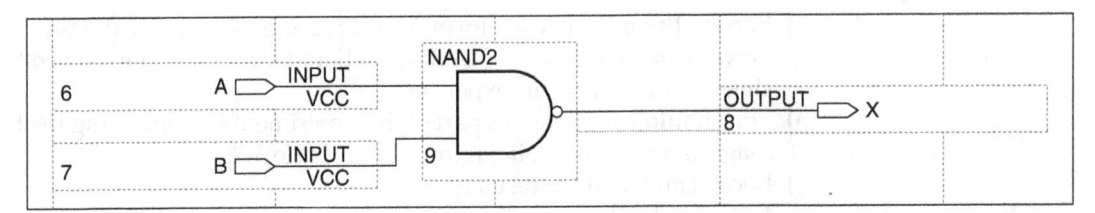

Figure S3-3

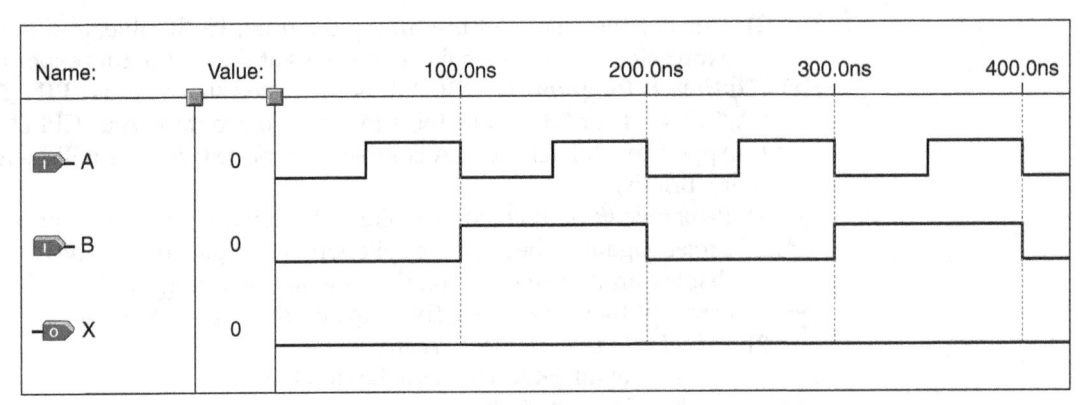

Figure S3-4

A	B	X
0	0	
0	1	
1	0	
1	1	

Table S3-2

DATA COLLECTION SHEET

Data Collection Sheets for Lab S3   Name _____

Date _____

Data collection for Part 1:

q)

    1) Write the Boolean expression for X:

        X = _____

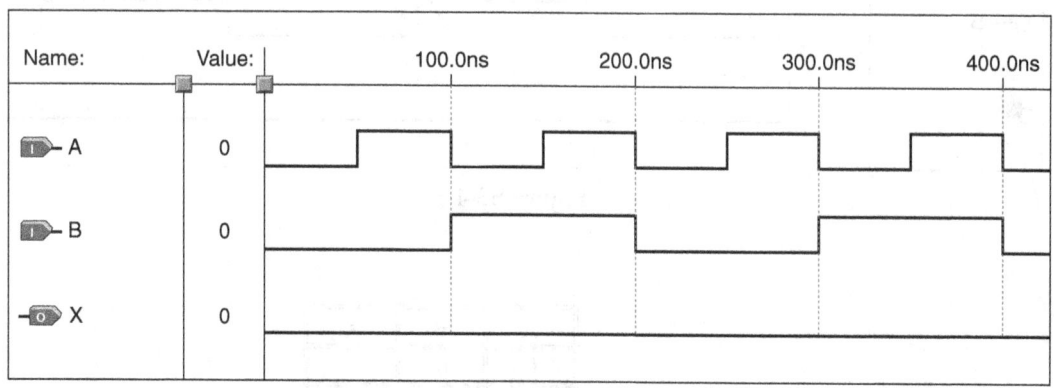

Figure S3-2

A	B	X
0	0	
0	1	
1	0	
1	1	

Table S3-1

Data Collection for Part 2:

q)

    1) Write the Boolean expression for X:

        X = _____

Demonstrated To
Instructor/FA _____ Date _____

Part 1 [ ]   Part 2 [ ]

Data Collection Sheets for Lab S3   Name _____

Date _____

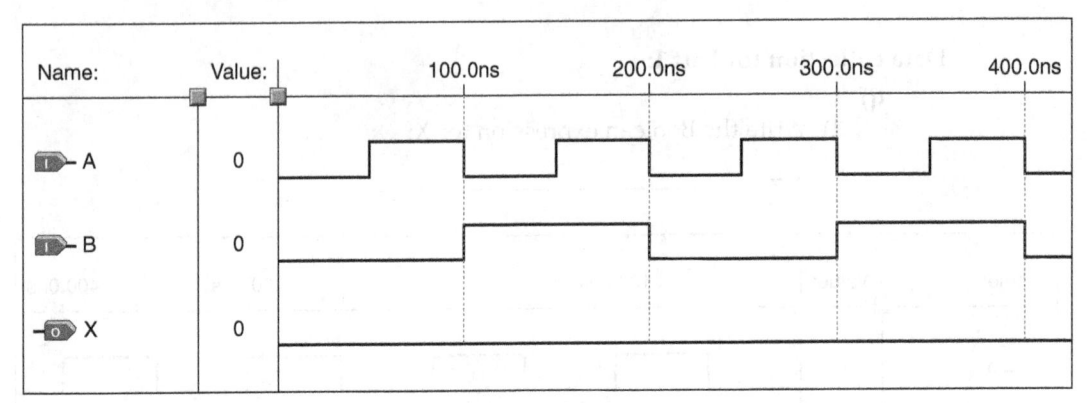

| Name: | Value: | 100.0ns | 200.0ns | 300.0ns | 400.0ns |

(waveform diagram showing signals A, B, and X over time)

Figure S3-4

A	B	X
0	0	
0	1	
1	0	
1	1	

Table S3-2

Demonstrated To
Instructor/FA _____ Date _____

Part 1 [ ]   Part 2 [ ]

Experiment S4

Name _____

BOOLEAN THEOREMS

OBJECTIVES

1. To practice using QUARTUS® II software.
2. To practice using QUARTUS® II software to create schematics and waveforms.
3. To practice using QUARTUS® II software to compile and simulate digital circuits.
4. To verify experimentally some of the Boolean theorems.
5. *Optional:* To program the EPM7128SLC84 CPLD.

TEXT REFERENCES

Read section 3.10. Review Appendix D of this manual.

EQUIPMENT NEEDED

Components

Blank floppy, zip disk, or memory stick for storing projects;

QUARTUS® II software (Altera Corporation);

Optional: Program the EPM7128SLC84 CPLD;

Desktop computer with minimum of INTEL Pentium PC @ 400MHz CPU running Microsoft Windows NT4SP1, 2000, or XP or better; Pentium III or 4 PC @ 400MHz running Red Hat Linux v7.3 or 8.0 or Red Hat Linux Enterprise 3; Sun Ultra running Solaris v8 or 9; HP9000 Series 700/800 running HP-UX v11.0 with ACE dated 11/1999 or later; or equivalent.

DISCUSSION

Boolean theorems are useful in simplifying expressions. You have already seen that logic circuits can be constructed from an expression, so it should follow, naturally, that simpler expressions mean simpler circuits. In this experiment, you will verify some of these theorems, discuss others, and demonstrate their impact on the physical circuit.

PROCEDURE

PART 1 – Eight Boolean Theorems

a) Start the QUARTUS® II program. Set up a new project and use **boolean1** for the name of the project. The diagram of the circuit you are to construct and its input waveform is given in Figures S4-1 and S4-2.
b) Create a new graphic file for the schematic, *boolean1.bdf*.
c) Place the logic symbols onto the schematic given in Figure S4-1.
d) Connect the symbols together according to the schematic.
e) Place the input and output connectors on the schematic. Label the input X; label the eight different outputs A through H.
f) Wire the input and output connectors to the logic symbols.
g) Select **Project | Top-Level Entity**.
h) Compile the project. If there are errors, fix them and save the file before recompiling the project. If there are no errors (or perhaps a warning concerning timing at most), then go on to the next step.
i) Create a new waveform file *boolean1.vwf* and define the input for simulation as shown in Figure S4-2.
j) Select **Assignments | Settings** to access the Settings Window.
k) Click on *Simulator* at the left.
l) Choose Functional from the *Simulator Mode* pull-down menu.
m) Find and use for the *Simulator Input boolean1.vwf* you just created above.
n) Select **Processing | Generate Functional Simulation Netlist**.
o) Simulate the project. If there are any errors and you change your circuit, remember to save and compile the circuit before re-simulating.
p) Record the output waveforms using Figure S4-1 or print out the Simulation Waveforms report from QUARTUS® II and attach it to the data collection sheet at the end of this experiment:
  1) Compare each output waveform (A through H) with the right-hand side of its corresponding Boolean theorem.
  2) You should observe that output A is a constant LOW or Boolean zero. Thus output A does correspond to the right-hand side of the Boolean theorem $x \cdot 0 = 0$.
  3) You should observe that output B is varying from 0 to 1 and is identical to the input X. Thus output B does correspond to the right-side of the Boolean theorem $x \cdot 1 = 1$.
  4) Finally, you should observe that output F is a constant HIGH or Boolean one. Thus output F corresponds to the right-side of the Boolean theorem $x + 1 = 1$.

PROCEDURE

5) Record your results in Table S4-1.
6) Transfer the results of the waveform and truth table to the data collection sheets at the end of this experiment. Alternately, print the Simulation Waveforms report from QUARTUS® II and attach it to the data collection sheet at the end of this experiment.

q) Save and close the project.

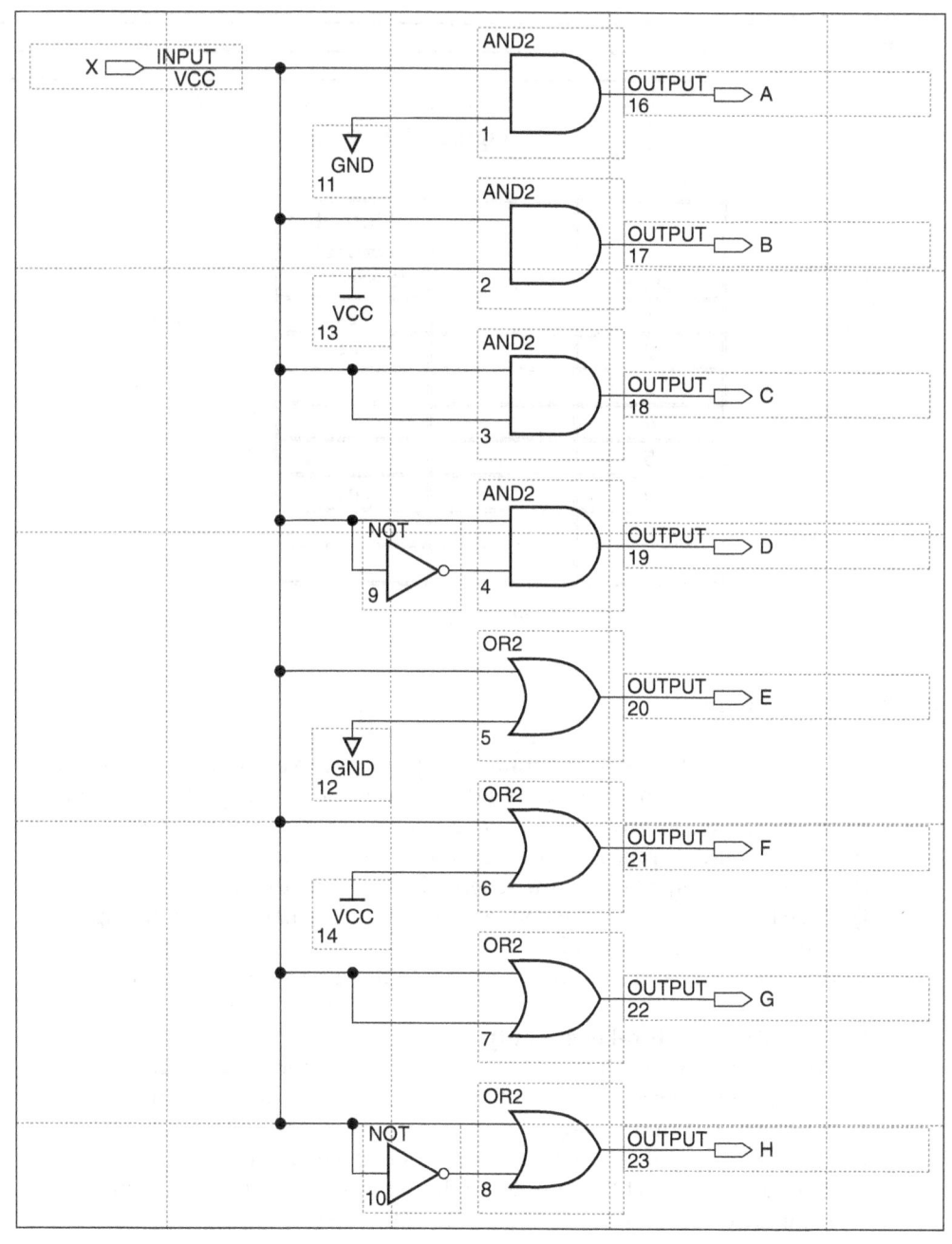

Figure S4-1

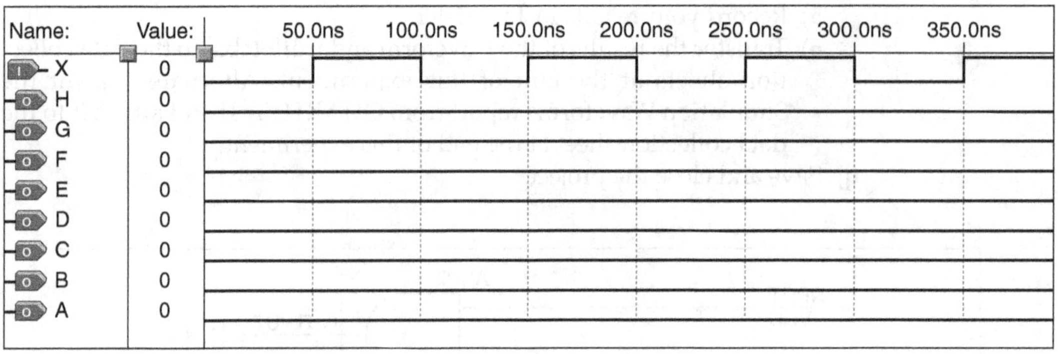

Name:	Value:	50.0ns	100.0ns	150.0ns	200.0ns	250.0ns	300.0ns	350.0ns
X	0							
H	0							
G	0							
F	0							
E	0							
D	0							
C	0							
B	0							
A	0							

Figure S4-2

Theorem	Output	Output verifies theorem?
1	0	yes
2	X	yes
3		
4		
5		
6	1	yes
7		
8		

Table S4-1

PART 2 – Theorem 14

a) Start the QUARTUS® II program. Set up a new project and use **boolean2** for the name of the project. The diagram of the circuit you are to construct and its input waveforms are given in Figures S4-3 and S4-4.
b) Create a new graphic file *boolean2.bdf* for the schematic.
c) Place the gate symbols on the schematic according to Figure S4-3.
d) Place the input and output connectors on the schematic. Label the inputs X and Y; label the output Z.
e) Wire the input and output connectors to the gate symbols according to Figure S4-3.
f) Select **Project | Top-Level Entity**.
g) Compile the project. If there are errors, fix them and save the file before recompiling the project. You will get a warning: "Design contains 1 input pin(s) that do not drive logic."
h) This warning could mean that Y is truly not necessary or you could have forgotten to connect Y to the circuit. In this case, this means the output Z is not dependent on Y and so Y can be eliminated.
i) Create a new waveform file *boolean2.vwf* and define the inputs for simulation. The inputs should look like those in Figure S4-4.

PROCEDURE

j) Select **Assignments | Settings** to access the Settings Window.
k) Click on *Simulator* at the left.
l) Choose Functional from the *Simulator Mode* pull-down menu.
m) Find and use for the *Simulator Input* boolean2.vwf you just created in step i.
n) Select **Processing | Generate Functional Simulation Netlist.**
o) Simulate the project. If there are any errors and you change your circuit, remember to save and compile the circuit before re-simulating.
p) Since the compiler did not include the input Y, you will get another Warning: Project does not include input node 'Y.' You can ignore this warning.
q) Record and analyze results:
  1) Record the output waveform X on Figure S4-4 or print the Simulation Waveforms report and attach it to the data collection sheets at the end of this experiment.
  2) Compare input X and output Z. Are they identical logically? _____ (If so, this verifies Theorem 14: $x + xy = x$.)
  3) Write the Boolean expression for Z:

  $$Z = \underline{\hspace{5cm}}$$

  4) Transfer the results of the waveform, truth table, and Boolean expression to the data collection sheets at the end of this experiment.
r) **(Optional) Program the CPLD:** It is suggested that you use **PIN_50** for "X," **PIN_51** for "Y," and **PIN_4** for "Z." To program your CPLD, refer to Appendix D, Exercise 3. A complete set of instructions will be found there. Briefly,
  1) *Recompile the project*: After making the pin assignments, compile the project again. When the compilation is completed, the software will display an information box showing any warnings and any detected errors. If there are errors, fix them and recompile the project.
  2) *Download the project to the board:*
      a. Set all eSOC DIP switches to OFF.
      b. Connect the board to the PC's parallel printer port and apply power.
      c. Select **Tools | Programmer**.
      d. Click on [Auto Detect] in the Chain Description File.
      e. When the eSOC board is recognized, press the [Add File...] button in the Chain Description File, then locate and Add the *boolean2.pof* file.
      f. Select the file and device line and scroll (horizontally) to the available programming option. Check the **Program/Configure** box.

g. Press the [Start] button. Observe the progress bar.
h. If the program downloads successfully, test the program. For each eSOC switch used by the program, turn its corresponding DIP switch on.

s) Save and close the project.

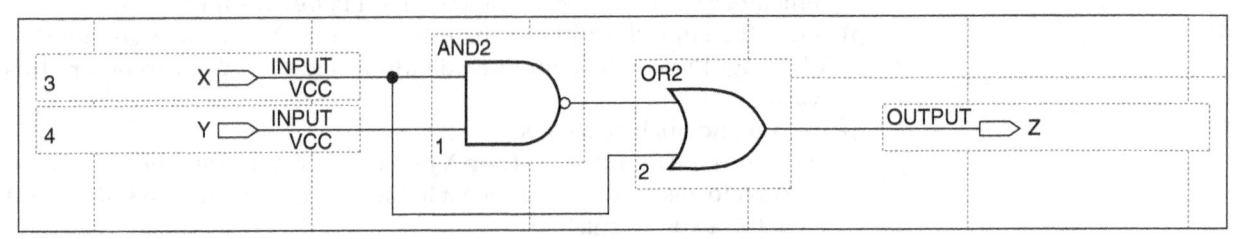

Figure S4-3

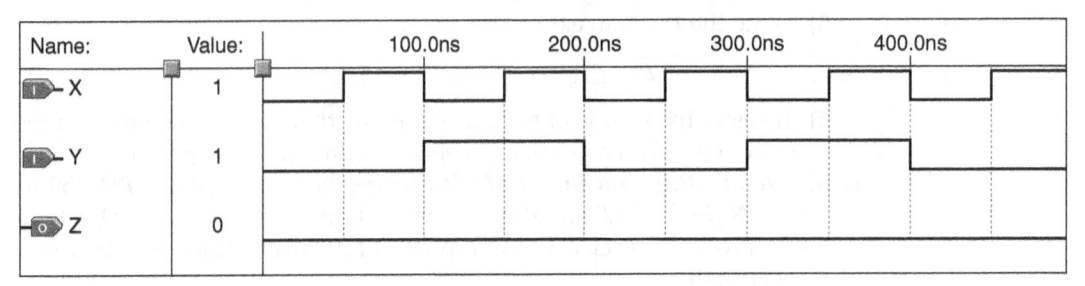

Figure S4-4

X	Y	Z
0	0	
0	1	
1	0	
1	1	

Table S4-2

PART 3 – Theorem 15

a) Start the QUARTUS® II program. Set up a new project and use **boolean3** for the name of the project. The diagram of the circuit you are to construct and its input waveforms are given in Figures S4-5 and S4-6.
b) Create a new graphic file for the schematic, *boolean3.bdf*.
c) Place the gate symbols on the schematic according to Figure S4-5.
d) Place the input and output connectors on the schematic. Label the inputs X and Y; label the output Z.
e) Wire the input and output connectors to the gate symbols according to Figure S4-5.
f) Select **Project | Top-Level Entity**.
g) Compile the project. If there are errors, fix them and save the file before recompiling the project. If there are no errors (or perhaps a warning concerning timing at most), then go on to the next step.

PROCEDURE

h) Create a new waveform file, *boolean3.vwf*, and define the inputs for simulation. The waveforms should look similar to those in Figure S4-6.
i) Simulate the project. If there are any errors and you change your circuit, remember to save and compile the circuit before re-simulating.
j) Record the output waveform X on Figure S4-4 or print the Simulation Waveforms report and attach it to the data collection sheets at the end of this experiment.
k) Compare this waveform with the waveform for a 2-input OR gate. Are the two waveforms identical? _____
l) Record and analyze results:
   1) You should complete this part of the experiment by converting the waveform information in Figure S4-6 to Table S4-3. Compare this truth table with that of a 2-input OR gate. Are the two tables the same? _____
   2) Based on your conclusions above, write the Boolean expression for Z:

      Z = _____

   3) Transfer the results of the waveform, truth table, and Boolean expression to the data collection sheets at the end of this experiment.
m) *(Optional) Program the CPLD:* If your instructor wants you to program your CPLD, refer to Appendix D, Exercise 3. A complete set of instructions will be found there. It is suggested that you use **PIN_50** for "X," **PIN_51** for "Y," and **PIN_4** for "Z." To program your CPLD, refer to Appendix D, Exercise 3. A complete set of instructions will be found there. Briefly,
   1) *Recompile the project*: After making the pin assignments, compile the project again. When the compilation is completed, the software will display an information box showing any warnings and any detected errors. If there are errors, fix them and recompile the project.
   2) *Download the project to the board:*
      a. Set all eSOC DIP switches to OFF.
      b. Connect the board to the PC's parallel printer port and apply power.
      c. Select **Tools | Programmer**.
      d. Click on **Auto Detect** in the Chain Description File.
      e. When the eSOC board is recognized, press the **Add File...** button in the Chain Description File, then locate and Add the *boolean3.pof* file.
      f. Select the file and device line and scroll (horizontally) to the available programming option. Check the **Program/Configure** box.

g. Press the **Start** button. Observe the progress bar.
h. If the program downloads successfully, test the program. For each eSOC switch used by the program, turn its corresponding DIP switch on.

n) Save and close the project.

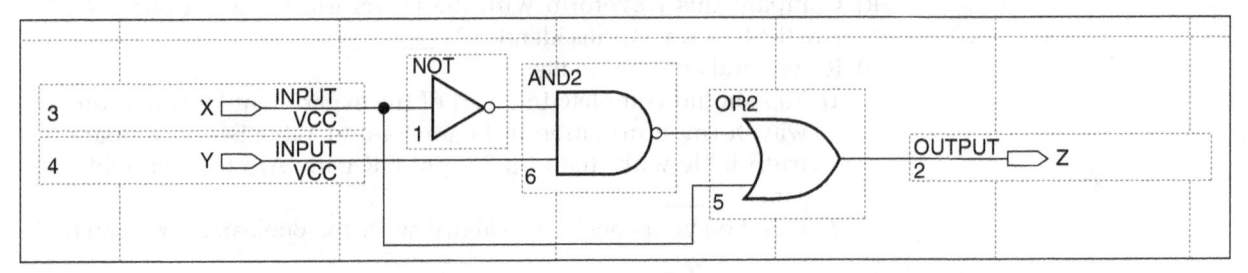

Figure S4-5

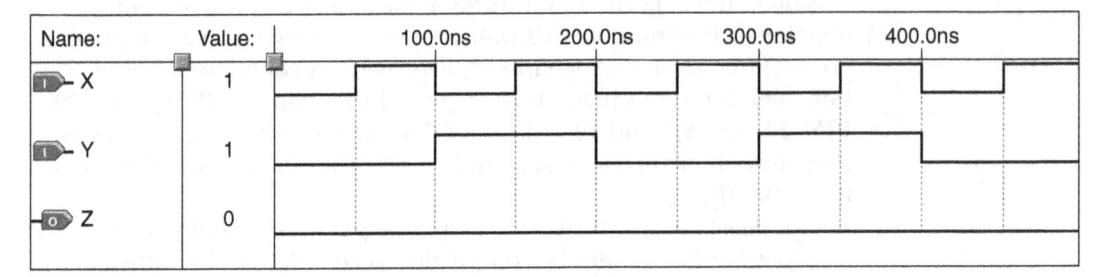

Figure S4-6

X	Y	Z
0	0	
0	1	
1	0	
1	1	

Table S4-3

DATA COLLECTION SHEET

Data Collection Sheets for Lab S4   Name _____
                                     Date _____

Data collection for Part 1:

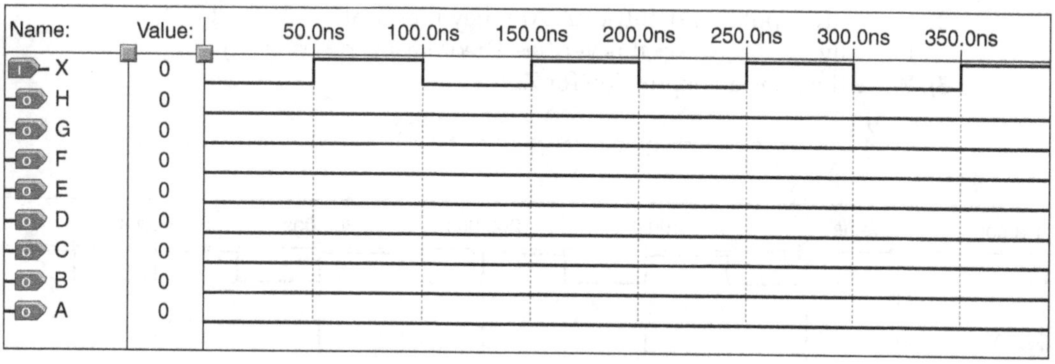

Name:	Value:	50.0ns	100.0ns	150.0ns	200.0ns	250.0ns	300.0ns	350.0ns
X	0							
H	0							
G	0							
F	0							
E	0							
D	0							
C	0							
B	0							
A	0							

Figure S4-2

Theorem	Output	Output verifies theorem?
1	0	yes
2	X	yes
3		
4		
5		
6	1	yes
7		
8		

Table S4-1

Demonstrated To
Instructor/FA _____ Date _____

Part 1 [ ]   Part 2 [ ]   Part 3 [ ]

Data Collection Sheets for Lab S4 Name _____
Date _____

Data collection for Part 2:

q)
  2) Compare input X and output Z. Are they identical logically? _____ (If so, this verifies Theorem 14: $x + xy = x$.)
  3) Write the Boolean expression for Z:

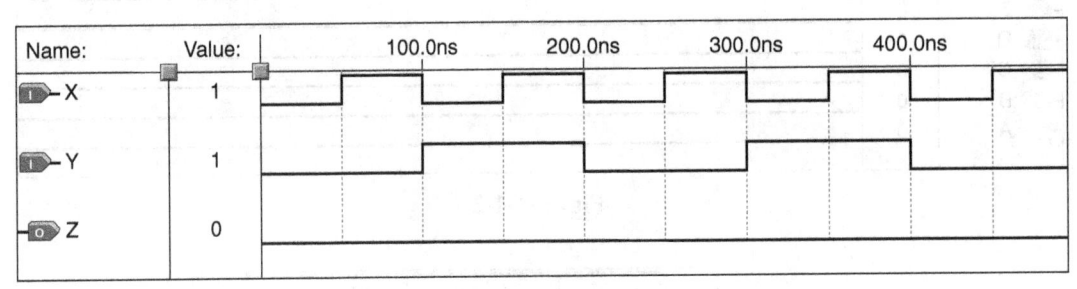

  Z = _____

Figure S4-4

X	Y	Z
0	0	
0	1	
1	0	
1	1	

Table S4-2

Demonstrated To
Instructor/FA _____ Date _____

Part 1 [ ]   Part 2 [ ]   Part 3 [ ]

PROCEDURE

321

Data Collection Sheets for Lab S4   Name _____
                                      Date _____

Data collection for Part 3:

1)

1) You should complete this part of the experiment by converting the waveform information in Figure S4-6 to Table S4-3. Compare this truth table with that of a 2-input OR gate. Are the two tables the same? _____

2) Based on your conclusions above, write the Boolean expression for Z:

   Z = _____

Name:	Value:	100.0ns	200.0ns	300.0ns	400.0ns
X	1				
Y	1				
Z	0				

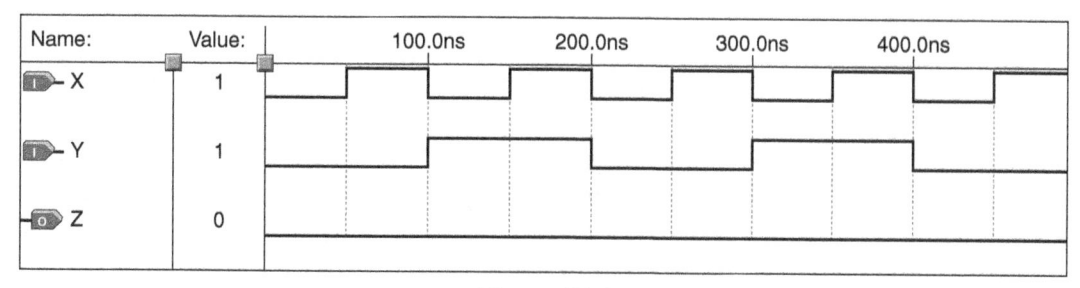

Figure S4-6

X	Y	Z
0	0	
0	1	
1	0	
1	1	

Table S4-3

Demonstrated To
Instructor/FA _____ Date _____

Part 1 [ ]   Part 2 [ ]   Part 3 [ ]

Experiment S5

Name _____

SIMPLIFICATION USING BOOLEAN THEOREMS

OBJECTIVES

1. To practice using QUARTUS® II software to create schematics and waveforms.
2. To use Boolean theorems to simplify circuits.
3. *Optional:* To program the EPM7128SLC84 CPLD.

TEXT REFERENCES

Read section 3.10. Review Appendix D of this manual.

EQUIPMENT NEEDED

Components

Blank floppy, zip disk, or memory stick for storing projects;
QUARTUS® II software (Altera Corporation);
Optional: DeVry University Board eSOC with EPM7128SLC84 CPLD (or Altera University Board with EPM7128SLC84 CPLD or any other equivalent board);
Desktop computer with minimum of INTEL Pentium PC @ 400MHz CPU running Microsoft Windows NT4SP1, 2000, or XP or better; Pentium III or 4 PC @ 400MHz running Red Hat Linux v7.3 or 8.0 or Red Hat Linux Enterprise 3; Sun Ultra running Solaris v8 or 9; HP9000 Series 700/800 running HP-UX v11.0 with ACE dated 11/1999 or later; or equivalent.

DISCUSSION

In Experiment S4, you were introduced to the Boolean theorems and their usefulness in explaining digital circuits and their simplification. In this experiment, you will continue your investigation of the application of Boolean theorems to simplification.

PROCEDURE

a) *Unsimplified version.* Construct the truth table for the circuit illustrated in Figure S5-1 using Table S5-1.

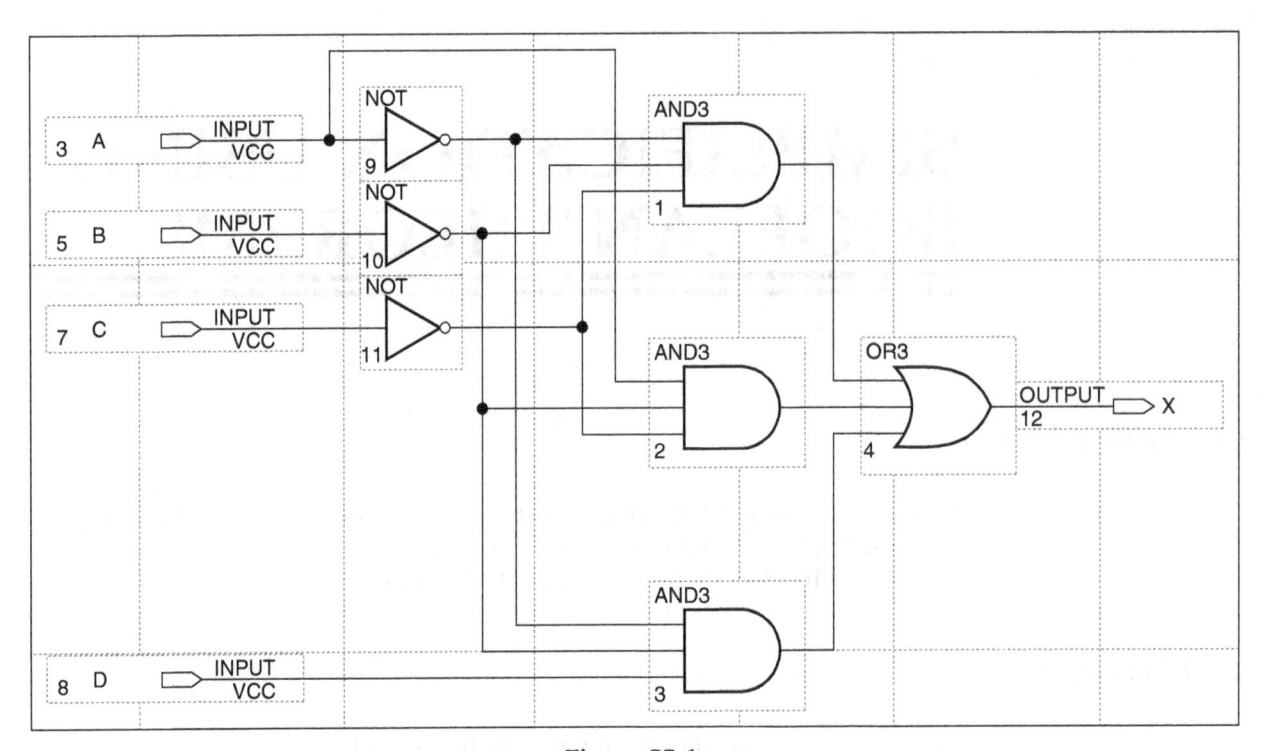

Figure S5-1

A	B	C	D	X
0	0	0	0	
0	0	0	1	
0	0	1	0	
0	0	1	1	
0	1	0	0	
0	1	0	1	
0	1	1	0	
0	1	1	1	
1	0	0	0	
1	0	0	1	
1	0	1	0	
1	0	1	1	
1	1	0	0	
1	1	0	1	
1	1	1	0	
1	1	1	1	

Table S5-1

PROCEDURE

b) Write the Boolean expression for X:

X = _____

c) Start the QUARTUS® II program. Set up a new project and use **simple1** for the name of the project.
d) Create a new graphic file *simple1.bdf* for the schematic of Figure S5-1 and place the symbols as shown.
e) Connect the symbols together according to the schematic.
f) Place the input and output connectors on the schematic. Label the inputs A, B, C, and D; label the output X.
g) Wire the input and output connectors to the logic symbols.
h) Select **Project | Top-Level Entity**.
i) Compile the project. If there are errors, fix them and save the file before recompiling the project. If there are no errors (or perhaps a warning concerning timing at most), then go on to the next step.
j) Create a new waveform file *simple1.vwf* and define the input for simulation as shown in Figure S5-2.
k) Select **Assignments | Settings** to access the Settings Window.
l) Click on *Simulator* at the left.
m) Choose Functional from the *Simulator Mode* pull-down menu.
n) Find and use for the *Simulator Input simple1.vwf* you just created above.
o) Select **Processing | Generate Functional Simulation Netlist**.

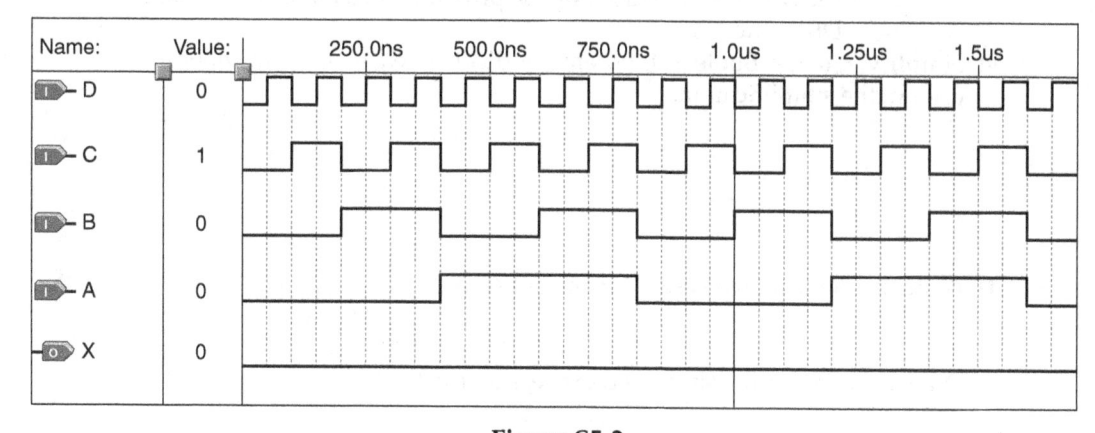

Figure S5-2

p) Simulate the project. If there are any errors and you change your circuit, remember to save and compile the circuit before re-simulating.
q) Record the output waveform X on Figure S5-2 or print the Simulation Waveforms report and attach it to the data collection sheets at the end of this experiment.
r) *(Optional) Program the CPLD:* It is suggested that you use **PIN_50** for "A," **PIN_51** for "B," **PIN_52** for "C," **PIN_54** for "D," and **PIN_4** for "X." To program your CPLD, refer to Appendix D, Exercise 3. A complete set of instructions will be found there. Briefly,

1) *Recompile the project*: After making the pin assignments, compile the project again. When the compilation is completed, the software will display an information box showing any warnings and any detected errors. If there are errors, fix them and recompile the project.
2) *Download the project to the board:*
   a. Set all eSOC DIP switches to OFF.
   b. Connect the board to the PC's parallel printer port and apply power.
   c. Select **Tools | Programmer**.
   d. Click on [Auto Detect] in the Chain Description File.
   e. When the eSOC board is recognized, press the [Add File...] button in the Chain Description File, then locate and Add the *simple1.pof* file.
   f. Select the file and device line and scroll (horizontally) to the available programming option. Check the **Program/Configure** box.
   g. Press the [Start] button. Observe the progress bar.
   h. If the program downloads successfully, test the program. For each eSOC switch used by the program, turn its corresponding DIP switch on.
s) Simplify X using Boolean theorems, indicating each Boolean theorem used in the simplification:

t) Write the Boolean expression for simplified X:

   X = _____

u) Draw a circuit for the simplified expression for X:

v) Save and close the project.
a) *Simplified version.* Create a new project using QUARTUS® II. Name the project **simple2**.
b) Enter the circuit for simplified X from step u using the Graphic Editor. Name the circuit *simple2.bdf*.
c) Select **Project | Top-Level Entity**.
d) Compile the project. If there are any errors, fix them and recompile the project.
e) Create a new waveform file *simple2.vwf* for the project using Figure S5-3.

PROCEDURE

f) Select **Assignments | Settings** to access the Settings Window.
g) Click on *Simulator* at the left.
h) Choose Functional from the *Simulator Mode* pull-down menu.
i) Find and use for the *Simulator Input* simple2.vwf you just created.
j) Select **Processing | Generate Functional Simulation Netlist**.
k) Simulate the project for *simple2*. If there are any errors and you change your circuit, remember to save and compile the circuit before re-simulating.
l) Use the output waveform generated by the simulator to complete Figure S5-3 (or print out the report from the Simulator Waveforms report) and Table S5-2.
m) Record and analyze results:
   1) Compare the output waveforms in Figures S5-2 and S5-3. Can you conclude that the original circuit and its simplified version are equivalent? _____
   2) Transfer the results of all the waveforms and truth tables to the data collection sheets at the end of this experiment.
n) *(Optional) Program the CPLD:* It is suggested that you use **PIN_50** for "A," **PIN_51** for "B," **PIN_52** for "C," **PIN_54** for "D," and **PIN_4** for "X." To program your CPLD, refer to Appendix D, Exercise 3. A complete set of instructions will be found there. Briefly,
   1) *Recompile the project*: After making the pin assignments, compile the project again. When the compilation is completed, the software will display an information box showing any warnings and any detected errors. If there are errors, fix them and recompile the project.
   2) *Download the project to the board:*
      a. Set all eSOC DIP switches to OFF.
      b. Connect the board to the PC's parallel printer port and apply power.
      c. Select **Tools | Programmer**.
      d. Click on [Auto Detect] in the Chain Description File.
      e. When the eSOC board is recognized, press the [Add File...] button in the Chain Description File, then locate and Add the *simple2.pof* file.
      f. Select the file and device line and scroll (horizontally) to the available programming option. Check the **Program/Configure** box.
      g. Press the [Start] button. Observe the progress bar.
      h. If the program downloads successfully, test the program. For each eSOC switch used by the program, turn its corresponding DIP switch on.
o) Save and close the project.

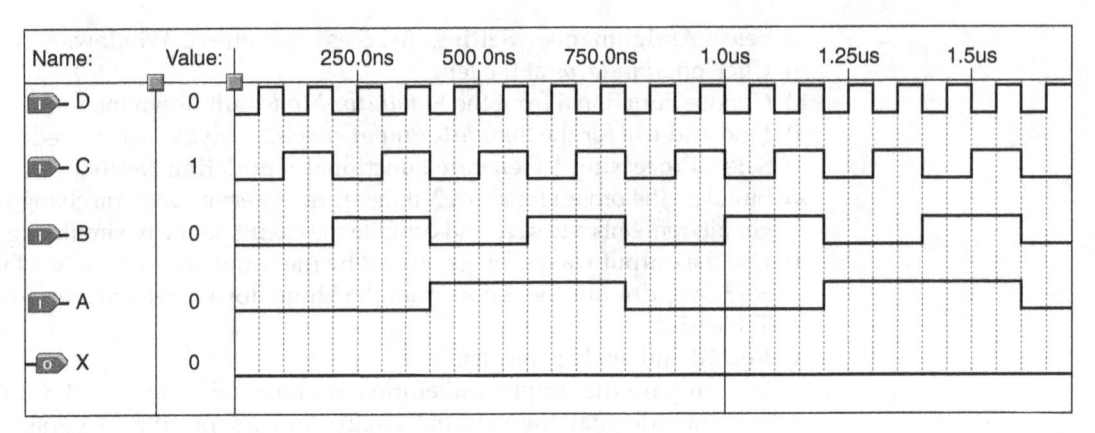

Name:	Value:	250.0ns	500.0ns	750.0ns	1.0us	1.25us	1.5us
D	0						
C	1						
B	0						
A	0						
X	0						

Figure S5-3

A	B	C	D	X
0	0	0	0	
0	0	0	1	
0	0	1	0	
0	0	1	1	
0	1	0	0	
0	1	0	1	
0	1	1	0	
0	1	1	1	
1	0	0	0	
1	0	0	1	
1	0	1	0	
1	0	1	1	
1	1	0	0	
1	1	0	1	
1	1	1	0	
1	1	1	1	

Table S5-2

DATA COLLECTION SHEET

Data Collection Sheets for Lab S5   Name _____

Date _____

Data collection for Part 1:

A	B	C	D	X
0	0	0	0	
0	0	0	1	
0	0	1	0	
0	0	1	1	
0	1	0	0	
0	1	0	1	
0	1	1	0	
0	1	1	1	
1	0	0	0	
1	0	0	1	
1	0	1	0	
1	0	1	1	
1	1	0	0	
1	1	0	1	
1	1	1	0	
1	1	1	1	

Table S5-1

b) Write the Boolean expression for X:

X = _____

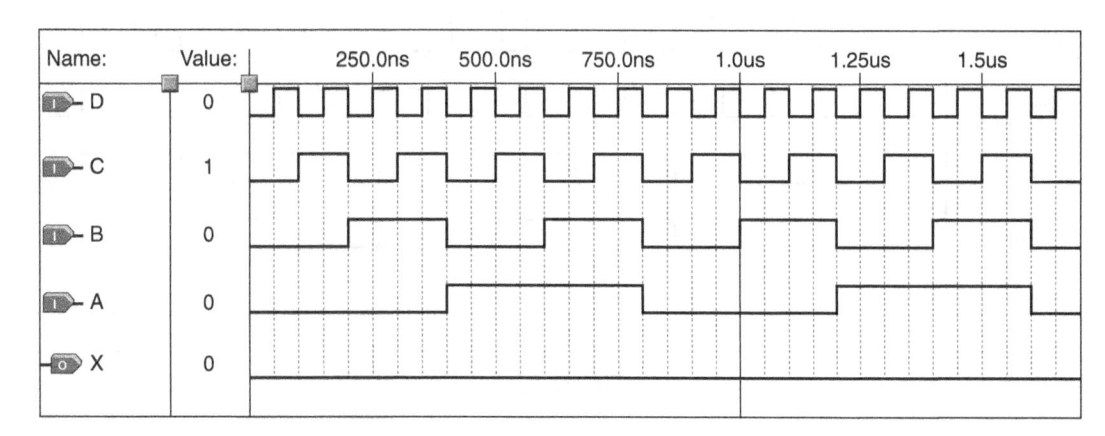

Figure S5-2

Demonstrated To
Instructor/FA _____ Date _____

Data Collection Sheets for Lab S5   Name _____

 Date _____

s) Simplify X using Boolean theorems, indicating each Boolean theorem used in the simplification:

t) Write the Boolean expression for simplified X:

   X = _____

u) Draw a circuit for the simplified expression for X:

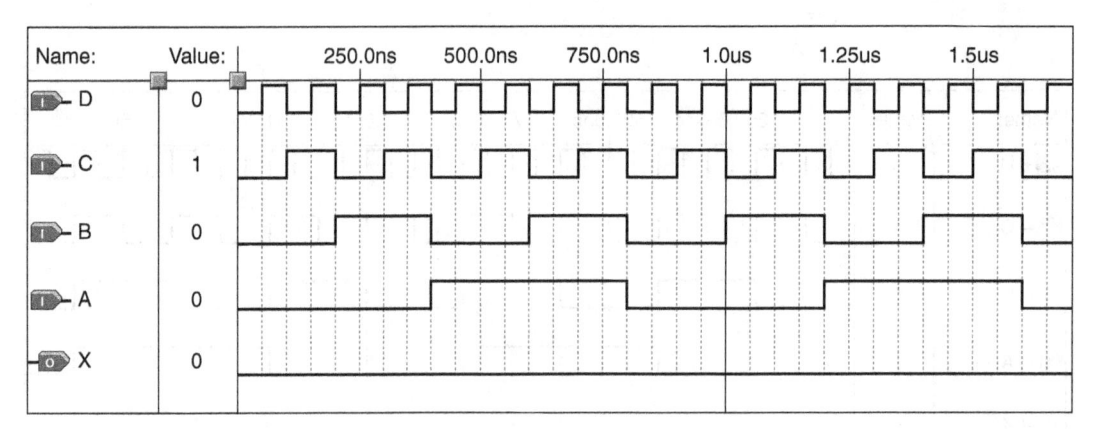

Figure S5-3

Demonstrated To
Instructor/FA _____ Date _____

Data Collection Sheets for Lab S5 Name _____

Date _____

m)

1) Compare the output waveforms in Figures S5-2 and S5-3. Can you conclude that the original circuit and its simplified version are equivalent?

A	B	C	D	X
0	0	0	0	
0	0	0	1	
0	0	1	0	
0	0	1	1	
0	1	0	0	
0	1	0	1	
0	1	1	0	
0	1	1	1	
1	0	0	0	
1	0	0	1	
1	0	1	0	
1	0	1	1	
1	1	0	0	
1	1	0	1	
1	1	1	0	
1	1	1	1	

Table S5-2

Demonstrated To
Instructor/FA _____ Date _____

Experiment S6

Name _____

DEMORGAN'S THEOREMS

OBJECTIVES

1. To practice using QUARTUS® II software to create schematics and waveforms.
2. To practice using QUARTUS® II software to compile and simulate digital circuits.
3. To verify experimentally DeMorgan's two theorems.
4. To investigate the use of DeMorgan's theorems in circuit simplification.
5. To demonstrate the extension of DeMorgan's theorems to three variables.
6. *Optional:* To program the EPM7128SLC84 CPLD.

TEXT REFERENCES

Read section 3.11. Review Appendix D of this manual.

EQUIPMENT NEEDED

Components

Blank floppy, zip disk, or memory stick for storing projects;

QUARTUS® II software (Altera Corporation);

Optional: DeVry University Board eSOC with EPM7128SLC84 CPLD (or Altera University Board with EPM7128SLC84 CPLD or any other equivalent board);

Desktop computer with minimum of INTEL Pentium PC @ 400MHz CPU running Microsoft Windows NT4SP1, 2000, or XP or better; Pentium III or 4 PC @ 400MHz running Red Hat Linux v7.3 or 8.0 or Red Hat Linux Enterprise 3; Sun Ultra running Solaris v8 or 9; HP9000 Series 700/800 running HP-UX v11.0 with ACE dated 11/1999 or later; or equivalent.

DISCUSSION

In Experiments S4 and S5, you investigated several rules of Boolean algebra. Now, in the current experiment, you are introduced to two more rules of Boolean algebra, known collectively as DeMorgan's theorems. The two theorems are:

1) $\overline{(x + y)} = \overline{x} \cdot \overline{y}$
2) $\overline{(x \cdot y)} = \overline{x} + \overline{y}$

You should note that each theorem permits you to simplify expressions involving inverted sums or products.

Example: Simplify $\overline{X + YZ}$.

Solution:

1) Using the first DeMorgan theorem, we can write $\overline{X + YZ} = \overline{X} \cdot \overline{YZ}$.
2) Using the second DeMorgan theorem, we have $\overline{X} \cdot \overline{YZ} = \overline{X} \cdot (\overline{Y} + \overline{Z})$.
3) Using the distributive rule we can write the right-hand side of the equation as $\overline{X} \cdot \overline{Y} + \overline{X} \cdot \overline{Z}$. Here the inverter signs invert only single variables.

PROCEDURE

PART 1 – Verifying DeMorgan's Theorems

a) Start the QUARTUS® II program. Set up a new project and use **demorgan1** for the name of the project. The diagram of the circuit you are to construct and its input waveforms are given in Figures S6-1 and S6-2. Note that both DeMorgan's theorems will be verified in this part.
b) Create a new graphic file for the schematic, *demorgan1.bdf*.
c) Place the gate symbols on the schematic according to Figure S6-1.
d) Place the input and output connectors on the schematic. Label the inputs A and B; label the outputs W, X, Y, and Z. The outputs W and X will be used to verify one of the DeMorgan's theorems, _____, while the outputs Y and Z will be used to verify the other of the DeMorgan's theorems, _____.
e) Wire the input and output connectors to the circuit as shown in Figure S6-1.
f) Select **Project | Top-Level Entity**.
g) Compile the project. If there are errors, fix them and save the file before recompiling the project. If there are no errors (or perhaps a warning concerning timing at most), then go on to the next step.
h) Create a new waveform *demorgan1.vwf* and define the inputs for the simulation file using Figure S6-2.
i) Select **Assignments | Settings** to access the Settings Window shown in Figure D-28 (Appendix D).
j) Click on *Simulator* at the left.

PROCEDURE

k) Choose Functional from the *Simulator Mode* pull-down menu.
l) Find and use for the *Simulator Input demorgan1.vwf* you just created.
m) Select **Processing | Generate Functional Simulation Netlist.**
n) Simulate the project. If there are any errors and you change your circuit, remember to save and compile the circuit before re-simulating.
o) Record the output waveforms X, Y, and Z on Figure S6-2 or print the Simulator Waveforms report from QUARTUS® II and attach it to the data collection sheets at the end of this experiment.
p) You should complete this part of the experiment by converting the waveform information in Figure S6-2 to Table S6-1 and Table S6-2.
q) Record and analyze results:
   1) Recall that two circuits are logically the same if they have the same truth table. According to your results in Truth Table S6-1, are W and X equivalent? _____ If so, give the DeMorgan theorem that explains this result: _____
   2) According to your results in Truth Table S6-2, are Y and Z equivalent? _____ If so, give the DeMorgan theorem that explains this result: _____
r) *(Optional) Program the CPLD:* It is suggested that you use **PIN_50** for "A," **PIN_51** for "B," **PIN_4** for "W," **PIN_5** for "X," **PIN_8** for "Y," and **PIN_4** for "Z." To program your CPLD, refer to Appendix D, Exercise 3. A complete set of instructions will be found there. Briefly,
   1) *Recompile the project*: After making the pin assignments, compile the project again. When the compilation is completed, the software will display an information box showing any warnings and any detected errors. If there are errors, fix them and recompile the project.
   2) *Download the project to the board:*
      a. Set all eSOC DIP switches to OFF.
      b. Connect the board to the PC's parallel printer port and apply power.
      c. Select **Tools | Programmer**.
      d. Click on [Auto Detect] in the Chain Description File.
      e. When the eSOC board is recognized, press the [Add File...] button in the Chain Description File, then locate and Add the *demorgan1.pof* file.
      f. Select the file and device line and scroll (horizontally) to the available programming option. Check the **Program/Configure** box.
      g. Press the [Start] button. Observe the progress bar.

h. If the program downloads successfully, test the program. For each eSOC switch used by the program, turn its corresponding DIP switch on.

s) Save and close the project.

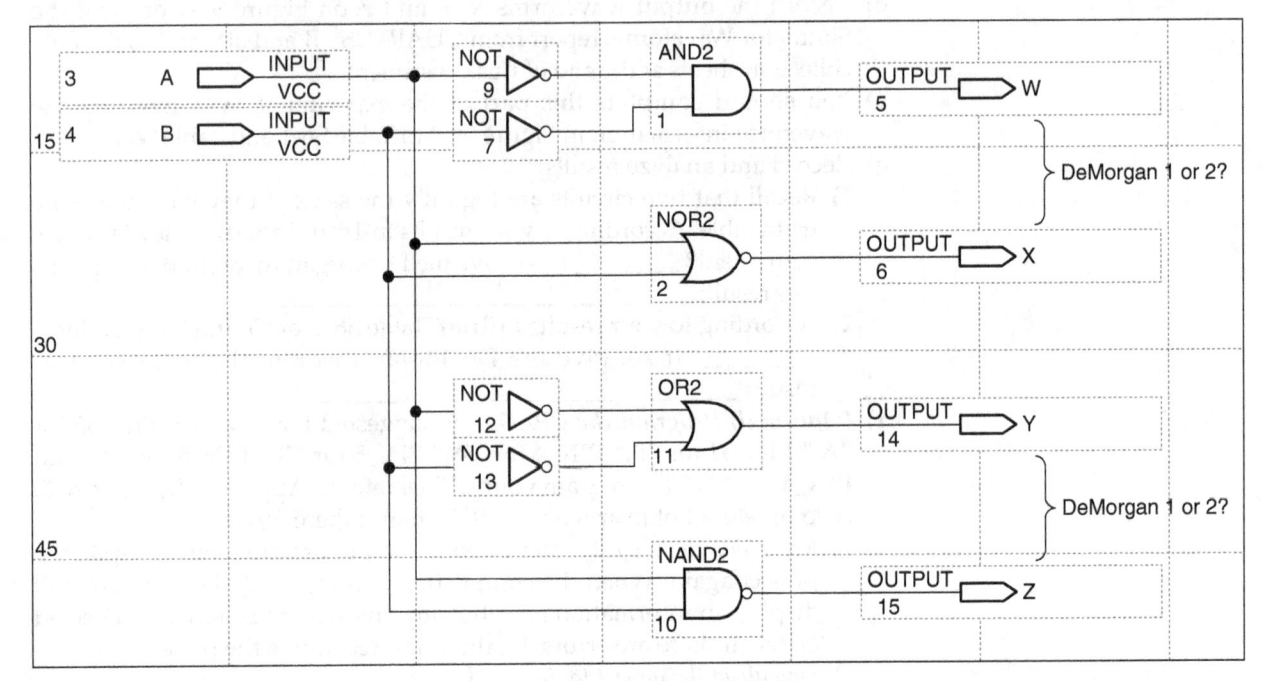

Figure S6-1

Figure S6-2

A	B	W	X
0	0		
0	1		
1	0		
1	1		

Table S6-1

PROCEDURE

A	B	Y	Z
0	0		
0	1		
1	0		
1	1		

Table S6-2

PART 2 – Simplification using DeMorgan's Theorems

a) Construct the truth table for the circuit illustrated in Figure S6-3 using Table S6-3. Use the X column for your output.

b) Write the Boolean expression for X:

$$X = \underline{\hspace{4cm}}$$

c) Start the QUARTUS® II program. Set up a new project and use **demorgan2** for the name of the project. The diagram of the circuit you are to construct and its input waveforms are given in Figures S6-3 and S6-4.

d) Create a new graphic file *demorgan2.bdf* for the schematic of Figure S6-3 and place the symbols as shown.

e) Connect the symbols together according to the schematic.

f) Place the input and output connectors on the schematic. Label the inputs A, B, C, and D; label the output X.

g) Wire the input and output connectors to the logic symbols.

h) Select **Project | Top-Level Entity**.

i) Compile the project. If there are errors, fix them and save the file before recompiling the project. You should have gotten three warnings:

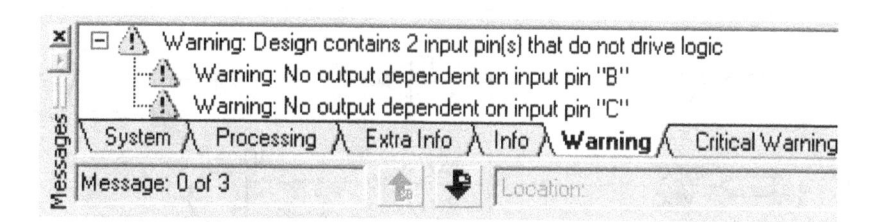

```
Warning: Design contains 2 input pin(s) that do not drive logic
  Warning: No output dependent on input pin "B"
  Warning: No output dependent on input pin "C"
\ System \ Processing \ Extra Info \ Info \ Warning \ Critical Warning
Message: 0 of 3                                    Location:
```

What do the warnings mean? When the compiler discovers inputs that are irrelevant to the operation of the circuit, they are eliminated as well as any connection to them. In our experiment, there will be only one variable, "A," remaining after compilation is completed. In other words, simplification takes place inside the compiler.

j) Create a new waveform file *demorgan2.vwf* and define the input for simulation as shown in Figure S6-4.

k) Select **Assignments | Settings** to access the Settings Window shown in Figure D-28 (Appendix D).

l) Click on *Simulator* at the left.

m) Choose Functional from the *Simulator Mode* pull-down menu.

n) Find and use for the **Simulator Input** *demorgan2.vwf* you just created above.

o) Select **Processing | Generate Functional Simulation Netlist**.

p) Simulate the project. If there are any errors and you change your circuit, remember to save and compile the circuit before re-simulating.
q) Use the output waveform generated by the simulator to verify your completed Table S6-3. Since "A" is the only input variable remaining, the values for "B" and "C" on the input side of the truth table do not matter. It is only the values in the A column that matter.
r) Note: You could have created a waveform that included inputs "B" and "C."
s) Simplify X using DeMorgan's theorems, indicating each DeMorgan theorem used in the simplification:

t) Write the Boolean expression for simplified X:

$X =$ _____

u) Does the expression for simplified X agree with the results you obtained for the original circuit? _____
v) Transfer the results of the waveform, truth table, and Boolean expression to the data collection sheets at the end of this experiment.
w) Save and close the project.

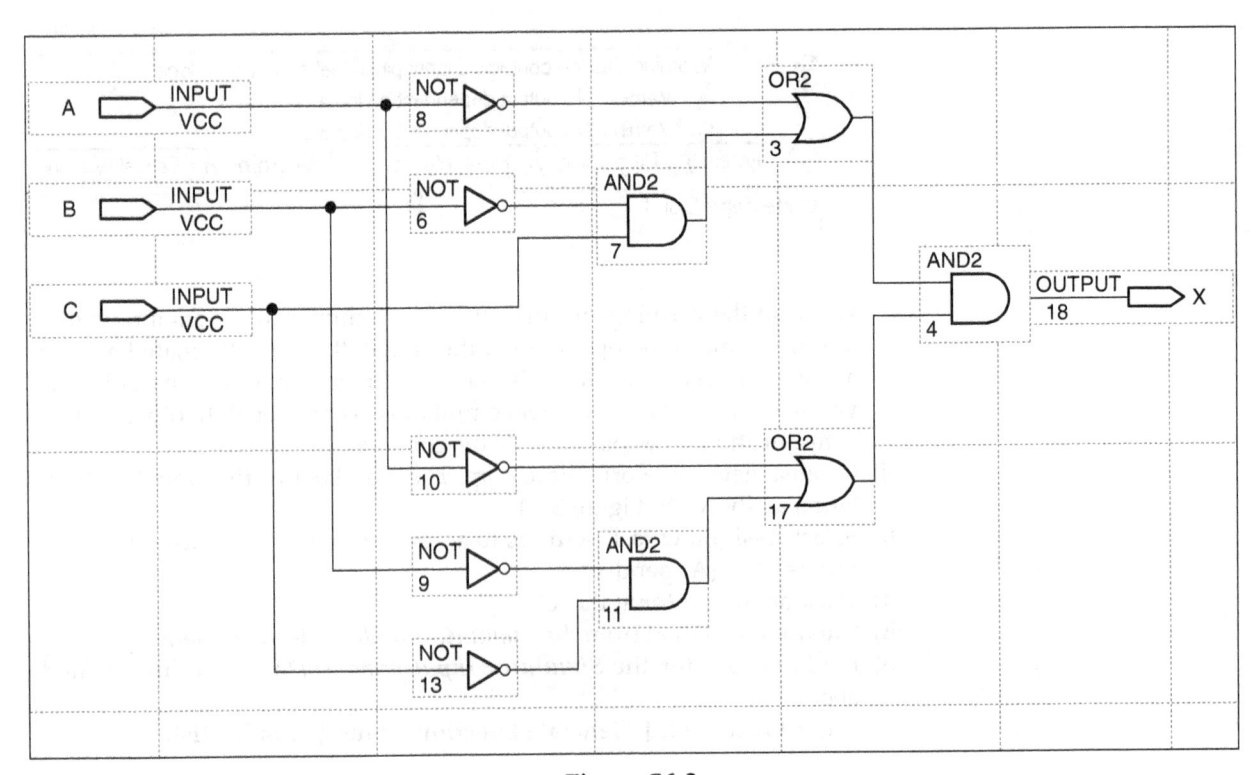

Figure S6-3

PROCEDURE

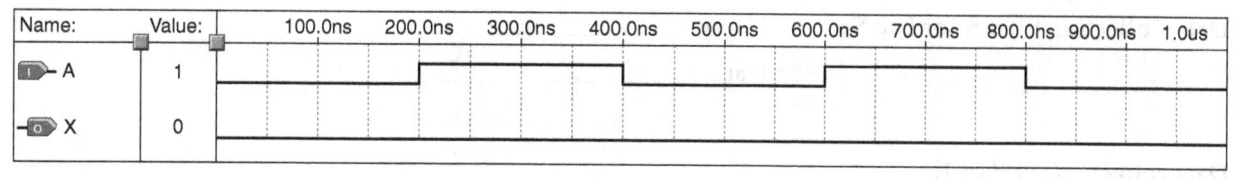

Figure S6-4

A	B	C	X	Simplified X
0	0	0		
0	0	1		
0	1	0		
0	1	1		
1	0	0		
1	0	1		
1	1	0		
1	1	1		

Table S6-3

Data Collection Sheets for Lab S6   Name _____

                                            Date _____

Data collection for Part 1:
  q)
   1) Recall that two circuits are logically the same if they have the same truth table. According to your results in Truth Table S6-1, are W and X equivalent? _____ If so, give the DeMorgan theorem that explains this result: _____
   2) According to your results in Truth Table S6-2, are Y and Z equivalent? _____ If so, give the DeMorgan theorem that explains this result: _____

Name:	Value:	50.0ns	100.0ns	150.0ns	200.0ns	250
B	1					
A	0					
W	0					
X	0					
Y	0					
Z	0					

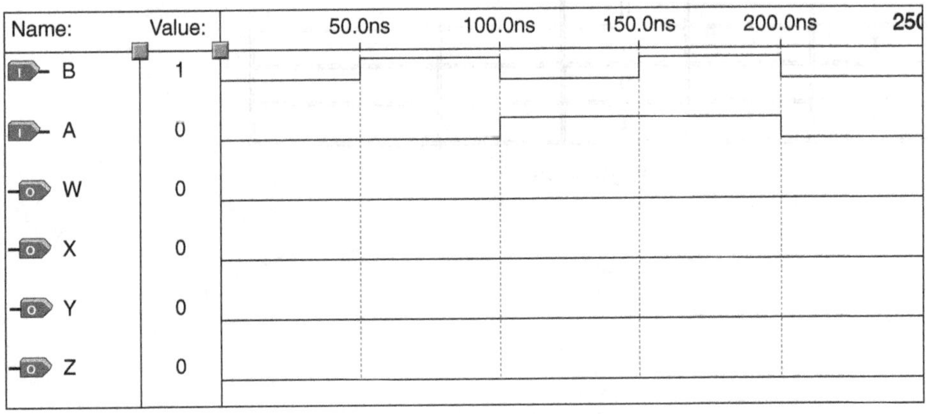

Figure S6-2

A	B	W	X
0	0		
0	1		
1	0		
1	1		

Table S6-1

Demonstrated To
Instructor/FA _____ Date _____

Part 1 [ ]   Part 2 [ ]

DATA COLLECTION SHEET 341

Data Collection Sheets for Lab S6   Name _____

Date _____

A	B	Y	Z
0	0		
0	1		
1	0		
1	1		

Table S6-2

Data collection for Part 2:

    b) Write the Boolean expression for X:

        X = _____

    t) Write the Boolean expression for simplified X:

        X = _____

    u) Does the expression for simplified X agree with the results you obtained for the original circuit? _____

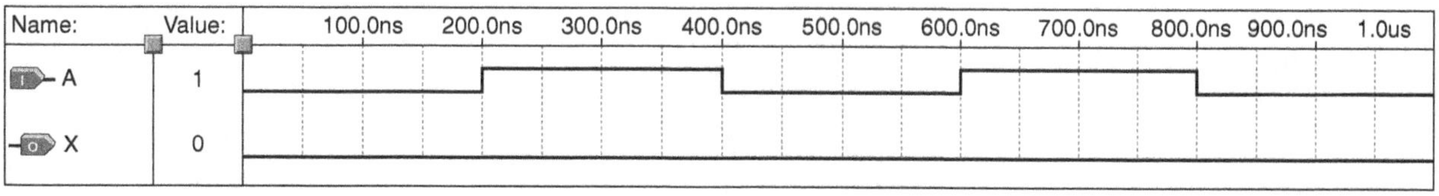

Name:	Value:	100.0ns	200.0ns	300.0ns	400.0ns	500.0ns	600.0ns	700.0ns	800.0ns	900.0ns	1.0us
A	1										
X	0										

Figure S6-4

Demonstrated To
Instructor/FA _____ Date _____

Part 1 [ ]   Part 2 [ ]

Data Collection Sheets for Lab S6 Name _____

Date _____

A	B	C	X	Simplified X
0	0	0		
0	0	1		
0	1	0		
0	1	1		
1	0	0		
1	0	1		
1	1	0		
1	1	1		

Table S6-3

Demonstrated To
Instructor/FA _____ Date _____

Part 1 [ ]   Part 2 [ ]

Experiment S7

Name _____

IMPLEMENTING LOGIC GATES AND CIRCUITS USING VHDL

OBJECTIVES

1. To learn how to use QUARTUS® II software to create logic circuits with VHDL.
2. To practice using QUARTUS® II software to compile and simulate digital circuits.
3. To create AND, OR, and NOT gates with VHDL.
4. To create a simple combinatorial circuit using VHDL.
5. *Optional:* Program the EPM7128SLC84 CPLD.

TEXT REFERENCES

Read sections 3.18 through 3.20. Review Appendix D of this manual.

EQUIPMENT NEEDED

Components

Blank floppy, zip disk, or memory stick for storing projects;
QUARTUS® II software (Altera Corporation);
Optional: DeVry University Board eSOC with EPM7128SLC84 CPLD (or Altera University Board with EPM7128SLC84 CPLD or any other equivalent board);
Desktop computer with minimum of INTEL Pentium PC @ 400MHz CPU running Microsoft Windows NT4SP1, 2000, or XP or better; Pentium III or 4 PC @ 400MHz running Red Hat Linux v7.3 or 8.0 or Red Hat Linux Enterprise 3; Sun Ultra running Solaris v8 or 9; HP9000 Series 700/800 running HP-UX v11.0 with ACE dated 11/1999 or later; or equivalent.

DISCUSSION

Up to this point, you have learned how to implement logic gates and circuits using the QUARTUS® II Graphic Editor. You made use of QUARTUS® II's extensive library of pre-defined symbols. Symbols for these gates and circuits can also be created using Hardware Definition Language (HDL). In particular you will use VHDL (Very High Speed Integrated Circuit HDL) to create a few gates and simple circuits with the QUARTUS® II Text Editor and compiler.

PROCEDURE

PART 1 – OR Gate using VHDL

a) The OR Gate using VHDL is covered in the QUARTUS® II tutorial (Appendix D) as Exercise 2 on page 607. Do this exercise to familiarize yourself with QUARTUS® II and the OR gate function. Also do Exercise 3 in Appendix D using the *example3.gdf* design file from Exercise 2.

b) When you have completed the exercises in Appendix D, close the current project and go on to Part 2.

PART 2 – AND Gate using VHDL

a) Start the QUARTUS® II program. Set up a new project and use **and_gate** for the name of the project.

b) Create a new text file for writing the *and_gate* VHDL description. Save this file with a vhd extension.

c) Type in the VHDL program from the listing in Figure S7-1.

```
Library IEEE;
use IEEE.std_logic_1164.all;

ENTITY   and_gate IS
    PORT (     a, b      : IN BIT;
               y         : OUT BIT);
END and_gate;

ARCHITECTURE ckt OF and_gate IS
    BEGIN
        y <= a AND b;
    END ckt;
```

Figure S7-1

d) Select **Project | Top-Level Entity**.

e) This project should be already set to use the EPM7128SLC84-15 CPLD. If necessary you can change the CPLD with the **Assignments | Device** window.

PROCEDURE

f) If you are using another board, check Appendix D for your device family, device, and speed.
g) Select **Processing | Start | Start Analysis & Synthesis** to synthesize a circuit that implements the given VHDL code. If there are errors, fix them and recompile the project.
h) Create a new Waveform Editor file with the name *and_gate.vwf*. Use the same waveforms created in Exercise 2 of Appendix D.
i) On the Settings page, choose Functional Simulation mode and select *and_gate.vwf* for the Simulation Input.
j) Select **Processing | Generate Functional Simulation Netlist**.
k) Start the simulation.
l) If there are no errors when the simulation is finished, go on to the next step. If there are errors and you change your circuit, remember to save and compile the circuit before re-simulating.
m) *Program the CPLD:* It is suggested that you use **PIN_50** for "a," **PIN_51** for "b," and **PIN_4** for "y." To program your CPLD, refer to Appendix D, Exercise 3. A complete set of instructions will be found there. Briefly,
   1) *Recompile the project*: After making the pin assignments, compile the project again. When the compilation is completed, the software will display an information box showing any warnings and any detected errors. If there are errors, fix them and recompile the project.
   2) *Download the project to the board:*
      a. Set all eSOC DIP switches to OFF.
      b. Connect the board to the PC's parallel printer port and apply power.
      c. Select **Tools | Programmer**.
      d. Click on [Auto Detect] in the Chain Description File.
      e. When the eSOC board is recognized, press the [Add File...] button in the Chain Description File, then locate and Add the *and_gate.pof* file.
      f. Select the file and device line and scroll (horizontally) to the available programming option. Check the **Program/Configure** box.
      g. Press the [Start] button. Observe the progress bar.
      h. If the program downloads successfully, test the program. For each eSOC switch used by the program, turn its corresponding DIP switch on.
n) *Test the program:* Toggle switches A and B and fill out Table S7-1 with your observations. Transfer this information to the data collection sheets at the end of this experiment.
o) Save and close the project.

a	b	y
0	0	
0	1	
1	0	
1	1	

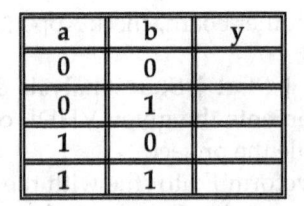

Table S7-1

PART 3 – NOT Gate using VHDL

a) Start the QUARTUS® II program. Set up a new project and use **not_gate** for the name of the project.
b) Create a new text file for writing the *not_gate* VHDL description. Save this file with a vhd extension.
c) Type in the VHDL program from the listing in Figure S7-2.

```
Library IEEE;
use IEEE.std_logic_1164.all;

ENTITY    not_gate IS
    PORT (    a    :IN BIT;
              y            : OUT BIT);
END   not_gate;

ARCHITECTURE ckt OF not_gate IS
    BEGIN
        y <= NOT a;
    END ckt;
```

Figure S7-2

d) Select **Project | Top-Level Entity**.
e) This project should be already set to use the EPM7128SLC84-15 CPLD. If necessary you can change the CPLD with the **Assignments | Device** window.
f) If you are using another board, check Appendix D for your device family, device, and speed.
g) Select **Processing | Start | Start Analysis & Synthesis** to synthesize a circuit that implements the given VHDL code. If there are errors, fix them and recompile the project.
h) Create a new Waveform Editor file with the name *not_gate.vwf*.
i) On the Settings page, choose Functional Simulation mode and select *not_gate.vwf* for the Simulation Input.
j) Select **Processing | Generate Functional Simulation Netlist**.
k) Start the simulation.
l) If there are no errors when the simulation is finished, go on to the next step. If there are errors and you change your circuit, remember to save and compile the circuit before re-simulating.

PROCEDURE

m) *Program the CPLD:* It is suggested that you use **PIN_50** for **"a,"** and **PIN_4** for **"y."** To program your CPLD, refer to Appendix D, Exercise 3. A complete set of instructions will be found there. Briefly,
   1) *Recompile the project*: After making the pin assignments, compile the project again. When the compilation is completed, the software will display an information box showing any warnings and any detected errors. If there are errors, fix them and recompile the project.
   2) *Download the project to the board:*
      a. Set all eSOC DIP switches to OFF.
      b. Connect the board to the PC's parallel printer port and apply power.
      c. Select **Tools | Programmer**.
      d. Click on [Auto Detect] in the Chain Description File.
      e. When the eSOC board is recognized, press the [Add File...] button in the Chain Description File, then locate and Add the *not_gate.pof* file.
      f. Select the file and device line and scroll (horizontally) to the available programming option. Check the **Program/Configure** box.
      g. Press the [Start] button. Observe the progress bar.
      h. If the program downloads successfully, test the program. For each eSOC switch used by the program, turn its corresponding DIP switch on.
n) *Test the program:* Toggle switch A and fill out Table S7-2 with your observations. Transfer this information to the data collection sheets at the end of this experiment.
o) Save and close the project.

a	y
0	
1	

Table S7-2

PART 4 – Simple Combinatorial Circuit using VHDL

a) Start the QUARTUS® II program. Set up a new project and use **comb1_ckt** for the name of the project.
b) Create a new text file for writing the *comb1_ckt* VHDL description. Save this file with a vhd extension.
c) Type in the VHDL program from the listing in Figure S7-3. The listing describes the circuit in Figure S7-4.

```
Library IEEE;
use IEEE.std_logic_1164.all;

ENTITY   comb1_ckt IS
    PORT (      a,b,c           : IN BIT;        -- inputs to block
                y               : OUT BIT);      -- output of block
END   comb1_ckt;

ARCHITECTURE ckt OF comb1_ckt IS
    SIGNAL m            :BIT;                    -- intermediate signal
    BEGIN
        m <= a AND b;
        y <= m or c;
    END ckt;
```

Figure S7-3

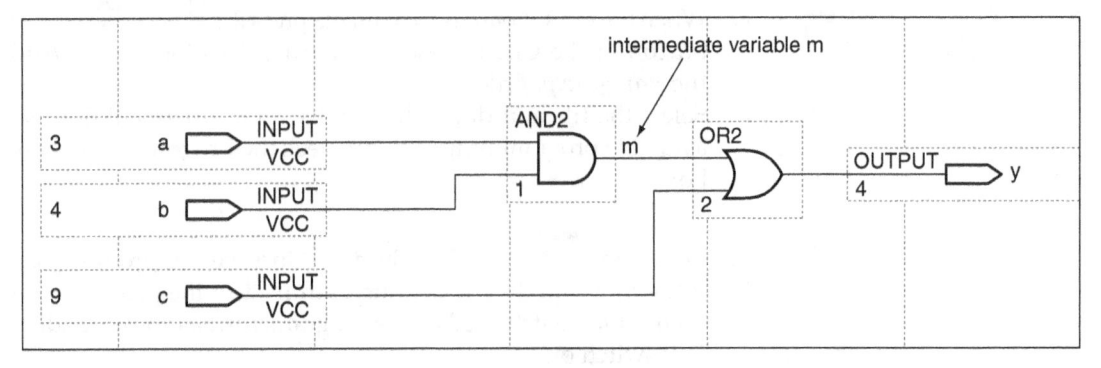

Figure S7-4

d) Select **Project | Top-Level Entity**.
e) This project should be already set to use the EPM7128SLC84-15 CPLD. If necessary you can change the CPLD with the **Assignments | Device** window.
f) If you are using another board, check Appendix D for your device family, device, and speed.
g) Select **Processing | Start | Start Analysis & Synthesis** to synthesize a circuit that implements the given VHDL code. If there are errors, fix them and recompile the project.
h) Create a new Waveform Editor file with the name *comb1_ckt.vwf*.
i) On the Settings page, choose Functional Simulation mode and select *comb1_ckt.vwf* for the Simulation Input.
j) Select **Processing | Generate Functional Simulation Netlist**.
k) Start the simulation.
l) If there are no errors when the simulation is finished, go on to the next step. If there are errors and you change your circuit, remember to save and compile the circuit before re-simulating.

PROCEDURE

m) *Program the CPLD:* It is suggested that you use **PIN_50** for "a," **PIN_51** for "b," **PIN_52** for "c," and **PIN_4** for "y." To program your CPLD, refer to Appendix D, Exercise 3. A complete set of instructions will be found there. Briefly,
   1) *Recompile the project:* After making the pin assignments, compile the project again. When the compilation is completed, the software will display an information box showing any warnings and any detected errors. If there are errors, fix them and recompile the project.
   2) *Download the project to the board:*
      a. Set all eSOC DIP switches to OFF.
      b. Connect the board to the PC's parallel printer port and apply power.
      c. Select **Tools | Programmer**.
      d. Click on [Auto Detect] in the Chain Description File.
      e. When the eSOC board is recognized, press the [Add File...] button in the Chain Description File, then locate and Add the *implement1.pof* file.
      f. Select the file and device line and scroll (horizontally) to the available programming option. Check the **Program/Configure** box.
      g. Press the [Start] button. Observe the progress bar.
      h. If the program downloads successfully, test the program. For each eSOC switch used by the program, turn its corresponding DIP switch on.
n) *Test the program:* Toggle switches a, b, and c and fill out Table S7-3 with your observations. Transfer this information to the data collection sheets at the end of this experiment.
o) Save and close the project.

a	b	c	y
0	0	0	
0	0	1	
0	1	0	
0	1	1	
1	0	0	
1	0	1	
1	1	0	
1	1	1	

Table S7-3

Data Collection Sheets for Lab S7   Name _____

Date _____

Data collection for Part 2:

a	b	y
0	0	
0	1	
1	0	
1	1	

Table S7-1

Data collection for Part 3:

a	y
0	
1	

Table S7-2

Data collection for Part 4:

a	b	c	y
0	0	0	
0	0	1	
0	1	0	
0	1	1	
1	0	0	
1	0	1	
1	1	0	
1	1	1	

Table S7-3

Demonstrated To
Instructor/FA _____ Date _____

Part 1 [ ]   Part 2 [ ]   Part 3 [ ]   Part 4 [ ]

Experiment S8

Name _____

IMPLEMENTING LOGIC CIRCUIT DESIGNS

OBJECTIVES

1. To practice using QUARTUS® II software to create schematics and waveforms.
2. To practice using QUARTUS® II software to compile and simulate digital circuits.
3. To design and implement logic circuits given either a truth table or a set of statements that describes the circuit's behavior.
4. *Optional:* To program the EPM7128SLC84 CPLD.

TEXT REFERENCES

Read sections 4.1 through 4.4. Review Appendix D of this manual.

EQUIPMENT NEEDED

Components

Blank floppy, zip disk, or memory stick for storing projects;
QUARTUS® II software (Altera Corporation);
Optional: DeVry University Board eSOC with EPM7128SLC84 CPLD (or Altera University Board with EPM7128SLC84 CPLD or any other equivalent board);
Desktop computer with minimum of INTEL Pentium PC @ 400MHz CPU running Microsoft Windows NT4SP1, 2000, or XP or better; Pentium III or 4 PC @ 400MHz running Red Hat Linux v7.3 or 8.0 or Red Hat Linux Enterprise 3; Sun Ultra running Solaris v8 or 9; HP9000 Series 700/800 running HP-UX v11.0 with ACE dated 11/1999 or later; or equivalent.

DISCUSSION

In this experiment, you are asked to design two combinatorial circuits. A combinatorial circuit is, as the name suggests, a circuit composed of a combination of logic gates. It acts on its inputs and gives an output based on its logic function. A combinatorial circuit can be designed directly from a truth table or from a description of its logic function. In either case, you are called upon to derive the simplest circuit possible. You should ask your instructor which circuits you will be required to build and demonstrate.

PROCEDURE

PART 1 – Truth Table to Implementation of a Combinatorial Circuit

a) Examine the truth table in Table S8-1. Draw a logic circuit in the space below. Do not be concerned with using a minimum number of gates.

A	B	C	X	X after simulation
0	0	0	1	
0	0	1	0	
0	1	0	1	
0	1	1	1	
1	0	0	1	
1	0	1	0	
1	1	0	0	
1	1	1	1	

Table S8-1

b) Write the Boolean expression for X:

X = _____

c) Start the QUARTUS® II program. Set up a new project and use **implement1** for the name of the project.
d) Create a new graphic file *implement1.bdf* for the schematic you drew above and place the symbols as shown.
e) Connect the symbols together according to the schematic you drew.
f) Place the input and output connectors on the schematic. Label the inputs A, B, and C; label the output X.
g) Wire the input and output connectors to the logic symbols.
h) Select **Project | Top-Level Entity**.
i) Compile the project. If there are errors, fix them and save the file before recompiling the project. If there are no errors (or perhaps a warning concerning timing at most), then go on to the next step.
j) Create a new waveform file *implement1.vwf* and define the inputs for simulation. The input waveforms should look similar to those in Figure S8-1.

PROCEDURE

k) Select **Assignments | Settings** to access the Settings Window.
l) Click on *Simulator* at the left.
m) Choose Functional from the *Simulator Mode* pull-down menu.
n) Find and use for the *Simulator Input* file *implement1.vwf* you just created.
o) Select **Processing | Generate Functional Simulation Netlist.**
p) Simulate the project. If there are any errors and you change your circuit, remember to save and compile the circuit before re-simulating.
q) Record the output waveform X on Figure S8-1 or print the Simulation Waveforms report from QUARTUS® II and attach it to the data collection sheet at the end of this experiment.
r) You should complete this part of the experiment by converting the waveform information in Figure S8-1 to Table S8-1 under "X after simulation."
s) Recall that two circuits are logically the same if they have the same truth table.
t) According to your results in Truth Table S8-1, are X and "X after simulation" equivalent? _____ If they are not, review your schematic that you derived from Table S8-1. Also check to see if *implement1.bfd* is the same as your schematic. Fix any errors found and repeat steps j-t.
u) Record and analyze results:
   1) Write the Boolean expression for X:

   $X = $ _____

   2) Transfer the results of the timing diagram, truth table, and Boolean expression to the data collection sheets at the end of this experiment.
v) *(Optional) Program the CPLD:* It is suggested that you use **PIN_50** for "A," **PIN_51** for "B," **PIN_52** for "C," and **PIN_4** for "X." To program your CPLD, refer to Appendix D, Exercise 3. A complete set of instructions will be found there. Briefly,
   1) *Recompile the project*: After making the pin assignments, compile the project again. When the compilation is completed, the software will display an information box showing any warnings and any detected errors. If there are errors, fix them and recompile the project.
   2) *Download the project to the board:*
      a. Set all eSOC DIP switches to OFF.
      b. Connect the board to the PC's parallel printer port and apply power.
      c. Select **Tools | Programmer**.
      d. Click on [Auto Detect] in the Chain Description File.
      e. When the eSOC board is recognized, press the [Add File...] button in the Chain Description File, then locate and Add the *implement1.pof* file.

f. Select the file and device line and scroll (horizontally) to the available programming option. Check the **Program/Configure** box.

g. Press the [Start] button. Observe the progress bar.

h. If the program downloads successfully, test the program. For each eSOC switch used by the program, turn its corresponding DIP switch on.

w) Save and close the project.

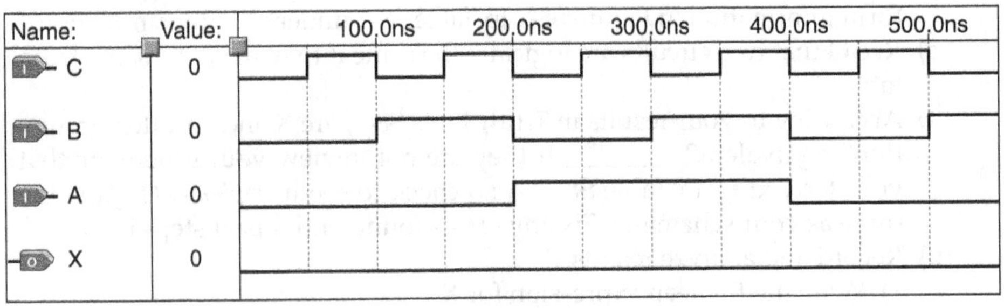

Figure S8-1

PART 2 – Simple Majority Tester

Design a circuit whose output is HIGH only when a majority of inputs A, B, and C are LOW. Use Table S8-2 to make a truth table for the circuit, then draw a logic diagram that corresponds to the truth table. Show your circuit design to the instructor before constructing it. Test the circuit using the truth table you made for it.

A	B	C	X	X after simulation
0	0	0		
0	0	1		
0	1	0		
0	1	1		
1	0	0		
1	0	1		
1	1	0		
1	1	1		

Table S8-2

PROCEDURE

a) Write the Boolean expression for X:

   X = _____

b) Start the QUARTUS® II program. Set up a new project and use **implement2** for the name of the project.
c) Create a new graphic file *implement2.bdf* for the schematic you drew and place the symbols as shown.
d) Connect the symbols together according to the schematic you drew.
e) Place the input and output connectors on the schematic. Label the inputs A, B, and C; label the output X.
f) Wire the input and output connectors to the logic symbols.
g) Select **Project | Top-Level Entity**.
h) Compile the project. If there are errors, fix them and save the file before recompiling the project. If there are no errors (or perhaps a warning concerning timing at most), then go on to the next step.
i) Create a new waveform file *implement2.vwf* and define the inputs for simulation. The input waveforms should look similar to those in Figure S8-2.
j) Select **Assignments | Settings** to access the Settings Window.
k) Click on *Simulator* at the left.
l) Choose Functional from the *Simulator Mode* pull-down menu.
m) Find and use for the *Simulator Input* file *implement1.vwf* you just created above.
n) Select **Processing | Generate Functional Simulation Netlist**.
o) Simulate the project. If there are any errors and you change your circuit, remember to save and compile the circuit before re-simulating.
p) Record the output waveform X on Figure S8-2 or print the Simulation Waveforms report from QUARTUS® II and attach it to the data collection sheet at the end of this experiment.
q) Record and analyze results:
   1) Record the output waveform X on Figure S8-2. You should complete this part of the experiment by converting the waveform information in Figure S8-2 to Table S8-2 under "X after simulation."
   2) Recall that two circuits are logically the same if they have the same truth table.
   3) According to your results in Truth Table S8-2, are X and "X after simulation" equivalent? _____ If they are not, review your schematic that you derived from Table S8-2. Also check to see if *implement1.gfd* is the same as your schematic. Fix any errors found and repeat steps i-q.
r) *(Optional) Program the CPLD:* It is suggested that you use **PIN_50** for "A," **PIN_51** for "B," **PIN_52** for "C," and **PIN_4** for "X." To program your CPLD, refer to Appendix D, Exercise 3. A complete set of instructions will be found there. Briefly,
   1) *Recompile the project*: After making the pin assignments, compile the project again. When the compilation is completed, the software will display an information box showing any warnings and any detected errors. If there are errors, fix them and recompile the project.

2) *Download the project to the board:*
   a. Set all eSOC DIP switches to OFF.
   b. Connect the board to the PC's parallel printer port and apply power.
   c. Select **Tools | Programmer**.
   d. Click on **Auto Detect** in the Chain Description File.
   e. When the eSOC board is recognized, press the **Add File...** button in the Chain Description File, then locate and Add the *implement2.pof* file.
   f. Select the file and device line and scroll (horizontally) to the available programming option. Check the **Program/Configure** box.
   g. Press the **Start** button. Observe the progress bar.
   h. If the program downloads successfully, test the program. For each eSOC switch used by the program, turn its corresponding DIP switch on.
s) Save and close the project.

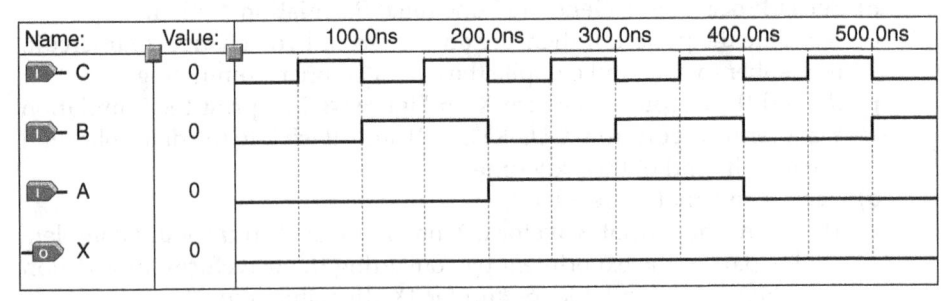

Figure S8-2

DATA COLLECTION SHEET

Data Collection Sheets for Lab S8   Name _____

Date _____

Data collection for Part 1:

a) Examine the truth table in Table S8-1. Draw a logic circuit in the space below. Do not be concerned with using a minimum number of gates.

A	B	C	X	X after simulation
0	0	0	1	
0	0	1	0	
0	1	0	1	
0	1	1	1	
1	0	0	1	
1	0	1	0	
1	1	0	0	
1	1	1	1	

Table S8-1

Demonstrated To
Instructor/FA _____ Date _____

Part 1 [ ]   Part 2 [ ]

Data Collection Sheets for Lab S8   Name _____

                                          Date _____

b) Write the Boolean expression for X:

   X = _____

t) According to your results in Truth Table S8-1, are X and "X after simulation" equivalent? _____

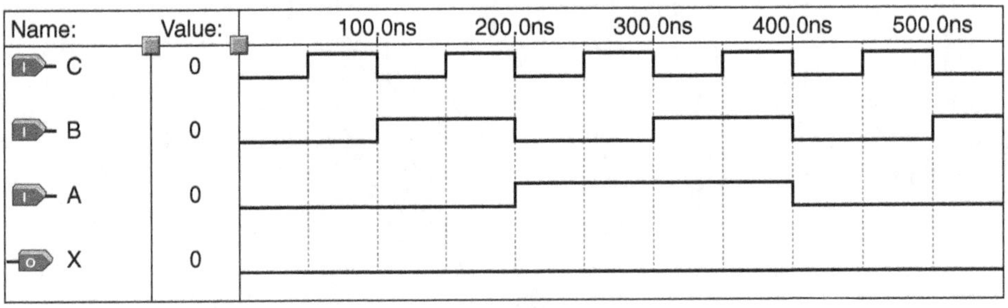

Figure S8-1

Data collection for Part 2:

A	B	C	X	X after simulation
0	0	0		
0	0	1		
0	1	0		
0	1	1		
1	0	0		
1	0	1		
1	1	0		
1	1	1		

Table S8-2

Demonstrated To
Instructor/FA _____ Date _____

Part 1 [ ]   Part 2 [ ]

DATA COLLECTION SHEET

Data Collection Sheets for Lab S8   Name _____

Date _____

a) Write the Boolean expression for X:

X = _____

Name:	Value:	100.0ns	200.0ns	300.0ns	400.0ns	500.0ns
C	0					
B	0					
A	0					
X	0					

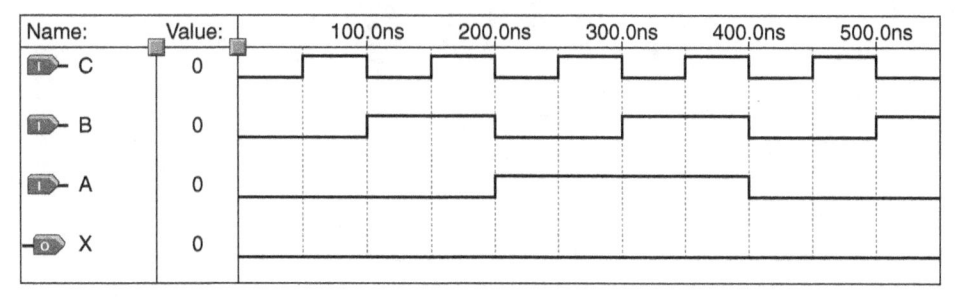

Figure S8-2

q)

3) According to your results in Truth Table S8-2, are X and "X after simulation" equivalent? _____

Demonstrated To
Instructor/FA _____ Date _____

Part 1 [ ]   Part 2 [ ]

Experiment S9

Name _____

EXCLUSIVE-OR AND EXCLUSIVE-NOR CIRCUITS

OBJECTIVES

1. To practice using QUARTUS® II software.
2. To practice using QUARTUS® II software to create schematics and waveforms.
3. To practice using QUARTUS® II software to compile and simulate digital circuits.
4. To investigate the truth tables and associated waveforms of the exclusive-OR (EX-OR or XOR) and exclusive-NOR (EX-NOR or XNOR) circuits.
5. *Optional:* To program the EPM7128SLC84 CPLD.

TEXT REFERENCE

Read section 4.6.

EQUIPMENT NEEDED

Components

Blank floppy, zip disk, or memory stick for storing projects;

QUARTUS® II software (Altera Corporation);

Optional: DeVry University Board eSOC with EPM7128SLC84 CPLD (or Altera University Board with EPM7128SLC84 CPLD or any other equivalent board);

Desktop computer with minimum of INTEL Pentium PC @ 400MHz CPU running Microsoft Windows NT4SP1, 2000, or XP or better; Pentium III or 4 PC @ 400MHz running Red Hat Linux v7.3 or 8.0 or Red Hat Linux Enterprise 3; Sun Ultra running Solaris v8 or 9; HP9000 Series 700/800 running HP-UX v11.0 with ACE dated 11/1999 or later; or equivalent.

DISCUSSION

Two Boolean expressions occur quite frequently in designing combinatorial circuits:

1) $x = \overline{A}B + A\overline{B}$

and

2) $x = \overline{AB} + AB$

Expression 1 defines the exclusive-OR function to be one that yields an output that is HIGH whenever its inputs are different. Similarly, expression 2 defines the exclusive-NOR function to be one that yields an output that is HIGH whenever its inputs are the same. While these circuits are combinatorial circuits, they have been given their own symbols and therefore are classified as gates. Both have been implemented with integrated circuits. In this experiment, you will investigate the behavior of both circuits.

PROCEDURE

PART 1 – The EX-OR (XOR) Gate and Its Combinatorial Equivalent

a) Start the QUARTUS® II program. Set up a new project and use **xor1** for the name of the project. The diagram of the circuit you are to construct and its input waveforms are given in Figures S9-1 and S9-2.

b) Create a new graphic file *xor1.bdf* for the schematic.

c) Place the XOR gate symbol on the schematic. Also place the XOR circuit on the schematic. Your schematic should look like Figure S9-1.

d) Place the input and output connectors on the schematic. Label the inputs A and B; label the output of the XOR gate X and the output of the XOR circuit Y.

e) Wire the input and output connectors to the gate and circuit as shown.

f) Select **Project | Top-Level Entity**.

g) Compile the project. If there are errors, fix them and save the file before recompiling the project. If there are no errors (or perhaps a warning concerning timing at most), then go on to the next step.

h) Create a new waveform file *xor1.vwf* and define the inputs for simulation. The input waveforms should look similar to those in Figure S9-2.

i) Select **Assignments | Settings** to access the Settings Window.

j) Click on *Simulator* at the left.

k) Choose Functional from the *Simulator Mode* pull-down menu.

l) Find and use for the *Simulator Input* file *xor1.vwf* you just created above.

m) Select **Processing | Generate Functional Simulation Netlist.**

n) Simulate the project. If there are any errors and you change your circuit, remember to save and compile the circuit before re-simulating.

o) Record the output waveform X on Figure S9-2 or print the Simulation Waveforms report from QUARTUS® II and attach it to the data collection sheet at the end of this experiment.

PROCEDURE

p) Record and analyze results:
   1) You should complete this part of the experiment by converting the waveform information in Figure S9-2 to Table S9-1.
   2) Are the values for X and Y the same? _____
   3) Write the Boolean expressions for X and Y:

      X = _____   Y = _____

   4) Transfer the results of the timing diagram, truth table, and Boolean expression to the data collection sheets at the end of this experiment.

q) *(Optional) Program the CPLD:* It is suggested that you use **PIN_50** for "**A**," **PIN_51** for "**B**," **PIN_4** for "**X**," and **PIN_** for "**Y**." To program your CPLD, refer to Appendix D, Exercise 3. A complete set of instructions will be found there. Briefly,
   1) *Recompile the project*: After making the pin assignments, compile the project again. When the compilation is completed, the software will display an information box showing any warnings and any detected errors. If there are errors, fix them and recompile the project.
   2) *Download the project to the board:*
      a. Set all eSOC DIP switches to OFF.
      b. Connect the board to the PC's parallel printer port and apply power.
      c. Select **Tools | Programmer**.
      d. Click on [Auto Detect] in the Chain Description File.
      e. When the eSOC board is recognized, press the [Add File...] button in the Chain Description File, then locate and Add the *xor1.pof* file.
      f. Select the file and device line and scroll (horizontally) to the available programming option. Check the **Program/Configure** box.
      g. Press the [Start] button. Observe the progress bar.
      h. If the program downloads successfully, test the program.

r) Save and close the project.

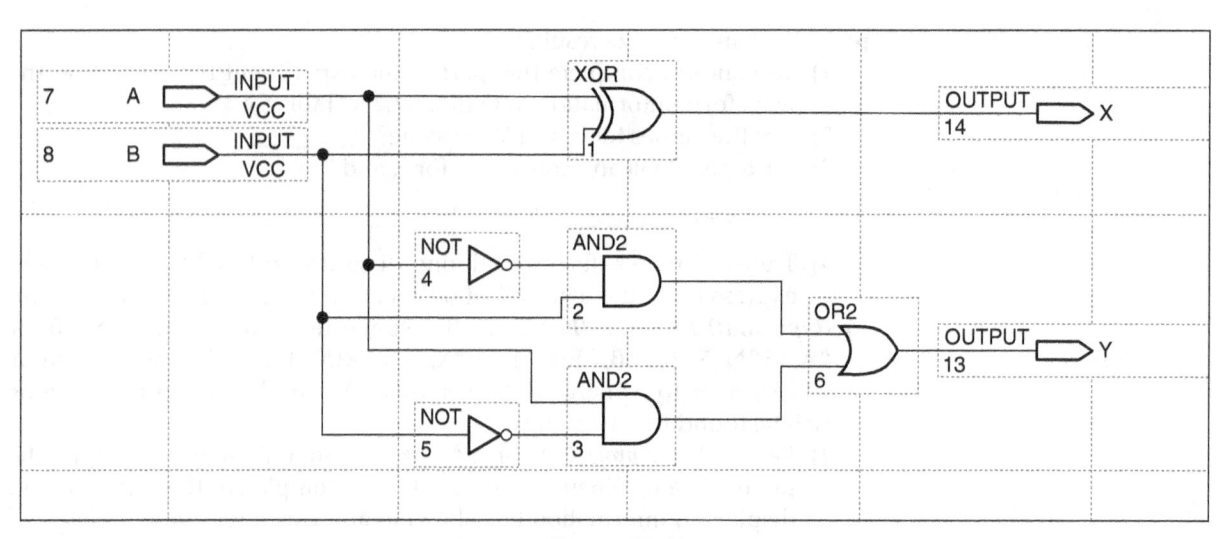

Figure S9-1

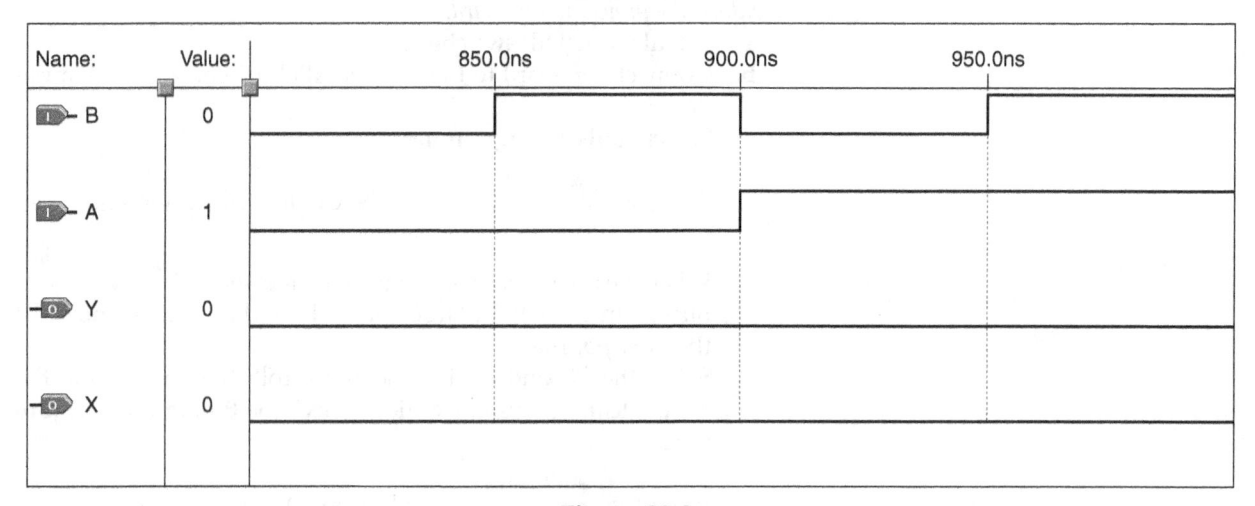

Figure S9-2

A	B	X	Y
0	0		
0	1		
1	0		
1	1		

Table S9-1

PART 2 – The EX-NOR (XNOR) Gate and Its Combinatorial Equivalent

a) Start the QUARTUS® II program. Set up a new project and use **xnor1** for the name of the project. The diagram of the circuit you are to construct and its input waveforms are given in Figures S9-3 and S9-4.

b) Create a new graphic file *xnor1.bdf* for the schematic.

c) Place the XNOR gate symbol on the schematic. Also place the XNOR circuit on the schematic. Your schematic should look like Figure S9-3.

d) Place the input and output connectors on the schematic. Label the inputs A and B; label the output of the XNOR gate X and the output of the XNOR circuit Y.

PROCEDURE

e) Wire the input and output connectors to the gate and circuit as shown.
f) Select **Project | Top-Level Entity**.
g) Compile the project. If there are errors, fix them and save the file before recompiling the project. If there are no errors (or perhaps a warning concerning timing at most), then go on to the next step.
h) Create a new waveform file *xnor1.vwf* and define the inputs for simulation. The input waveforms should look similar to those in Figure S9-4.
i) Select **Assignments | Settings** to access the Settings Window.
j) Click on *Simulator* at the left.
k) Choose Functional from the *Simulator Mode* pull-down menu.
l) Find and use for the *Simulator Input* file *xnor1.vwf* you just created above.
m) Select **Processing | Generate Functional Simulation Netlist**.
n) Simulate the project. If there are any errors and you change your circuit, remember to save and compile the circuit before re-simulating.
o) Record the output waveform X on Figure S9-4 or print the Simulation Waveforms report from QUARTUS® II and attach it to the data collection sheet at the end of this experiment.
p) Record and analyze results:
   1) You should complete this part of the experiment by converting the waveform information in Figure S9-4 to Table S9-2.
   2) Are the values for X and Y the same? _____
   3) Write the Boolean expressions for X and Y:

      X = _____   Y = _____

   4) Transfer the results of the timing diagram, truth table, and Boolean expression to the data collection sheets at the end of this experiment.
q) *(Optional) Program the CPLD:* It is suggested that you use **PIN_50** for "A," **PIN_51** for "B," **PIN_4** for "X," and **PIN_** for "Y." To program your CPLD, refer to Appendix D, Exercise 3. A complete set of instructions will be found there. Briefly,
   1) *Recompile the project*: After making the pin assignments, compile the project again. When the compilation is completed, the software will display an information box showing any warnings and any detected errors. If there are errors, fix them and recompile the project.
   2) *Download the project to the board:*
      a. Set all eSOC DIP switches to OFF.
      b. Connect the board to the PC's parallel printer port and apply power.
      c. Select **Tools | Programmer**.
      d. Click on *Auto Detect* in the Chain Description File.
      e. When the eSOC board is recognized, press the *Add File...* button in the Chain Description File, then locate and Add the *xnor1_ctr.pof* file.

f. Select the file and device line and scroll (horizontally) to the available programming option. Check the **Program/Configure** box.
g. Press the [Start] button. Observe the progress bar.
h. If the program downloads successfully, test the program.

r) Save and close the project.

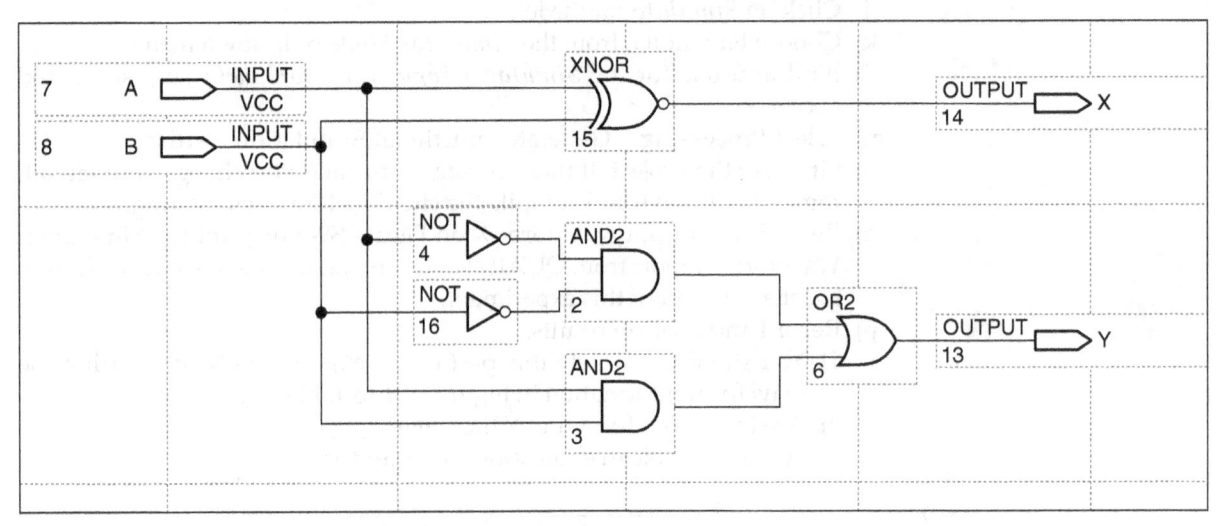

Figure S9-3

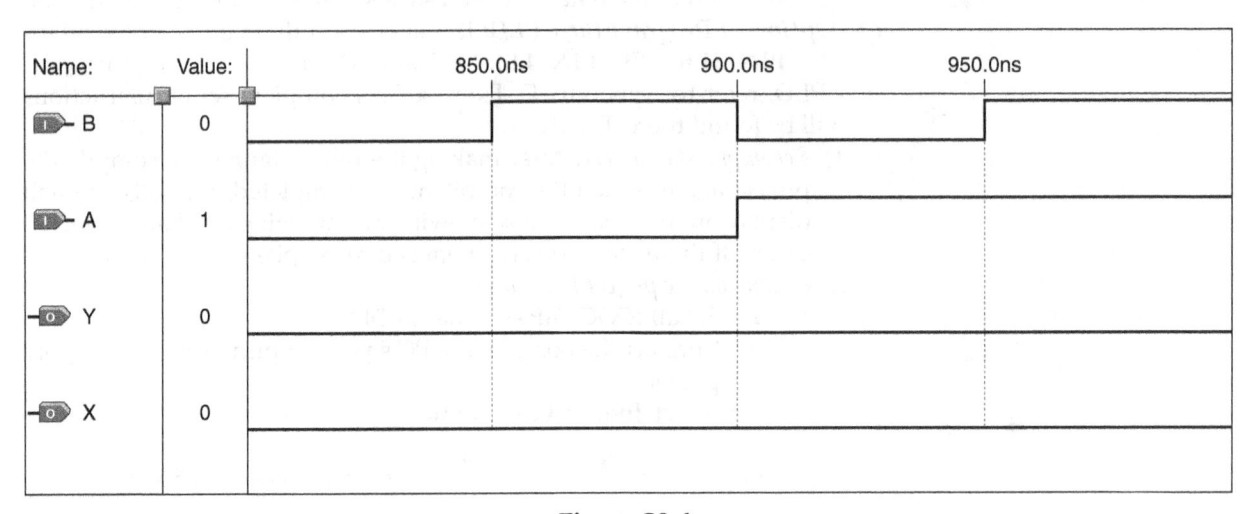

Figure S9-4

A	B	X	Y
0	0		
0	1		
1	0		
1	1		

Table S9-2

DATA COLLECTION SHEET

Data Collection Sheets for Lab S9   Name _____

Date _____

Data collection for Part 1:

p)

2) Are the values for X and Y the same? _____

3) Write the Boolean expressions for X and Y:

X = _____   Y = _____

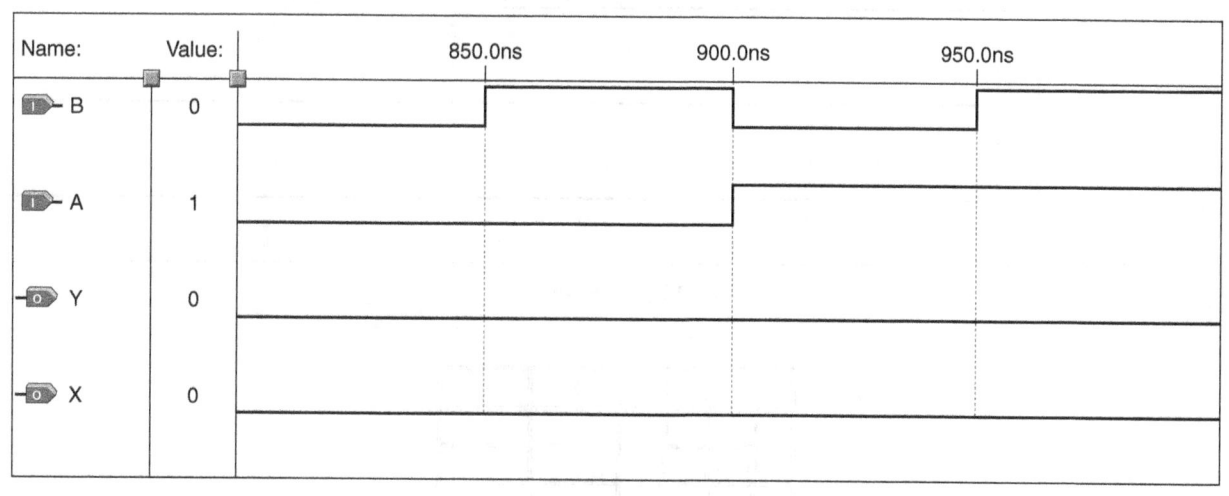

Name:	Value:	850.0ns	900.0ns	950.0ns
B	0			
A	1			
Y	0			
X	0			

Figure S9-2

A	B	X	Y
0	0		
0	1		
1	0		
1	1		

Table S9-1

Data collection for Part 2:

p)

2) Are the values for X and Y the same? _____

3) Write the Boolean expressions for X and Y:

X = _____   Y = _____

Demonstrated To
Instructor/FA _____ Date _____

Part 1 [ ]   Part 2 [ ]

Data collection Sheets for Lab S9   Name _____

Date _____

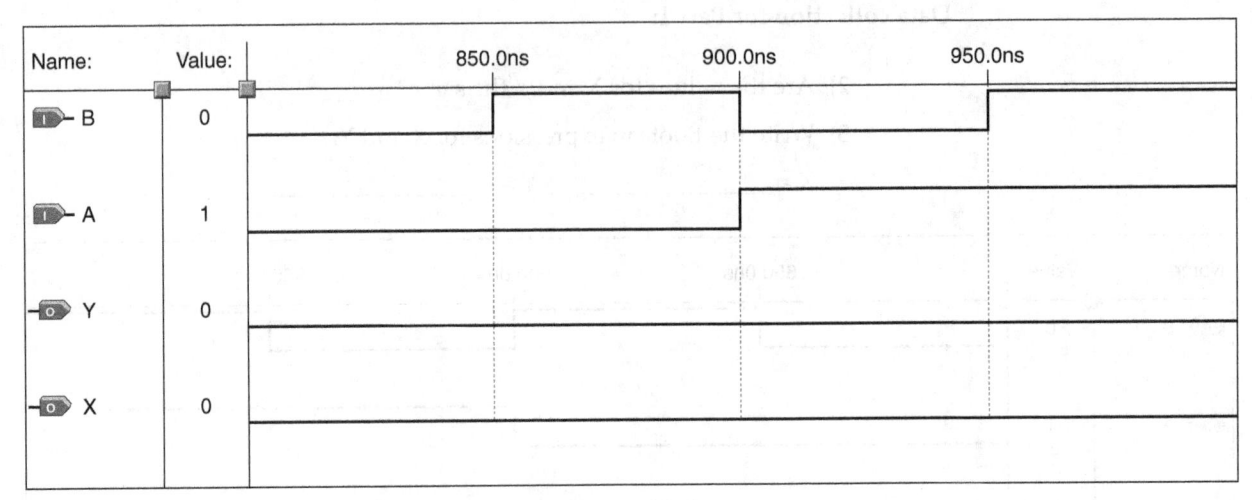

Name:	Value:	850.0ns	900.0ns	950.0ns

- B 0
- A 1
- Y 0
- X 0

Figure S9-4

A	B	X	Y
0	0		
0	1		
1	0		
1	1		

Table S9-2

Demonstrated To
Instructor/FA _____ Date _____

Part 1 [ ]   Part 2 [ ]

Experiment S10

Name _____

DESIGNING WITH EXCLUSIVE-OR AND EXCLUSIVE-NOR CIRCUITS

OBJECTIVES

1. To practice using QUARTUS® II software to create schematics and waveforms.
2. To practice using QUARTUS® II software to compile and simulate digital circuits.
3. To investigate the application of an exclusive-OR circuit in a parity generator circuit.
4. To design a 4-bit parity generator.
5. To investigate the application of an exclusive-NOR circuit in a digital comparator circuit.
6. To program the EPM7128SLC84 CPLD.

TEXT REFERENCES

Read sections 2.9 and 4.6. Review Appendix D of this manual.

EQUIPMENT NEEDED

Components

Blank floppy, zip disk, or memory stick for storing projects;
QUARTUS® II software (Altera Corporation);
Optional: DeVry University Board eSOC with EPM7128SLC84 CPLD (or Altera University Board with EPM7128SLC84 CPLD or any other equivalent board);
Desktop computer with minimum of INTEL Pentium PC @ 400MHz CPU running Microsoft Windows NT4SP1, 2000, or XP or better; Pentium III or 4 PC @ 400MHz running Red Hat Linux v7.3 or 8.0 or Red Hat Linux Enterprise 3; Sun Ultra running Solaris v8 or 9; HP9000 Series 700/800 running HP-UX v11.0 with ACE dated 11/1999 or later; or equivalent.

DISCUSSION

In the previous experiment, you were introduced to the exclusive-OR and exclusive-NOR circuits. You discovered that these circuits can be used to compare the level of two inputs. Indeed, the exclusive-OR and exclusive-NOR are basically digital comparators. In the current experiment, you will design two circuits using the exclusive-OR and exclusive-NOR circuits:

1) <u>Parity generator</u>. When binary information (**data**) is transmitted from one place to another, the data sometimes gets corrupted for one reason or another. That is, a 1 may get changed to a 0, or vice versa. That will, of course, change the meaning or value of the data. To permit detection of a single bit error on the receiving end, the transmitter adds an extra bit to the data for the receiver to use to predict if the data received has been corrupted or not. If an error is detected, the receiver usually makes a request for the transmitter to re-transmit the corrupted data or causes an error message to be generated by the operating system. This extra bit is called the **parity bit**. It is generated by a **parity generator**.

In odd parity, if the total number of 1s in the data is odd, the parity bit generated will have a value 0 while the value will be 1 if the total number of 1s in the data is even. In even parity, if the total number of 1s in the data is odd, the parity bit generated will have a value 1 while the value will be 0 if the total number of 1s in the data is even. The receiver uses a **parity checker** to detect any data errors. Its function is to compare the number of 1s in the data to the value of the parity bit according to the type of parity being used by BOTH the transmitter and receiver (they must both use the same type of parity). Both the parity generator and parity checker can be implemented with XOR and XNOR circuits.

2) <u>Binary comparator</u>. Digital arithmetic units contain devices that compare two binary numbers, say A and B, to determine if they are the same or different. If A and B are the same, then the device indicates "A = B". If A and B are different, then the device indicates that "A < B" or "A > B". Such a unit can be implemented with XOR and XNOR circuits.

PROCEDURE

PART 1 – Parity Generator for 3-bit Data

a) Start the QUARTUS® II program. Set up a new project and use **parity1** for the name of the project. The diagram of the circuit you are to construct and its input waveforms are given in Figures S10-1 and S10-2.
b) Compute the EVEN and ODD parity bit values for all 3-bit data and place the values in Tables S10-1a and S10-1b, respectively. Don't change these tables later even if your experimental results do not bear you out.
c) Create a new graphic file for the schematic, *parity1.bdf*.
d) Place the gate symbols on the schematic according to Figure S10-1.

PROCEDURE

e) Place the input and output connectors on the schematic. Label the inputs A, B, and C; label the outputs X, Y, D2, D1, and D0. A, B, and C are the original data and D2, D1, and D0 are the data bits transmitted along with either X or Y, the generated parity bits for EVEN and ODD parity, respectively.

f) Wire the input and output connectors to the circuit as shown in Figure S10-1.

g) Select **Project | Top-Level Entity**.

h) Compile the project. If there are errors, fix them and save the file before recompiling the project. If there are no errors (or perhaps a warning concerning timing at most), then go on to the next step.

i) Create a new waveform file *parity1.vwf* and define the inputs for simulation. The input waveforms should look similar to those in Figure S10-2.

j) Select **Assignments | Settings** to access the Settings Window.

k) Click on *Simulator* at the left.

l) Choose Functional from the *Simulator Mode* pull-down menu.

m) Find and use for the *Simulator Input* file *parity1.vwf* you just created above.

n) Select **Processing | Generate Functional Simulation Netlist**.

o) Simulate the project. If there are any errors and you change your circuit, remember to save and compile the circuit before re-simulating.

p) Record the output waveform X on Figure S10-2 or print the Simulation Waveforms report from QUARTUS® II and attach it to the data collection sheet at the end of this experiment.

q) Drag the **Time Bar** to the middle of interval 0-50ns and copy the D0, D1, D2, and X data from the Values column to Table S10-2a, 0-50ns time interval. Repeat this for all time intervals in the table.

r) Repeat **step q** for D0, D1, D2, and Y data from the Values column for all of the time intervals in Table S10-2b.

s) Analyze Tables S10-2a and S10-2b and verify that X and Y are the generated parity bits for an EVEN parity generator and an ODD parity generator, respectively. Do Tables S10-2a and S10-2b agree with your computations recorded in Tables S10-1a and S10-1b? _____

t) Write the Boolean expression for X:

$$X = \underline{\hspace{5cm}}$$

u) Transfer the results of the timing diagram, truth tables, and Boolean expression to the data collection sheets at the end of this experiment.

v) *(Optional) Program the CPLD*: It is suggested that you use **PIN_50** for "A," **PIN_51** for "B," **PIN_52** for "C," **PIN_4** for "Y," **PIN_5** for "X," **PIN_8** for "D2," **PIN_9** for "D1," and **PIN_10** for "D0." To program your CPLD, refer to Appendix D, Exercise 3. A complete set of instructions will be found there. Briefly,

   1) *Recompile the project*: After making the pin assignments, compile the project again. When the compilation is completed, the software will display an information box showing any warnings and any detected errors. If there are errors, fix them and recompile the project.

2) *Download the project to the board:*
   a. Set all eSOC DIP switches to OFF.
   b. Connect the board to the PC's parallel printer port and apply power.
   c. Select **Tools | Programmer**.
   d. Click on [Auto Detect] in the Chain Description File.
   e. When the eSOC board is recognized, press the [Add File...] button in the Chain Description File, then locate and Add the *parity1.pof* file.
   f. Select the file and device line and scroll (horizontally) to the available programming option. Check the **Program/Configure** box.
   g. Press the [Start] button. Observe the progress bar.
   h. If the program downloads successfully, test the program.

w) Save and close the project.

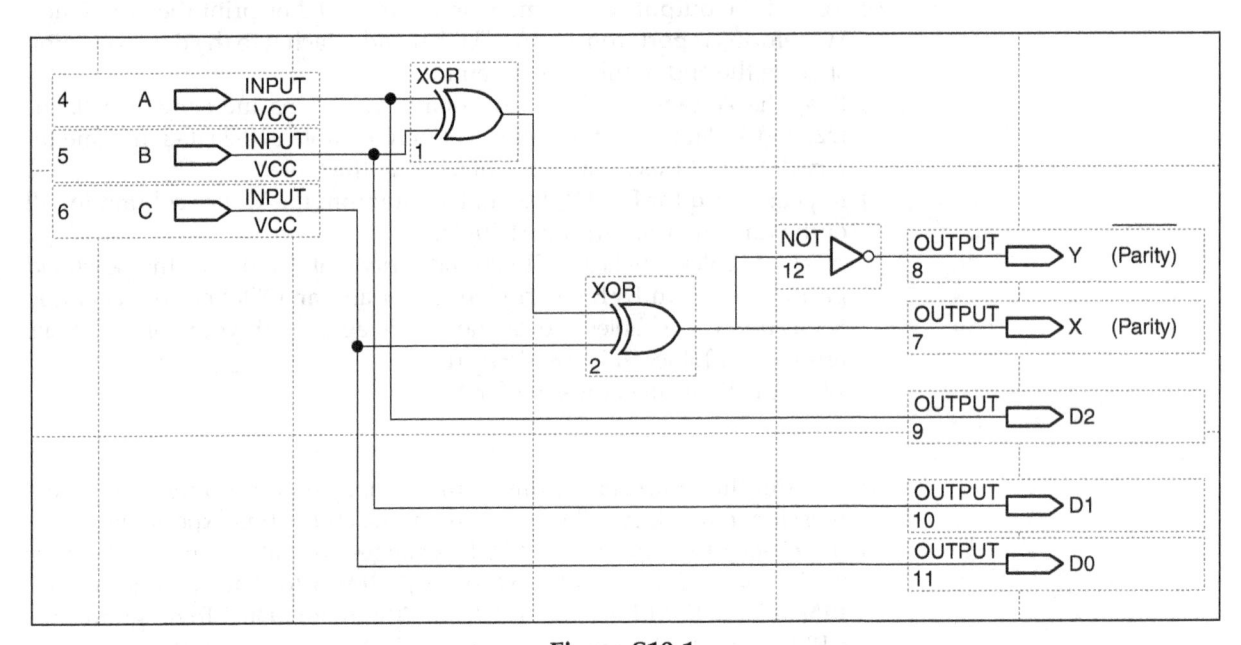

Figure S10-1

PROCEDURE

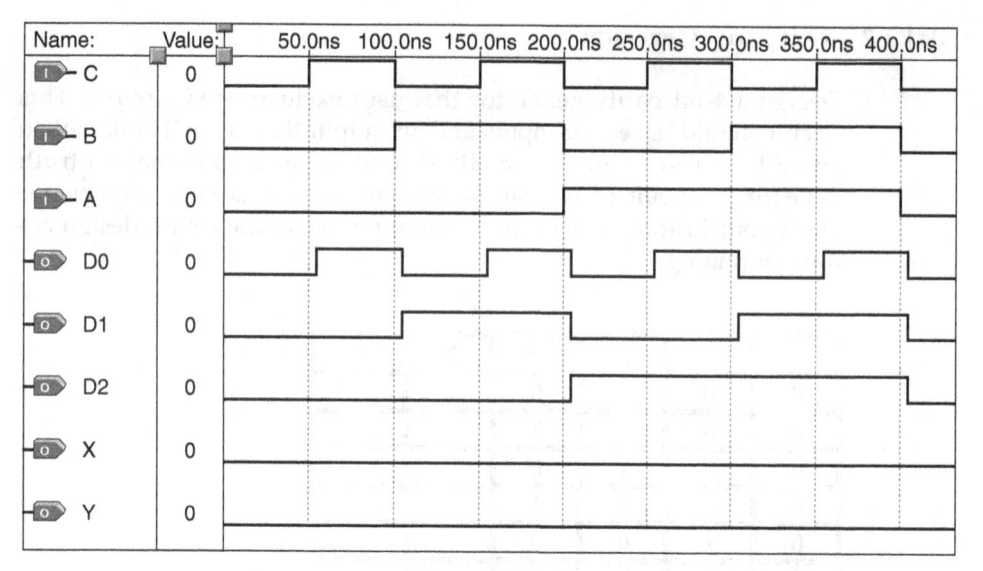

Figure S10-2

X	D2	D1	D0
	0	0	0
	0	0	1
	0	1	0
	0	1	1
	1	0	0
	1	0	1
	1	1	0
	1	1	1

Table S10-1a: EVEN Parity

Y	D2	D1	D0
	0	0	0
	0	0	1
	0	1	0
	0	1	1
	1	0	0
	1	0	1
	1	1	0
	1	1	1

Table S10-1b: ODD Parity

Time Intervals	X	D2	D1	D0
0-50ns				
50ns-100ns				
100ns-150ns				
150ns-200ns				
200ns-250ns				
250ns-300ns				
300ns-350ns				
350ns-400ns				

Table S10-2a

Time Intervals	Y	D2	D1	D0
0-50ns				
50ns-100ns				
100ns-150ns				
150ns-200ns				
200ns-250ns				
250ns-300ns				
300ns-350ns				
350ns-400ns				

Table S10-2b

PART 2 – 4-bit Parity Generator

a) Design a 4-bit parity generator that uses exclusive-OR circuits. This circuit should have four inputs and an output that is HIGH only when an odd number of inputs are HIGH. Use Table S10-3 to make a truth table for the circuit. Draw your logic circuit in the space provided below. Have your instructor or faculty assistant approve the circuit design before continuing.

A	B	C	D	X	Y
0	0	0	0		
0	0	0	1		
0	0	1	0		
0	0	1	1		
0	1	0	0		
0	1	0	1		
0	1	1	0		
0	1	1	1		
1	0	0	0		
1	0	0	1		
1	0	1	0		
1	0	1	1		
1	1	0	0		
1	1	0	1		
1	1	1	0		
1	1	1	1		

Table S10-3

b) Draw a logic circuit in the space below. Do not be concerned with using a minimum number of gates.

Instructor/FA _____

PROCEDURE

c) Write the Boolean expression for X:

X = _____

d) Start the QUARTUS® II program. Set up a new project and use **parity2** for the name of the project.
e) Create a new graphic file *parity2.bdf* for the schematic you drew and place the symbols as shown.
f) Connect the symbols together according to the schematic you drew.
g) Place the input and output connectors on the schematic. Label the inputs A, B, C, and D; label the outputs X and Y.
h) Wire the input and output connectors to the logic symbols.
i) Select **Project | Top-Level Entity**.
j) Compile the project. If there are errors, fix them and save the file before recompiling the project. If there are no errors (or perhaps a warning concerning timing at most), then go on to the next step.
k) Create a new waveform file *parity2.vwf* and define the inputs for simulation. The input waveforms should look similar to those in Figure S10-3.
l) Select **Assignments | Settings** to access the Settings Window.
m) Click on *Simulator* at the left.
n) Choose Functional from the *Simulator Mode* pull-down menu.
o) Find and use for the *Simulator Input* file *parity2.vwf* you just created above.
p) Select **Processing | Generate Functional Simulation Netlist**.
q) Simulate the project. If there are any errors and you change your circuit, remember to save and compile the circuit before re-simulating.
r) Record the output waveform X on Figure S10-3 or print the Simulation Waveforms report from QUARTUS® II and attach it to the data collection sheet at the end of this experiment.
s) Drag the **Time Bar** to the middle of interval 0-50ns and copy the D0, D1, D2, D3, and X data from the Values column to Table S10-4a, 0-50ns time interval. Repeat this for all time intervals in the table.
t) Repeat **step s** for D0, D1, D2, D3, and Y data from the Values column for all of the time intervals in Table S10-4b.
u) Analyze Tables S10-4a and S10-4b and verify that X and Y are the generated parity bits for an EVEN parity generator and an ODD parity generator, respectively. Do Tables S10-4a and S10-4b agree with your computations recorded in Table S10-3? _____
v) Write the Boolean expression for X:

X = _____

w) Transfer the results of the waveform information, truth tables, and Boolean expression to the data collection sheets at the end of this experiment.
x) *(Optional) Program the CPLD:* It is suggested that you use **PIN_50** for "A," **PIN_51** for "B," **PIN_52** for "C," **PIN_54** for "D," **PIN_4** for "Y," **PIN_5** for "X," **PIN_8** for "D3," **PIN_9** for "D2," **PIN_10** for "D1," and **PIN_11** for "D0." To program your CPLD, refer to Appendix D, Exercise 3. A complete set of instructions will be found there. Briefly,
   1) *Recompile the project*: After making the pin assignments, compile the project again. When the compilation is completed, the software will display an information box showing any warnings and any detected errors. If there are errors, fix them and recompile the project.
   2) *Download the project to the board:*
      a. Set all eSOC DIP switches to OFF.
      b. Connect the board to the PC's parallel printer port and apply power.
      c. Select **Tools | Programmer**.
      d. Click on **Auto Detect** in the Chain Description File.
      e. When the eSOC board is recognized, press the **Add File...** button in the Chain Description File, then locate and Add the *parity2.pof* file.
      f. Select the file and device line and scroll (horizontally) to the available programming option. Check the **Program/Configure** box.
      g. Press the **Start** button. Observe the progress bar.
      h. If the program downloads successfully, test the program.
y) Save and close the project.

Time Intervals	X	D3	D2	D1	D0
0-50ns		0	0	0	0
50ns-100ns		0	0	0	1
100ns-150ns		0	0	1	0
150ns-200ns		0	0	1	1
200ns-250ns		0	1	0	0
250ns-300ns		0	1	0	1
300ns-350ns		0	1	1	0
350ns-400ns		0	1	1	1
400ns-450ns		1	0	0	0
450ns-500ns		1	0	0	1
500ns-550ns		1	0	1	0
550ns-600ns		1	0	1	1
600ns-650ns		1	1	0	0
650ns-700ns		1	1	0	1
700ns-750ns		1	1	1	0
750ns-800ns		1	1	1	1

Table S10-4a

Time Intervals	Y	D3	D2	D1	D0
0-50ns		0	0	0	0
50ns-100ns		0	0	0	1
100ns-150ns		0	0	1	0
150ns-200ns		0	0	1	1
200ns-250ns		0	1	0	0
250ns-300ns		0	1	0	1
300ns-350ns		0	1	1	0
350ns-400ns		0	1	1	1
400ns-450ns		1	0	0	0
450ns-500ns		1	0	0	1
500ns-550ns		1	0	1	0
550ns-600ns		1	0	1	1
600ns-650ns		1	1	0	0
650ns-700ns		1	1	0	1
700ns-750ns		1	1	1	0
750ns-800ns		1	1	1	1

Table S10-4b

PROCEDURE

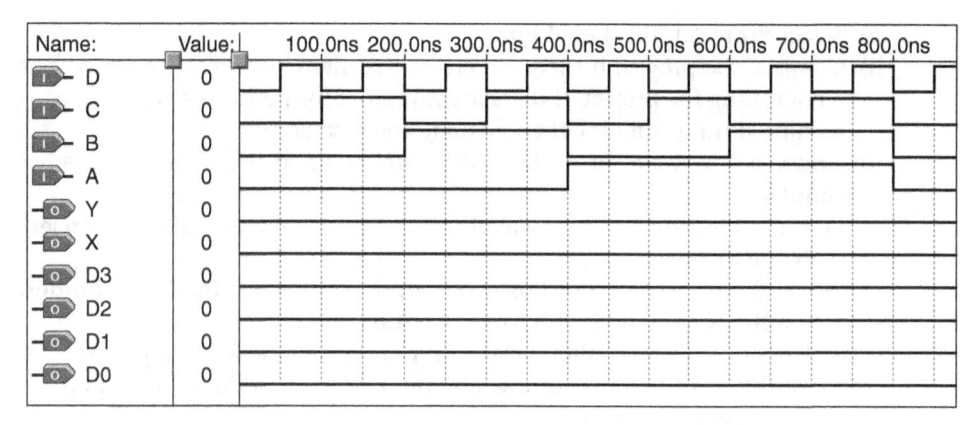

Figure S10-3

PART 3 – 2-Bit Binary Equality Detector

a) Start the QUARTUS® II program. Set up a new project and use **combcircuit1** for the name of the project. The diagram of the circuit you are to construct and its input waveforms are given in Figures S10-4 and S10-5.

b) Create a new graphic file for the schematic *combcircuit4.bdf*.

c) Place the gate symbols on the schematic according to Figure S10-4.

d) Place the input and output connectors on the schematic. Label the inputs A[1..0] and B[1..0]; label the output X.

e) **Note:** The symbol A[1..0] is a **bus representation** for A1 and A0 collectively. It must be written precisely as shown in order for the program to understand what you are trying to convey. B[1..0] is a bus representation for B1 and B0. Use only two dots between the 1 and the 0. To represent an 8-bit bus G, you would write G[7..0].

f) *Construct and label two input buses:* Wire the input connectors to the circuit as shown in Figure S10-4:

  1) Place the mouse pointer on the right side of the A[1..0] symbol. Using the bus tool, click and drag the mouse about an inch to the right and then down an inch or two before releasing the mouse button.

  2) Connect the XOR inputs as shown using the node line (wiring) tool, starting your line from the input of the gate back to the bus.

  3) To label the gate inputs, click on a thin input connecting wire. This will highlight the thin connecting wire to the bus and you will see the insertion point flashing on the thin line. Now type in the symbol according to Figure S10-4. Do this for each XOR input.

  4) Repeat the above steps for the B[1..0] bus.

g) Select **Project | Top-Level Entity**.
h) Compile the project. If there are errors, fix them and save the file before recompiling the project. If there are no errors (or perhaps a warning concerning timing at most), then go on to the next step.
i) Create a new waveform file *combcircuit1.vwf* and define the inputs for simulation:
   1) Right-click on the new waveform file under name and select **Insert Node or Bus...**
   2) On the Insert Node or Bus... pop-up box, press on the **Node Finder.**
   3) On the Node Finder window, press **List.**
   4) Choose A[0], A[1], B[0], B[1], and X and press the **>** button. Press **OK** on the Node Finder window and on the Insert Node or Bus... pop-up box, press **OK** again.
   5) On the waveform window, click on A[1] to highlight the node name, and press the [X₀] button. In the Clock pop-up box, enter 400ns for the Period and press **OK**. Make sure the Duty Cycle is 50%.
   6) Repeat the last step for A[0], B[1], and B[0] while using 200ns, 100ns, and 50ns for their respective Periods.
   7) Highlight A[0] and A[1] together and right-click on their node names. From the pop-up menu, select **Group**. Select *Hexadecimal* for the **Radix** on the Group pop-up box.
   8) Repeat the previous step for B[0] and B[1].
   9) You have created bus signals for A[1..0] and B[1..0] (indicated by just ⊞ A and ⊞ B) in *combcircuit4.vwf*.
j) The waveforms, after grouping, should look similar to those in Figure S10-5. You should note that the two bus signals have Values in hexadecimal and for each time interval there is a hexadecimal number value for the bus. For example, if B1 = 1 and B0 = 0, the hexadecimal value for the signal on the B[1..0] bus is 2. Note the Time Bar is at 270ns. In the Value column, B[1..0] = H 2, the H standing for hexadecimal. A[1..0] is also H 2.
k) Note that if you click on the "+" symbol next to each node name, you will expand each bus so that you can see the individual bus lines. This is shown in Figure S10-6.
l) Simulate the project. If there are any errors and you change your circuit, remember to save and compile the circuit before re-simulating.
m) Record and analyze results:
   1) Record the output waveform X on Figure S10-5. Transfer the output X waveform information to Table S10-5.
   2) Write the Boolean expression for X:

   $$X = \underline{}$$

   3) Transfer the results of the waveform information, truth table, and Boolean expression to the data collection sheets at the end of this experiment.

PROCEDURE

n) *(Optional) Program the CPLD:* It is suggested that you use **PIN_50** for "A0," **PIN_51** for "A1," **PIN_52** for "B0," **PIN_54** for "B1," and **PIN_4** for "X." To program your CPLD, refer to Appendix D, Exercise 3. A complete set of instructions will be found there. Briefly,

1) *Recompile the project*: After making the pin assignments, compile the project again. When the compilation is completed, the software will display an information box showing any warnings and any detected errors. If there are errors, fix them and recompile the project.
2) *Download the project to the board:*
   a. Set all eSOC DIP switches to OFF.
   b. Connect the board to the PC's parallel printer port and apply power.
   c. Select **Tools | Programmer**.
   d. Click on [Auto Detect] in the Chain Description File.
   e. When the eSOC board is recognized, press the [Add File...] button in the Chain Description File, then locate and Add the *combcircuit4.pof* file.
   f. Select the file and device line and scroll (horizontally) to the available programming option. Check the **Program/Configure** box.
   g. Press the [Start] button. Observe the progress bar.
   h. If the program downloads successfully, test the program.

o) Save and close the project.

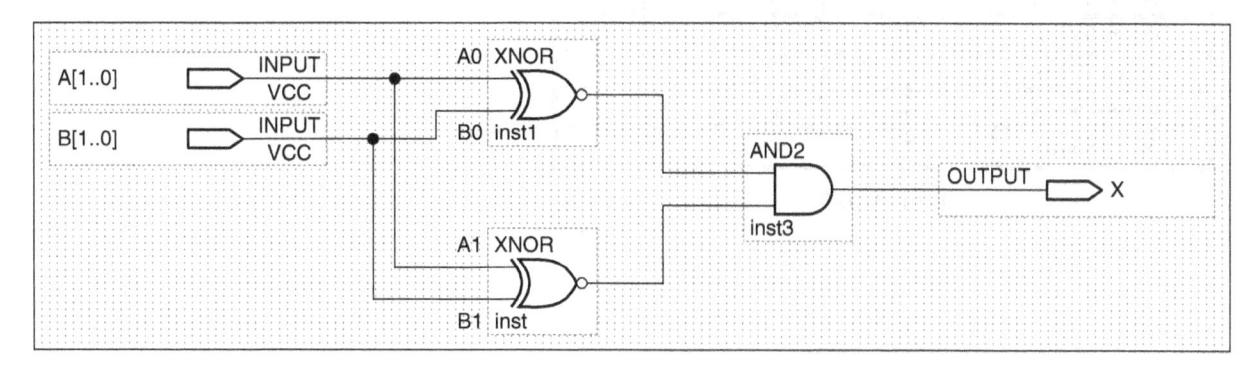

Figure S10-4

Name	Value at 270.0 ns	0 ps	80.0 ns	160.0 ns	240.0 ns	320.0 ns	400.0 ns	480.0 ns	560.0 ns
A	H 2	0		1		2	3	0	1
B	H 2	0 1 2 3	0 1 2 3	0 1 2 3	0 1 2 3	0 1 2 3	0 1 2 3	0 1 2 3	
X	B 0								

Figure S10-5

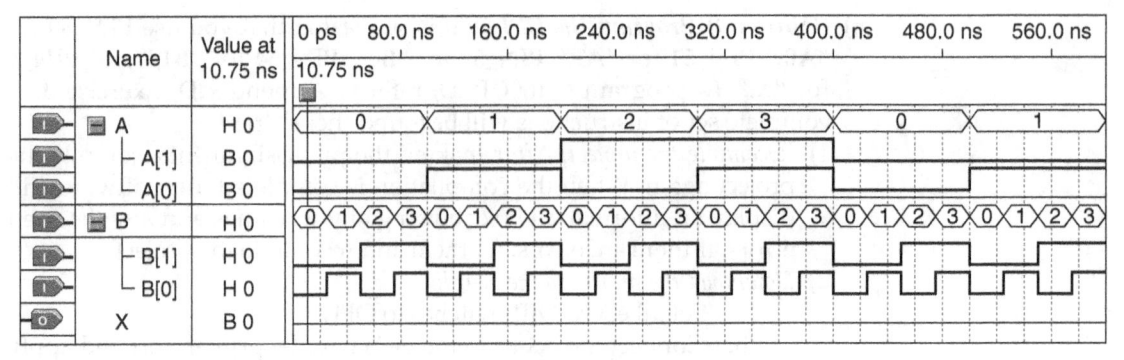

Name	Value at 10.75 ns	Waveform
A	H 0	0, 1, 2, 3, 0, 1
A[1]	B 0	
A[0]	B 0	
B	H 0	0,1,2,3 repeating
B[1]	H 0	
B[0]	H 0	
X	B 0	

Figure S10-6

Time Intervals	X	A1	A0	B1	B0
0-50ns		0	0	0	0
50ns-100ns		0	0	0	1
100ns-150ns		0	0	1	0
150ns-200ns		0	0	1	1
200ns-250ns		0	1	0	0
250ns-300ns		0	1	0	1
300ns-350ns		0	1	1	0
350ns-400ns		0	1	1	1
400ns-450ns		1	0	0	0
450ns-500ns		1	0	0	1
500ns-550ns		1	0	1	0
550ns-600ns		1	0	1	1
600ns-650ns		1	1	0	0
650ns-700ns		1	1	0	1
700ns-750ns		1	1	1	0
750ns-800ns		1	1	1	1

Table S10-5

DATA COLLECTION SHEET

Data Collection Sheets for Lab S10   Name _____
                                      Date _____

Data collection for Part 1:

    s) Analyze Tables S10-2a and S10-2b and verify that X and Y are the generated parity bits for an EVEN parity generator and an ODD parity generator, respectively. Do Tables S10-2a and S10-2b agree with your computations recorded in Tables S10-1a and S10-1b? _____

    t) Write the Boolean expression for X:

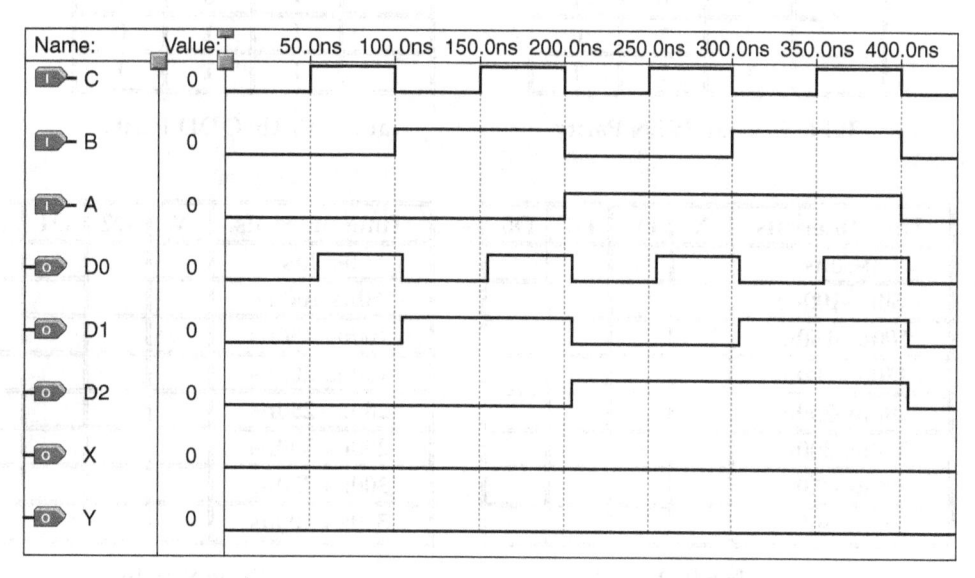

$$X = \text{\underline{\hspace{5cm}}}$$

Figure S10-2

Demonstrated To
Instructor/FA _____ Date _____

Part 1 [ ]   Part 2 [ ]   Part 3 [ ]

Data Collection Sheets for Lab S10 Name _____

Date _____

X	D2	D1	D0
	0	0	0
	0	0	1
	0	1	0
	0	1	1
	1	0	0
	1	0	1
	1	1	0
	1	1	1

Table S10-1a: EVEN Parity

Y	D2	D1	D0
	0	0	0
	0	0	1
	0	1	0
	0	1	1
	1	0	0
	1	0	1
	1	1	0
	1	1	1

Table S10-1b: ODD Parity

Time Intervals	X	D2	D1	D0
0-50ns				
50ns-100ns				
100ns-150ns				
150ns-200ns				
200ns-250ns				
250ns-300ns				
300ns-350ns				
350ns-400ns				

Table S10-2a

Time Intervals	Y	D2	D1	D0
0-50ns				
50ns-100ns				
100ns-150ns				
150ns-200ns				
200ns-250ns				
250ns-300ns				
300ns-350ns				
350ns-400ns				

Table S10-2b

Demonstrated To
Instructor/FA _____ Date _____

Part 1 [ ]   Part 2 [ ]   Part 3 [ ]

DATA COLLECTION SHEET

Data Collection Sheets for Lab S10   Name _____

Date _____

Data collection for Part 2:

A	B	C	D	X	Y
0	0	0	0		
0	0	0	1		
0	0	1	0		
0	0	1	1		
0	1	0	0		
0	1	0	1		
0	1	1	0		
0	1	1	1		
1	0	0	0		
1	0	0	1		
1	0	1	0		
1	0	1	1		
1	1	0	0		
1	1	0	1		
1	1	1	0		
1	1	1	1		

Table S10-3

c) Write the Boolean expression for X:

X = _____

u) Analyze Tables S10-4a and S10-4b and verify that X and Y are the generated parity bits for an EVEN parity generator and an ODD parity generator, respectively. Do Tables S10-4a and S10-4b agree with your computations recorded in Table S10-3? _____

v) Write the Boolean expression for X:

X = _____

Demonstrated To
Instructor/FA _____ Date _____

Part 1 [ ]   Part 2 [ ]   Part 3 [ ]

Data Collection Sheets for Lab S10   Name _____

Date _____

Time Intervals	X	D3	D2	D1	D0
0-50ns		0	0	0	0
50ns-100ns		0	0	0	1
100ns-150ns		0	0	1	0
150ns-200ns		0	0	1	1
200ns-250ns		0	1	0	0
250ns-300ns		0	1	0	1
300ns-350ns		0	1	1	0
350ns-400ns		0	1	1	1
400ns-450ns		1	0	0	0
450ns-500ns		1	0	0	1
500ns-550ns		1	0	1	0
550ns-600ns		1	0	1	1
600ns-650ns		1	1	0	0
650ns-700ns		1	1	0	1
700ns-750ns		1	1	1	0
750ns-800ns		1	1	1	1

Table S10-4a

Time Intervals	Y	D3	D2	D1	D0
0-50ns		0	0	0	0
50ns-100ns		0	0	0	1
100ns-150ns		0	0	1	0
150ns-200ns		0	0	1	1
200ns-250ns		0	1	0	0
250ns-300ns		0	1	0	1
300ns-350ns		0	1	1	0
350ns-400ns		0	1	1	1
400ns-450ns		1	0	0	0
450ns-500ns		1	0	0	1
500ns-550ns		1	0	1	0
550ns-600ns		1	0	1	1
600ns-650ns		1	1	0	0
650ns-700ns		1	1	0	1
700ns-750ns		1	1	1	0
750ns-800ns		1	1	1	1

Table S10-4b

Demonstrated To
Instructor/FA _____ Date _____

Part 1 [ ]   Part 2 [ ]   Part 3 [ ]

DATA COLLECTION SHEET

Data Collection Sheets for Lab S10 Name _____

Date _____

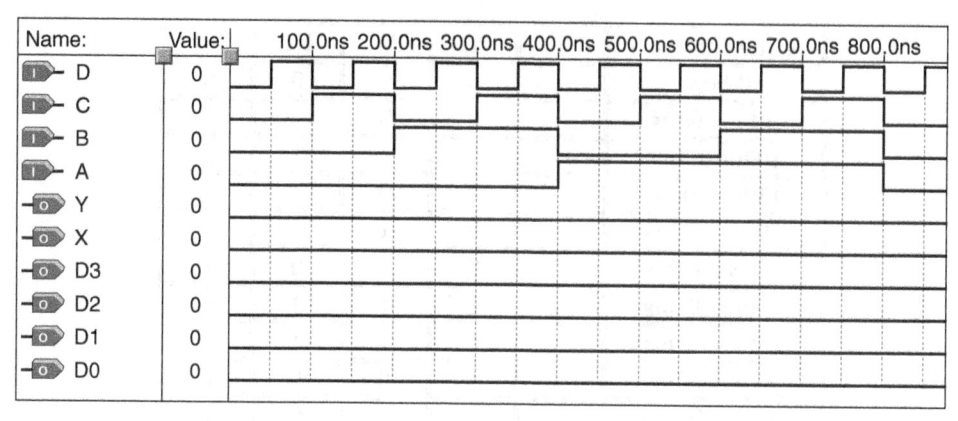

Figure S10-3

Data collection for Part 3:

m)

2) Write the Boolean expression for X:

X = _____

Figure S10-5

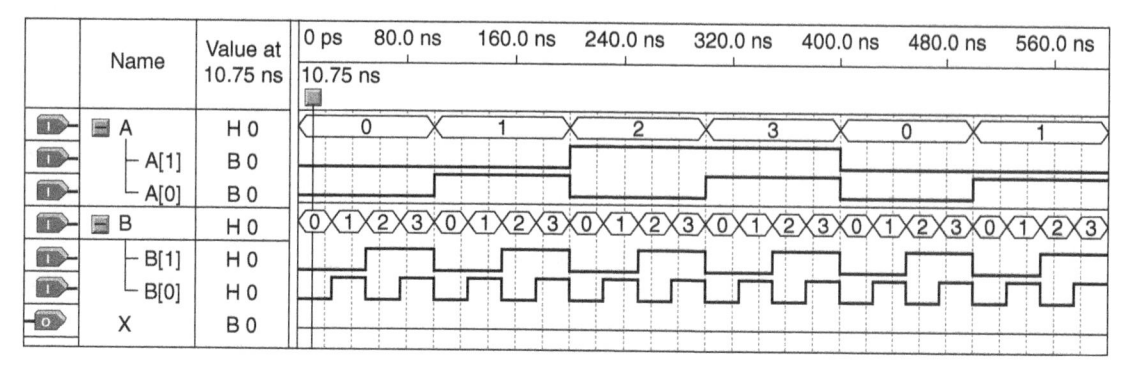

Figure S10-6

Demonstrated To
Instructor/FA _____ Date _____

Part 1 [ ] Part 2 [ ] Part 3 [ ]

Data Collection Sheets for Lab S10 Name _____
                                       Date _____

Time Intervals	X	A1	A0	B1	B0
0-50ns		0	0	0	0
50ns-100ns		0	0	0	1
100ns-150ns		0	0	1	0
150ns-200ns		0	0	1	1
200ns-250ns		0	1	0	0
250ns-300ns		0	1	0	1
300ns-350ns		0	1	1	0
350ns-400ns		0	1	1	1
400ns-450ns		1	0	0	0
450ns-500ns		1	0	0	1
500ns-550ns		1	0	1	0
550ns-600ns		1	0	1	1
600ns-650ns		1	1	0	0
650ns-700ns		1	1	0	1
700ns-750ns		1	1	1	0
750ns-800ns		1	1	1	1

Table S10-5

Demonstrated To
Instructor/FA _____ Date _____

Part 1 [ ]   Part 2 [ ]   Part 3 [ ]

Experiment S11

Name _____

IMPLEMENTING EXCLUSIVE-OR AND EXCLUSIVE-NOR CIRCUITS USING VHDL

OBJECTIVES

1. To practice using QUARTUS® II software to create schematics and waveforms.
2. To practice using QUARTUS® II software to compile and simulate digital circuits.
3. To implement exclusive-OR and exclusive-NOR circuits with VHDL.
4. To implement a digital comparator circuit with VHDL.
5. To program the EPM7128SLC84 CPLD.

TEXT REFERENCES

Read sections 2.9 and 4.6. Review Appendix D of this manual.

EQUIPMENT NEEDED

Components

Blank floppy, zip disk, or memory stick for storing projects;
QUARTUS® II software (Altera Corporation);
Optional: DeVry University Board eSOC with EPM7128SLC84 CPLD
   (or Altera University Board with EPM7128SLC84 CPLD or any other equivalent board);
Desktop computer with minimum of INTEL Pentium PC @ 400MHz CPU
   running Microsoft Windows NT4SP1, 2000, or XP or better; Pentium III
   or 4 PC @ 400MHz running Red Hat Linux v7.3 or 8.0 or Red Hat Linux
   Enterprise 3; Sun Ultra running Solaris v8 or 9; HP9000 Series 700/800
   running HP-UX v11.0 with ACE dated 11/1999 or later; or equivalent.

DISCUSSION

In Experiments S9 and S10, you were introduced to the exclusive-OR and exclusive-NOR circuits and their applications. In the current experiment, you will implement the exclusive-OR and exclusive-NOR circuits with VHDL. You will also implement a 4-bit magnitude comparator with VHDL.

PROCEDURE

PART 1 – Implementing the Exclusive-OR with VHDL

a) Start the QUARTUS® II program. Set up a new project and use **xor_gate** for the name of the project.
b) Create a new text file *xor_gate.vhd* for writing the XOR gate VHDL description.
c) Type in the VHDL program from the listing in Figure S11-1.

```
Library IEEE;
use IEEE.std_logic_1164.all;

ENTITY    xor_ckt IS
    PORT (      a, b        : IN BIT;
                y           : OUT BIT);
END xor_ckt;

ARCHITECTURE ckt OF xor_ckt IS
    BEGIN
        y <= a XOR b;
    END ckt;
```

Figure S11-1

d) Select **Project | Top-Level Entity**.
e) This project should be already set to use the EPM7128SLC84-15 CPLD. If necessary you can change the CPLD with the **Assignments | Device** window.
f) If you are using another board, check Appendix D for your device family, device, and speed.
g) Select **Processing | Start | Start Analysis & Synthesis** to synthesize a circuit that implements the given VHDL code. If there are errors, fix them and recompile the project.
h) Create a new Waveform Editor file with the name *xor_gate.vwf*. Use the same waveforms created in Exercise 2 of Appendix D.
i) On the Settings page, choose Functional Simulation mode and select *xor_gate.vwf* for the Simulation Input.
j) Select **Processing | Generate Functional Simulation Netlist**.
k) Start the simulation.
l) If there are no errors when the simulation is finished, go on to the next step. If there are errors and you change your circuit, remember to save and compile the circuit before re-simulating.

PROCEDURE

m) *(Optional) Program the CPLD:* It is suggested that you use **PIN_50** for "**a**," **PIN_51** for "**b**," and **PIN_4** for "**y**." To program your CPLD, refer to Appendix D, Exercise 3. A complete set of instructions will be found there. Briefly,
  1) *Recompile the project*: After making the pin assignments, compile the project again. When the compilation is completed, the software will display an information box showing any warnings and any detected errors. If there are errors, fix them and recompile the project.
  2) *Download the project to the board:*
      a. Set all eSOC DIP switches to OFF.
      b. Connect the board to the PC's parallel printer port and apply power.
      c. Select **Tools | Programmer**.
      d. Click on [Auto Detect] in the Chain Description File.
      e. When the eSOC board is recognized, press the [Add File...] button in the Chain Description File, then locate and Add the *xor_gate.pof* file.
      f. Select the file and device line and scroll (horizontally) to the available programming option. Check the **Program/Configure** box.
      g. Press the [Start] button. Observe the progress bar.
      h. If the program downloads successfully, test the program.
n) *Test the program:* Toggle switches A and B and fill out Table S11-1 with your observations. Transfer this information to the data collection sheets at the end of this experiment.
o) Save and close the project.

a	b	y
0	0	
0	1	
1	0	
1	1	

Table S11-1

PART 2 – Implementing the Exclusive-NOR with VHDL

a) Start the QUARTUS® II program. Set up a new project and use **xnor_gate** for the name of the project.
b) Create a new text file for writing the XNOR gate VHDL description. Save this file with a vhd extension.
c) Type in the VHDL program from the listing in Figure S11-2.

```
Library IEEE;
use IEEE.std_logic_1164.all;

ENTITY   xnor_ckt IS
    PORT (     a, b      : IN BIT;
               y         : OUT BIT);
END  xnor_ckt;

ARCHITECTURE ckt OF xnor_ckt IS
    BEGIN
        y <= a XNOR b;
    END ckt;
```

Figure S11-2

d) Select **Project | Top-Level Entity**.
e) This project should be already set to use the EPM7128SLC84-15 CPLD. If necessary you can change the CPLD with the **Assignments | Device** window.
f) If you are using another board, check Appendix D for your device family, device, and speed.
g) Select **Processing | Start | Start Analysis & Synthesis** to synthesize a circuit that implements the given VHDL code. If there are errors, fix them and recompile the project.
h) Create a new Waveform Editor file with the name *xnor_gate.vwf*. Use the same waveforms created in Exercise 2 of Appendix D.
i) On the Settings page, choose Functional Simulation mode and select *xnor_gate.vwf* for the Simulation Input.
j) Select **Processing | Generate Functional Simulation Netlist**.
k) Start the simulation.
l) If there are no errors when the simulation is finished, go on to the next step. If there are errors and you change your circuit, remember to save and compile the circuit before re-simulating.
m) *(Optional) Program the CPLD:* Program your CPLD, referring to Appendix D, Exercise 3. A complete set of instructions will be found there. It is suggested that you use **PIN_50** for "a," **PIN_51** for "b," and **PIN_4** for "y." To program your CPLD, refer to Appendix D, Exercise 3. A complete set of instructions will be found there. Briefly,

PROCEDURE

1) *Recompile the project*: After making the pin assignments, compile the project again. When the compilation is completed, the software will display an information box showing any warnings and any detected errors. If there are errors, fix them and recompile the project.
2) *Download the project to the board:*
   a. Set all eSOC DIP switches to OFF.
   b. Connect the board to the PC's parallel printer port and apply power.
   c. Select **Tools | Programmer**.
   d. Click on [Auto Detect] in the Chain Description File.
   e. When the eSOC board is recognized, press the [Add File...] button in the Chain Description File, then locate and Add the *xnor_gate.pof* file.
   f. Select the file and device line and scroll (horizontally) to the available programming option. Check the **Program/Configure** box.
   g. Press the [Start] button. Observe the progress bar.
   h. If the program downloads successfully, test the program.
n) *Test the program:* Toggle switches A and B and fill out Table S11-2 with your observations. Transfer this information to the data collection sheets at the end of this experiment.
o) Save and close the project.

a	b	y
0	0	
0	1	
1	0	
1	1	

Table S11-2

PART 3 – Implementing a 3-bit Comparator with VHDL

a) Start the MAX+PLUS® II program. Set up a new project and use **comp3_ckt** for the name of the project.
b) Create a new text file for writing the 3-bit comparator's VHDL description. Save this file with a vhd extension.

```
Library IEEE;
use IEEE.std_logic_1164.all;

ENTITY   comp3_ckt IS
    PORT (a,b:          IN INTEGER RANGE 0 to 7; -- Bit arrays of 3 bits
          less, equal, greater: OUT BIT);
END ENTITY comp3_ckt;

ARCHITECTURE comp of comp3_ckt is
BEGIN
    less    <= '0' when (a < b) else '1';
    equal   <= '0' when (a = b) else '1';
    greater <= '0' when (a > b) else '1';
END comp;
```

Figure S11-3

c) Type in the VHDL program from the listing in Figure S11-3.
d) Select **Project | Top-Level Entity**.
e) This project should be already set to use the EPM7128SLC84-15 CPLD. If necessary you can change the CPLD with the **Assignments | Device** window.
f) If you are using another board, check Appendix D for your device family, device, and speed.
g) Select **Processing | Start | Start Analysis & Synthesis** to synthesize a circuit that implements the given VHDL code. If there are errors, fix them and recompile the project.
h) Create a new Waveform Editor file with the name *comp3_ckt.vwf*. Use the same waveforms created in Exercise 2 of Appendix D.
i) On the Settings page, choose Functional Simulation mode and select *comp3_ckt.vwf* for the Simulation Input.
j) Select **Processing | Generate Functional Simulation Netlist**.
k) Start the simulation.
l) If there are no errors when the simulation is finished, go on to the next step. If there are errors and you change your circuit, remember to save and compile the circuit before re-simulating.
m) *(Optional) Program the CPLD:* Program your CPLD, referring to Appendix D, Exercise 3. A complete set of instructions will be found there. It is suggested that you use **PIN_50** for "a," **PIN_51** for "b," and **PIN_4** for "y." To program your CPLD, refer to Appendix D, Exercise 3. A complete set of instructions will be found there. Briefly,
   1) *Recompile the project*: After making the pin assignments, compile the project again. When the compilation is completed, the software will display an information box showing any warnings and any detected errors. If there are errors, fix them and recompile the project.
   2) *Download the project to the board:*
      a. Set all eSOC DIP switches to OFF.
      b. Connect the board to the PC's parallel printer port and apply power.

PROCEDURE

    c. Select **Tools | Programmer**.

    d. Click on [Auto Detect] in the Chain Description File.

    e. When the eSOC board is recognized, press the [Add File...] button in the Chain Description File, then locate and Add the *comp3_ckt.pof* file.

    f. Select the file and device line and scroll (horizontally) to the available programming option. Check the **Program/Configure** box.

    g. Press the [Start] button. Observe the progress bar.

    h. If the program downloads successfully, test the program.

n) *Test the program:* Toggle switches A and B and fill out Table S11-3 with your observations. Transfer this information to the data collection sheets at the end of this experiment.

o) Save and close the project.

A	B	Less	Equal	Greater
0	0			
1	2			
2	1			
3	4			
4	2			
5	6			
6	5			
7	7			

Table S11-3

Exper. S11

Data Collection Sheets for Lab S11   Name _____

Date _____

Data collection for Part 1:

a	b	y
0	0	
0	1	
1	0	
1	1	

Table S11-1

Data collection for Part 1:

a	b	y
0	0	
0	1	
1	0	
1	1	

Table S11-2

Data collection for Part 1:

A	B	Less	Equal	Greater
0	0			
1	2			
2	1			
3	4			
4	2			
5	6			
6	5			
7	7			

Table S11-3

Demonstrated To
Instructor/FA _____ Date _____

Part 1 [ ]   Part 2 [ ]   Part 3 [ ]

Experiment S12

Name _____

LATCHES AND D-TYPE FLIP-FLOPS

OBJECTIVES

1. To practice using QUARTUS® II software to create schematics and waveforms.
2. To practice using QUARTUS® II software to compile and simulate digital circuits.
3. To investigate the operation of the NAND gate SET/RESET latch.
4. To investigate the operation of the D latch.
5. To investigate the operation of the edge-triggered D flip-flop using hardware switch debouncing.
6. To investigate the operation of the edge-triggered D flip-flop using software switch debouncing.
7. *Optional:* To program the EPM7128SLC84 CPLD.

TEXT REFERENCES

Read sections 5.1, 5.2, 5.5, 5.7, 5.8, 5.10, 5.11, and glance at 5.24. Review Appendix D of this manual.

EQUIPMENT NEEDED

Components

Blank floppy, zip disk, or memory stick for storing projects;

QUARTUS® II software (Altera Corporation);

Optional: DeVry University Board eSOC with EPM7128SLC84 CPLD (or Altera University Board with EPM7128SLC84 CPLD or any other equivalent board);

Optional: Hardware switch debouncer;

Desktop computer with minimum of INTEL Pentium PC @ 400MHz CPU running Microsoft Windows NT4SP1, 2000, or XP or better; Pentium III or 4 PC @ 400MHz running Red Hat Linux v7.3 or 8.0 or Red Hat Linux Enterprise 3; Sun Ultra running Solaris v8 or 9; HP9000 Series 700/800 running HP-UX v11.0 with ACE dated 11/1999 or later; or equivalent;

TTL compatible signal or function generator.

DISCUSSION

All of the previous experiments have been concerned with learning the fundamentals of logic gates and combinatorial circuits. Recall that an output of such a device or circuit responds to changes in its inputs and that when its inputs are removed, the output may not be sustained. In this experiment, you will be introduced to a device that can sustain a given output even when its inputs are removed. Such a device is said to possess memory. Examples of memory devices include latches and flip-flops, which are the topics for this experiment. Concentration will be on the D-type latch and flip-flop and some of their parameters such as set-up and hold times.

PRE-LAB

This lab makes use of the concept of switch debouncing. Read section 5.1 in the text, especially the material in Example 5-2.

If you are going to program your CPLD and your instructor wants you to use an external hardware debounced switch, refer to the circuit given in the Preface of this lab manual. You should construct the debouncer on a solderless circuit board that can be dedicated for this purpose.

If you are going to program your CPLD and your instructor wants you to use an on-board software debounced switch, refer to the procedure given in Appendix D of this lab manual.

In a departure from the normal flow of the text, you will investigate the effect of noise on certain inputs. Be sure to glance at section 5.24, especially the paragraph on **Open Inputs**.

PROCEDURE

PART 1 – The NAND SET/RESET Latch

a) Start the QUARTUS® II program. Set up a new project and use **nand_latch** for the name of the project. The diagram of the circuit you are to construct and its input waveforms are given in Figures S12-1 and S12-2.
b) Create a new graphic file for the schematic, *nand_latch.bdf*.
c) Place the symbols on the schematic according to Figure S12-1.
d) Place the input and output connectors on the schematic. Label the inputs SET and RESET; label the outputs Q and QNOT.
e) Wire the input and output connectors to the circuit as shown in Figure S12-1.
f) Select **Project | Top-Level Entity**.
g) Compile the project. If there are errors, fix them and save the file before recompiling the project. If there are no errors (or perhaps a warning concerning timing at most), then go on to the next step.
h) Create a new waveform file *nand_latch.vwf* and define the inputs for simulation. The input waveforms should look similar to those in Figure S12-2.
i) Select **Assignments | Settings** to access the Settings Window.
j) Click on *Simulator* at the left.
k) Choose Functional from the *Simulator Mode* pull-down menu.

PROCEDURE

l) Find and use for the *Simulator Input* file *parity1.vwf* you just created.
m) Select **Processing | Generate Functional Simulation Netlist**.
n) Simulate the project. If there are any errors and you change your circuit, remember to save and compile the circuit before re-simulating.
o) Record and analyze results:
   1) Record the output waveforms Q and QNOT on Figure S12-2. You should complete this part of the experiment by converting the waveform information in Figure S12-2 to Table S12-1a.
   2) Transfer the waveform information and truth table to the data collection sheets at the end of this experiment.
p) *(Optional) Program the CPLD:* It is suggested that you use **PIN_50** for "SET," **PIN_51** for "RESET," **PIN_4** for "Q," and **PIN_5** for "QNOT." To program your CPLD, refer to Appendix D, Exercise 3. A complete set of instructions will be found there. Briefly,
   1) *Recompile the project*: After making the pin assignments, compile your project again. When the compilation is completed, the software will display an information box showing any warnings and any detected errors. If there are errors, fix them and recompile the project.
   2) *Download the project to the board:*
      a. Set all eSOC DIP switches to OFF.
      b. Connect the board to the PC's parallel printer port and apply power.
      c. Select **Tools | Programmer**.
      d. Click on [Auto Detect] in the Chain Description File.
      e. When the eSOC board is recognized, press the [Add File...] button in the Chain Description File, then locate and Add the *nand_latch.pof* file.
      f. Select the file and device line and scroll (horizontally) to the available programming option. Check the **Program/Configure** box.
      g. Press the [Start] button. Observe the progress bar.
      h. If the program downloads successfully, go to the next step. Otherwise, fix the problem and try downloading again.
q) *Test the program:* Toggle switches SET and RESET, and verify that the LED responds in a manner consistent with a NAND SET/RESET latch. Place your observations in Table S12-1b.
r) Transfer your observations from the simulation and program test results to the data collection sheets at the end of this experiment.
s) Save and close the project.

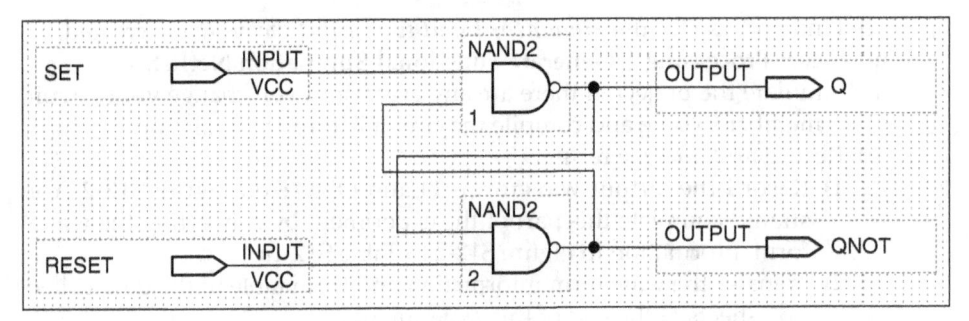

Figure S12-1

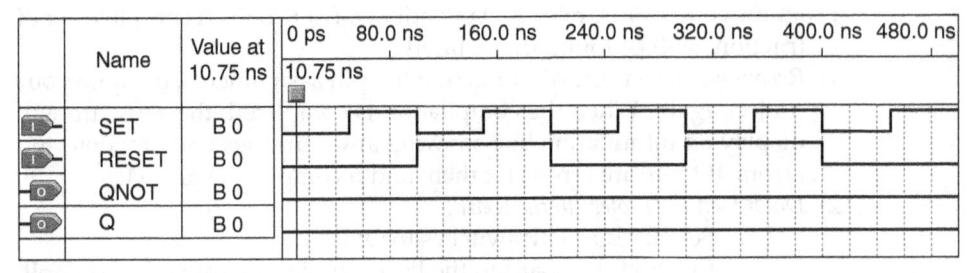

Figure S12-2

SET	RESET	Q	QNOT
0	0		
1	0		
0	1		
1	1		

Table S12-1a

SET	RESET	Q	QNOT
0	0		
1	0		
0	1		
1	1		

Table S12-1b

PART 2 – The D-type Latch

a) Start the QUARTUS® II program. Set up a new project and use **d_latch** for the name of the project. The diagram of the circuit you are to construct and its input waveforms are given in Figures S12-3 and S12-4.
b) Create a new graphic file for the schematic, *d_latch.bdf*.
c) Place the symbols on the schematic according to Figure S12-3.
d) Place the input and output connectors on the schematic. Label the inputs D and ENABLE; label the output Q.
e) Wire the input and output connectors to the circuit as shown in Figure S12-3.

PROCEDURE

f) Select **Project | Top-Level Entity**.
g) Compile the project. If there are errors, fix them and save the file before recompiling the project. If there are no errors (or perhaps a warning concerning timing at most), then go on to the next step.
h) Create a new waveform file *d_latch.vwf* and define the inputs for simulation. The input waveforms should look similar to those in Figure S12-4.
i) Select **Assignments | Settings** to access the Settings Window.
j) Click on *Simulator* at the left.
k) Choose *Functional* from the *Simulator Mode* pull-down menu.
l) Find and use for the *Simulator Input* file *d_latch.vwf* you just created above.
m) Select **Processing | Generate Functional Simulation Netlist**.
n) Simulate the project. If there are any errors and you change your circuit, remember to save and compile the circuit before re-simulating.
o) Record and analyze results. Record the output waveform on Figure S12-4. You should complete this part of the experiment by converting the waveform information in Figure S12-4 to Table S12-2.
p) *(Optional) Program the CPLD:* If your instructor wants you to program your CPLD, refer to Appendix D, Exercise 3. A complete set of instructions will be found there. It is suggested that you use **PIN_50** for "D," **PIN_51** for "ENABLE," and **PIN_4** for "Q."
q) *Test the program:* Toggle switches ENABLE and D, and verify that the LED responds in a manner consistent with a D-type latch. Place your observations in Table S12-2a.
r) Transfer your observations from the simulation and program test results to the data collection sheets at the end of this experiment.
s) Save and close the project.

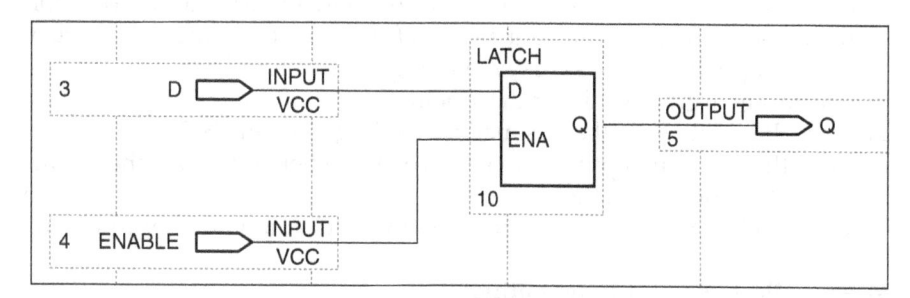

Figure S12-3

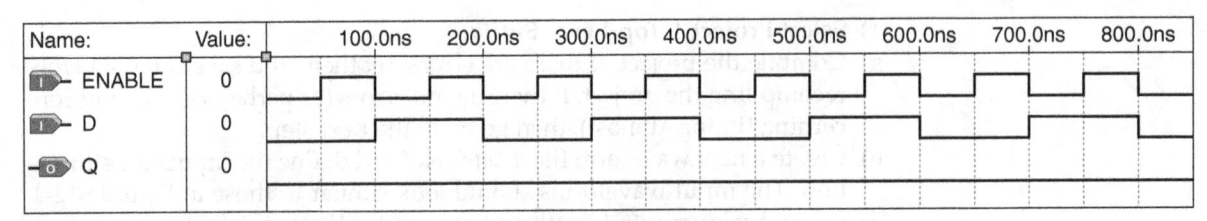

Name:	Value:	100.0ns	200.0ns	300.0ns	400.0ns	500.0ns	600.0ns	700.0ns	800.0ns
ENABLE	0								
D	0								
Q	0								

Figure S12-4

ENABLE	D	Q
0	0	
0	1	
1	0	
1	1	

Table S12-2

ENABLE	D	Q
0	0	
0	1	
1	0	
1	1	

Table S12-2a

PART 3 – The D-type Flip-Flop with Hardware Clock Debounce Circuit

a) In this part, you will investigate the positive edge-triggered D flip-flop. In order to clock edge-triggered flip-flops using switches, you will need to use an external, hardware debounced switch or on-board software debounced switch, preferably a pushbutton switch. In this part, an external hardware debounced switch is assumed. If your instructor prefers you to use an on-board switch with software debouncing, go on to Part 4 of this experiment.

b) Start the QUARTUS® II program. Set up a new project and use **d_ff** for the name of the project. The diagram of the circuit you are to construct and its input waveforms are given in Figures S12-5 and S12-6.

c) Create a new graphic file for the schematic, *d_ff.bdf*.

d) Place the symbols on the schematic according to Figure S12-5.

e) Place the input and output connectors on the schematic. Label the inputs D and ENABLE; label the output Q.

f) Wire the input and output connectors to the circuit as shown in Figure S12-5.

g) Select **Project | Top-Level Entity**.

h) Compile the project. If there are errors, fix them and save the file before recompiling the project. If there are no errors (or perhaps a warning concerning timing at most), then go on to the next step.

i) Create a new waveform file *d_ff.vwf* and define the inputs for simulation. The input waveforms should look similar to those in Figure S12-4. You should review the manual methods of creating waveforms in Exercise 1 of Appendix D before trying to create the "clr" waveform.

j) Select **Assignments | Settings** to access the Settings Window.

k) Click on *Simulator* at the left.

l) Choose *Functional* from the *Simulator Mode* pull-down menu.

m) Find and use for the *Simulator Input* file *d_ff.vwf* you just created above.

n) Select **Processing | Generate Functional Simulation Netlist**.

PROCEDURE

o) Simulate the project. If there are any errors and you change your circuit, remember to save and compile the circuit before re-simulating.
p) Record and analyze results. Record the output waveform on Figure S12-6. You should complete this part of the experiment by converting the waveform information in Figure S12-6 to Table S12-3.
q) *Program the CPLD:* It is suggested that you use **PIN_50** for "**D**," **PIN_73** for "**pre**," **PIN_74** for "**clr**," and **PIN_4** for "**Q**." You are going to use an external debounced switch or a TTL-compatible function generator at PLD pin 83 (Header J14 pin 5) for your global PLD clock, so this connection is through J14 pin 5. The jumper at JP1 must be removed. See Figure D-39 in Appendix D. To program your CPLD, refer to Appendix D, Exercise 3. A complete set of instructions will be found there. Briefly,
  1) *Recompile the project*: After making the pin assignments, compile your project again. When the compilation is completed, the software will display an information box showing any warnings and any detected errors. If there are errors, fix them and recompile the project.
  2) *Download the project to the board:*
     a. Set all eSOC DIP switches to OFF.
     b. Connect the board to the PC's parallel printer port and apply power.
     c. Select **Tools | Programmer**.
     d. Click on [Auto Detect] in the Chain Description File.
     e. When the eSOC board is recognized, press the [Add File...] button in the Chain Description File, then locate and Add the *d_ff.pof* file.
     f. Select the file and device line and scroll (horizontally) to the available programming option. Check the **Program/Configure** box.
     g. Press the [Start] button. Observe the progress bar.
     h. If the program downloads successfully, go to the next step. Otherwise, fix the problem and try downloading again.
r) *Test the program:* Toggle switch D, and verify that the LED # 1 responds in a manner consistent with a D-type FF when you pulse (push and release) the "**clk**" pushbutton. Alternately pulse the "**pre**" and "**clr**" pushbuttons and verify that the LED # 1 responds by indicating Q = 1 (LED is OFF) when you pulse the "**pre**" pushbutton and Q = 0 (LED is ON) when you pulse the "**clr**" pushbutton. Place your observations in Table S12-2a.
s) Transfer your observations from the simulation and program test results to the data collection sheets at the end of this experiment.
t) Save and close the project.

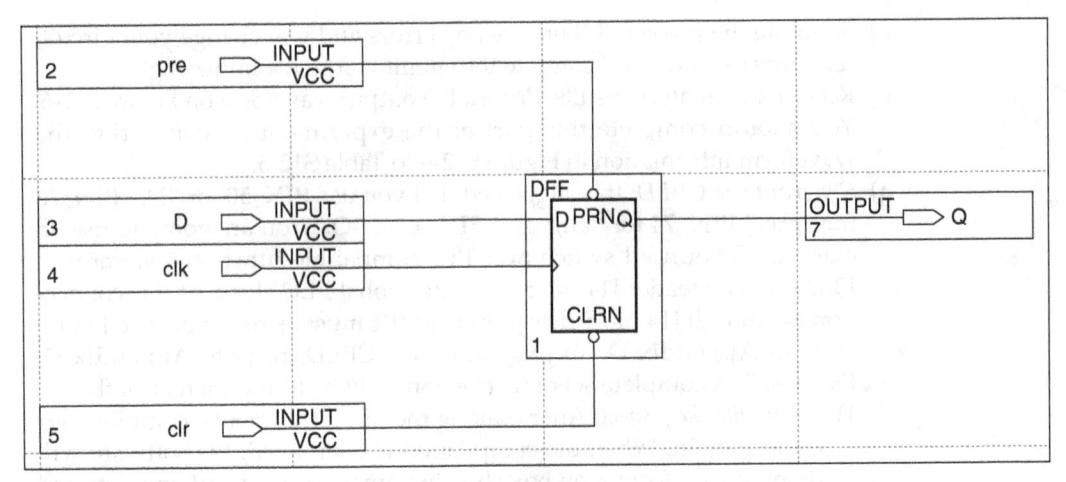

Figure S12-5

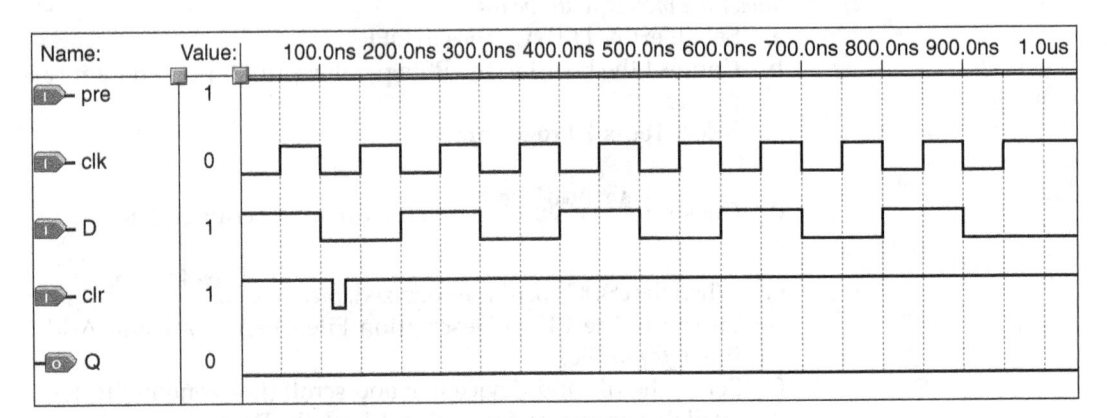

Figure S12-6

Preset	Reset	Clock	D	Q
1	1			
1	1			
0	1			
1	0			

Table S12-3

u) *The effect of noise on flip-flops:* The timing of the simulation in step k was pre-planned to demonstrate the response of the D flip-flop under ideal conditions. What if the flip-flop inputs are subjected to random noise "spikes"? Modify the waveform in Figure S12-6 by placing small pulses in the **"pre," "clr,"** and **"clk"** waveforms as shown in the suggested waveforms in Figure S12-7.

PROCEDURE

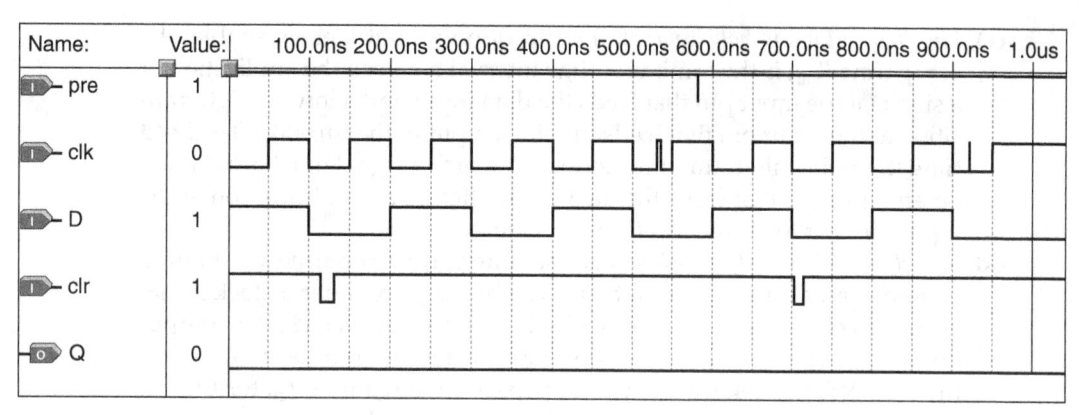

Figure S12-7

v) Select **Assignments | Settings** to access the Settings Window.
w) Click on *Simulator* at the left.
x) Choose *Timing* from the *Simulator Mode* pull-down menu.
y) Find and use for the *Simulator Input* file *d_ff.vwf* you just created.
z) Select **Processing | Generate Functional Simulation Netlist.**
aa) Simulate **Processing | Start Compilation** the project. If there are any errors and you change your circuit, remember to save and compile the circuit before re-simulating.
bb) Analyze the results:

1) You should have gotten several warnings from the simulator. In general, what were the warnings?

   _____
   _____
   _____
   _____

2) Compare the Q outputs for this simulation against that of the simulation done in step k (see Figure 12-6). Generally, what is different?

   _____
   _____

3) What did the noise spike on the "**clk**" signal (near 550ns) do to the Q waveform?

   _____

4) What did the noise spike on the "**pre**" signal (near 600ns) do to the Q waveform?

   _____

5) What did the noise spike on the "**clr**" signal (near 700ns) do to the Q waveform?

   _____

6) What did the noise spike on the "**clk**" signal (near 900ns) do to the Q waveform?

   _____

cc) *Flip-flop setup and hold times:* On a flip-flop without a clock enable, the **setup time (t_{su})** is the minimum time interval between the application of a signal at the input pin that feeds the data input and a low-to-high transition at the input pin that feeds the clock input of the flip-flop. The **hold time (t_h)** is the minimum time period for which a signal must be retained on an input pin that feeds the data input after an active transition at the input pin that feeds the flip-flop's clock input.

dd) *Flip-flop clock to output delay:* The maximum time required to obtain a valid output at an output pin that is fed by a register after a clock signal transition on an input pin that clocks the register is called clock to output delay (t_{co}). This time always represents an external pin-to-pin delay. In the QUARTUS® II software, you can specify the required t_{co} for the entire project and/or any clock signal, any register driving an output or bidirectional pin, or any output or bidirectional pin driven by a register. You can also specify a point-to-point (P2P) t_{co} requirement between a clock and a register, a clock and an output pin, or a register and an output pin.

ee) *Measure t_{su}, t_h, and t_{co}, (worst case):* After compilation, click on the Timing Analyzer Summary report (part of the Compilation Report) and display the report. Record the values from the Timing Analyzer Summary report in Table S12-4.

ff) *Measure t_{su}, t_h, and t_{co}, (required times):* To demonstrate the Timing Analyzer's function when there are requirements for t_{su} and t_h select **Assignments | Time Settings...** to display the Timing Requirements and Options page. Enter the values shown in Table S12-5 for t_{su}, t_h, and t_{co}. Recompile and record the values from the Timing Analyzer Summary report in Table S12-5. Compare these results with the worst case results in Table S12-4. You should note the t_{su}, t_h, and t_{co} meet requirements. To further validate this, check the QUARTUS® II Message window.

Note: Slack is the margin by which a timing requirement was met or not met. A positive slack value, displayed in black, indicates the margin by which a requirement was met. A negative slack value, displayed in red, indicates the margin by which a requirement was not met.

Type	Slack	Required Time	Actual Time	From	To	From Clock	To Clock	Failed paths
Worst-case tsu	N/A	None						
Worst-case tco	N/A	None						
Worst-case th	N/A	None						
Clock Setup: 'clk'	N/A	None						
Total number of failed paths								

Table S12-4

PROCEDURE

gg) *Program the CPLD:* To program your CPLD, refer to Appendix D, Exercise 3. A complete set of instructions will be found there. It is suggested that you use **PIN_50** for "**D**," **PIN_73** for "**pre**," **PIN_74** for "**clr**," and **PIN_4** for "**Q**."

Type	Slack Required Time	Actual Time	From	To	From Clock	To Clock	Failed paths
Worst-case tsu	20.000 ns						
Worst-case tco	20.000 ns						
Worst-case th	0.000 ns						
Clock Setup: 'clk'	30.00 MHz (period = 33.333 ns)						
Clock Hold: 'clk'	30.00 MHz (period = 33.333 ns)						
Total number of failed paths							

Table S12-5

hh) If you are going to connect an external function generator or external debounced switch to "**clk**," go on to step 3. Be sure to check the correct procedure for wiring this input to your board.

ii) If you are going to connect an on-board switch with software debouncing to "**clk**," add the clock divider and debounce symbols to your schematic and assign an on-board switch (suggestion: **PB1** at **PIN_70**) for the key-pressed input.

jj) To program your CPLD, refer to Appendix D, Exercise 3. A complete set of instructions will be found there. Briefly,

1) *Recompile the project*: After making the pin assignments, compile your project again. When the compilation is completed, the software will display an information box showing any warnings and any detected errors. If there are errors, fix them and recompile the project.

2) *Download the project to the board:*
   a. Set all eSOC DIP switches to OFF.
   b. Connect the board to the PC's parallel printer port and apply power.
   c. Select **Tools | Programmer**.
   d. Click on [Auto Detect] in the Chain Description File.
   e. When the eSOC board is recognized, press the [Add File...] button in the Chain Description File, then locate and Add the *d_ff.pof* file.
   f. Select the file and device line and scroll (horizontally) to the available programming option. Check the **Program/Configure** box.
   g. Press the [Start] button. Observe the progress bar.

      h. If the program downloads successfully, go to next the step. Otherwise, fix the problem and try downloading again.

kk) *Test the program:* Toggle switches **"pre," "clr,"** D, and **"clk"** (if using external or on-board switch) and verify that the LED responds in a manner consistent with a D-type flip-flop:

    1) If the Q output is HIGH, clear the flip-flop by momentarily pulsing the **"clr"** input LOW.
    2) Set the D input HIGH. Momentarily pulse the **"clk"** (if using external or on-board switch) input. Q = _____.
    3) Set the D input LOW. Momentarily pulse the **"clk"** (if using external or on-board switch) input. Q = _____.
    4) Momentarily pulse the **"pre"** input. Q = _____.
    5) Set the J input HIGH. Momentarily pulse the **"clk"** input (if using external or on-board switch).

ll) Fill out Table S12-6 with your observations in step kk and transfer this information to the data collection sheets at the end of this experiment.

mm) Save and close the project.

pre	clr	Clock	D	Q
1	1			
1	1			
0	1			
1	0			

Table S12-6

PART 4 – The D-type Flip-Flop with Software Clock Debouncing

a) In this part, you will investigate the positive edge-triggered D flip-flop, this time using an on-board software debounced switch.

b) Start the QUARTUS® II program. Set up a new project and use **d1_ff** for the name of the project. The circuit you are to construct is given in Figure S12-8.

c) If you haven't already done so, refer to Appendix D and create the symbols for a clock divider and switch debouncer.

d) Create a new graphic file for the schematic *d1_ff.bdf*.

e) Place the symbols on the schematic according to Figure S12-8.

f) Place the input and output connectors on the schematic. Label the inputs **"pre," "clr," "clk," and "pb3"**; label the output Q.

g) Wire the input and output connectors to the circuit as shown in Figure S12-8.

h) Set project to the current file.

i) Compile the project. If there are errors, fix them and save the file before recompiling the project. If there are no errors (or perhaps a warning concerning timing at most), then go on to the next step.

PROCEDURE 407

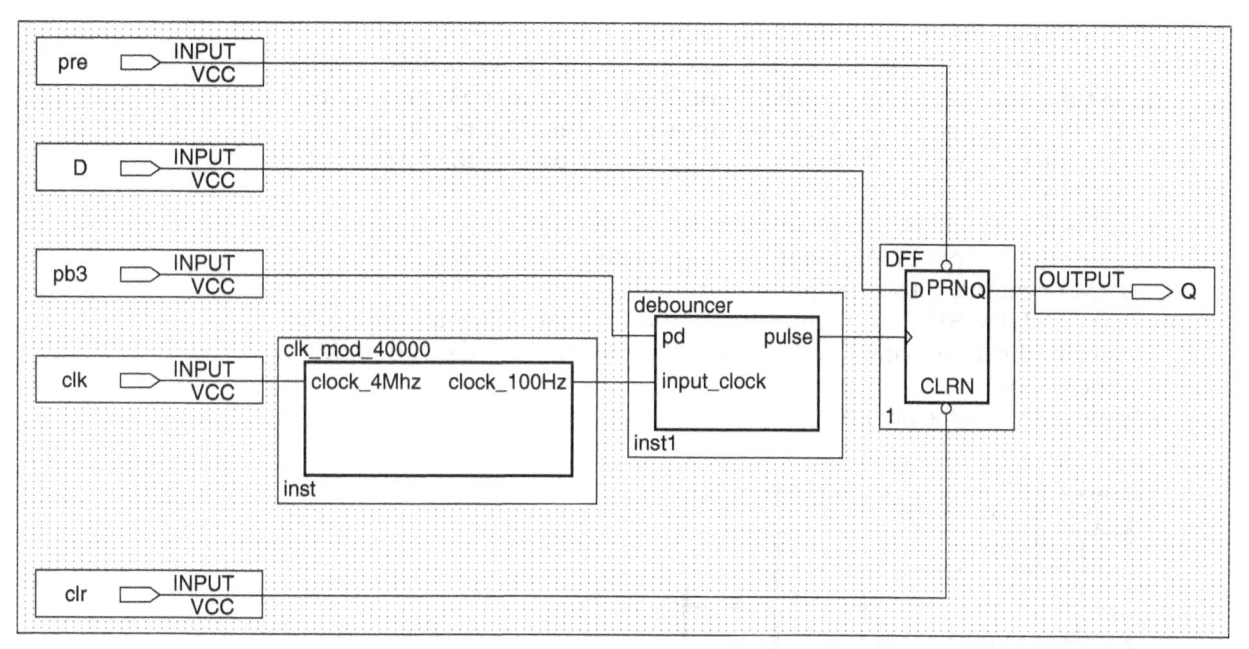

Figure S12-8

j) *Program the CPLD:* Program your CPLD with this project.
   1) Assign on-board pushbutton switches for **"pre," "clr,"** and **"pb3."** (If your board does not have pushbutton switches, toggle switches will do.)
   2) You will be using the board's clock signal (4MHz for the DeVry eSOC board) so be sure not to connect anything to pin 83.
   3) Assign a toggle switch for D and an LED for Q.
k) To program your CPLD, refer to Appendix D, Exercise 3. A complete set of instructions will be found there. Briefly,
   1) *Recompile the project*: After making the pin assignments, compile your project again. When the compilation is completed, the software will display an information box showing any warnings and any detected errors. If there are errors, fix them and recompile the project.
   2) *Download the project to the board:*
      a. Set all eSOC DIP switches to OFF.
      b. Connect the board to the PC's parallel printer port and apply power.
      c. Select **Tools | Programmer**.
      d. Click on *Auto Detect* in the Chain Description File.
      e. When the eSOC board is recognized, press the *Add File...* button in the Chain Description File, then locate and Add the *d1_ff.pof* file.
      f. Select the file and device line and scroll (horizontally) to the available programming option. Check the **Program/Configure** box.
      g. Press the *Start* button. Observe the progress bar.

h. If the program downloads successfully, test the program. Go to the next step. Otherwise, fix the problem and try downloading again.

l) *Test the program:* Pulse switches **"pre," "clr,"** D, and **"pb3"** and verify that the LED responds in a manner consistent with a D-type flip-flop:
   1) If the Q output is HIGH, reset the flip-flop by pulsing the **"clr"** input LOW.
   2) Set the D input HIGH. Momentarily pulse the **"pb3"** input. Q = _____.
   3) Set the D input LOW. Momentarily pulse the **"pb3"** input. Q = _____.
   4) Momentarily pulse the **"pre"** input. Q = _____.

m) Fill out Table S12-7 with your observations in the previous step and transfer this information to the data collection sheets at the end of this experiment.

n) Save and close the project.

pre	clr	clk	D	Q
1	1			
1	1			
0	1			
1	0			

Table S12-7

DATA COLLECTION SHEET

Data Collection Sheets for Lab S12   Name _____

                                        Date _____

Data collection for Part 1:

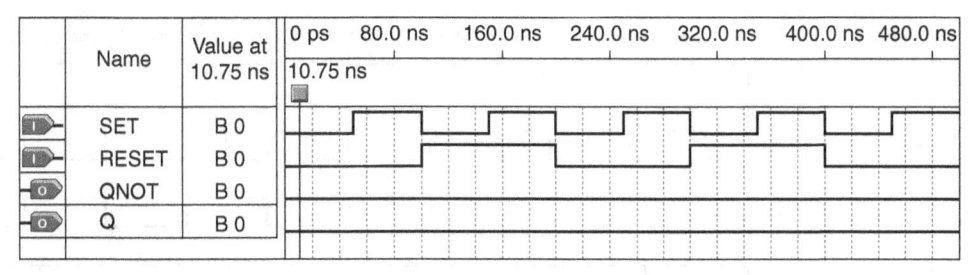

Name	Value at 10.75 ns
SET	B 0
RESET	B 0
QNOT	B 0
Q	B 0

Figure S12-2

SET	RESET	Q	QNOT
0	0		
1	0		
0	1		
1	1		

Table S12-1a

SET	RESET	Q	QNOT
0	0		
1	0		
0	1		
1	1		

Table S12-1b

Data collection for Part 2:

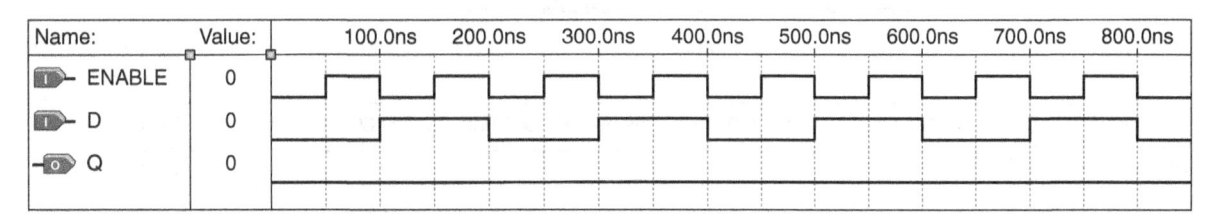

Name:	Value:
ENABLE	0
D	0
Q	0

Figure S12-4

ENABLE	D	Q
0	0	
0	1	
1	0	
1	1	

Table S12-2

ENABLE	D	Q
0	0	
0	1	
1	0	
1	1	

Table S12-2a

Demonstrated To
Instructor/FA _____ Date _____

Part 1 [ ]   Part 2 [ ]   Part 3 [ ]   Part 4 [ ]

Data Collection Sheets for Lab S12   Name _____

Date _____

Data collection for Part 3:

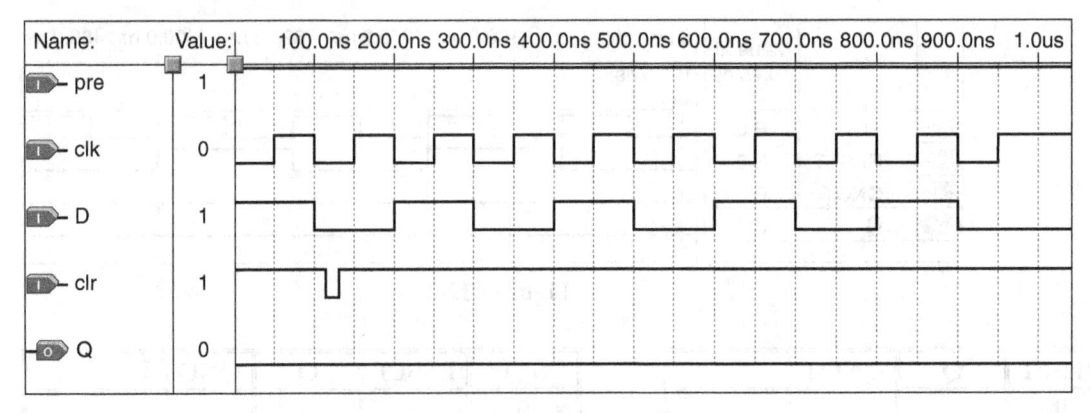

Figure S12-6

Preset	Reset	Clock	D	Q
1	1			
1	1			
0	1			
1	0			

Table S12-3

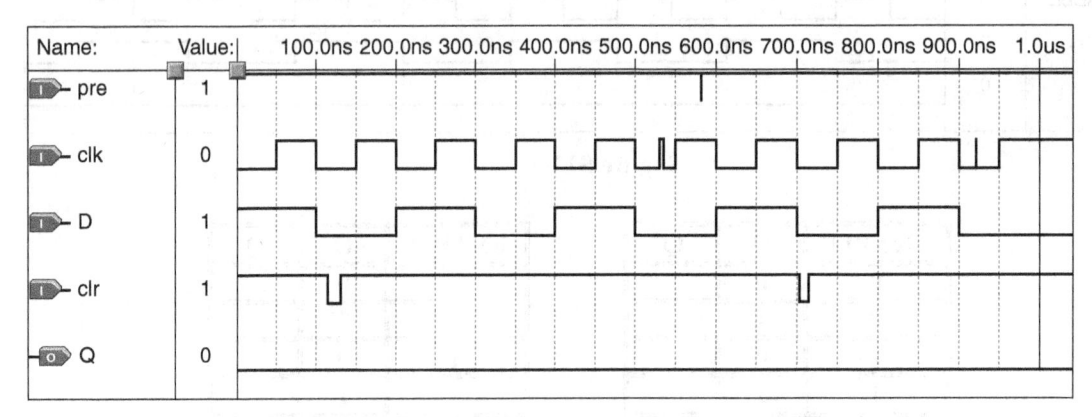

Figure S12-7

Demonstrated To
Instructor/FA _____ Date _____

Part 1 [ ]   Part 2 [ ]   Part 3 [ ]   Part 4 [ ]

DATA COLLECTION SHEET

Data collection Sheets for Lab S12   Name _____

Date _____

bb)

1) You should have gotten several warnings from the simulator. In general, what were the warnings?

   _____

   _____

2) Compare the Q outputs for this simulation against that of the simulation done in step k (see Figure 12-6). Generally, what is different?

   _____

3) What did the noise spike on the "**clk**" signal (near 550ns) do to the Q waveform?

   _____

4) What did the noise spike on the "**pre**" signal (near 600ns) do to the Q waveform?

   _____

5) What did the noise spike on the "**clr**" signal (near 700ns) do to the Q waveform?

   _____

6) What did the noise spike on the "**clk**" signal (near 900ns) do to the Q waveform?

   _____

Type	Slack	Required Time	Actual Time	From	To	From Clock	To Clock	Failed paths
Worst-case tsu	N/A	None						
Worst-case tco	N/A	None						
Worst-case th	N/A	None						
Clock Setup: 'clk'	N/A	None						
Total number of failed paths								

Table S12-4

Demonstrated To
Instructor/FA _____ Date _____

Part 1 [ ]   Part 2 [ ]   Part 3 [ ]   Part 4 [ ]

Data Collection Sheets for Lab S12   Name _____

Date _____

Type	Slack Required Time	Actual Time	From	To	From Clock	To Clock	Failed paths
Worst-case tsu	20.000 ns						
Worst-case tco	20.000 ns						
Worst-case th	0.000 ns						
Clock Setup: 'clk'	30.00 MHz (period = 33.333 ns)						
Clock Hold: 'clk'	30.00 MHz (period = 33.333 ns)						
Total number of failed paths							

Table S12-5

kk)
1) If the Q output is HIGH, clear the flip-flop by momentarily pulsing the "**clr**" input LOW.
2) Set the D input HIGH. Momentarily pulse the "**clk**" (if using external or on-board switch) input. Q = _____.
3) Set the D input LOW. Momentarily pulse the "**clk**" (if using external or on-board switch) input. Q = _____.
4) Momentarily pulse the "**pre**" input. Q = _____.
5) Set the J input HIGH. Momentarily pulse the "**clk**" input (if using external or on-board switch).

Preset	Clear	Clock	D	Q
1	1			
1	1			
0	1			
1	0			

Table S12-6

Demonstrated To
Instructor/FA _____ Date _____

Part 1 [ ]   Part 2 [ ]   Part 3 [ ]   Part 4 [ ]

DATA COLLECTION SHEET

Data Collection Sheets for Lab S12   Name _____

Date _____

Data collection for Part 4:

1)

    2) Set the D input HIGH. Momentarily pulse the **"pb3"** input. Q = _____.
    3) Set the D input LOW. Momentarily pulse the **"pb3"** input. Q = _____.
    4) Momentarily pulse the **"pre"** input. Q = _____.

pre	clr	Clock	D	Q
1	1			
1	1			
0	1			
1	0			

Table S12-7

Demonstrated To
Instructor/FA _____ Date _____

Part 1 [ ]   Part 2 [ ]   Part 3 [ ]   Part 4 [ ]

Experiment S13

Name _____

J-K AND T-TYPE FLIP-FLOPS

OBJECTIVES

1. To practice using QUARTUS® II software to create schematics and waveforms.
2. To practice using QUARTUS® II software to compile and simulate digital circuits.
3. To investigate the operation of the edge-triggered J-K flip-flop.
4. To investigate the operation of the edge-triggered T-type flip-flop.
5. To program the EPM7128SLC84 CPLD with flip-flop applications.

TEXT REFERENCES

Read sections 5.6, 5.8, and 5.13.

EQUIPMENT NEEDED

Components

Blank floppy, zip disk, or memory stick for storing projects;

QUARTUS® II software (Altera Corporation);

DeVry University Board eSOC with EPM7128SLC84 CPLD (or Altera University Board with EPM7128SLC84 CPLD or any other equivalent board);

Desktop computer with minimum of INTEL Pentium PC @ 400MHz CPU running Microsoft Windows NT4SP1, 2000, or XP or better; Pentium III or 4 PC @ 400MHz running Red Hat Linux v7.3 or 8.0 or Red Hat Linux Enterprise 3; Sun Ultra running Solaris v8 or 9; HP9000 Series 700/800 running HP-UX v11.0 with ACE dated 11/1999 or later; or equivalent;

TTL compatible signal or function generator.

DISCUSSION

In Experiment 12, you investigated latches and D-type flip-flops. You will complete your study of flip-flops by investigating the J-K and T-type (or toggle-type) flip-flops. The T-type flip-flop is essentially a J-K flip-flop with its J and K inputs tied HIGH. The output toggles whenever an active transition occurs at its clock input. The T-type flip-flop has application in counting and timing circuits of digital systems.

PROCEDURE

PART 1 – The J-K Flip-Flop

a) Start the QUARTUS® II program. Set up a new project and use **jk_ff** for the name of the project. The diagram of the circuit you are to construct and its input waveforms are given in Figures S13-1 and S13-2.
b) Create a new graphic file for the schematic, *jk_ff.bdf*.
c) Place the symbols on the schematic according to Figure S13-1.
d) Place the input and output connectors on the schematic. Label the inputs J, K, **"pre,"** **"clr,"** and **"clk"**; label the output Q.
e) Wire the input and output connectors to the circuit as shown in Figure S13-1.
f) Select **Project | Top-Level Entity**.
g) Compile the project. If there are errors, fix them and save the file before recompiling the project. If there are no errors (or perhaps a warning concerning timing at most), then go on to the next step.
h) Create a new waveform file *jk_ff.vwf* and define the inputs for simulation. The input waveforms should look similar to those in Figure S13-2. You should review the manual methods of creating waveforms in Exercise 1 of Appendix D before trying to create the **"clr"** waveform.
i) Select **Assignments | Settings** to access the Settings Window.
j) Click on *Simulator* at the left.
k) Choose *Functional* from the *Simulator Mode* pull-down menu.
l) Find and use for the *Simulator Input* file *jk_ff.vwf* you just created above.
m) Select **Processing | Generate Functional Simulation Netlist**.
n) Simulate the project. If there are any errors and you change your circuit, remember to save and compile the circuit before re-simulating.
o) Record and analyze results. Record the output waveform on Figure S13-2. You should complete this part of the experiment by converting the waveform information in Figure S13-2 to Table S13-1.
p) *Program the CPLD:* It is suggested that you use **PIN_50 for "J," PIN_51 for "K," PIN_73** for **"pre," PIN_74** for **"clr,"** and **PIN_4** for **"Q."** You are going to use an external debounced switch or a TTL-compatible function generator at PLD pin 83 (Header J14 pin 5) for your global PLD clock, so this connection is through J14 pin 5. The jumper at JP1 must be removed. See Figure D-39 in Appendix D. To program your CPLD, refer to Appendix D, Exercise 3. A complete set of instructions will be found there. Briefly,

PROCEDURE

1) *Recompile the project*: After making the pin assignments, compile your project again. When the compilation is completed, the software will display an information box showing any warnings and any detected errors. If there are errors, fix them and recompile the project.
2) *Download the project to the board:*
    a. Set all eSOC DIP switches to OFF.
    b. Connect the board to the PC's parallel printer port and apply power.
    c. Select **Tools | Programmer**.
    d. Click on [Auto Detect] in the Chain Description File.
    e. When the eSOC board is recognized, press the [Add File...] button in the Chain Description File, then locate and Add the *jk_ff.pof* file.
    f. Select the file and device line and scroll (horizontally) to the available programming option. Check the **Program/Configure** box.
    g. Press the [Start] button. Observe the progress bar.
    h. If the program downloads successfully, go to the next step. Otherwise, fix the problem and try downloading again.
q) *Test the program:* Toggle switches J and K, and verify that LED # 1 responds in a manner consistent with a J-K-type FF when **"clk"** is active. Alternately pulse the **"pre"** and **"clr"** pushbuttons and verify that LED # 1 responds by indicating Q = 1 (LED is OFF) when you pulse the **"pre"** pushbutton and Q = 0 (LED is ON) when you pulse the **clr** pushbutton.
r) Transfer your observations from the simulation and program test results to the data collection sheets at the end of this experiment.
s) Save and close the project.

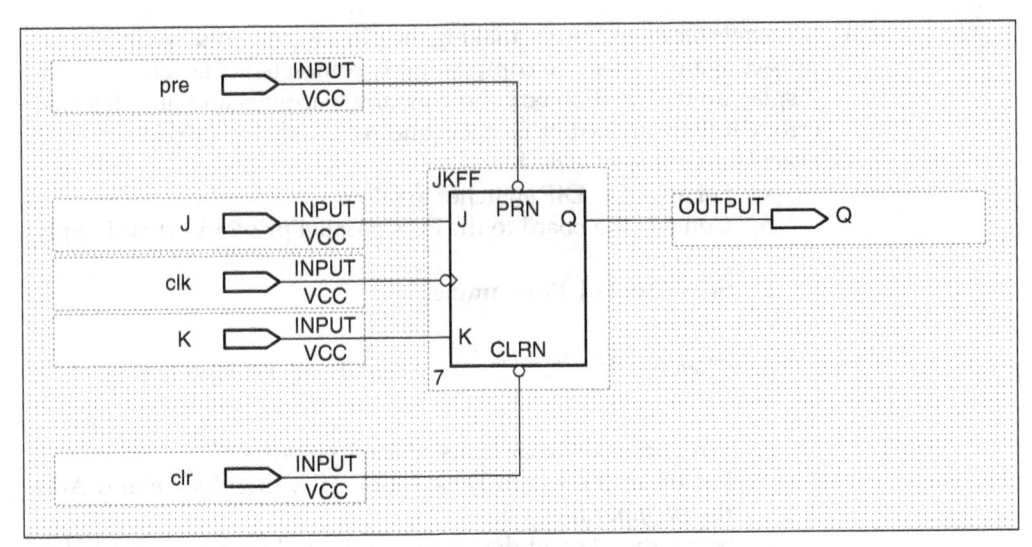

Figure S13-1

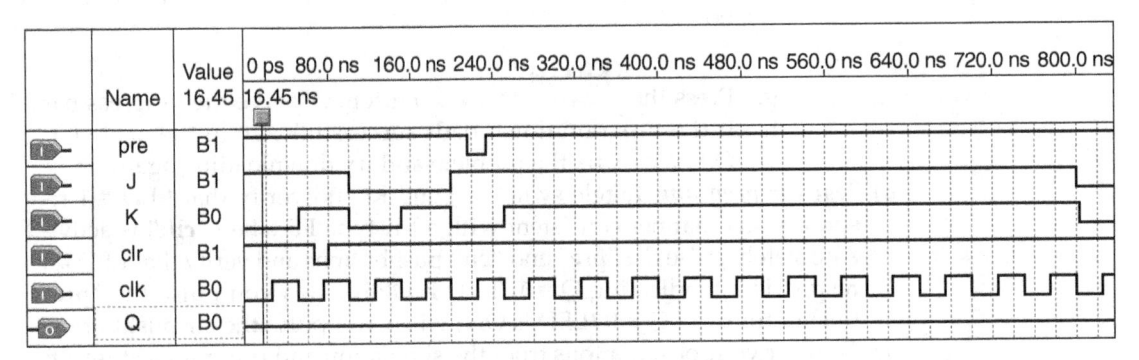

Figure S13-2

pre	clr	clk	J	K	Q
1	1	↑	0	0	
1	1	↑	1	0	
0	1	↑	0	1	
1	0	↑	1	1	

Table S13-1

t) *Flip-flop setup, hold times, and clock to output delay*: For definitions, refer to Experiment S12.

u) *Measure t_{su}, t_h, and t_{co} (worst case)*: After compilation, click on the Timing Analyzer Summary report (part of the Compilation Report) and display the report. Record the values from the Timing Analyzer Summary report in Table S13-2.

v) *Measure t_{su}, t_h, and t_{co} (required times)*: To demonstrate the Timing Analyzer's function when there are requirements for t_{su} and t_h select **Assignments | Time Settings...** to display the Timing Requirements and Options page. Enter the values shown in Table S13-3 for t_{su}, t_h, and t_{co}. Recompile and record the values from the Timing Analyzer Summary report in Table S13-3. Compare these results with the worst case results. You should note the t_{su}, t_h, and t_{co} meet requirements. To further validate this, check the QUARTUS® II Message window.

PROCEDURE

Note: Slack is the margin by which a timing requirement was met or not met. A positive slack value, displayed in black, indicates the margin by which a requirement was met. A negative slack value, displayed in red, indicates the margin by which a requirement was not met.

Type	Slack	Required Time	Actual Time	From	To	From Clock	To Clock	Failed paths
Worst-case tsu	N/A	None						
Worst-case tco	N/A	None						
Worst-case th	N/A	None						
Clock Setup: 'clk'	N/A	None						
Total number of failed paths								

Table S13-2

w) *Program the CPLD:* To program your CPLD, refer to Appendix D, Exercise 3. A complete set of instructions will be found there. It is suggested that you use **PIN_50** for "**J**," **PIN_51** for "**K**," **PIN_73** for "**pre**," **PIN_74** for "**clr**," and **PIN_4** for "**Q**."

Type	Slack	Required Time	Actual Time	From	To	From Clock	To Clock	Failed paths
Worst-case tsu		20.000 ns						
Worst-case tco		20.000 ns						
Worst-case th		0.000 ns						
Clock Setup: 'clk'		30.00 MHz (period = 33.333 ns)						
Clock Hold: 'clk'		30.00 MHz (period = 33.333 ns)						
Total number of failed paths								

Table S13-3

1) If you are going to connect an external function generator or external debounced switch to "**clk**," go on to step 3 below. Be sure to check the correct procedure for wiring this input to your board.
2) If you are going to connect an on-board switch with software debouncing to "**clk**," add the clock divider and debounce symbols to your schematic and assign an on-board switch (pushbutton preferably, suggested **PB1** at **PIN_70**) for the key-pressed input.
3) Recompile your project.
4) Refer to Appendix D for instructions to complete the programming of the CPLD.

x) *Test the program:* Toggle switches "**pre**," "**clr**," J, K, and "**clk**" (if using external or on-board switch) and verify that the LED responds in a manner consistent with a J-K-type flip-flop:
  1) If the Q output is HIGH, clear the flip-flop by momentarily pulsing the "**clr**" input LOW.
  2) Set the J input HIGH and K to LOW. Momentarily pulse the "**clk**" (if using external or on-board switch) input. Q = _____.

3) Set the J input LOW. Momentarily pulse the **"clk"** (if using external or on-board switch) input. Q = _____.
4) Set the K input HIGH. Momentarily pulse the **"clk"** (if using external or on-board switch) input.
5) Momentarily pulse the **"pre"** input. Q = _____.
6) Set the J input HIGH. Momentarily pulse the **"clk"** input (if using external or on-board switch).

y) Fill out Table S13-4 with your observations in step **x** and transfer this information to the data collection sheets at the end of this experiment.

z) Save and close the project.

pre	clr	clk	J	K	Q
1	1	↑	0	0	
1	1	↑	1	0	
0	1	↑	0	1	
1	0	↑	1	1	

Table S13-4

PART 2 – The T-type Flip-Flop

a) Start the QUARTUS® II program. Set up a new project and use **t_ff** for the name of the project. The diagram of the circuit you are to construct and its input waveforms are given in Figures S13-4 and S13-5.
b) Create a new graphic file for the schematic, *t_ff.bdf*.
c) Place the flip-flop symbol on the schematic according to Figure S13-4.
d) Place the input and output connectors on the schematic. Label the inputs T, **"pre," "clr,"** and **"clk"**; label the output Q.
e) Wire the input and output connectors to the circuit as shown in Figure S13-4.
f) Select **Project | Top-Level Entity**.
g) Compile the project. If there are errors, fix them and save the file before recompiling the project. If there are no errors (or perhaps a warning concerning timing at most), then go on to the next step.
h) Create a new waveform file *t_ff.vwf* and define the inputs for simulation. The input waveforms should look similar to those in Figure S13-5. Note that both **"pre"** and **"clr"** waveforms are a constant "1." To create the constant "1," select the node names **"pre"** and **"clr"** on the waveform window and press [1].
i) Select **Assignments | Settings** to access the Settings Window.
j) Click on *Simulator* at the left.
k) Choose *Functional* from the *Simulator Mode* pull-down menu.
l) Find and use for the *Simulator Input* file *t_ff.vwf* you just created above.
m) Select **Processing | Generate Functional Simulation Netlist.**
n) Simulate the project. If there are any errors and you change your circuit, remember to save and compile the circuit before re-simulating.

PROCEDURE

o) Record and analyze results. Record the output waveform on Figure S13-6. You should complete this part of the experiment by converting the waveform information in Figure S13-6 to Table S13-5.

p) *Program the CPLD.* It is suggested that you use **PIN_50 for "T," PIN_73 for "pre," PIN_74 for "clr,"** and **PIN_4 for "Q."** To program your CPLD, refer to Appendix D, Exercise 3. A complete set of instructions will be found there. Briefly,

1) *Recompile the project*: After making the pin assignments, compile your project again. When the compilation is completed, the software will display an information box showing any warnings and any detected errors. If there are errors, fix them and recompile the project.

2) *Download the project to the board:*
   a. Set all eSOC DIP switches to OFF.
   b. Connect the board to the PC's parallel printer port and apply power.
   c. Select **Tools | Programmer**.
   d. Click on [Auto Detect] in the Chain Description File.
   e. When the eSOC board is recognized, press the [Add File...] button in the Chain Description File, then locate and Add the *t_ff.pof* file.
   f. Select the file and device line and scroll (horizontally) to the available programming option. Check the **Program/Configure** box.
   g. Press the [Start] button. Observe the progress bar.
   h. If the program downloads successfully, go to the next step. Otherwise, fix the problem and try downloading again.

q) You are going to use an external TTL-compatible function generator at PLD pin 83 (Header J14 pin 5) for your global PLD clock, so this connection is through J14 pin 5. Connect PLD pin 4 to an oscilloscope set up to measure 5V DC vertical and 1 kHz horizontal. See Figure S13-3.

r) Jumper 1 at JP1 on the eSOC board must be removed. See Figure D-39 in Appendix D.

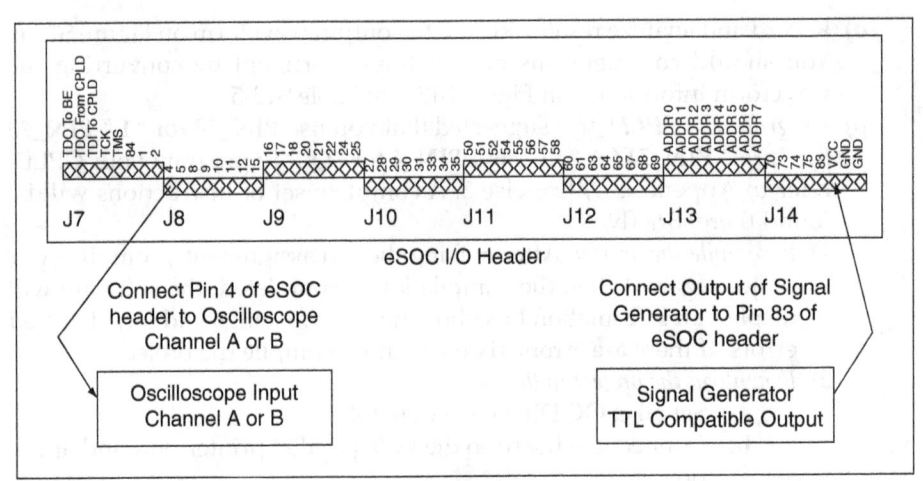

Figure S13-3

s) *Test the program:* With the eSOC board off, connect the setup in Figure S13-3 and then apply power to the eSOC board.
t) Observe the output at Pin 4 of the PLD on the scope.
u) Transfer your observations from the simulation and program test results to the data collection sheets at the end of this experiment.
v) *Analyze results:*
   1) What is the simulated frequency of "clk"? f_{CLK} = _____
   2) Compute the simulated frequency of Q: F_Q = _____
   3) The relationship between f_{CLK} and F_Q is F_Q = _____ x f_{CLK}
   4) Repeat 1) – 4) for the programmed CPLD.
   5) f_{CLK} = _____
   6) Measure the frequency of Q at PIN_4: F_Q = _____
   7) The relationship between f_{CLK} and F_Q is F_Q = _____ x f_{CLK}
   8) Are the results of 3) and 7) the same?
w) Save and close the project.

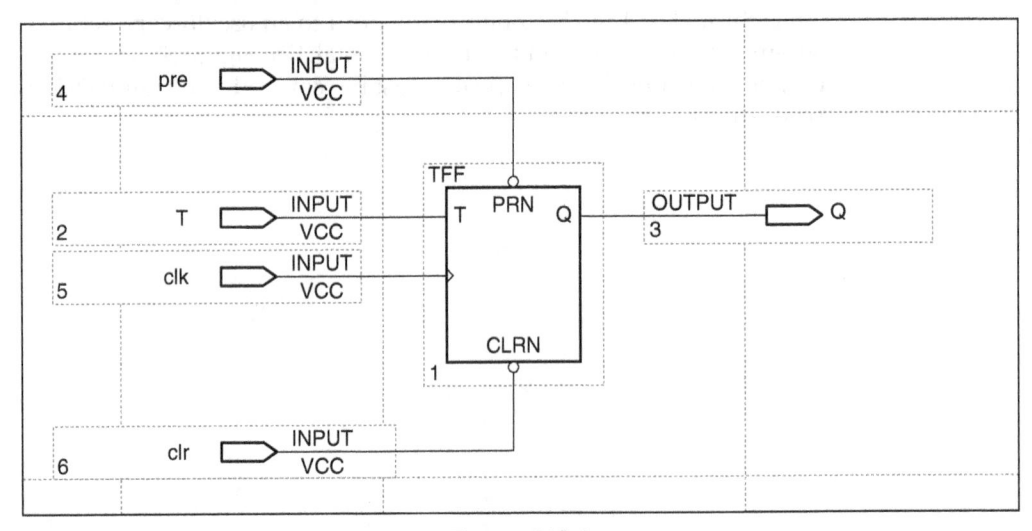

Figure S13-4

PROCEDURE

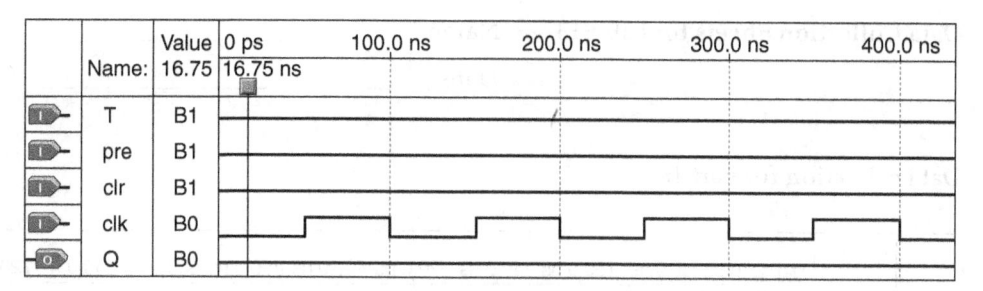

	Name:	Value 16.75	0 ps 16.75 ns	100.0 ns	200.0 ns	300.0 ns	400.0 ns
▷	T	B1					
▷	pre	B1					
▷	clr	B1					
▷	clk	B0		⎍⎍⎍⎍			
○	Q	B0					

Figure S13-5

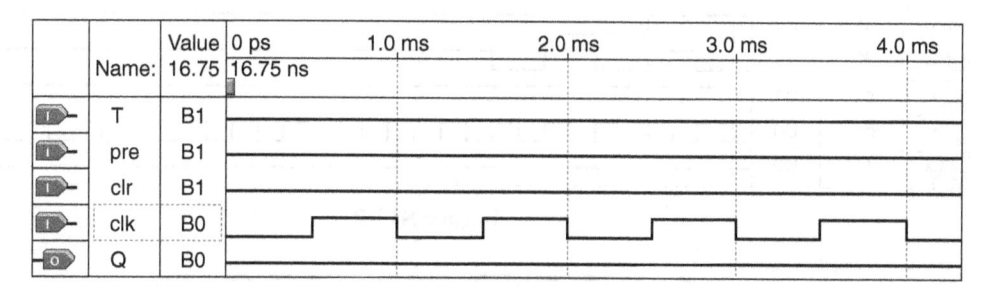

	Name:	Value 16.75	0 ps 16.75 ns	1.0 ms	2.0 ms	3.0 ms	4.0 ms
▷	T	B1					
▷	pre	B1					
▷	clr	B1					
▷	clk	B0		⎍⎍⎍⎍			
○	Q	B0					

Figure S13-6

pre	clr	clk	T	Q
1	1	↑	0	
1	1	↑	1	
0	1	↑	0	
1	0	↑	1	

Table S13-5

Data Collection Sheets for Lab S13   Name _____

Date _____

Data collection for Part 1:

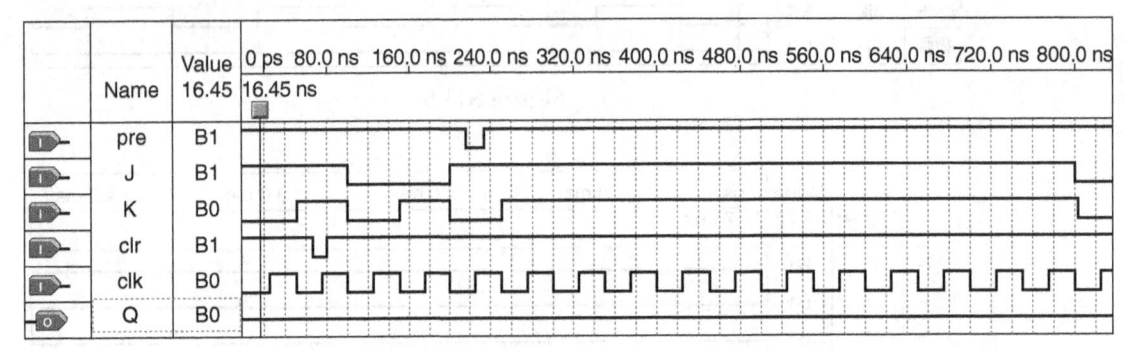

	Name	Value 16.45	0 ps 80.0 ns 160.0 ns 240.0 ns 320.0 ns 400.0 ns 480.0 ns 560.0 ns 640.0 ns 720.0 ns 800.0 ns
▷	pre	B1	
▷	J	B1	
▷	K	B0	
▷	clr	B1	
▷	clk	B0	
○	Q	B0	

Figure S13-2

pre	clr	clk	J	K	Q
1	1	↑	0	0	
1	1	↑	1	0	
0	1	↑	0	1	
1	0	↑	1	1	

Table S13-1

Type	Slack	Required Time	Actual Time	From	To	From Clock	To Clock	Failed paths
Worst-case tsu	N/A	None						
Worst-case tco	N/A	None						
Worst-case th	N/A	None						
Clock Setup: 'clk'	N/A	None						
Total number of failed paths								

Table S13-2

Demonstrated To
Instructor/FA _____ Date _____

Part 1 [ ]   Part 2 [ ]

DATA COLLECTION SHEET

Data Collection Sheets for Lab S13   Name _____

Date _____

Type	Slack	Required Time	Actual Time	From	To	From Clock	To Clock	Failed paths
Worst-case tsu		20.000 ns						
Worst-case tco		20.000 ns						
Worst-case th		0.000 ns						
Clock Setup: 'clk'		30.00 MHz (period = 33.333 ns)						
Clock Hold: 'clk'		30.00 MHz (period = 33.333 ns)						
Total number of failed paths								

Table S13-3

x)

1) If the Q output is HIGH, clear the flip-flop by momentarily pulsing the **"clr"** input LOW.
2) Set the J input HIGH and K to LOW. Momentarily pulse the **"clk"** (if using external or on-board switch) input. Q = ____.
3) Set the J input LOW. Momentarily pulse the **"clk"** (if using external or on-board switch) input. Q = ____.
4) Set the K input HIGH. Momentarily pulse the **"clk"** (if using external or on-board switch) input.
5) Momentarily pulse the **"pre"** input. Q = ____.

pre	clr	clk	J	K	Q
1	1	↑	0	0	
1	1	↑	1	0	
0	1	↑	0	1	
1	0	↑	1	1	

Table S13-4

Demonstrated To
Instructor/FA _____ Date _____

Part 1 [ ]   Part 2 [ ]

Data Collection Sheets for Lab S13   Name _____

Date _____

Data collection for Part 2:
  v)
   1) What is the simulated frequency of "clk"?
      f_{CLK} = _____
   2) Compute the simulated frequency of Q:
      F_Q = _____
   3) The relationship between f_{CLK} and F_Q is
      F_Q = _____ × f_{CLK}
   4) Repeat 1) – 4) for the programmed CPLD.
   5) f_{CLK} = _____
   6) Measure the frequency of Q at PIN_4:
      F_Q = _____
   7) The relationship between f_{CLK} and F_Q is
      F_Q = _____ × f_{CLK}
   8) Are the results of 3) and 7) the same?

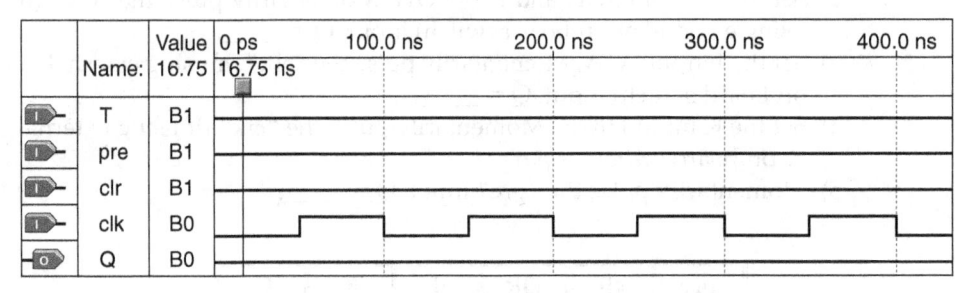

Figure S13-5

Figure S13-6

Demonstrated To	
Instructor/FA _____ Date _____	
Part 1 [ ] Part 2 [ ]	

DATA COLLECTION SHEET

Data Collection Sheets for Lab S13 Name _____

Date _____

pre	clr	clk	T	Q
1	1	↑	0	
1	1	↑	1	
0	1	↑	0	
1	0	↑	1	

Table S13-5

Demonstrated To
Instructor/FA _____ Date _____

Part 1 [ ]   Part 2 [ ]

Experiment S14

Name _____

FLIP-FLOP APPLICATIONS

OBJECTIVES

1. To practice using QUARTUS® II software to create schematics and waveforms.
2. To practice using QUARTUS® II software to compile and simulate digital circuits.
3. To investigate the application of J-K flip-flops in counting circuits.
4. To investigate the application of D flip-flops in data registers.
5. To investigate parallel data transfer.
6. To program the EPM7128SLC84 CPLD.

TEXT REFERENCES

Read sections 5-17 through 5-19.

EQUIPMENT NEEDED

Components

Blank floppy, zip disk or memory stick for storing projects;

QUARTUS® II software (Altera Corporation);

Optional: DeVry University Board eSOC with EPM7128SLC84 CPLD (or Altera University Board with EPM7128SLC84 CPLD or any other equivalent board);

Desktop computer with minimum of INTEL Pentium PC @ 400MHz CPU running Microsoft Windows NT4SP1, 2000, or XP or better; Pentium III or 4 PC @ 400MHz running Red Hat Linux v7.3 or 8.0 or Red Hat Linux Enterprise 3; Sun Ultra running Solaris v8 or 9; HP9000 Series 700/800 running HP-UX v11.0 with ACE dated 11/1999 or later; or equivalent;

TTL compatible signal or function generator.

DISCUSSION

Now that you have become acquainted with the fundamentals of flip-flops, you are ready to apply your knowledge in designing and constructing circuits that use flip-flops. Two very important applications of flip-flops are counters and registers.

Counters

A counter is a digital device capable of producing an output, which represents a sequence of numbers. The output count of the counter is binary and is usually triggered by a clock signal. Each bit of the binary count is produced by a flip-flop. Among the various types of counters used in digital circuits is the so-called MOD counter. The MOD counter has a count sequence that may start at 0 and end at some number $N-1$ and repeat. The number N is called the modulus (MOD-number) of the counter and is equal to 2^M, where M is the number of flip-flops. The number N can be changed for a given MOD counter. In the current experiment, you will investigate these principles and also be asked to design a counter whose MOD-number can be changed by "programming" the counter.

Registers

While counters are a very important application of flip-flops, probably the most common application is the register. A register is simply a collection of flip-flops used for storing information (data) temporarily. In the current experiment, you will construct a simple register for storing data, and then investigate a method of transferring data to and from such registers.

PROCEDURE

PART 1 – A MOD-8 Asynchronous Counter

a) Start the QUARTUS® II program. Set up a new project and use **amod8_ctr** for the name of the project. The diagram of the circuit you are to construct and its input waveforms are given in Figures S14-1 and S14-2.
b) Create a new graphic file *amod8_ctr.bdf* for the schematic.
c) Place the symbols on the schematic according to Figure S14-1.
d) Place the input and output connectors on the schematic. Label the inputs SET and CLEAR; label the outputs Q and QNOT.
e) Wire the input and output connectors to the circuit as shown in Figure S14-1.
f) Select **Project | Top-Level Entity**.
g) Compile the project. If there are errors, fix them and save the file before recompiling the project. If there are no errors (or perhaps a warning concerning timing at most), then go on to the next step.
h) *Measure worst-case* t_{co}: Without any design requirements assigned, open the tco report, which is part of Compilation Report (under Timing Analyzer). Input the appropriate data in Table S14-1.

PROCEDURE

		Actual			From
Slack	Required tco	**tco**	From	To	Clock
N/A	None				
N/A	None				
N/A	None				

Table S14-1

i) Create a new waveform file *amod8_ctr.vwf* and define the inputs for simulation. The input waveforms should look similar to those in Figure S14-2.
j) Select **Assignments | Settings** to access the Settings Window.
k) Click on *Simulator* at the left.
l) Choose *Functional* from the *Simulator Mode* pull-down menu.
m) Find and use for the *Simulator Input* file *amod8_ctr.vwf* you just created above.
n) Select **Processing | Generate Functional Simulation Netlist**.
o) Simulate the project. If there are any errors and you change your circuit, remember to save and compile the circuit before re-simulating.
p) Record and analyze results. Record the output waveform on Figure S14-2. You should complete this part of the experiment by converting the waveform information in Figure S14-2 to Table S14-1. This type of logic table is called a state table.
q) *Program the CPLD*. It is suggested that you use **PIN_4** for "QA," **PIN_5** for "QB," and **PIN_8** for "QC." You are going to use an external debounced switch or a TTL-compatible function generator at PLD pin 83 (Header J14 pin 5) for your global PLD clock, so this connection is through J14 pin 5. The jumper at JP1 must be removed. See Figure D-39 in Appendix D. Also refer to Figure S13-3 in Experiment S13 for wiring information. To program your CPLD, refer to Appendix D, Exercise 3. A complete set of instructions will be found there. Briefly,
  1) *Recompile the project*: After making the pin assignments, compile the project again. When the compilation is completed, the software will display an information box showing any warnings and any detected errors. If there are errors, fix them and recompile the project.
  2) *Download the project to the board:*
     a. Set all eSOC DIP switches to OFF.
     b. Connect the board to the PC's parallel printer port and apply power.
     c. Select **Tools | Programmer**.
     d. Click on [Auto Detect] in the Chain Description File.
     e. When the eSOC board is recognized, press the [Add File...] button in the Chain Description File, then locate and Add the *amod8_ctr.pof* file.

f. Select the file and device line and scroll (horizontally) to the available programming option. Check the **Program/Configure** box.

g. Press the [Start] button. Observe the progress bar.

h. If the program downloads successfully, test the program.

r) *Test the program:* After connections are made to your board, apply power and verify that the LEDs respond in a manner consistent with a MOD-8 counter. Place your observations in Table S14-2.

s) Transfer your observations from the simulation and program test results to the data collection sheets at the end of this experiment.

  1) The frequency of the clock signal is _____ MHz.
  2) The frequency of QC is _____ MHz.
     QC divides "**clk**" by _____.
  3) The frequency of QB is _____ MHz.
     QB divides "**clk**" by _____.
  4) The frequency of QA is _____ MHz.
     QA divides "**clk**" by _____.
  5) The MOD number of the counter is _____.
  6) Record the output waveforms QC, QB, and QA on Figure S14-2. You should complete this part of the experiment by converting the waveform information in Figure S14-2 to Table S14-2:
     1. Place the Time Bar in the 0 – 100ns time interval.
     2. Read the values for QA, QB, and QC in the Values column of the waveform diagram and place them in their corresponding columns in Table S14-3.
     3. Repeat steps 2 and 3 for each time interval.
     4. Tables S14-2 and S14-3 are referred to as the **state tables** for the counter.
  7) Does the counter count up or down? _____
  8) Transfer the answers to the above questions, the waveform information, state table, and propagation delay measurement to the data collection sheets at the end of this experiment.

t) Save and close your project.

PROCEDURE

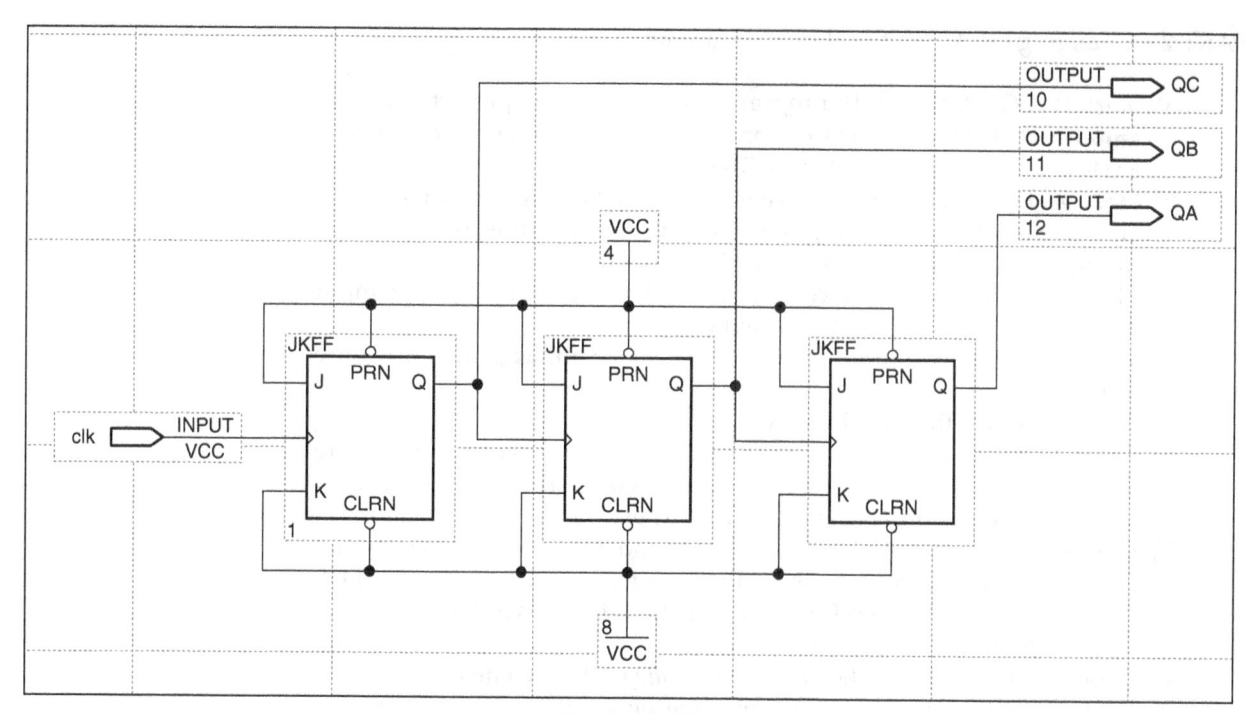

Figure S14-1

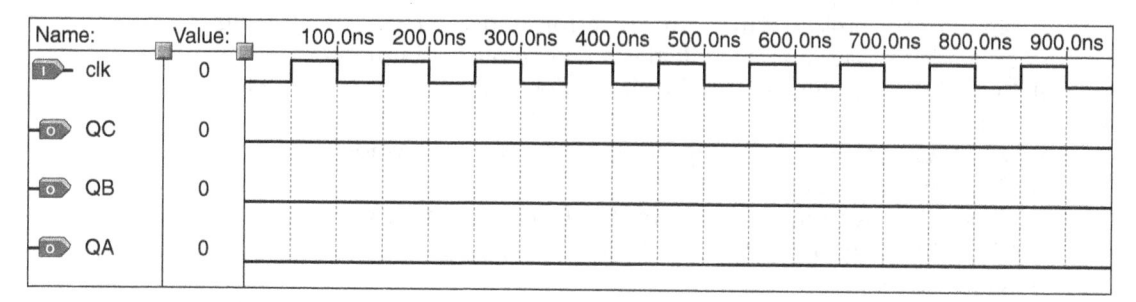

Figure S14-2

clk	Time Interval	QA	QB	QC
0	0 – 100ns			
1	100ns – 200ns			
2	200ns – 300ns			
3	300ns – 400ns			
4	400ns – 500ns			
5	500ns – 600ns			
6	600ns – 700ns			
7	700ns – 800ns			

Table S14-2

clk	Time Interval	QA	QB	QC
0	0 – 100ns			
1	100ns – 200ns			
2	200ns – 300ns			
3	300ns – 400ns			
4	400ns – 500ns			
5	500ns – 600ns			
6	600ns – 700ns			
7	700ns – 800ns			

Table S14-3

PART 2 – Changing the MOD Number of a Counter.

a) Start the QUARTUS® II program. Set up a new project and use **amod8r_ctr** for the name of the project. The diagram of the circuit you are to construct is given in Figure S14-3.

b) Open *amod8_ctr*. Choose **File - Save As...** from the program's main menu and save this file as *amod8r_ctr.gdf* in the new project directory.

c) Modify the schematic according to Figure S14-3.

d) Place the input and output connectors on the schematic. Label the input "clk"; label the outputs QA, QB, and QC.

e) Wire the input and output connectors to the circuit as shown in Figure S14-3.

f) Select **Project | Top-Level Entity**.

g) Compile the project. If there are errors, fix them and save the file before recompiling the project. If there are no errors (or perhaps a warning concerning timing at most), then go on to the next step.

h) Create a new waveform file *amod8r_ctr.vwf* and define the inputs for simulation. The input waveforms should look similar to those in Figure S14-4.

i) Select **Assignments | Settings** to access the Settings Window.

j) Click on *Simulator* at the left.

k) Choose *Functional* from the *Simulator Mode* pull-down menu.

l) Find and use for the *Simulator Input* file *amod8_ctr.vwf* you just created above.

m) Select **Processing | Generate Functional Simulation Netlist**.

n) Simulate the project. If there are any errors and you change your circuit, remember to save and compile the circuit before re-simulating.

o) Record the waveforms from the Simulation Waveforms report on Figure S14-4.

p) Select **Assignments | Settings** to access the Settings Window.

q) Click on *Simulator* at the left.

r) Choose *Timing* from the *Simulator Mode* pull-down menu.

s) Check the *Glitch Detection* box on the same page.

t) Simulate the project.

u) Record the waveforms from the Simulation Waveforms report on Figure S14-5.

v) Record and analyze results:

  1) A glitch is an extremely narrow pulse. It is not counted as a 1 except in determining the number that is decoded when the glitch occurs. [Note: You will cover glitches in Chapter 7.]

  2) Examine the waveforms in Figure S14-5. Did you get any glitches? _____

  3) If so, measure the time at which they occur with a Time Bar. _____, _____

  4) Counting the glitch as a 1, what counter numbers produce a glitch? _____, _____

  5) What is the count sequence of the counter? _____, _____

  6) The MOD number of the counter is _____.

  7) You should complete this part of the experiment by converting the waveform information in Figure S14-5 to Table S14-4.

PROCEDURE

8) Does the counter count up or down? _____
9) What is the frequency of QB? _____
10) Transfer the answers to the above questions, the waveform information, and state table to the data collection sheets at the end of this experiment.

w) Save and close your project.

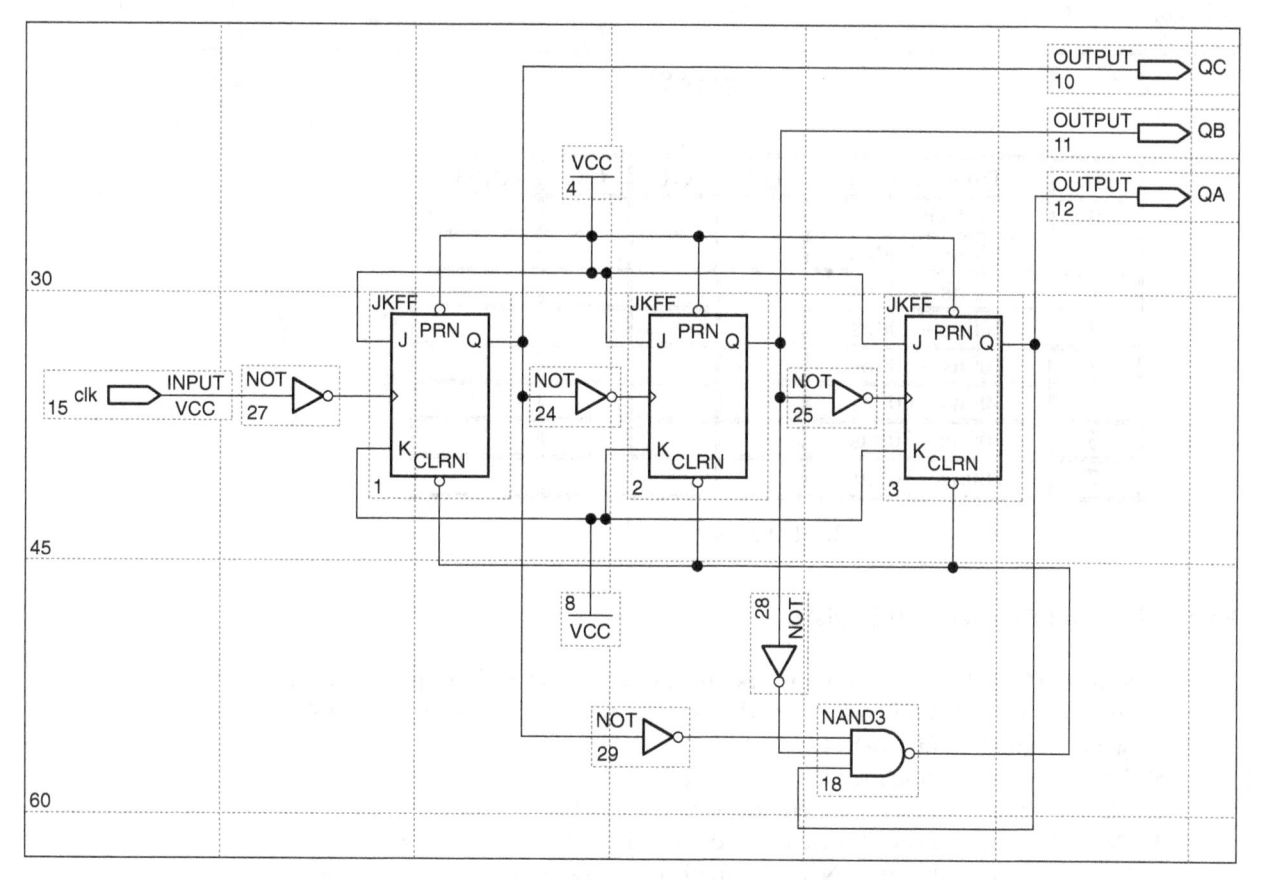

Figure S14-3

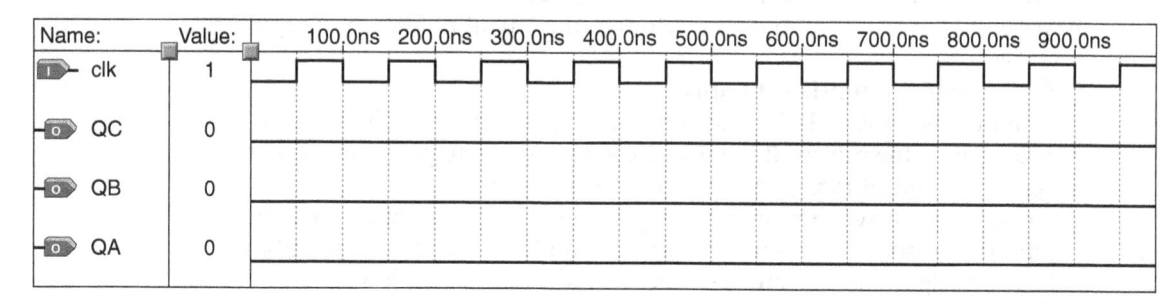

Figure S14-4

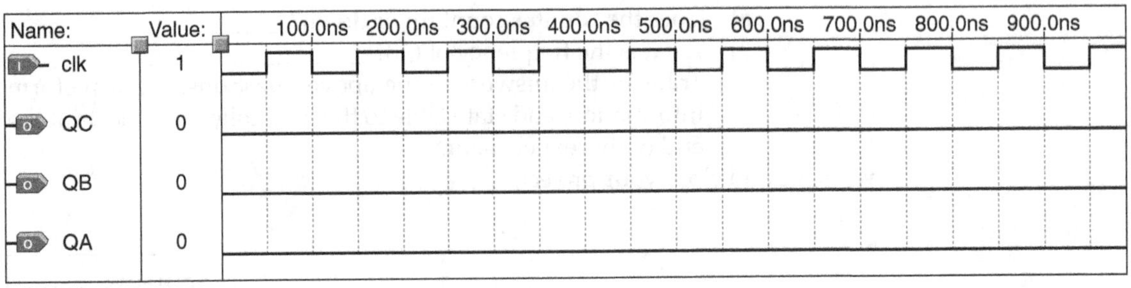

Name:	Value:	100.0ns 200.0ns 300.0ns 400.0ns 500.0ns 600.0ns 700.0ns 800.0ns 900.0ns
clk	1	⎍⎍⎍⎍⎍⎍⎍⎍⎍
QC	0	
QB	0	
QA	0	

Figure S14-5

clk	Time Interval	QA	QB	QC	Glitch?
0	0 – 100ns				
1	100ns – 200ns				
2	200ns – 300ns				
3	300ns – 400ns				
4	400ns – 500ns				
5	500ns – 600ns				
6	600ns – 700ns				
7	700ns – 800ns				

Table S14-4

PART 3 – A 3-bit Parallel Shift Register

a) Start the QUARTUS® II program. Set up a new project and use **para3_reg** for the name of the project. The diagram of the circuit you are to construct is given in Figure S14-6.
b) Create a new graphic file for the schematic, *para3_reg.bdf*.
c) Place the symbols on the schematic according to Figure S14-6.
d) Place the input and output connectors on the schematic. Create a 3-bit bus input and label it DIN[2..0]. Label the other two inputs **"clr"** and **"clk."** Create a 3-bit output bus and label it Q[2..0]. If you need to review buses, refer to Experiment S10 or use the QUARTUS® II online help.
e) Wire the input and output connectors to the circuit as shown in Figure S14-6.
f) Select **Project | Top-Level Entity**.
g) Compile the project. If there are errors, fix them and save the file before recompiling the project. If there are no errors (or perhaps a warning concerning timing at most), then go on to the next step.
h) Create a new waveform file *para3_reg.vwf* and define the inputs for simulation. The input waveforms should look similar to those in Figure S14-7.
i) Select **Assignments | Settings** to access the Settings Window.
j) Click on *Simulator* at the left.
k) Choose *Functional* from the *Simulator Mode* pull-down menu.

PROCEDURE

l) Find and use for the *Simulator Input* file *para3_reg.vwf* you just created.
m) Select **Processing | Generate Functional Simulation Netlist.**
n) Simulate the project. If there are any errors and you change your circuit, remember to save and compile the circuit before re-simulating.
o) Compare the simulation outputs to those in Figure S14-7. If they are the same, go on to the next step. If not, check your circuit and make sure that you have completed all steps.

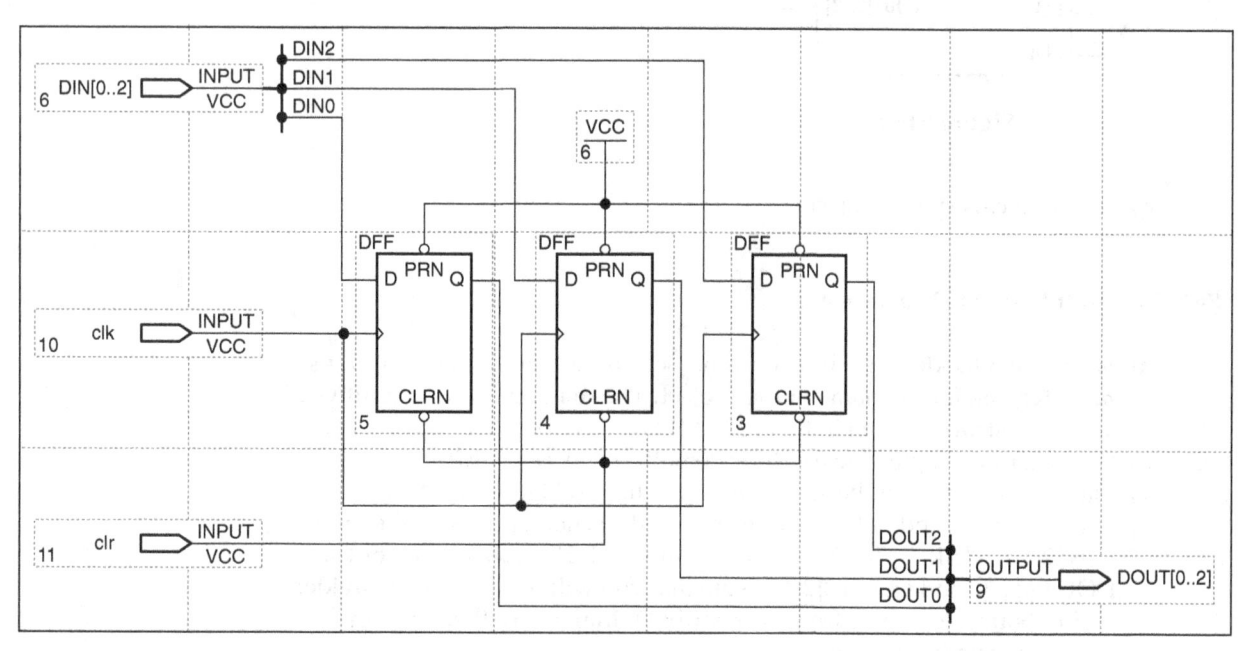

Figure S14-6

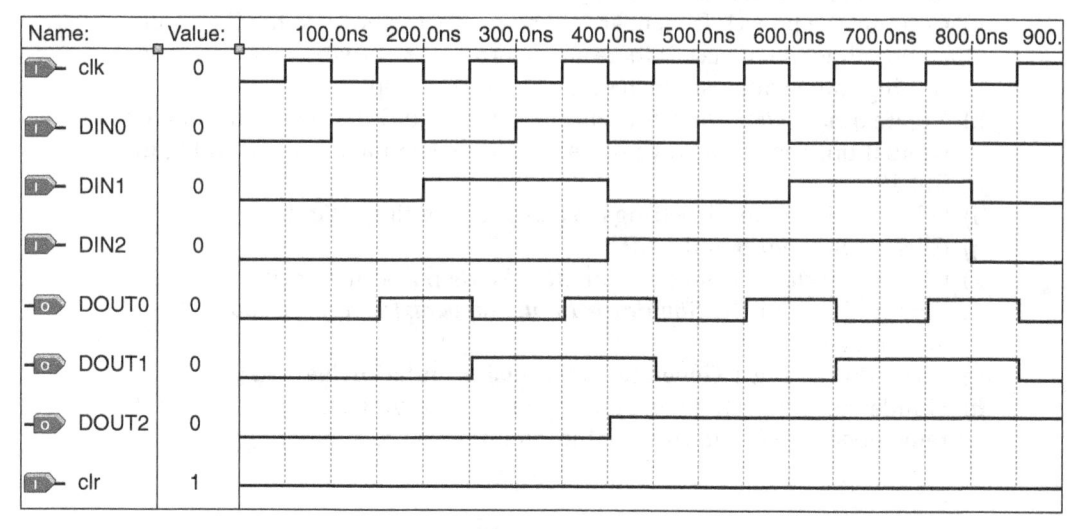

Figure S14-7

p) With the schematic on top, choose **File | Create/Update | Create Symbol Files for Current File**. This will create a symbol for the circuit, which we will use in Part 4 of this experiment.

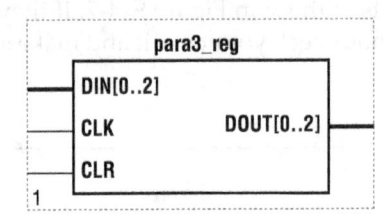

```
        para3_reg
   ──│ DIN[0..2]
   ──│ CLK    DOUT[0..2] │──
   ──│ CLR
      1
```

Figure S14-8

q) Save and close your project.

PART 4 – 3-bit Parallel Data Transfer

a) Start the QUARTUS® II program. Set up a new project and use **paraxfer_reg** for the name of the project. The diagram of the circuit you are to construct is given in Figure S14-9.
b) Create a new graphic file for the schematic, *paraxfer_reg.bdf*.
c) Place the symbols on the schematic according to Figure S14-9.
d) Place the input and output connectors on the schematic. Label the inputs DIN[2..0], "clk," "pb3," "clr1," and "clr2"; label the outputs DOUTA[2..0] and DOUTB[2..0]. Note that you will need the clock divider and debounce symbols from Appendix D. Make sure their design files are in your project folder.
e) Wire the input and output connectors to the circuit as shown in Figure S14-9.
f) Select **Project | Top-Level Entity**.
g) Compile the project. If there are errors, fix them and save the file before recompiling the project. If there are no errors (or perhaps a warning concerning timing at most), then go on to the next step.
h) Create a new waveform file *paraxfer_reg.vwf* and define the inputs for simulation. The input waveforms should look similar to those in Figure S14-10.
i) Select **Assignments | Settings** to access the Settings Window.
j) Click on *Simulator* at the left.
k) Choose *Functional* from the *Simulator Mode* pull-down menu.
l) Find and use for the *Simulator Input* file *paraxfer_reg.vwf* you just created above.
m) Select **Processing | Generate Functional Simulation Netlist.**
n) Simulate the project. If there are any errors and you change your circuit, remember to save and compile the circuit before re-simulating.

PROCEDURE

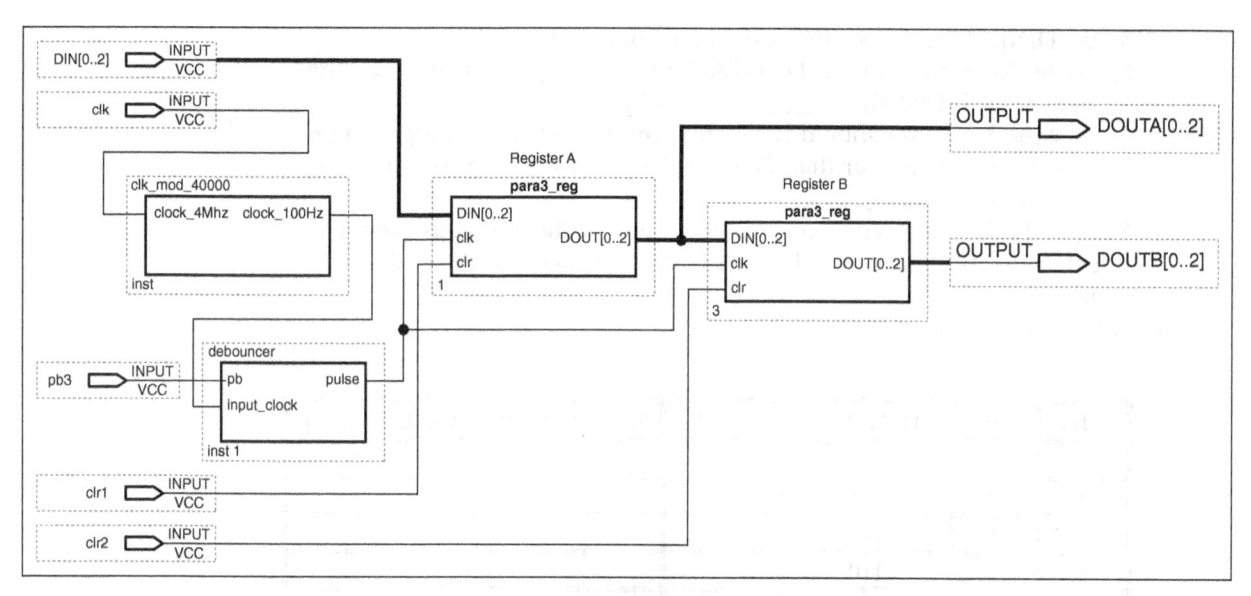

Figure S14-9

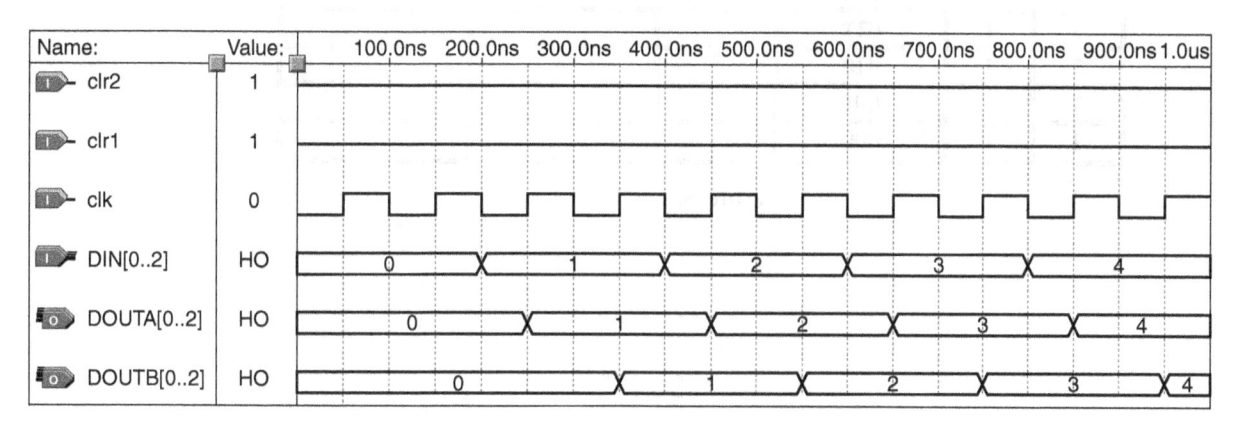

Figure S14-10

o) *Program the CPLD:* To program your CPLD, refer to Appendix D, Exercise 3. A complete set of instructions will be found there. It is suggested that you use **PIN_4** for "**DOUTA0**," **PIN_5** for "**DOUTA1**," and **PIN_8** for "**DOUTA2**." Also, use **PIN_16** for "**DOUTB0**," **PIN_17** for "**DOUTB1**," and **PIN_18** for "**DOUTB2**." For DIN, use **PIN_50** for "**DIN0**," **PIN_51** for "**DIN1**," **PIN_52** for "**DIN2**," **PIN_70** for "**pb3**," **PIN_73** for "**clr1**," and **PIN_74** for "**clr2**."

p) *Test the program:*
   1) Write down the state of DOUTA[2..0]: _____. If this value is not 000, press "**clr1**." Does DOUTA[2..0] clear? If not, then you most likely have a logical error. Recheck your circuit, procedure, and assignments. If you have to change your circuit or perhaps a logic option, don't forget to re-compile the project.
   2) Write down the state of DOUTB[2..0]: _____. If this value is not 000, press "**clr2**." Take the same precautions as in step 1.
   3) Set DIN[2..0] to 111 and clock the data transfer system by pressing "**pb3**" momentarily.
   4) Write down the state of DOUTA[2..0]: _____. Write down the state of DOUTB[2..0]: _____.

5) Set DIN[2..0] and clock the system one more time.
6) Write down the state of DOUTA[2..0]: _____. Write down the state of DOUTB[2..0]: _____.
7) In steps 3 – 6, you should have observed that DOUTA[2..0] changes one clock cycle earlier than DOUTB[2..0]. Does this agree with your simulation results? _____

q) Fill out Table S14-5 with your observations in the previous step and transfer this information to the data collection sheets at the end of this experiment.

r) Save and close your project.

Clr1	Clr2	DIN[2..0]	Clock	DOUTA[2..0]	DOUTB[2..0]
0	1	111	x		
1	0	111	x		
1	1	111	↑		
1	1	110	↑		
1	1	101	↑		
1	1	100	↑		
1	1	011	↑		
1	1	010	↑		
1	1	001	↑		
1	1	000	↑		

Table S14-5

DATA COLLECTION SHEET

Data Collection Sheets for Lab S14 Name _____

 Date _____

Data collection for Part 1:

h) *Measure worst-case* t_{co}: Input the appropriate data in Table S14-1.

Slack	Required tco	Actual tco	From	To	From Clock
N/A	None				
N/A	None				
N/A	None				

Table S14-1

s)
1) The frequency of the clock signal is _____ MHz.
2) The frequency of QC is _____ MHz.
   QC divides "clk" by _____.
3) The frequency of QB is _____ MHz.
   QB divides "clk" by _____.
4) The frequency of QA is _____ MHz.
   QA divides "clk" by _____.
5) The MOD number of the counter is _____.
7) Does the counter count up or down? _____

| Demonstrated To |
| Instructor/FA _____ Date _____ |
| Part 1 [ ]   Part 2 [ ]   Part 3 [ ]   Part 4 [ ] |

Data Collection Sheets for Lab S14   Name _____

Date _____

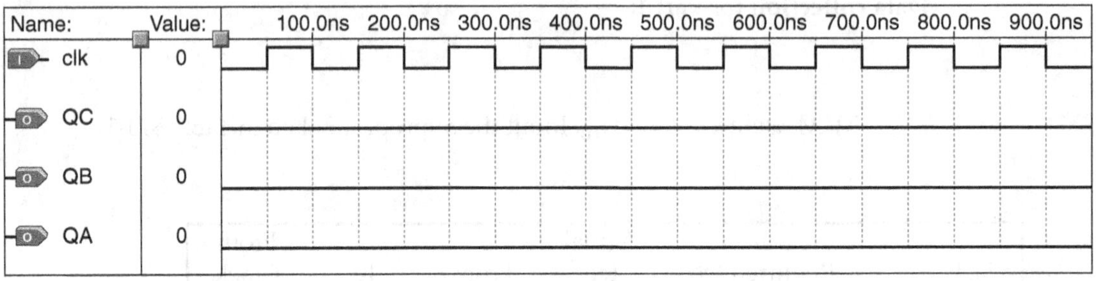

Name:	Value:	clk waveform 0–900 ns
clk	0	
QC	0	
QB	0	
QA	0	

Figure S14-2

clk	Time Interval	QA	QB	QC
0	0 – 100ns			
1	100ns – 200ns			
2	200ns – 300ns			
3	300ns – 400ns			
4	400ns – 500ns			
5	500ns – 600ns			
6	600ns – 700ns			
7	700ns – 800ns			

Table S14-2

clk	Time Interval	QA	QB	QC
0	0 – 100ns			
1	100ns – 200ns			
2	200ns – 300ns			
3	300ns – 400ns			
4	400ns – 500ns			
5	500ns – 600ns			
6	600ns – 700ns			
7	700ns – 800ns			

Table S14-3

Demonstrated To
Instructor/FA _____ Date _____

Part 1 [ ]   Part 2 [ ]   Part 3 [ ]   Part 4 [ ]

DATA COLLECTION SHEET

Data Collection Sheets for Lab S14   Name _____

Date _____

Data collection for Part 2:

v)

2) Examine the waveforms in Figure S14-5. Did you get any glitches? _____

3) If so, measure the time at which they occur with a Time Bar. _____, _____

4) Counting the glitch as a 1, what counter numbers produce a glitch? _____, _____

5) What is the count sequence of the counter? _____, _____.

6) The MOD number of the counter is _____.

8) Does the counter count up or down? _____

9) What is the frequency of QB? _____

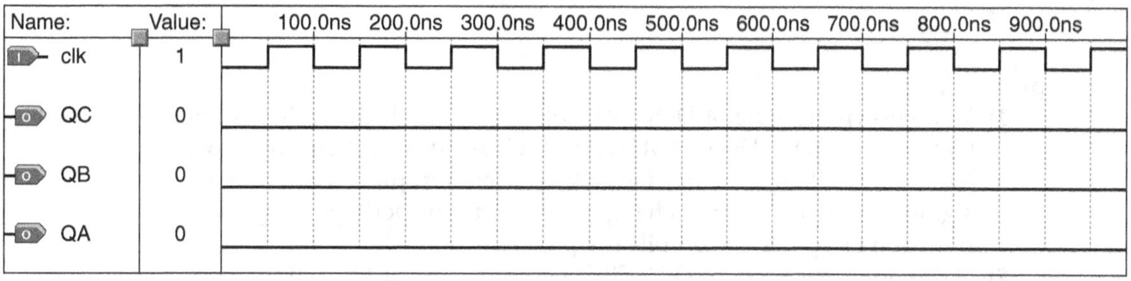

Name:	Value:	100.0ns 200.0ns 300.0ns 400.0ns 500.0ns 600.0ns 700.0ns 800.0ns 900.0ns
clk	1	
QC	0	
QB	0	
QA	0	

Figure S14-4

| Demonstrated To |
| Instructor/FA _____ Date _____ |
| Part 1 [ ]   Part 2 [ ]   Part 3 [ ]   Part 4 [ ] |

Data Collection Sheets for Lab S14   Name _____

Date _____

clk	Time Interval	QA	QB	QC	Glitch?
0	0 – 100ns				
1	100ns – 200ns				
2	200ns – 300ns				
3	300ns – 400ns				
4	400ns – 500ns				
5	500ns – 600ns				
6	600ns – 700ns				
7	700ns – 800ns				

Table S14-4

Data collection for Part 4:

p)
1) Write down the state of DOUTA[2..0]: _____. If this value is not 000, press **"clr1."** Does DOUTA[2..0] clear? If not, then you most likely have a logical error. Recheck your circuit, procedure, and assignments. If you have to change your circuit or perhaps a logic option, don't forget to re-compile the project.
2) Write down the state of DOUTB[2..0]: _____. If this value is not 000, press **"clr2."** Take the same precautions as in step 1.

Demonstrated To
Instructor/FA _____ Date _____

Part 1 [ ]   Part 2 [ ]   Part 3 [ ]   Part 4 [ ]

DATA COLLECTION SHEET

Data Collection Sheets for Lab S14   Name _____

Date _____

3) Set DIN[2..0] to 111 and clock the data transfer system by pressing "**pb3**" momentarily.
4) Write down the state of DOUTA[2..0]: _____. Write down the state of DOUTB[2..0]: _____.
5) Set DIN[2..0] and clock the system one more time.
6) Write down the state of DOUTA[2..0]: _____. Write down the state of DOUTB[2..0]: _____.
7) In steps 3 – 6, you should have observed that DOUTA[2..0] changes one clock cycle earlier than DOUTB[2..0]. Does this agree with your simulation results? _____

Clr1	Clr2	DIN[2..0]	Clock	DOUTA[2..0]	DOUTB[2..0]
0	1	111	x		
1	0	111	x		
1	1	111	↑		
1	1	110	↑		
1	1	101	↑		
1	1	100	↑		
1	1	011	↑		
1	1	010	↑		
1	1	001	↑		
1	1	000	↑		

Table S14-5

Demonstrated To
Instructor/FA _____ Date _____

Part 1 [ ]   Part 2 [ ]   Part 3 [ ]   Part 4 [ ]

Experiment S15

Name _____

IMPLEMENTING FLIP-FLOPS AND FLIP-FLOP DEVICES WITH VHDL

OBJECTIVES

1. To practice using QUARTUS® II software to create schematics and waveforms.
2. To practice using QUARTUS® II software to compile and simulate digital circuits.
3. To investigate the implementation of D flip-flops using VHDL.
4. To investigate the implementation of J-K flip-flops using VHDL.
5. To investigate the implementation of a MOD-8 counter using VHDL.

TEXT REFERENCES

Read sections 5.17 through 5.19. Review Appendix D of this manual.

EQUIPMENT NEEDED

Components

QUARTUS® II software (Altera Corporation);

DeVry University Board eSOC with EPM7128SLC84 CPLD (or Altera University Board with EPM7128SLC84 CPLD or any other equivalent board);

Desktop computer with minimum of INTEL Pentium PC @ 400MHz CPU running Microsoft Windows NT4SP1, 2000, or XP or better; Pentium III or 4 PC @ 400MHz running Red Hat Linux v7.3 or 8.0 or Red Hat Linux Enterprise 3; Sun Ultra running Solaris v8 or 9; HP9000 Series 700/800 running HP-UX v11.0 with ACE dated 11/1999 or later; or equivalent.

DISCUSSION

Now that you have become acquainted with the programming sequential devices such as latches, flip-flops, counters, and registers using schematics, you are ready to apply your knowledge in creating the same devices with VHDL.

PROCEDURE

PART 1 – D-type Flip-Flop Using VHDL

a) Start the QUARTUS® II program. Set up a new project and use **dff_vhdl** for the name of the project.
b) Create a new text file *dff_vhdl.vhd* for writing the D-type flip-flop VHDL description.
c) Type in the VHDL program from the listing in Figure S15-1.
d) Select **Project | Top-Level Entity**.
e) Compile the project. If there are errors, fix them and save the file before recompiling the project. If there are no errors (or perhaps a warning concerning timing at most), then go on to the next step.
f) Create a new waveform file *dff_vhdl.vwf* and define the inputs for simulation. The input waveforms should look similar to those in Figure S15-2. If necessary, review the manual methods of creating waveforms in Exercise 1 of Appendix D before trying to create the **"clr"** waveform.
g) Select **Assignments | Settings** to access the Settings Window.
h) Click on *Simulator* at the left.
i) Choose *Functional* from the *Simulator Mode* pull-down menu.
j) Find and use for the *Simulator Input* file *dff_vhdl.vwf* you just created above.
k) Select **Processing | Generate Functional Simulation Netlist**.
l) Simulate the project. If there are any errors and you change your circuit, remember to save and compile the circuit before re-simulating.
m) Record and analyze results. Record the output waveform on Figure S15-2. You should complete this part of the experiment by converting the waveform information in Figure S15-2 to Table S15-1.
n) Compare these results with the results of Experiment S12 Part 3 (Figure S12-6).
o) Transfer the results of the waveform and truth table to the data collection sheets at the end of this experiment.
p) Save and close the project.

PROCEDURE

```
-- dff_vhd1.vhd
-- MAX+plus II VHDL Example of D flip-flop
LIBRARY ieee;
USE ieee.std_logic_1164.all;
                                -- Positive Edged Triggered D-type FF
ENTITY dff_vhd1 IS
PORT (
    d,clk,prn,clrn: IN BIT;    -- defines the inputs
    q: OUT BIT);    -- defines the outputs
END dff_vhd1;

ARCHITECTURE  a OF dff_vhd1 IS
SIGNAL qstate: BIT;
BEGIN
    PROCESS(clk, prn, clrn) -- respond to any of these signals
    BEGIN
        If prn = '0' THEN qstate <= '1'; -- async preset
         ELSIF clrn = '0' THEN qstate <= '0'; -- async clear
         ELSIF clrn = '1' AND clk'EVENT THEN -- on PGT clock edge
            qstate <= d;
        END IF;
    END PROCESS;
    q <= qstate;                                -- update output pin
END a;
```

Figure S15-1

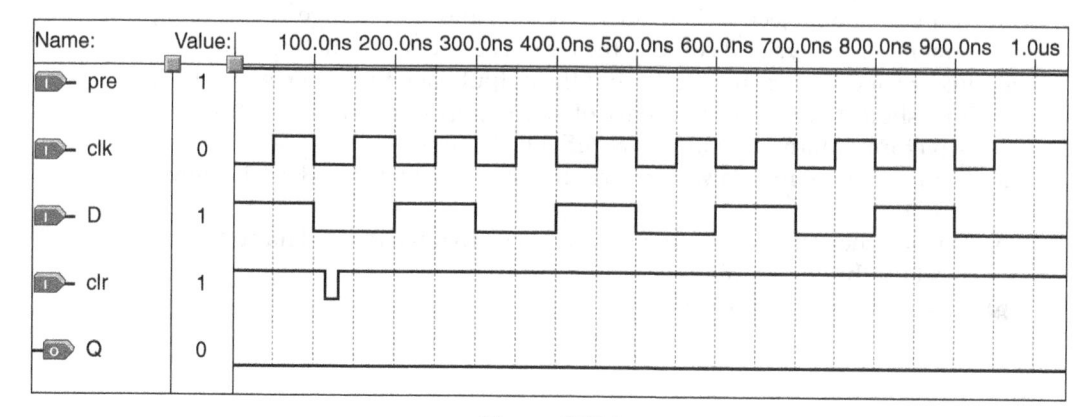

Figure S15-2

pre	clr	clk	D	Q
1	1			
1	1			
0	1			
1	0			

Table S15-1

PART 2 – J-K Flip-Flop Using VHDL

a) Start the QUARTUS® II program. Set up a new project and use **jkff_vhdl** for the name of the project.
b) Create a new text file *jkff_vhdl.vhd* for writing the J-K flip-flop VHDL description.
c) Type in the VHDL program from the listing in Figure S15-3.
d) Select **Project | Top-Level Entity**.
e) Compile the project. If there are errors, fix them and save the file before recompiling the project. If there are no errors (or perhaps a warning concerning timing at most), then go on to the next step.
f) Create a new waveform file *jkff_vhdl.vwf* and define the inputs for simulation. The input waveforms should look similar to those in Figure S15-4. If necessary, review the manual methods of creating waveforms in Exercise 1 of Appendix D before trying to create the **"clr"** and the **"pre"** waveforms.
g) Select **Assignments | Settings** to access the Settings Window.
h) Click on *Simulator* at the left.
i) Choose *Functional* from the *Simulator Mode* pull-down menu.
j) Find and use for the *Simulator Input* file *jkff_vhdl.vwf* you just created above.
k) Select **Processing | Generate Functional Simulation Netlist.**
l) Simulate the project. If there are any errors and you change your circuit, remember to save and compile the circuit before re-simulating.
m) Record and analyze results. Record the output waveform on Figure S15-4. You should complete this part of the experiment by converting the waveform information in Figure S15-4 to Table S15-2.
n) Compare these results with the results of Experiment S13 Part 1 (Figure S13-2).
o) Transfer the results of the waveform and truth table to the data collection sheets at the end of this experiment.
p) Save and close the project.

PROCEDURE

```
-- MAX+plus II VHDL Example of J-K flip-flop
LIBRARY ieee;
USE ieee.std_logic_1164.all;
                                -- Positive Edged Triggered J-K-type FF
ENTITY jkff_vhdl IS
POST (
    j,k,clk,prn,clrn: IN BIT;  -- defines the inputs
    q:  OUT BIT);  -- defines the output
END jkff_vhdl;

ARCHITECTURE a OF jkff_vhdl IS
SIGNAL qstate: BIT;
BEGIN
    PROCESS(clk, prn, clrn) -- respond to any of these signals
    BEGIN
        If prn = '0' THEN qstate <= '1'; -- async preset
        ELSIF clrn = '0' THEN qstate <= '0'; -- async clear
        ELSIF clk = '1' AND clk'EVENT THEN -- on PGT clock edge
            IF j = '1' AND k = '1' THEN qstate <= NOT qstate;
            ELSIF j = '1' AND k = '0' THEN qstate <= '1';
            ELSIF j = '0' AND k = '1' THEN qstate <= '0';
            END IF;
        END IF;
    END PROCESS;
    q <= qstate;                                    -- update output pin
END a;
```

Figure S15-3

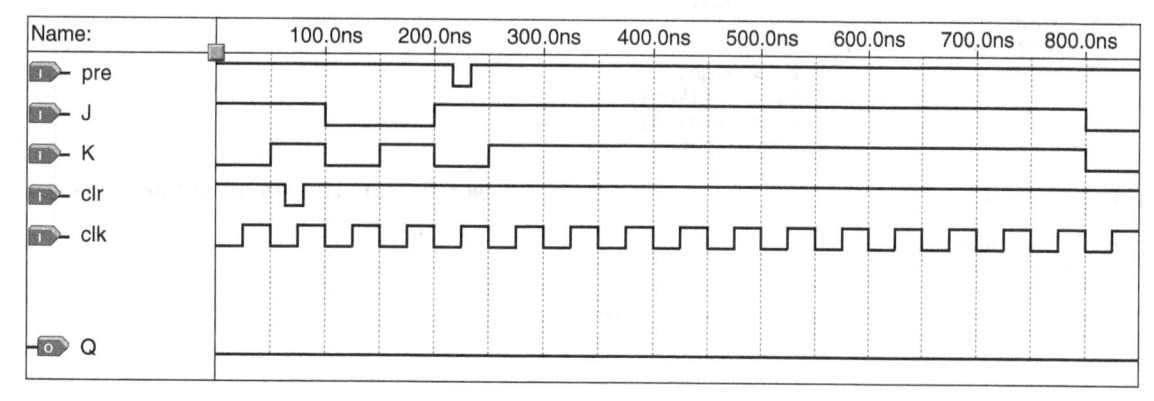

Figure S15-4

pre	clr	clk	J	K	Q
1	1	↑	0	0	
1	1	↑	1	0	
0	1	↑	0	1	
1	0	↑	1	1	

Table S15-2

```vhdl
-- From Tocci/Widmer fig 5-72
LIBRARY ieee;
USE ieee.std_logic_1164.ALL;
LIBRARY altera;
USE altera.maxplus2.ALL;
ENTITY mod8_ctr_vhdl IS
PORT(
        clock                                   :IN std_logic;
        qout0, qout1, qout2                     :OUT std_logic);
END mod8_ctr_vhdl;
ARCHITECTURE a of mod8_ctr_vhdl IS
SIGNAL clknot, q0not, q1not, q0, q1, q2, high: std_logic;
BEGIN
     high     <= '1';                   -- connection for Vcc
     clknot   <= NOT clock;    -- invert clock signal
     q0not    <= NOT q0;                -- invert each q to emulate neg edge trig
     q1not    <= NOT q1;
ff2: JKFF
     PORT MAP( j => high,     -- toggle mode
                       k => high,
                       clk => q1not,     -- ripple clock connection
                       clrn => high,     -- asynch inputs inactive
                       prn => high,
                       q => q2);                 -- connect to output signal
ff1: JKFF
     PORT MAP( j => high,
                       k => high,
                       clk => q0not,
                       clrn => high,
                       prn => high,
                       q => q1);
ff0: JKFF
     PORT MAP( j => high,
                       k => high,
                       clk => clknot,
                       clrn => high,
                       prn => high,
                       q => q0);
qout0    <= q0;                          -- connect ff out signals to output pins
qout1    <= q1;
qout2    <= q2;
END a;
```

Figure S15-5

PROCEDURE

PART 3 – MOD-8 Counter Using VHDL

a) Start the QUARTUS® II program. Set up a new project and use **mod8_ctr_vhdl** for the name of the project.
b) Create a new text file for writing the counter's VHDL description. Save this file as *mod8_ctr_vhdl.bdf*.
c) Type in the VHDL program from the listing in Figure S15-5.
d) Select **Project | Top-Level Entity**.
e) Compile the project. If there are errors, fix them and save the file before recompiling the project. If there are no errors (or perhaps a warning concerning timing at most), then go on to the next step.
f) Create a new waveform file *mod8_ctr_vhdl.vwf* and define the inputs for simulation. The input waveforms should look similar to those in Figure S15-6.
g) Select **Assignments | Settings** to access the Settings Window.
h) Click on *Simulator* at the left.
i) Choose *Functional* from the *Simulator Mode* pull-down menu.
j) Find and use for the *Simulator Input* file *mod8_ctr_vhdl.vwf* you just created above.
k) Select **Processing | Generate Functional Simulation Netlist**.
l) Simulate the project. If there are any errors and you change your circuit, remember to save and compile the circuit before re-simulating.
m) Record and analyze results. Record the output waveforms on Figure S15-6. Compare these results with the results of Experiment S14 Part 1 (Figure S14-2).
n) Transfer the results of the waveform to the data collection sheets at the end of this experiment.
o) Save and close the project.

Name:	Value:	100.0ns	200.0ns	300.0ns	400.0ns	500.0ns	600.0ns	700.0ns	800.0ns
clock	0								
qout2	0								
qout1	0								
qout0	0								

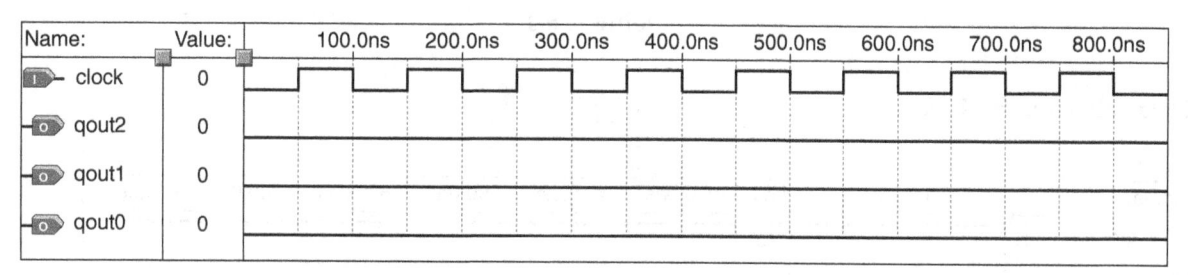

Figure S15-6

Data Collection Sheets for Lab S15

Name _____

Date _____

Data collection for Part 1:

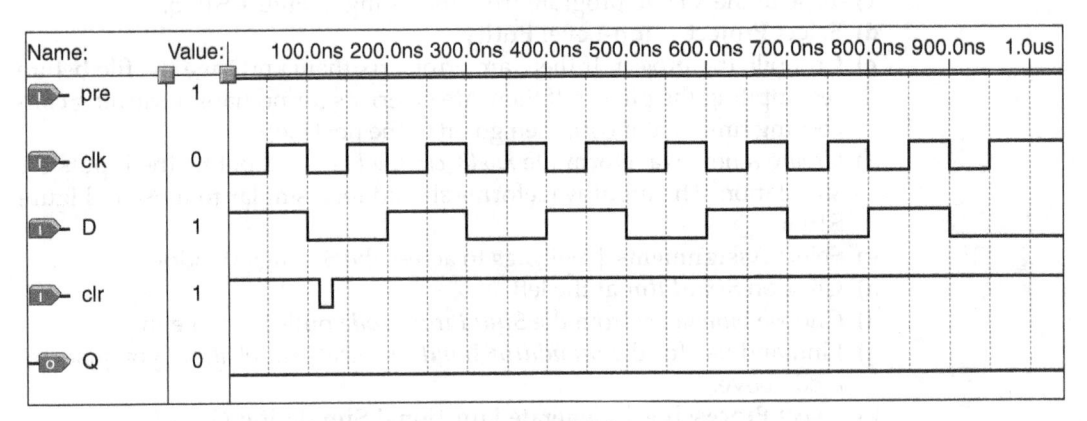

Figure S15-2

pre	clr	clk	D	Q
1	1			
1	1			
0	1			
1	0			

Table S15-1

Data collection for Part 2:

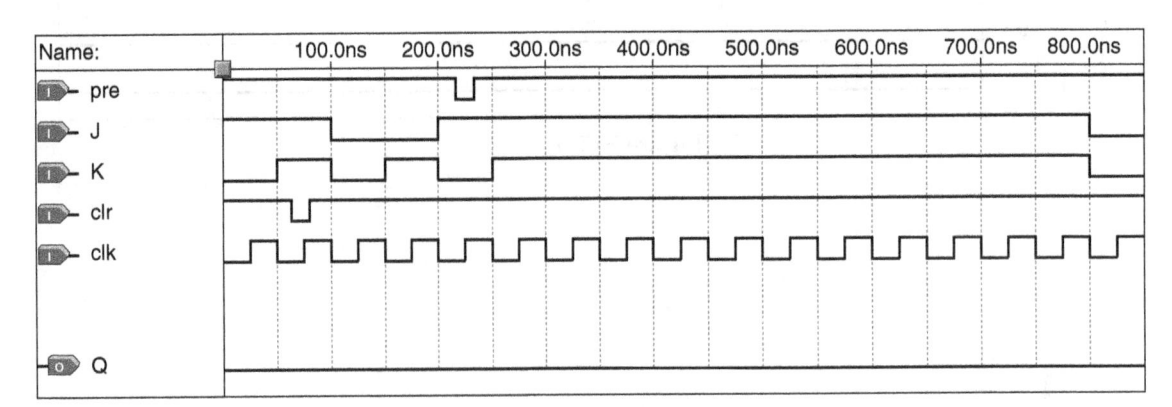

Figure S15-4

Demonstrated To
Instructor/FA _____ Date _____

Part 1 [ ]   Part 2 [ ]   Part 3 [ ]

DATA COLLECTION SHEET

Data Collection Sheets for Lab S15   Name _____

Date _____

pre	clr	clk	J	K	Q
1	1	↑	0	0	
1	1	↑	1	0	
0	1	↑	0	1	
1	0	↑	1	1	

Table S15-2

Data collection for Part 3:

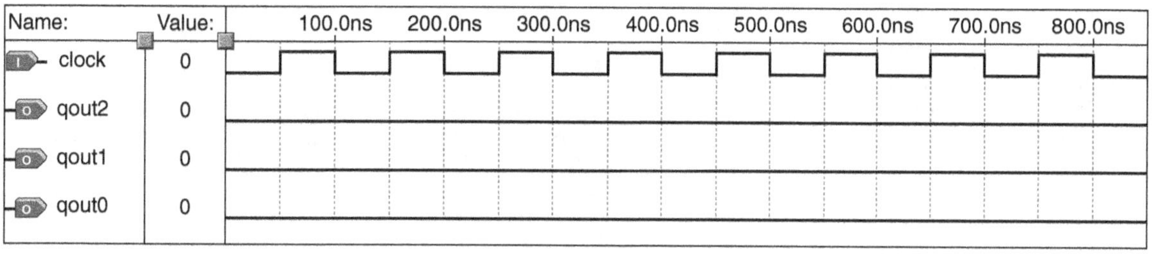

Figure S15-6

Demonstrated To
Instructor/FA _____ Date _____

Part 1 [ ]   Part 2 [ ]   Part 3 [ ]

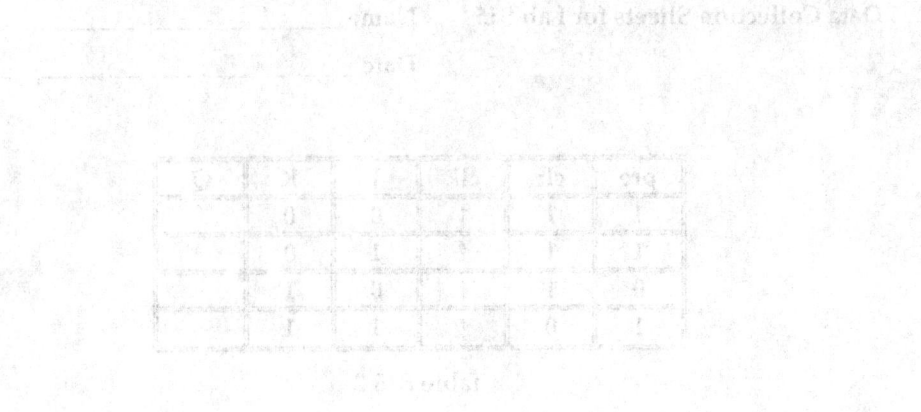

Experiment S16

Name _____

BINARY ADDERS AND 2'S COMPLEMENT SYSTEM

OBJECTIVES

1. To practice using QUARTUS® II software to create schematics and waveforms.
2. To practice using QUARTUS® II software to compile and simulate digital circuits.
3. To investigate the half and full adder.
4. To investigate a 2's complement adder/subtractor system.
5. To program the EPM7128SLC84 CPLD with the 2's complement adder/subtractor.

TEXT REFERENCES

Read sections 6.1 through 6.4 and sections 6.9 through 6.15. Review Appendix D of this manual.

EQUIPMENT NEEDED

Components

Blank floppy, zip disk or memory stick for storing projects;

QUARTUS® II software (Altera Corporation);

DeVry University Board eSOC with EPM7128SLC84 CPLD (or Altera University Board with EPM7128SLC84 CPLD or any other equivalent board);

Desktop computer with minimum of INTEL Pentium PC @ 400MHz CPU running Microsoft Windows NT4SP1, 2000, or XP or better; Pentium III or 4 PC @ 400MHz running Red Hat Linux v7.3 or 8.0 or Red Hat Linux Enterprise 3; Sun Ultra running Solaris v8 or 9; HP9000 Series 700/800 running HP-UX v11.0 with ACE dated 11/1999 or later; or equivalent.

DISCUSSION

In this experiment, you will investigate the parallel binary adder. You will first investigate simple adders like the half adder and full adder. These are the basic building blocks of more complex adders such as the 4-bit parallel adder. Finally you will investigate the operation of a 2's complement adder/subtractor system that makes use of the 7483A parallel adder.

PROCEDURE

PART 1 – Half Adder

a) Start the QUARTUS® II program. Set up a new project and use **half_adder** for the name of the project. The diagram of the circuit you are to construct and its input waveforms are given in Figures S16-1 and S16-2.
b) Create a new graphic file for the schematic and save it as *half_adder.bdf*.
c) Place the symbols on the schematic according to Figure S16-1.
d) Place the input and output connectors on the schematic. Label the inputs A0 and B0; label the outputs Sum0 and Carry0.
e) Wire the input and output connectors to the circuit as shown in Figure S16-1.
f) Select **Project | Top-Level Entity**.
g) Compile the project. If there are errors, fix them and save the file before recompiling the project. If there are no errors (or perhaps a warning concerning timing at most), then go on to the next step.
h) Create a new waveform *half_adder.vwf* and define the inputs for simulation. The input waveforms should look similar to those in Figure S16-2.
i) Select **Assignments | Settings** to access the Settings Window.
j) Click on *Simulator* at the left.
k) Choose *Functional* from the *Simulator Mode* pull-down menu.
l) Find and use for the *Simulator Input* file *half_adder.vwf* you just created above.
m) Select **Processing | Generate Functional Simulation Netlist**.
n) Simulate the project. If there are any errors and you change your circuit, remember to save and compile the circuit before re-simulating.
o) Record and analyze results. Record the output waveforms on Figure S16-2. You should complete this part of the experiment by converting the waveform information in Figure S16-2 to Table S16-1.
Program the CPLD: It is suggested that you use **PIN_50** for "A0," **PIN_51** for "B0," **PIN_4** for "Sum0," and **PIN_5** for "Carry0." To program your CPLD, refer to Appendix D, Exercise 3. A complete set of instructions will be found there. Briefly,
1) *Recompile the project*: After making the pin assignments, compile your project again. When the compilation is completed, the software will display an information box showing any warnings and any detected errors. If there are errors, fix them and recompile the project.

PROCEDURE

2) *Download the project to the board:*
   a. Set all eSOC DIP switches to OFF.
   b. Connect the board to the PC's parallel printer port and apply power.
   c. Select **Tools | Programmer**.
   d. Click on [Auto Detect] in the Chain Description File.
   e. When the eSOC board is recognized, press the [Add File...] button in the Chain Description File, then locate and Add the *half_adder.pof* file.
   f. Select the file and device line and scroll (horizontally) to the available programming option. Check the **Program/Configure** box.
   g. Press the [Start] button. Observe the progress bar.
   h. If the program downloads successfully, go to the next step. Otherwise, fix the problem and try downloading again.

p) *Test the program:* After connections are made to your board, apply power and verify that the LEDs respond in a manner consistent with a half adder. Place your observations in Table S16-2.

q) Transfer your observations from the simulation and program test results to the data collection sheets at the end of this experiment.

r) Save and close your project.

s) Record and analyze results:
   1) Write the Boolean expression for the sum bit:

      Sum0 = _____

   2) Write the Boolean expression for the carry bit:

      Carry0 = _____

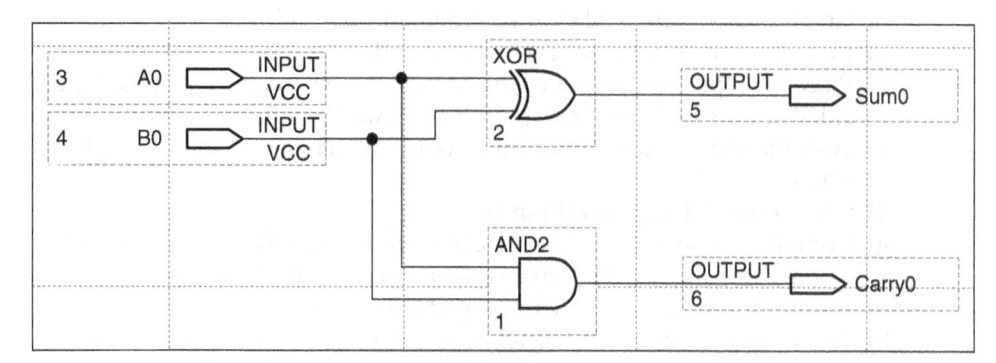

Figure S16-1

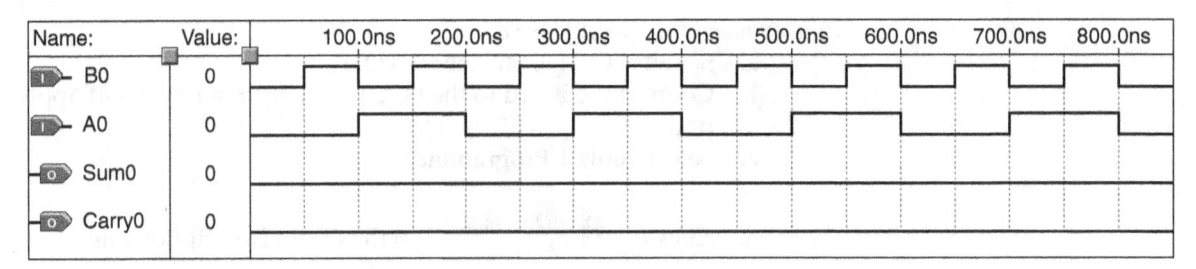

Figure S16-2

A0	B0	Sum0	Carry0
0	0		
0	1		
1	0		
1	1		

Table S16-1

A0	B0	Sum0	Carry0
0	0		
0	1		
1	0		
1	1		

Table S16-2

PART 2 – The Full Adder

a) Start the QUARTUS® II program. Set up a new project and use **full_adder** for the name of the project. The diagram of the circuit you are to construct and its input waveforms are given in Figures S16-3 and S16-4.
b) Create a new graphic file for the schematic *full_adder.bdf*.
c) Place the symbols on the schematic according to Figure S16-3.
d) Place the input and output connectors on the schematic. Label the inputs A0, B0, and Cin; label the outputs Sum0 and Carry0.
e) Wire the input and output connectors to the circuit as shown in Figure S16-3.
f) Select **Project | Top-Level Entity**.
g) Compile the project. If there are errors, fix them and save the file before recompiling the project. If there are no errors (or perhaps a warning concerning timing at most), then go on to the next step.
h) Create a new waveform *full_adder.vwf* and define the inputs for simulation. The input waveforms should look similar to those in Figure S16-4.
i) Select **Assignments | Settings** to access the Settings Window.
j) Click on *Simulator* at the left.
k) Choose *Functional* from the *Simulator Mode* pull-down menu.
l) Find and use for the *Simulator Input* file *full_adder.vwf* you just created above.
m) Select **Processing | Generate Functional Simulation Netlist**.

PROCEDURE

n) Simulate the project. If there are any errors and you change your circuit, remember to save and compile the circuit before re-simulating.

o) Record and analyze results. Record the output waveforms on Figure S16-4. You should complete this part of the experiment by converting the waveform information in Figure S16-4 to Table S16-3.

p) *Program the CPLD:* It is suggested that you use **PIN_50** for "**A0**," **PIN_51** for "**B0**," **PIN_52** for "**Cin**," **PIN_4** for "**Sum0**," and **PIN_5** for "**Carry0**." To program your CPLD, refer to Appendix D, Exercise 3. A complete set of instructions will be found there. Briefly,

  1) *Recompile the project:* After making the pin assignments, compile your project again. When the compilation is completed, the software will display an information box showing any warnings and any detected errors. If there are errors, fix them and recompile the project.

  2) *Download the project to the board:*
     a. Set all eSOC DIP switches to OFF.
     b. Connect the board to the PC's parallel printer port and apply power.
     c. Select **Tools | Programmer**.
     d. Click on [Auto Detect] in the Chain Description File.
     e. When the eSOC board is recognized, press the [Add File...] button in the Chain Description File, then locate and Add the *full_adder.pof* file.
     f. Select the file and device line and scroll (horizontally) to the available programming option. Check the **Program/Configure** box.
     g. Press the [Start] button. Observe the progress bar.
     h. If the program downloads successfully, go to the next step. Otherwise, fix the problem and try downloading again.

q) *Test the program:* After connections are made to your board, apply power and verify that the LEDs respond in a manner consistent with a full adder. Place your observations in Table S16-4.

r) Transfer your observations from the simulation and program test results to the data collection sheets at the end of this experiment.

s) Save and close your project.

t) Analyze results:
  1) Write the Boolean expression for the sum bit:

  Sum0 = _____

  2) Write the Boolean expression for the carry bit:

  Carry0 = _____

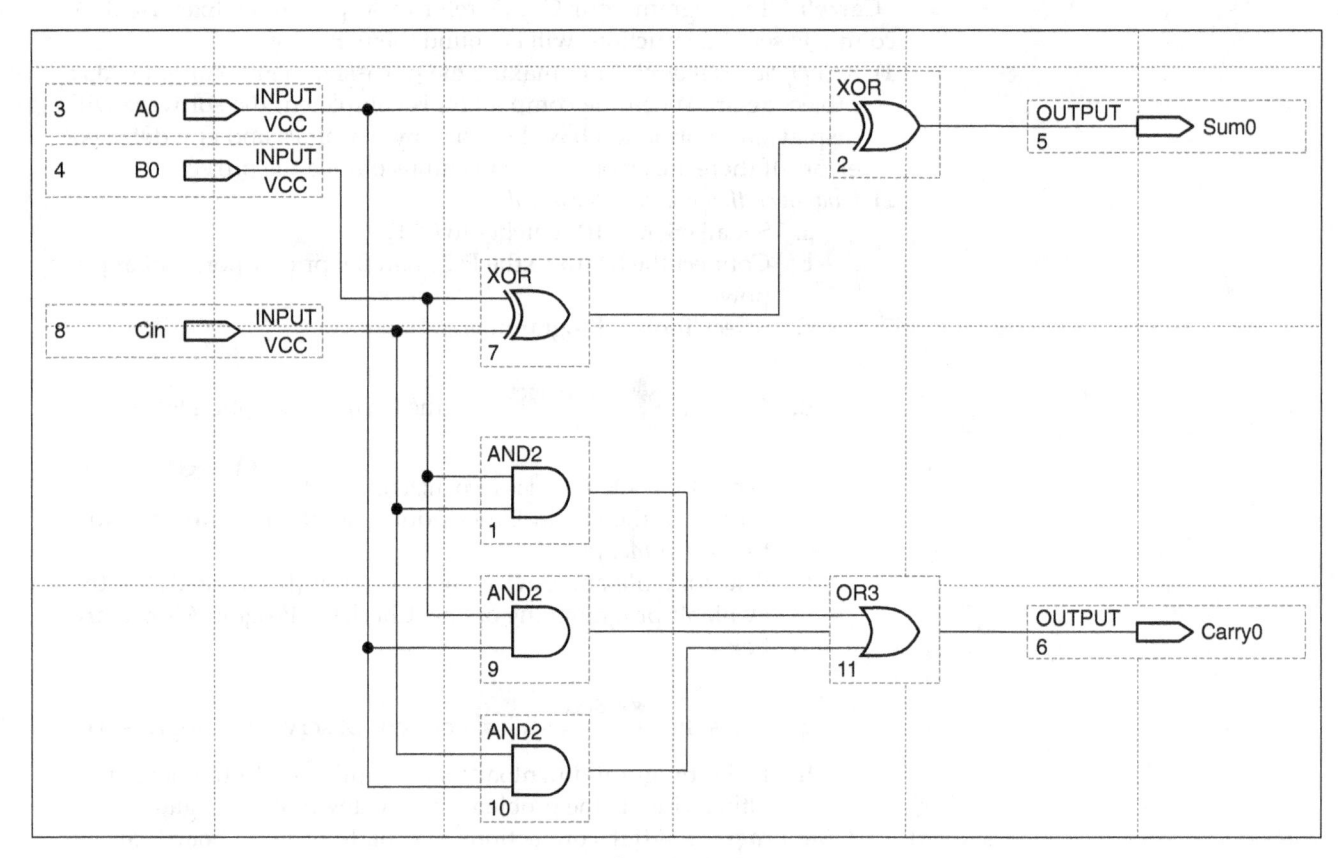

Figure S16-3

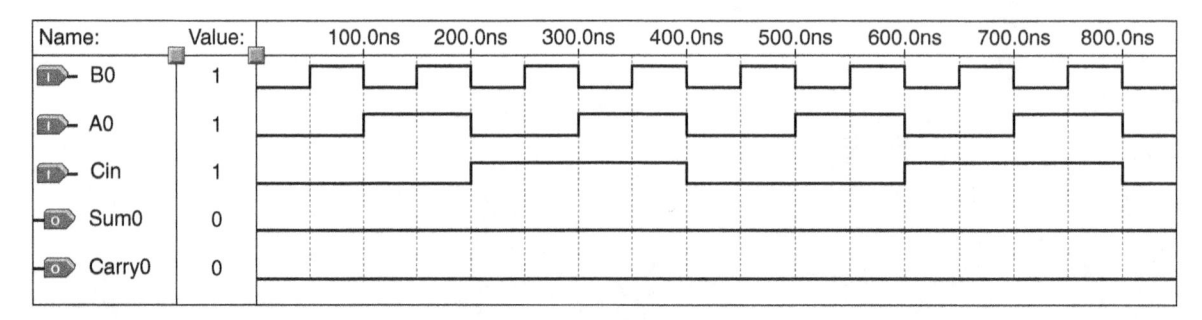

Figure S16-4

PROCEDURE

Cin	A0	B0	Sum0	Carry0
0	0	0		
0	0	1		
0	1	0		
0	1	1		
1	0	0		
1	0	1		
1	1	0		
1	1	1		

Table S16-3

Cin	A0	B0	Sum0	Carry0
0	0	0		
0	0	1		
0	1	0		
0	1	1		
1	0	0		
1	0	1		
1	1	0		
1	1	1		

Table S16-4

PART 3 – A 2's Complement Adder/Subtractor

a) Start the QUARTUS® II program. Set up a new project and use **addsub** for the name of the project. The diagram of the circuit you are to construct is given in Figure S16-5.
b) Create a new graphic file for the schematic *addsub.bdf*.
c) Place the symbols on the schematic according to Figure S16-5.
d) Place the input and output connectors on the schematic. Create a 4-bit bus input and label it DIN[3..0]. Label the other two inputs clr and clk. Create a 4-bit output bus and label it S[3..0]. You will need two toggle inputs, ADD and SUB. You will also need a clock input INCLK for the input register and another clock input ADDSUBCLK to clock the actual operation of addition or subtraction. Finally, you will need a 1-bit output for CARRY.
e) Wire the input and output connectors to the circuit as shown in Figure S16-5.
f) Select **Project | Top-Level Entity**.

g) Compile the project. If there are errors, fix them and save the file before recompiling the project. If there are no errors (or perhaps a warning concerning timing at most), then go on to the next step.

h) *Program the CPLD:* It is suggested that you use **PIN_50** for "**DIN0**," **PIN_51** for "**DIN1**," **PIN_52** for "**DIN2**," **PIN_54** for "**DIN3**," **PIN_55** for "**ADD**," **PIN_56** for "**SUB**," **PIN_4** for "**S0**," **PIN_5** for "**S1**," **PIN_8** for "**S2**," **PIN_9** for "**S3**," and **PIN_10** for "**CARRY**." You can also use **PIN_70** for "**INCLK**," **PIN_73** for "**ADDSUBCLK**," and **PIN_74** for "**clr**." To program your CPLD, refer to Appendix D, Exercise 3. A complete set of instructions will be found there. Briefly,

1) *Recompile the project*: After making the pin assignments, compile your project again. When the compilation is completed, the software will display an information box showing any warnings and any detected errors. If there are errors, fix them and recompile the project.

2) *Download the project to the board:*
   a. Set all eSOC DIP switches to OFF.
   b. Connect the board to the PC's parallel printer port and apply power.
   c. Select **Tools | Programmer**.
   d. Click on [Auto Detect] in the Chain Description File.
   e. When the eSOC board is recognized, press the [Add File...] button in the Chain Description File, then locate and Add the *full_adder.pof* file.
   f. Select the file and device line and scroll (horizontally) to the available programming option. Check the **Program/Configure** box.
   g. Press the [Start] button. Observe the progress bar.
   h. If the program downloads successfully, go to the next step. Otherwise, fix the problem and try downloading again.

i) *Test the program:*
   1) Set the ADD toggle switch ON and the SUB toggle switch OFF.
   2) Press CLR to flush the accumulator.
   3) Use the toggle switches DIN0-DIN3 to enter 0.
   4) Press the INCLK pushbutton to clock 0 in. This will clear the input register.
   5) Use the toggle switches DIN0-DIN3 to enter 3.
   6) Press the INCLK pushbutton to clock 3 into the input register.
   7) The LEDs indicate _____.

PROCEDURE

8) Press the ADDSUBCLK pushbutton to add whatever is in the accumulator to the value entered. The LEDs indicate _____.
9) Use the toggle switches DIN0-DIN3 to enter 5.
10) Press the INCLK pushbutton to clock the 5 into the input register.
11) Press the ADDSUBCLK pushbutton to add whatever is in the accumulator to the value entered. The LEDs indicate _____.
12) Use the toggle switches DIN0-DIN3 to enter 0.
13) Press the INCLK pushbutton to clock the 0 in.
14) The LEDs now indicate _____.
15) Set the ADD toggle switch OFF and the SUB toggle switch ON.
16) Use the toggle switches DIN0-DIN3 to enter 2.
17) Press the INCLK pushbutton to clock the 2 into the input register.
18) The LEDs indicate _____.
19) Use the toggle switches DIN0-DIN3 to enter 0.
20) Press the INCLK pushbutton to clock the 0 in.
21) The LEDs indicate _____.
22) Complete Table S16-5 by practicing adding and subtracting numbers with the adder/subtractor.

Operations	S	CARRY
1–2 + 3+4+0		
5+4+1+2-7+0		
1+3+5+6+0		
-1-2-4-6-0		

Table S16-5

j) Transfer the information in Table S16-5 to the data collection sheets at the end of this experiment.

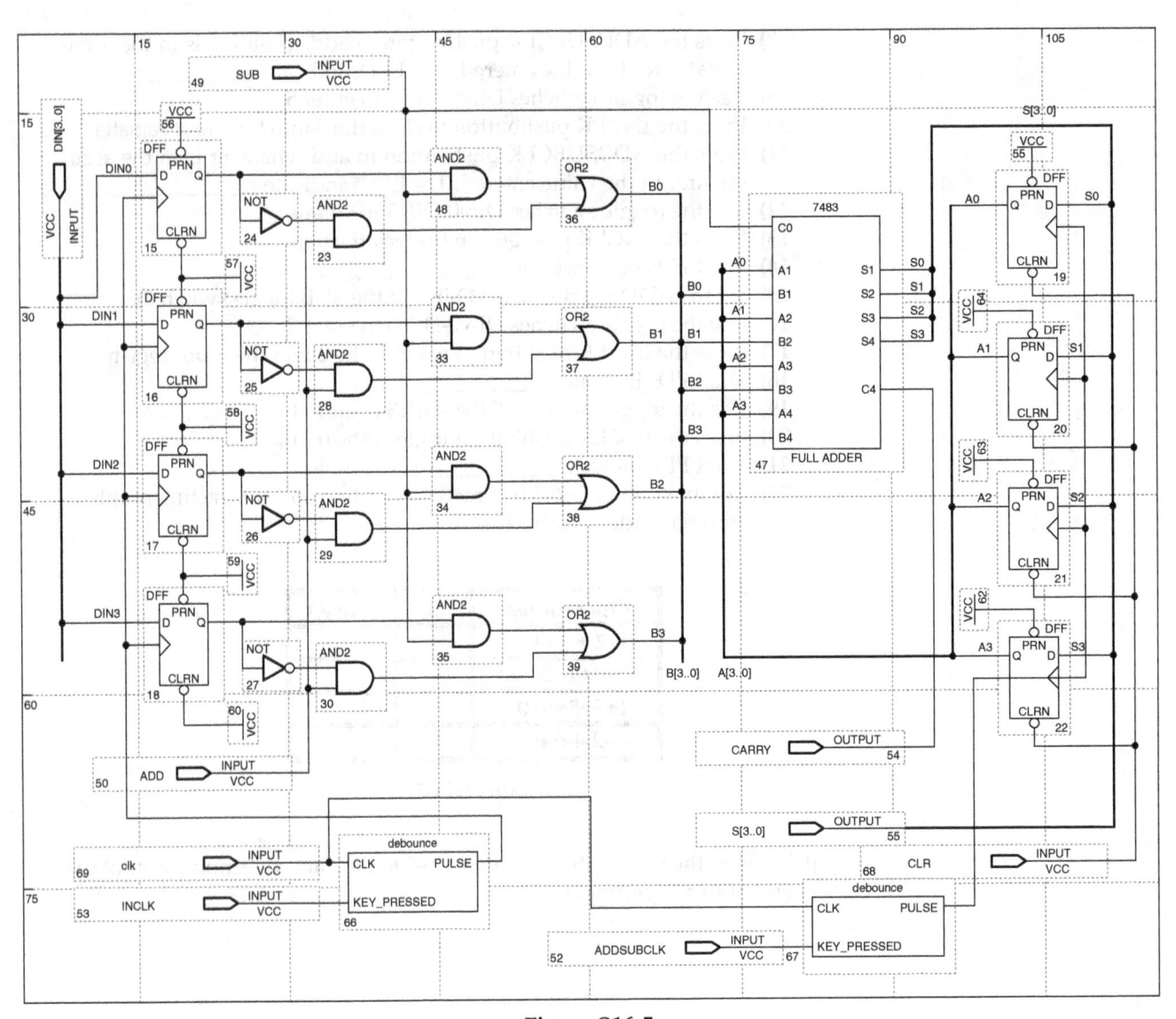

Figure S16-5

DATA COLLECTION SHEET

Data Collection Sheets for Lab S16   Name _____

Date _____

Data collection for Part 1:

    s) Record and analyze results:
       1) Write the Boolean expression for the sum bit:

          $Sum0 = $ _____

       2) Write the Boolean expression for the carry bit:

          $Carry0 = $ _____

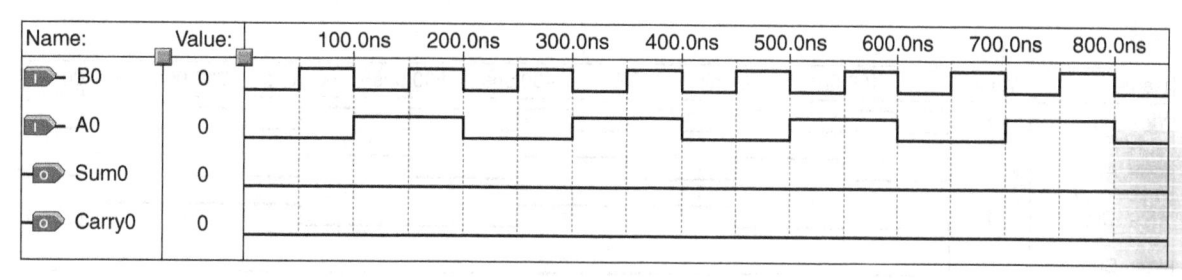

Figure S16-2

A0	B0	Sum0	Carry0
0	0		
0	1		
1	0		
1	1		

Table S16-1

A0	B0	Sum0	Carry0
0	0		
0	1		
1	0		
1	1		

Table S16-2

Demonstrated To
Instructor/FA _____ Date _____

Part 1 [ ]   Part 2 [ ]   Part 3 [ ]

Exper. S16

Data Collection Sheets for Lab S16 Name _____

Date _____

Data collection for Part 2:

   t) Analyze results:
      1) Write the Boolean expression for the sum bit:

         $Sum0 =$ _____

      2) Write the Boolean expression for the carry bit:

         $Carry0 =$ _____

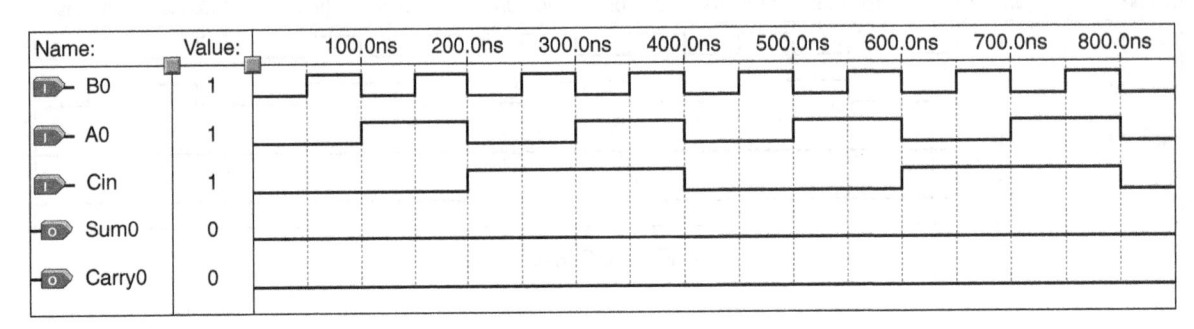

Figure S16-4

Cin	A0	B0	Sum0	Carry0
0	0	0		
0	0	1		
0	1	0		
0	1	1		
1	0	0		
1	0	1		
1	1	0		
1	1	1		

Table S16-3

Demonstrated To
Instructor/FA _____ Date _____

Part 1 [ ]   Part 2 [ ]   Part 3 [ ]

DATA COLLECTION SHEET

Data Collection Sheets for Lab S16   Name _____

Date _____

Cin	A0	B0	Sum0	Carry0
0	0	0		
0	0	1		
0	1	0		
0	1	1		
1	0	0		
1	0	1		
1	1	0		
1	1	1		

Table S16-4

Data collection for Part 3:

i)
7) The LEDs indicate _____.
8) The LEDs indicate _____.
11) The LEDs indicate _____.
14) The LEDs now indicate _____.
18) The LEDs indicate _____.
21) The LEDs indicate _____.
22) Complete Table S16-5 by practicing adding and subtracting numbers with the adder/subtractor.

Operations	S	CARRY
1–2 + 3+4+0		
5+4+1+2-7+0		
1+3+5+6+0		
-1-2-4-6-0		

Table S16-5

Demonstrated To
Instructor/FA _____ Date _____

Part 1 [ ]   Part 2 [ ]   Part 3 [ ]

Experiment S17

Name _____

ASYNCHRONOUS COUNTERS

OBJECTIVES

1. To practice using QUARTUS® II software to create schematics and waveforms.
2. To practice using QUARTUS® II software to compile and simulate digital circuits.
3. To investigate the 74293 IC counter.
4. To investigate the cascading of two 74293 IC counters.
5. To program the EPM7128SLC84 CPLD with a circuit having an asynchronous counter as a component.

TEXT REFERENCES

Read sections 7.1 and 7.2. Review Appendix D of this manual.

EQUIPMENT NEEDED

Components

Blank floppy, zip disk, or memory stick for storing projects;

QUARTUS® II software (Altera Corporation);

DeVry University Board eSOC with EPM7128SLC84 CPLD (or Altera University Board with EPM7128SLC84 CPLD or any other equivalent board);

Desktop computer with minimum of INTEL Pentium PC @ 400MHz CPU running Microsoft Windows NT4SP1, 2000, or XP or better; Pentium III or 4 PC @ 400MHz running Red Hat Linux v7.3 or 8.0 or Red Hat Linux Enterprise 3; Sun Ultra running Solaris v8 or 9; HP9000 Series 700/800 running HP-UX v11.0 with ACE dated 11/1999 or later; or equivalent;

TTL compatible signal or function generator.

DISCUSSION

In this experiment, you will investigate asynchronous IC counters, the 74LS293 counter in particular. You will become familiar with their properties, especially their MOD number capabilities, and various ways to connect them. You will then investigate the cascading of such counters to make counters with larger MOD numbers. You will also discover why asynchronous counters are not particularly useful in modern digital systems.

PROCEDURE

PART 1 – The 74293 IC Asynchronous Counter – MOD-16 Configuration

a) Start the QUARTUS® II program. Set up a new project and use *a74293_ctr* for the name of the project. The diagram of the circuit you are to construct and its input waveforms are given in Figures S17-1 and S17-2.
b) Create a new graphic file for the schematic *a74293_ctr.bdf*.
c) Place the symbols on the schematic according to Figure S17-1.
d) Place the input and output connectors on the schematic. Label the input clk and CLEAR; label the outputs QA, QB, QC, and QD.
e) Wire the input and output connectors to the circuit as shown in Figure S17-1. This configuration is for a MOD-16 counter.
f) Select **Project | Top-Level Entity**.
g) Compile the project. If there are errors, fix them and save the file before recompiling the project. If there are no errors (or perhaps a warning concerning timing at most), then go on to the next step.
h) *Measure worst-case* t_{co}: Without any design requirements assigned, open the tco report, which is part of Compilation Report (under Timing Analyzer). Input the appropriate data in Table S17-1.

Slack	Required **tco**	Actual **tco**	From	To	From Clock
N/A	None				
N/A	None				
N/A	None				

Table S17-1

i) Create a new waveform *a74293_ctr.vwf* and define the inputs for simulation. The input waveforms should look similar to those in Figure S17-2.
j) Select **Assignments | Settings** to access the Settings Window.
k) Click on *Simulator* at the left.
l) Choose *Functional* from the *Simulator Mode* pull-down menu.
m) Find and use for the *Simulator Input* file *a74293_ctr.vwf* you just created above.
n) Select **Processing | Generate Functional Simulation Netlist**.
o) Simulate the project. If there are any errors and you change your circuit, remember to save and compile the circuit before re-simulating.

PROCEDURE

p) Record and analyze results. Record the output waveforms on Figure S17-2 or print out the Simulation Waveforms report from QUARTUS® II and attach it to the data collection sheet at the end of this experiment.

q) You should complete this part of the experiment by converting the waveform information in Figure S17-2 to Table S17-2.
   1) The frequency of the clock signal is _____ MHz.
   2) The frequency of QD is _____ MHz.
      QD divides clk by _____.
   3) The frequency of QC is _____ MHz.
      QC divides clk by _____.
   4) The frequency of QB is _____ MHz.
      QB divides clk by _____.
   5) The frequency of QA is _____ MHz.
      QA divides clk by _____.
   6) The MOD number of the counter is _____.
   7) Place the Time Bar in the 25–50ns time interval.
   8) Read the values for QA, QB, and QC in the Values column of the waveform diagram and place them in their corresponding columns in Table S17-2.
   9) Repeat steps 2 and 3 for each time interval.
   10) Does the counter count up or down? _____

r) **Program the CPLD:** It is suggested that you use **PIN_4** for "**QA**," **PIN_5** for "**QB**," **PIN_8** for "**QC**," and **PIN_9** for "**QD**." Also, use an external TTL compatible clock input at 1 Hz. If you choose to use an on-board or external switch to do your clocking, make sure it is debounced. To program your CPLD, refer to Appendix D, Exercise 3. A complete set of instructions will be found there. Briefly,
   1) *Recompile the project*: After making the pin assignments, compile your project again. When the compilation is completed, the software will display an information box showing any warnings and any detected errors. If there are errors, fix them and recompile the project.
   2) *Download the project to the board:*
      a. Set all eSOC DIP switches to OFF.
      b. Connect the board to the PC's parallel printer port and apply power.
      c. Select **Tools | Programmer**.
      d. Click on [Auto Detect] in the Chain Description File.
      e. When the eSOC board is recognized, press the [Add File...] button in the Chain Description File, then locate and Add the *a74293_ctr.pof* file.

f. Select the file and device line and scroll (horizontally) to the available programming option. Check the **Program/Configure** box.

g. Press the [Start] button. Observe the progress bar.

h. If the program downloads successfully, go to the next step. Otherwise, fix the problem and try downloading again.

s) Transfer the answers to the above questions, the waveform information, and state table to the data collection sheets at the end of this experiment.

t) Save and close the project.

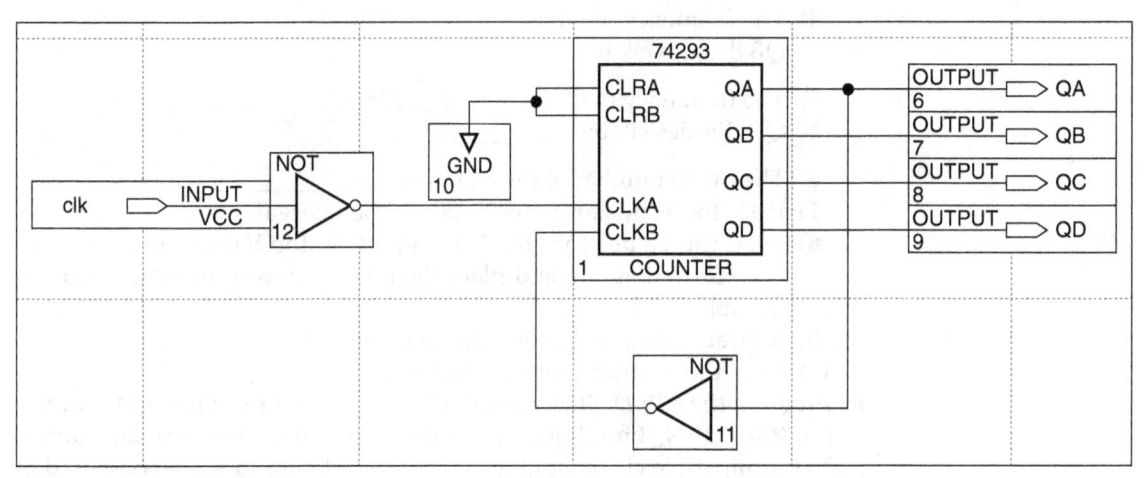

Figure S17-1

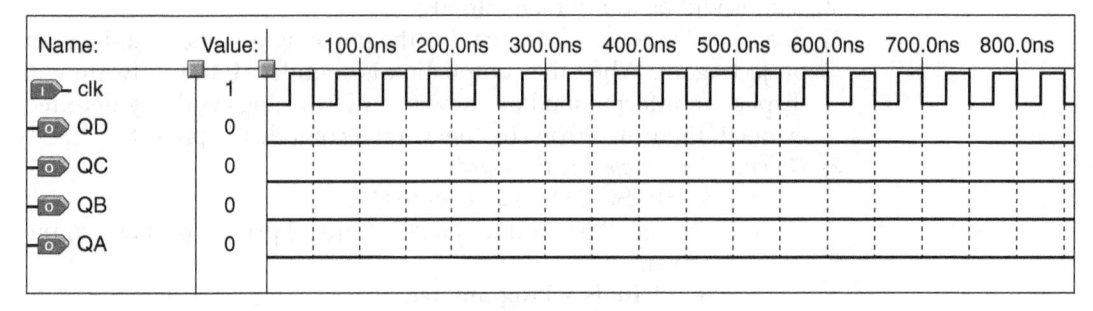

Figure S17-2

PROCEDURE

clk	Time Interval	QA	QB	QC	QD
0	25ns – 50ns				
1	75ns – 100ns				
2	125ns – 150ns				
3	175ns – 200ns				
4	225ns – 250ns				
5	275ns – 300ns				
6	325ns – 350ns				
7	375ns – 400ns				
8	425ns – 450ns				
9	475ns – 500ns				
10	525ns – 550ns				
11	575ns – 600ns				
12	625ns – 650ns				
13	675ns – 700ns				
14	725ns – 750ns				
15	775ns – 800ns				

Table S17-2

PART 2 – Cascading the 74293 IC Asynchronous Counter

a) Start the QUARTUS® II program. Set up a new project and use **amod256_ctr** for the name of the project. The diagram of the circuit you are to construct and its input waveforms are given in Figures S17-3 and S17-4.

b) Create a new graphic file for the schematic *amod256_ctr.bdf*.

c) Place the symbols on the schematic according to Figure S17-3.

d) Place the input and output connectors on the schematic. Label the input clk; label the outputs QA, QB, QC, QD, QE, QF, QG, and QH.

e) Wire the input and output connectors to the circuit as shown in Figure S17-3. This configuration is for a MOD-256 counter.

f) Select **Project | Top-Level Entity**.

g) Compile the project. If there are errors, fix them and save the file before recompiling the project. If there are no errors (or perhaps a warning concerning timing at most), then go on to the next step.

h) *Measure worst-case* t_{co}: Without any design requirements assigned, open the tco report, which is part of Compilation Report (under Timing Analyzer). Input the appropriate data in Table S17-3.

Slack	Required **tco**	Actual **tco**	From	To	From Clock
N/A	None				
N/A	None				
N/A	None				

Table S17-3

i) Create a new waveform *amod256_ctr.vwf* and define the inputs for simulation. The input waveforms should look similar to those in Figure S17-4.
j) Select **Assignments | Settings** to access the Settings Window.
k) Click on *Simulator* at the left.
l) Choose *Functional* from the *Simulator Mode* pull-down menu.
m) Find and use for the *Simulator Input* file *amod256_ctr.vwf* you just created above.
n) Select **Processing | Generate Functional Simulation Netlist**.
o) Simulate the project. If there are any errors and you change your circuit, remember to save and compile the circuit before re-simulating.
p) Record the output waveforms using Figure S17-4 or print out the Simulation Waveforms report from QUARTUS® II and attach it to the data collection sheet at the end of this experiment.
q) *Program the CPLD:* It is suggested that you use **PIN_4** for "**QA**," **PIN_5** for "**QB**," **PIN_8** for "**QC**," **PIN_9** for "**QD**," **PIN_10** for "**QE**," **PIN_11** for "**QF**," **PIN_12** for "**QG**," **and PIN_15** for "**QH**." Also, use an external TTL compatible clock input at 1 Hz. If you choose to use an on-board or external switch to do your clocking, make sure it is debounced. To program your CPLD, refer to Appendix D, Exercise 3. A complete set of instructions will be found there. Briefly,
   1) *Recompile the project*: After making the pin assignments, compile your project again. When the compilation is completed, the software will display an information box showing any warnings and any detected errors. If there are errors, fix them and recompile the project.
   2) *Download the project to the board:*
      a. Set all eSOC DIP switches to OFF.
      b. Connect the board to the PC's parallel printer port and apply power.
      c. Select **Tools | Programmer**.
      d. Click on [Auto Detect] in the Chain Description File.
      e. When the eSOC board is recognized, press the [Add File...] button in the Chain Description File, then locate and Add the *amod256_ctr.pof* file.
      f. Select the file and device line and scroll (horizontally) to the available programming option. Check the **Program/Configure** box.
      g. Press the [Start] button. Observe the progress bar.

PROCEDURE

       h. If the program downloads successfully, go to the next step. Otherwise, fix the problem and try downloading again.
- **r)** Record and analyze results:
  1) The frequency of the clock signal is _____MHz.
  2) The frequency of QH is _____MHz.
     QH divides clk by _____.
  3) The frequency of QG is _____MHz.
     QG divides clk by _____.
  4) The frequency of QF is _____MHz.
     QF divides clk by _____.
  5) The frequency of QE is _____MHz.
     QE divides clk by _____.
  6) The frequency of QD is _____MHz.
     QD divides clk by _____.
  7) The frequency of QC is _____MHz.
     QC divides clk by _____.
  8) The frequency of QB is _____MHz.
     QB divides clk by _____.
  9) The frequency of QA is _____MHz.
     QA divides clk by _____.
  10) The MOD number of the counter is _____.
  11) Does the counter count up or down? _____
- **s)** Transfer the answers to the above questions to the data collection sheets at the end of this experiment.
- **t)** Save and close the project.

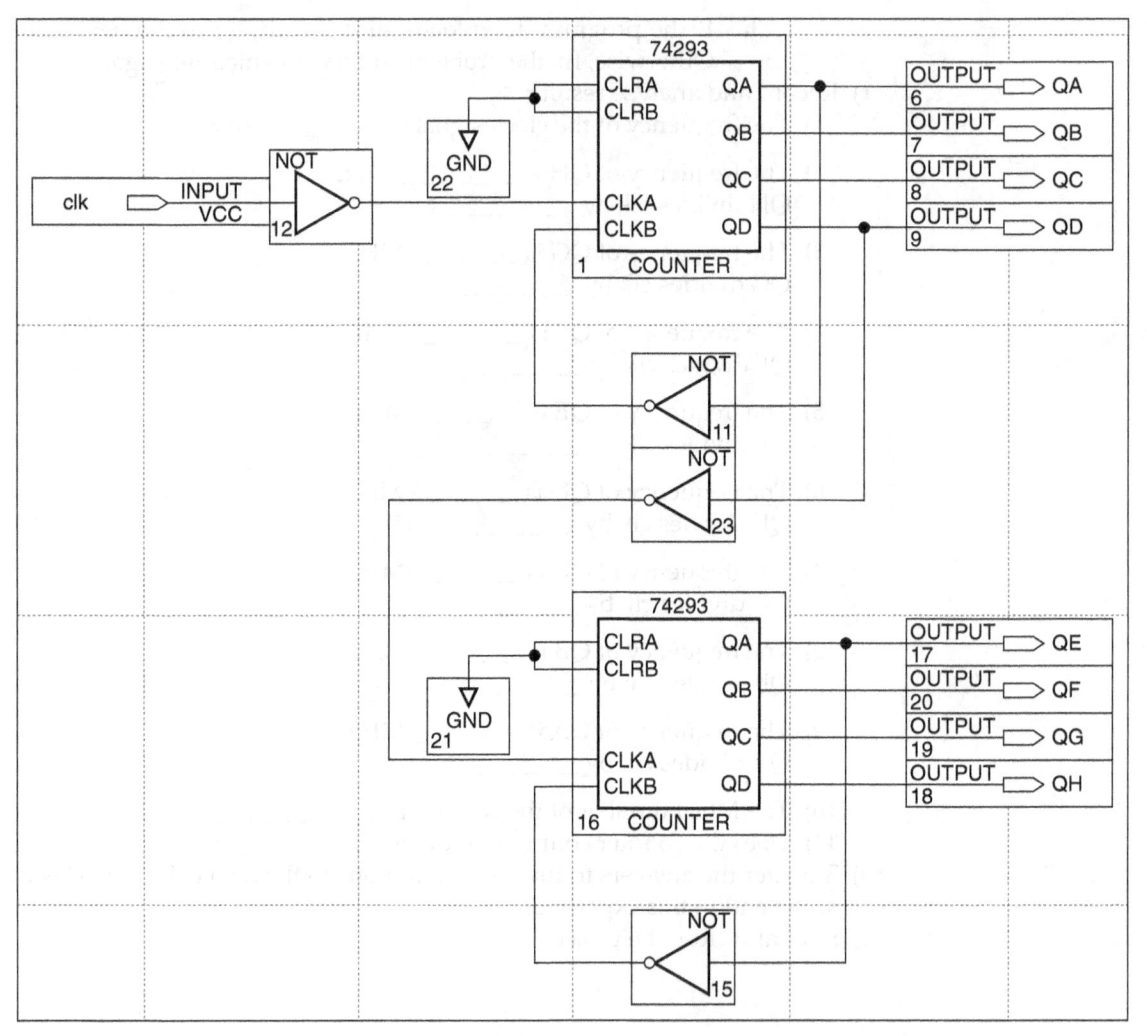

Figure S17-3

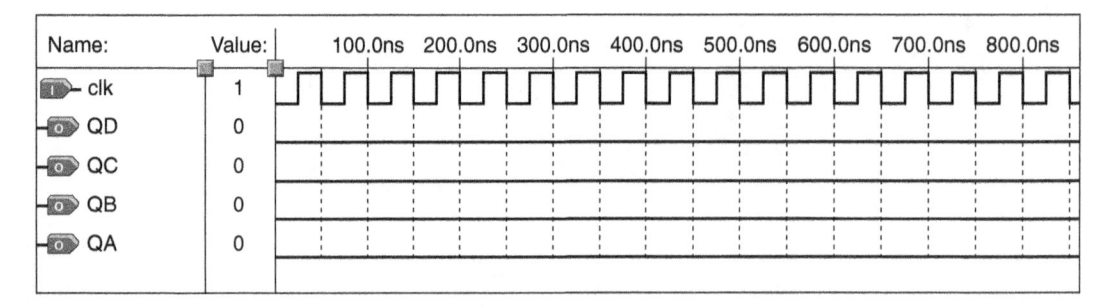

Figure S17-4

DATA COLLECTION SHEET

Data Collection Sheets for Lab S17   Name _____

Date _____

Data collection for Part 1:

h) Examine the **Timing Analyzer Summary** report. What is the worst-case t_{co}? _____

Slack	Required **tco**	Actual **tco**	From	To	From Clock
N/A	None				
N/A	None				
N/A	None				

Table S17-1

q) 1) The frequency of the clock signal is _____ MHz.

2) The frequency of QD is _____ MHz. QD divides clk by _____.

3) The frequency of QC is _____ MHz. QC divides clk by _____.

4) The frequency of QB is _____ MHz. QB divides clk by _____.

5) The frequency of QA is _____ MHz. QA divides clk by _____.

6) The MOD number of the counter is _____.

10) Does the counter count up or down? _____

Demonstrated To Instructor/FA _____ Date _____

Part 1 [ ]   Part 2 [ ]   Part 3 [ ]

Data Collection Sheets for Lab S17

Name _____

Date _____

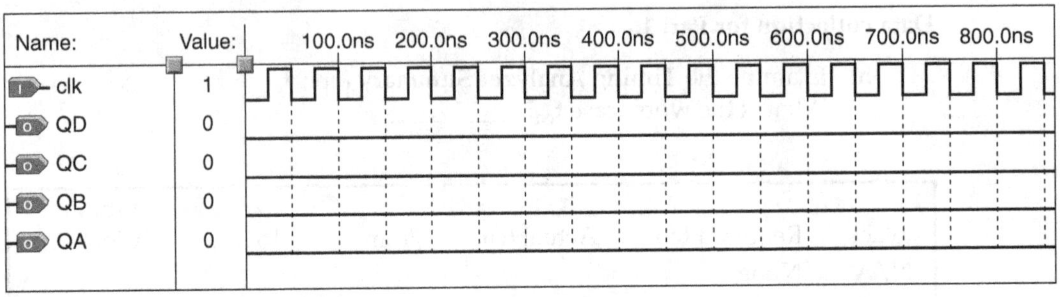

Figure S17-2

clk	Time Interval	QA	QB	QC	QD
0	25ns – 50ns				
1	75ns – 100ns				
2	125ns – 150ns				
3	175ns – 200ns				
4	225ns – 250ns				
5	275ns – 300ns				
6	325ns – 350ns				
7	375ns – 400ns				
8	425ns – 450ns				
9	475ns – 500ns				
10	525ns – 550ns				
11	575ns – 600ns				
12	625ns – 650ns				
13	675ns – 700ns				
14	725ns – 750ns				
15	775ns – 800ns				

Table S17-2

Demonstrated To
Instructor/FA _____ Date _____

Part 1 [ ]   Part 2 [ ]   Part 3 [ ]

DATA COLLECTION SHEET

Data Collection Sheets for Lab S17   Name _____

                                      Date _____

Data collection for Part 2:

    h) Examine the **Timing Analyzer Summary** report. What is the worst-case t_{co}? _____

Slack	Required tco	Actual tco	From	To	From Clock
N/A	None				
N/A	None				
N/A	None				

Table S17-3

    r) Record and analyze results:

        1) The frequency of the clock signal is _____ MHz.

        2) The frequency of QH is _____ MHz.
           QH divides clk by _____.

        3) The frequency of QG is _____ MHz.
           QG divides clk by _____.

        4) The frequency of QF is _____ MHz.
           QF divides clk by _____.

        5) The frequency of QE is _____ MHz.
           QE divides clk by _____.

        6) The frequency of QD is _____ MHz.
           QD divides clk by _____.

        7) The frequency of QC is _____ MHz.
           QC divides clk by _____.

Demonstrated To
Instructor/FA _____ Date _____

Part 1 [ ]   Part 2 [ ]   Part 3 [ ]

Data Collection Sheets for Lab S17   Name _____

Date _____

8) The frequency of QB is _____ MHz.
   QB divides clk by _____.

9) The frequency of QA is _____ MHz.
   QA divides clk by _____.

10) The MOD number of the counter is _____.

11) Does the counter count up or down? _____

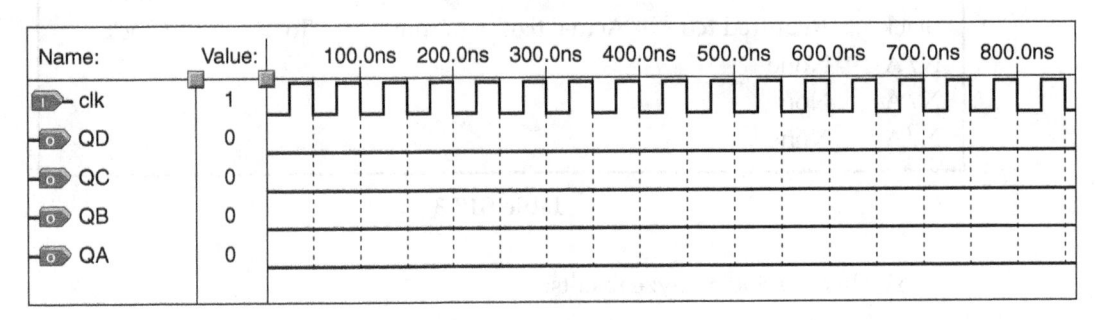

Name:	Value:	100.0ns 200.0ns 300.0ns 400.0ns 500.0ns 600.0ns 700.0ns 800.0ns
clk	1	
QD	0	
QC	0	
QB	0	
QA	0	

Figure S17-4

Demonstrated To
Instructor/FA _____ Date _____

Part 1 [ ]   Part 2 [ ]   Part 3 [ ]

Experiment S18

Name _____

SYNCHRONOUS COUNTERS

OBJECTIVES

1. To practice using QUARTUS® II software to create schematics and waveforms.
2. To practice using QUARTUS® II software to compile and simulate digital circuits.
3. To investigate the operation of a synchronous 4-bit flip-flop counter.
4. To investigate the operation of the 74193 IC counter.
5. To program the EPM7128SLC84 CPLD with a circuit having an asynchronous counter as a component.

TEXT REFERENCES

Review Appendix D of this manual.
Read sections 7.3, 7.6, 7.7, and 7.10.

EQUIPMENT NEEDED

Components

Blank floppy, zip disk, or memory stick for storing projects;
QUARTUS® II software (Altera Corporation);
DeVry University Board eSOC with EPM7128SLC84 CPLD (or Altera University Board with EPM7128SLC84 CPLD or any other equivalent board);
Desktop computer with minimum of INTEL Pentium PC @ 400MHz CPU running Microsoft Windows NT4SP1, 2000, or XP or better; Pentium III or 4 PC @ 400MHz running Red Hat Linux v7.3 or 8.0 or Red Hat Linux Enterprise 3; Sun Ultra running Solaris v8 or 9; HP9000 Series 700/800 running HP-UX v11.0 with ACE dated 11/1999 or later; or equivalent;
TTL compatible signal or function generator.

DISCUSSION

In Experiments S14 and S17, you investigated flip-flop and IC asynchronous counters. One undesirable characteristic of these counters is the accumulation of propagation delay from clock input to clock output. Asynchronous counters have clock signals that are serialized, thus the clock for the second flip-flop onward is generated by the previous flip-flop's output. Another undesirable characteristic is the unwanted glitches at the output of decoders that are used to decode the counter.

Flip-flops in synchronous counters all have a common clock. This not only reduces the propagation delay to virtually zero, but it also significantly reduces the number of glitches that appear at the output of decoders. In this experiment, you will investigate synchronous IC counters, the 74LS193 counter in particular. You will become familiar with their properties, especially their MOD number capabilities, and various ways to connect them. You will then investigate the cascading of such counters to make counters with larger MOD numbers.

PROCEDURE

PART 1 – A 3-bit Synchronous Flip-Flop Counter

a) Start the QUARTUS® II program. Set up a new project and use **s3bit_ctr** for the name of the project. The diagram of the circuit you are to construct and its input waveforms are given in Figures S18-1 and S18-2.

b) Create a new graphic file for the schematic *s3bit_ctr.bdf*.

c) Place the symbols on the schematic according to Figure S18-1.

d) Place the input and output connectors on the schematic. Label the input clk and clr; label the outputs QA, QB, and QC.

e) Wire the input and output connectors to the circuit as shown in Figure S18-1. This configuration is for a MOD-8 counter.

f) Select **Project | Top-Level Entity**.

g) Compile the project. If there are errors, fix them and save the file before recompiling the project. If there are no errors (or perhaps a warning concerning timing at most), then go on to the next step.

h) *Measure worst-case* t_{co}: Without any design requirements assigned, open the tco report, which is part of Compilation Report (under Timing Analyzer). Input the appropriate data in Table S18-1. Examine the **Timing Analyzer Summary** report. What is the worst-case t_{co}? _____

Slack	Required **tco**	Actual **tco**	From	To	From Clock
N/A	None				
N/A	None				
N/A	None				

Table S18-1

PROCEDURE

i) In a few well-chosen words, compare these results with those in Part 1 of Experiment 14 (Table S14-1).

_____

_____

j) Create a new waveform *s3bit_ctr.vwf* and define the inputs for simulation. The input waveforms should look similar to those in Figure S18-2.
k) Select **Assignments | Settings** to access the Settings Window.
l) Click on *Simulator* at the left.
m) Choose *Functional* from the *Simulator Mode* pull-down menu.
n) Find and use for the *Simulator Input* file *s3bit_ctr.vwf* you just created above.
o) Select **Processing | Generate Functional Simulation Netlist**.
p) Simulate the project. If there are any errors and you change your circuit, remember to save and compile the circuit before re-simulating.
q) Record and analyze results:
   1) The frequency of the clock signal is _____ MHz.
   2) The frequency of QA is _____ MHz.
      QA divides clk by _____.
   3) The MOD number of the counter is _____.
r) Record the output waveforms QC, QB, and QA on Figure S18-2. You should complete this part of the experiment by converting the waveform information in Figure S18-2 to Table S18-1:
   1) Place the Time Bar in the 25–50ns time interval.
   2) Read the values for QA, QB, and QC in the Values column of the waveform diagram and place them in their corresponding columns in Table S18-2.
   3) Repeat steps 1 and 2 for each time interval.
s) Does the counter count up or down? _____
t) Transfer the answers to the above questions, the waveform information, and state table to the data collection sheets at the end of this experiment.
u) Save and close the project.

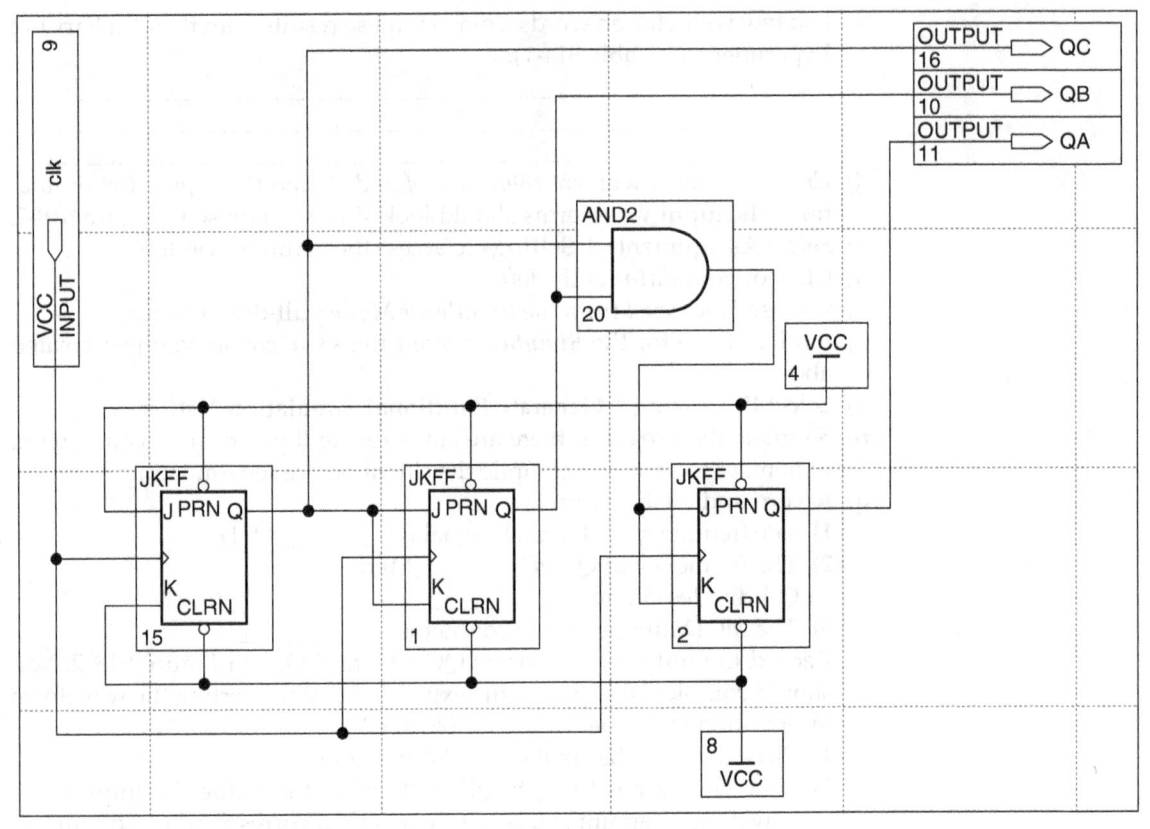

Figure S18-1

Figure S18-2

clk	Time Interval	QA	QB	QC
0	0ns – 25ns			
1	25ns – 75ns			
2	75ns – 125ns			
3	175ns – 200ns			
4	225ns – 250ns			
5	275ns – 300ns			
6	325ns – 350ns			
7	375ns – 400ns			
8	425ns – 450ns			

Table S18-2

PROCEDURE

PART 2 – The 74193 IC Synchronous Counter – MOD-16 Configuration

a) Start the QUARTUS® II program. Set up a new project and use **s74193_ctr** for the name of the project. The diagram of the circuit you are to construct and its input waveforms are given in Figures S18-3 and S18-4.
b) Create a new graphic file for the schematic.
c) Place the symbols on the schematic according to Figure S18-3.
d) Place the input and output connectors on the schematic. Label the inputs clk and CLEAR; label the outputs QA, QB, QC, and QD.
e) Wire the input and output connectors to the circuit as shown in Figure S18-3. This configuration is for a MOD-16 counter.
f) Select **Project | Top-Level Entity**.
g) Compile the project. If there are errors, fix them and save the file before recompiling the project. If there are no errors (or perhaps a warning concerning timing at most), then go on to the next step.
h) Create a new waveform file *s74193_ctr.vwf* and define the inputs for simulation. The waveforms should look similar to those in Figure S18-4.
i) Select **Assignments | Settings** to access the Settings Window.
j) Click on *Simulator* at the left.
k) Choose *Functional* from the *Simulator Mode* pull-down menu.
l) Find and use for the *Simulator Input* file *s74193_ctr.vwf* you just created above.
m) Select **Processing | Generate Functional Simulation Netlist**.
n) Simulate the project. If there are any errors and you change your circuit, remember to save and compile the circuit before re-simulating.
o) Record and analyze results:
  1) Record the output waveforms QD, QC, QB, and QA on Figure S18-4 or print out the Simulation Waveforms report from QUARTUS® II and attach it to the data collection sheet at the end of this experiment.
  2) The frequency of the clock signal is _____ MHz.
  3) The frequency of QD is _____ MHz.
     QD divides clk by _____.
  4) The MOD number of the counter is _____.
p) You should complete this part of the experiment by converting the waveform information in Figure S18-4 to Table S18-3:
  1) Place the Time Bar in the 25–50ns time interval.
  2) Read the values for QA, QB, QC, and QD in the Values column of the waveform diagram and place them in their corresponding columns in Table S18-3.
  3) Repeat steps 1 and 2 for each time interval.
  4) Does the counter count up or down? _____
  5) What must you do to make the counter count in the opposite direction? _____

q) Transfer the answers to the above questions, the waveform information, and state table, to the data collection sheets at the end of this experiment.
r) Save and close the project.

Figure S18-3

```
         29  PL  >INPUT/VCC    o|LDN  74193  QA|--OUTPUT 6 --> QA
         25  a   >INPUT/VCC     |A           QB|--OUTPUT 7 --> QB
         26  b   >INPUT/VCC     |B           QC|--OUTPUT 8 --> QC
         27  c   >INPUT/VCC     |C           QD|--OUTPUT 9 --> QD
         28  d   >INPUT/VCC     |D          CON|o-OUTPUT 33 --> TCU
                                |DN         BON|o
  31    30  clk >INPUT/VCC      |UP
  VCC   34  clr >INPUT/VCC      |CLR
                                24  COUNTER
```

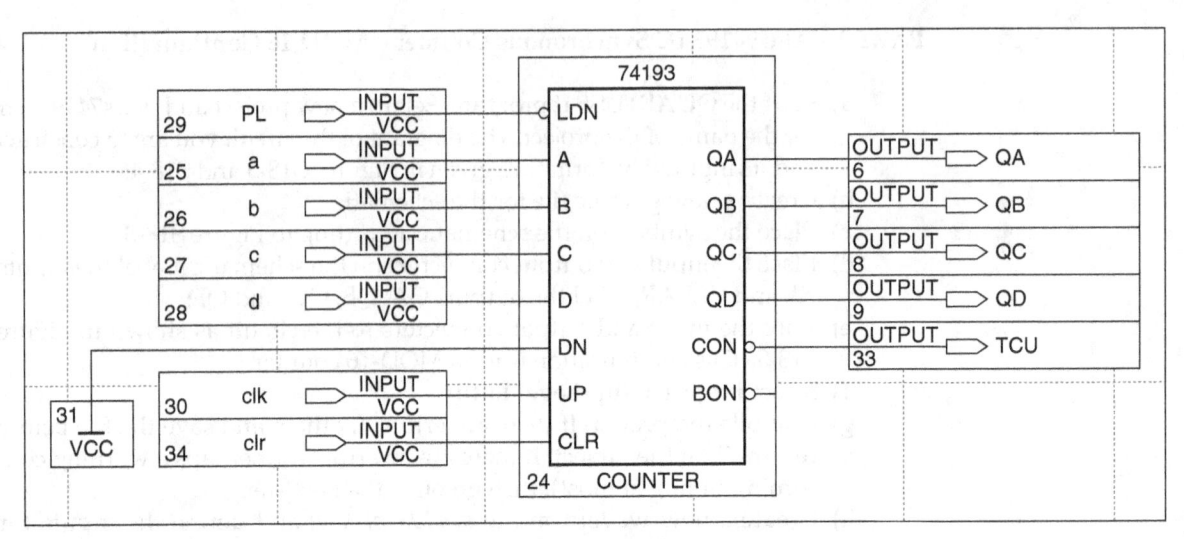

Figure S18-3

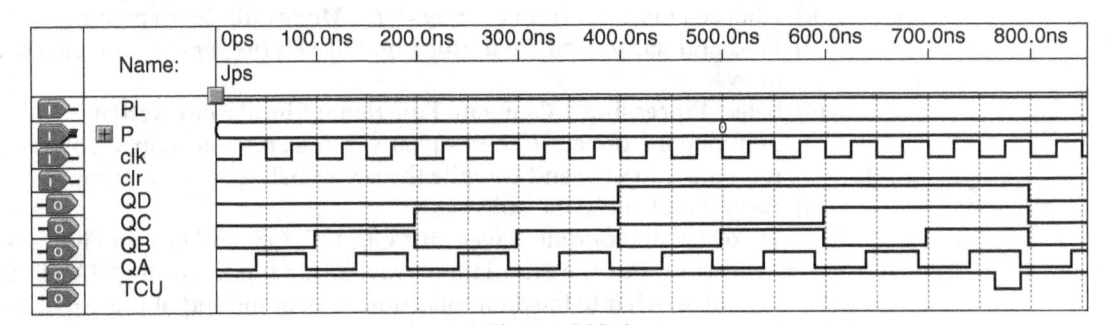

Figure S18-4

clk	Time Interval	QA	QB	QC	QD
0	25ns – 50ns				
1	75ns – 100ns				
2	125ns – 150ns				
3	175ns – 200ns				
4	225ns – 250ns				
5	275ns – 300ns				
6	325ns – 350ns				
7	375ns – 400ns				
8	425ns – 450ns				
9	475ns – 500ns				
10	525ns – 550ns				
11	575ns – 600ns				
12	625ns – 650ns				
13	675ns – 700ns				
14	725ns – 750ns				
15	775ns – 800ns				

Table S18-3

PROCEDURE

PART 3 – Multistage Arrangement

a) Start the QUARTUS® II program. Set up a new project and use **multistage_ctr** for the name of the project. The diagram of the circuit you are to construct is given in Figure S18-5.
b) Create a new graphic file for the schematic *multistage_ctr.bdf*.
c) Place the symbols on the schematic according to Figure S18-5.
d) Place the input and output connectors on the schematic. Label the input clk; label the outputs Q0, Q1, Q2, Q3, Q4, Q5, Q6, and Q7.
e) Wire the input and output connectors to the circuit as shown in Figure S18-5. This configuration is for a multistage counter using the 74193 IC. From this configuration you can obtain an extended count range of 0 – 255 and with the parallel inputs, P0-P7, you can start the counter at some number other than zero. This permits you to change the MOD number of the counter.
f) Select **Project | Top-Level Entity**.
g) Compile the project. If there are errors, fix them and save the file before recompiling the project. If there are no errors (or perhaps a warning concerning timing at most), then go on to the next step.
h) Create a new waveform file *multistage_ctr.vwf* (see Figure 18-6). Set the clock period to 100ns. Set the end time to 21µs. Group the parallel inputs and outputs into two groups of 4 each.
i) In the 50-100ns time interval, create a negative-going pulse on the Load input. For the remainder of the time Load should be HIGH.
j) Set both down clock inputs HIGH.
k) Select **Assignments | Settings** to access the Settings Window.
l) Click on *Simulator* at the left.
m) Choose *Functional* from the *Simulator Mode* pull-down menu.
n) Find and use for the *Simulator Input* file *multistage_ctr.vwf* you just created above.
o) Select **Processing | Generate Functional Simulation Netlist**.
p) Simulate the project. If there are any errors and you change your circuit, remember to save and compile the circuit before re-simulating.
q) Now set P0-P3 to 8 and P4-P7 to 3. This will make the count sequence of the counter go from 38H-FFH (56_{10}-255_{10}). What is the MOD number with this setting? _____
r) On the Waveform Editor window, insert node Q7. The waveform for this node is generated without further simulation.
s) Use the Time Bar and its movement buttons and measure the period of Q7: _____
t) Transfer the answers to the above questions to the data collection sheets at the end of this experiment.

u) *Program the CPLD:* It is suggested that you use **PIN_4** for "Q0," **PIN_5** for "Q1," **PIN_8** for "Q2," **PIN_9** for "Q3," **PIN_10** for "Q4," **PIN_11** for "Q5," **PIN_12** for "Q6," and **PIN_15** for "Q7." Use **PIN_50** for "P0," **PIN_51** for "P1," **PIN_52** for "P2," **PIN_54** for "P3," **PIN_55** for "P4," **PIN_56** for "P5," **PIN_57** for "P6," and **PIN_58** for "P7." Use **PIN_60** for "DN1," **PIN_61** for "DN2," and **PIN_62** for "Clear." Use **PIN_70** for "Load." Also, use an external TTL compatible clock input at 1 Hz. If you choose to use an on-board or external switch to do your clocking, make sure it is debounced. To program your CPLD, refer to Appendix D, Exercise 3. A complete set of instructions will be found there. Briefly,

1) *Recompile the project*: After making the pin assignments, compile your project again. When the compilation is completed, the software will display an information box showing any warnings and any detected errors. If there are errors, fix them and recompile the project.
2) *Download the project to the board:*
    a. Set all eSOC DIP switches to OFF.
    b. Connect the board to the PC's parallel printer port and apply power.
    c. Select **Tools | Programmer**.
    d. Click on [Auto Detect] in the Chain Description File.
    e. When the eSOC board is recognized, press the [Add File...] button in the Chain Description File, then locate and Add the *multistage_ctr.pof* file.
    f. Select the file and device line and scroll (horizontally) to the available programming option. Check the **Program/Configure** box.
    g. Press the [Start] button. Observe the progress bar.
    h. If the program downloads successfully, go to the next step. Otherwise, fix the problem and try downloading again.
3) *Test the program:* After connections are made to your board, apply power and verify that the LEDs respond in a manner consistent with a MOD-256 counter.
    a. Toggle DN1 and DN2 to HIGH.
    b. Set Clear to LOW.
    c. Set P[7..0] = 38H.
    d. Push Clear momentarily and release.
    e. Observe the count's direction and sequence.

v) Transfer your observations from the simulation and program test results to the data collection sheets at the end of this experiment.

w) Save and close the project.

PROCEDURE

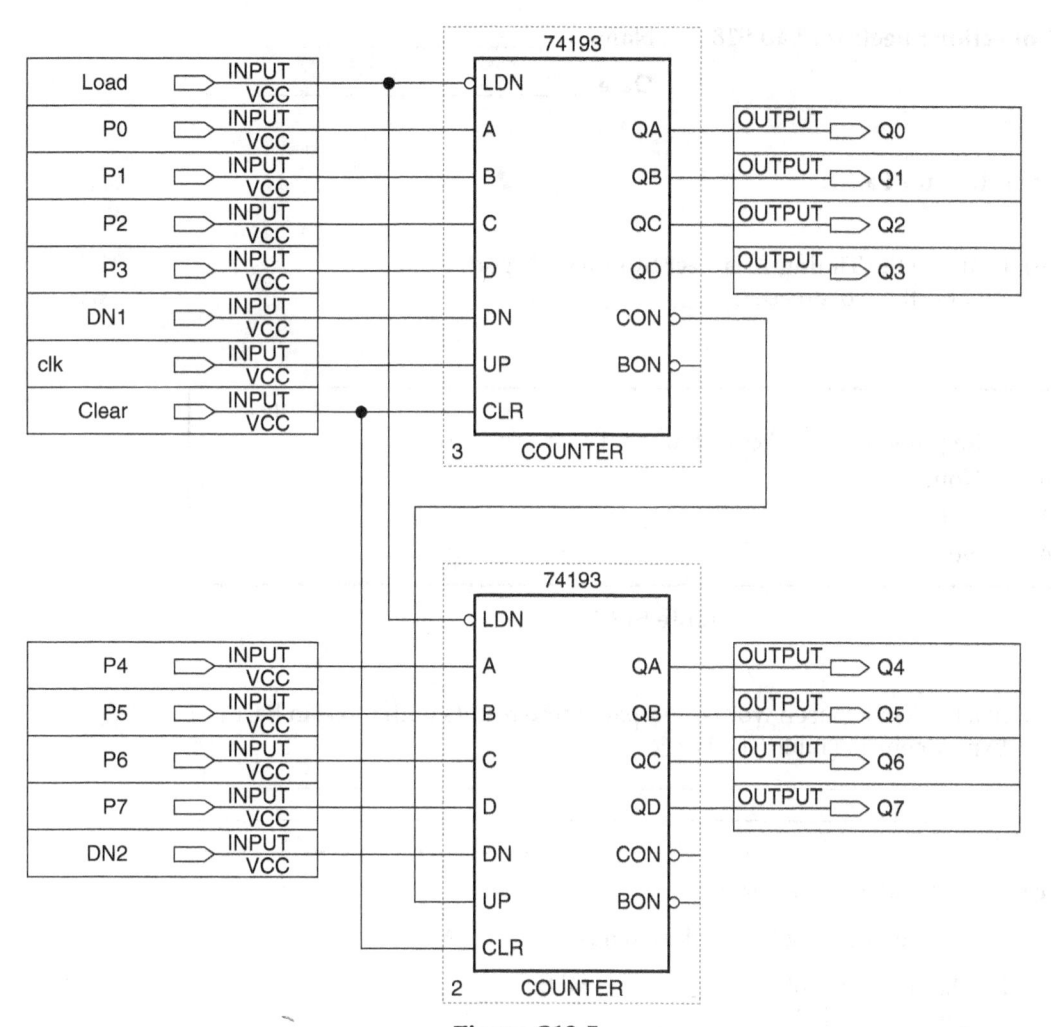

Figure S18-5

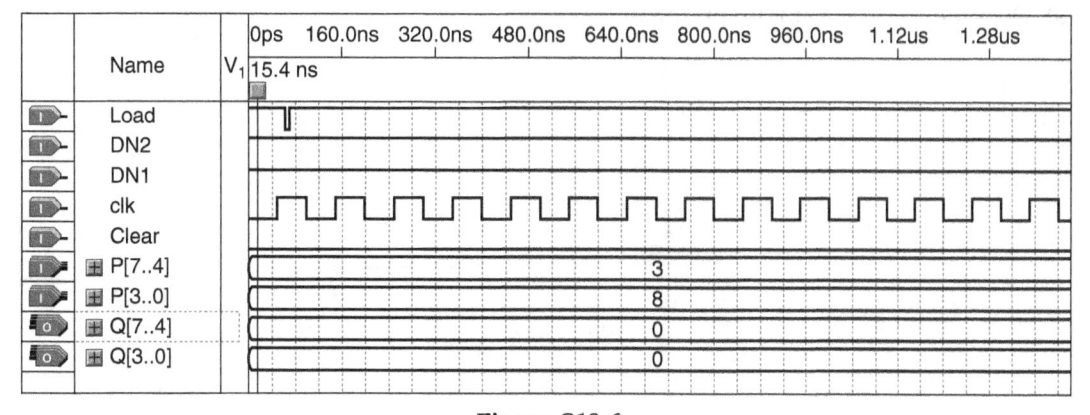

Figure S18-6

Data Collection Sheets for Lab S18 Name _____

Date _____

Data collection for Part 1:

    h) Examine the **Timing Analyzer Summary** report.
        What is the worst-case t_{co}? _____

Slack	Required **tco**	Actual **tco**	From	To	From Clock
N/A	None				
N/A	None				
N/A	None				

Table S18-1

    i) In a few well-chosen words, compare these results with those in Part 1 of Experiment 14 (Table S14-1).
        _____
        _____
        _____

    q) Record and analyze results:

        1) The frequency of the clock signal is _____MHz.

        2) The frequency of QA is _____MHz.
           QA divides clk by _____.

        3) The MOD number of the counter is _____.

    s) Does the counter count up or down? _____

Demonstrated To
Instructor/FA _____ Date _____

Part 1 [ ]   Part 2 [ ]   Part 3 [ ]

DATA COLLECTION SHEET

Data Collection Sheets for Lab S18 Name _____

Date _____

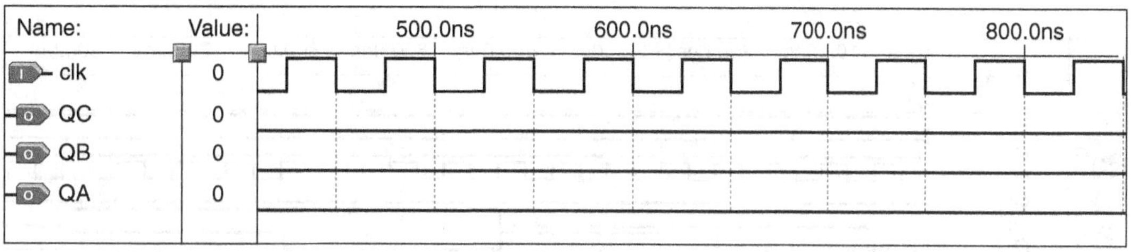

```
Name:    Value:    500.0ns      600.0ns      700.0ns      800.0ns
clk      0
QC       0
QB       0
QA       0
```

Figure S18-2

clk	Time Interval	QA	QB	QC
0	0ns – 25ns			
1	25ns – 75ns			
2	75ns – 125ns			
3	175ns – 200ns			
4	225ns – 250ns			
5	275ns – 300ns			
6	325ns – 350ns			
7	375ns – 400ns			
8	425ns – 450ns			

Table S18-2

Data collection for Part 2:

o)

2) The frequency of the clock signal is _____ MHz.

3) The frequency of QD is _____ MHz.
   QD divides clk by _____.

4) The MOD number of the counter is _____.

Demonstrated To
Instructor/FA _____ Date _____

Part 1 [ ]   Part 2 [ ]   Part 3 [ ]

Data Collection Sheets for Lab S18

Name _____

Date _____

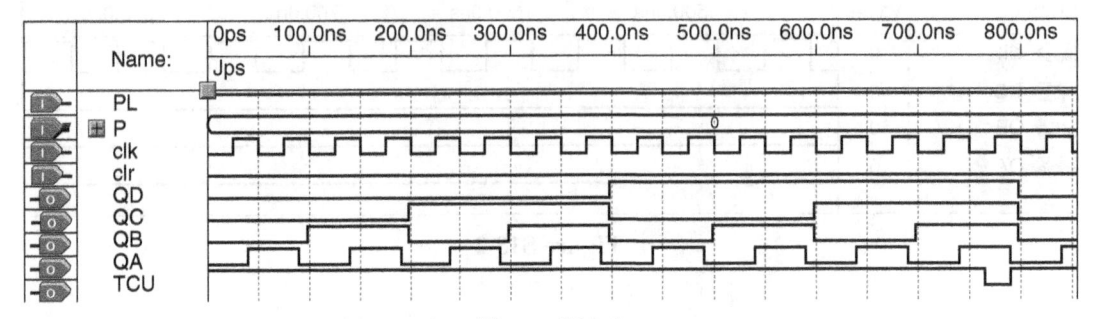

Figure S18-4

clk	Time Interval	QA	QB	QC	QD
0	25ns – 50ns				
1	75ns – 100ns				
2	125ns – 150ns				
3	175ns – 200ns				
4	225ns – 250ns				
5	275ns – 300ns				
6	325ns – 350ns				
7	375ns – 400ns				
8	425ns – 450ns				
9	475ns – 500ns				
10	525ns – 550ns				
11	575ns – 600ns				
12	625ns – 650ns				
13	675ns – 700ns				
14	725ns – 750ns				
15	775ns – 800ns				

Table S18-3

Demonstrated To
Instructor/FA _____ Date _____

Part 1 [ ]   Part 2 [ ]   Part 3 [ ]

DATA COLLECTION SHEET

Data Collection Sheets for Lab S18   Name _____

Date _____

p)
4) Does the counter count up or down? _____

5) What must you do to make the counter count in the opposite direction?
_____
_____
_____

Data collection for Part 3:

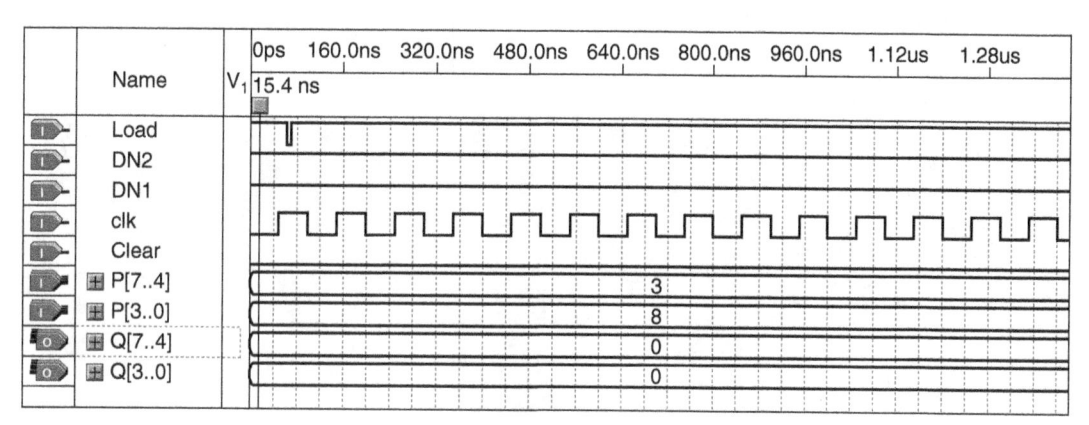

	Name	V₁	0ps 160.0ns 320.0ns 480.0ns 640.0ns 800.0ns 960.0ns 1.12us 1.28us
▷	Load		
▷	DN2		
▷	DN1		
▷	clk		⊓⊔⊓⊔⊓⊔⊓⊔⊓⊔⊓⊔⊓⊔⊓⊔
▷	Clear		
▷	⊞ P[7..4]		3
▷	⊞ P[3..0]		8
◁	⊞ Q[7..4]		0
◁	⊞ Q[3..0]		0

Figure S18-6

q) Now set P0-P3 to 8 and P4-P7 to 3. This will make the count sequence of the counter go from 38H-FFH (56_{10}-255_{10}). What is the MOD number with this setting? _____

s) Use the Time Bar and its movement buttons and measure the period of Q7: _____

Demonstrated To		
Instructor/FA _____ Date _____		
Part 1 [ ]	Part 2 [ ]	Part 3 [ ]

Experiment S19

Name _____

BCD COUNTERS

OBJECTIVES

1. To practice using QUARTUS® II software to create schematics and waveforms.
2. To practice using QUARTUS® II software to compile and simulate digital circuits.
3. To investigate the operation of the 74160 IC counter.
4. To demonstrate that BCD counters may be displayed directly with seven-segment display units.
5. To investigate the cascading of 74160 IC counters.
6. To program the EPM7128SLC84 CPLD with a circuit having an asynchronous counter as a component.

TEXT REFERENCES

Read sections 7.7 and section 9.2. Review Appendix D of this manual.

EQUIPMENT NEEDED

Components

Blank floppy, zip disk, or memory stick for storing projects;

QUARTUS® II software (Altera Corporation);

DeVry University Board eSOC with EPM7128SLC84 CPLD (or Altera University Board with EPM7128SLC84 CPLD or any other equivalent board);

Desktop computer with minimum of INTEL Pentium PC @ 400MHz CPU running Microsoft Windows NT4SP1, 2000, or XP or better; Pentium III or 4 PC @ 400MHz running Red Hat Linux v7.3 or 8.0 or Red Hat Linux Enterprise 3; Sun Ultra running Solaris v8 or 9; HP9000 Series 700/800 running HP-UX v11.0 with ACE dated 11/1999 or later; or equivalent;

TTL compatible signal or function generator.

DISCUSSION

In this experiment, you will investigate synchronous BCD IC counters, the 74160 counter in particular. You will become familiar with their properties and the use of seven-segment devices to display their outputs. You will then investigate the cascading of such counters to make counters with larger MOD numbers.

PROCEDURE

PART 1 – The 74160 IC BCD Counter

a) Start the QUARTUS® II program. Set up a new project and use **syn_bcd_ctr** for the name of the project. The diagram of the circuit you are to construct and its input waveforms are given in Figures S19-1 and S19-2.
b) Create a new graphic file for the schematic *syn_bcd_ctr.bdf*.
c) Place the symbols on the schematic according to Figure S19-1.
d) Place the input and output connectors on the schematic. Label the clock inputs CP0 and CP1. Label the resets MR1 and MR2. Label the outputs Q0, Q1, Q2, and Q3.
e) Wire the input and output connectors to the circuit as shown in Figure S19-1.
f) Select **Project | Top-Level Entity**.
g) Compile the project. If there are errors, fix them and save the file before recompiling the project. If there are no errors (or perhaps a warning concerning timing at most), then go on to the next step.
h) Create a new waveform *syn_bcd_ctr.vwf* and define the inputs for simulation. The input waveforms should look similar to those in Figure S19-2.
i) Select **Assignments | Settings** to access the Settings Window.
j) Click on **Simulator** at the left.
k) Choose *Functional* from the **Simulator Mode** pull-down menu.
l) Find and use for the **Simulator Input** file *syn_bcd_ctr.vwf* you just created above.
m) Select **Processing | Generate Functional Simulation Netlist**.
n) Simulate the project. If there are any errors and you change your circuit, remember to save and compile the circuit before re-simulating.
o) Record and analyze results:
  1) Record the output waveforms on Figure S19-2 or print out the Simulation Waveforms report from QUARTUS® II and attach it to the data collection sheet at the end of this experiment.
  2) You should complete this part of the experiment by converting the waveform information in Figure S19-2 to Table S19-1:
     a. Select **View | Snap To Transition**. Check its box if necessary.
     b. Place the Time Bar at the first positive-edge transition of the clock at 125 ns.

PROCEDURE

  c. Read the values for Q0, Q1, Q2, and Q3 in the Values column of the waveform diagram and place them in their corresponding columns in Table S19-1.
  d. Repeat steps 2 and 3 for each positive-edge transition of the clock.
3) What is the count sequence for this counter? _____

4) Does the counter count up or down? _____
5) Transfer the answers to the above questions, the waveform information, and state table to the data collection sheets at the end of this experiment.

p) **Program the CPLD:** It is suggested that you use **PIN_50** for "**Enable**," **PIN_4** for "**Q0**," **PIN_5** for "**Q1**," **PIN_8** for "**Q2**," **PIN_9** for "**Q3**," and **PIN_10** for "**Carry**." Also, use an external TTL compatible clock input at 1 Hz. If you choose to use an on-board or external switch to do your clocking, make sure it is debounced. To program your CPLD, refer to Appendix D, Exercise 3. A complete set of instructions will be found there. Briefly,

  1) *Recompile the project*: After making the pin assignments, compile your project again. When the compilation is completed, the software will display an information box showing any warnings and any detected errors. If there are errors, fix them and recompile the project.
  2) *Download the project to the board:*
      a. Set all eSOC DIP switches to OFF.
      b. Connect the board to the PC's parallel printer port and apply power.
      c. Select **Tools | Programmer**.
      d. Click on [Auto Detect] in the Chain Description File.
      e. When the eSOC board is recognized, press the [Add File...] button in the Chain Description File, then locate and Add the *syn_bcd_ctr.pof* file.
      f. Select the file and device line and scroll (horizontally) to the available programming option. Check the **Program/Configure** box.
      g. Press the [Start] button. Observe the progress bar.
      h. If the program downloads successfully, go to the next step. Otherwise, fix the problem and try downloading again.

q) *Test the program:*
   1) Toggle the Enable switch to HIGH. What value is displayed on the LEDs? _____
   2) Pulse the clock pushbutton through the counter's counting sequence.
   3) What is the counter's sequence?
      _____

r) Save and close the project.

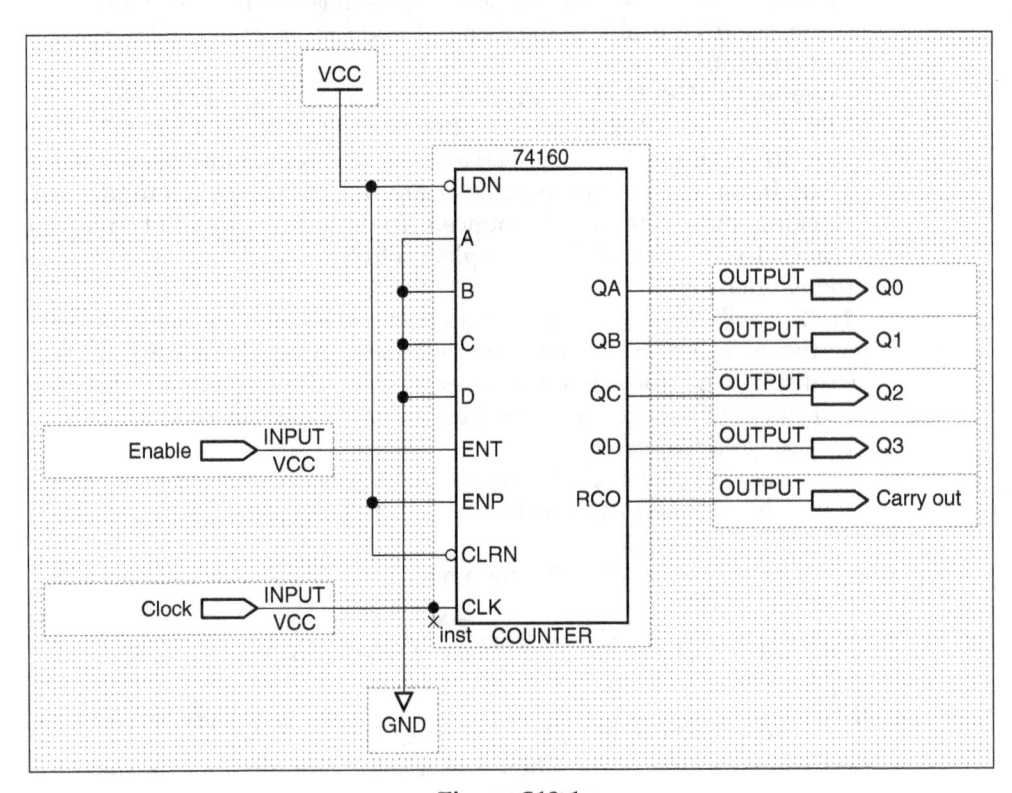

Figure S19-1

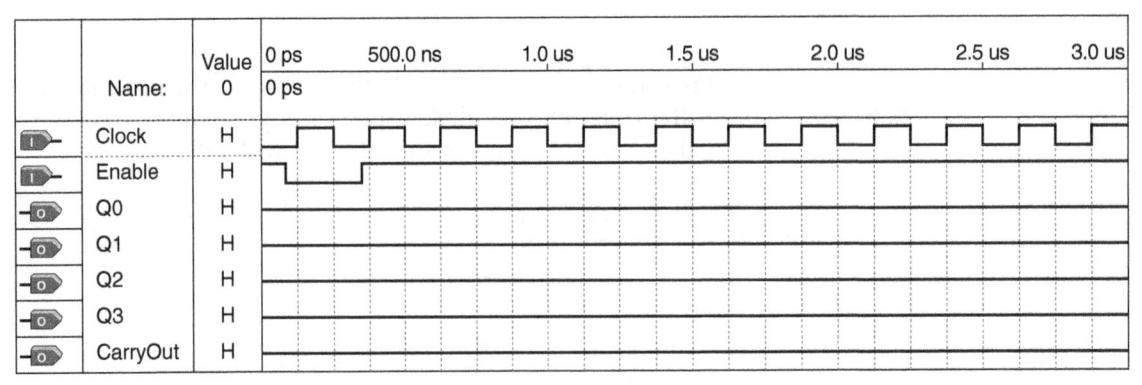

Figure S19-2

PROCEDURE

clk	Time Interval	Carry Out	Q3	Q2	Q1	Q0
0	375ns – 425ns					
1	425ns – 675ns					
2	675ns – 925ns					
3	925ns – 1.175us					
4	1.175us – 1.425us					
5	1.425us – 1.675us					
6	1.675us – 1.925us					
7	1.925us – 2.175us					
8	2.175us – 2.425us					
9	2.425us – 2.675us					
10	2.675us – 2.925us					

Table S19-1

PART 2 – Cascading BCD Counters & Using the 7-segment Display Unit

a) Start the QUARTUS® II program. Set up a new project and use **bcd_ctr_display** for the name of the project. The diagram of the circuit you are to construct is given in Figure S19-3. It is suggested that you connect an external TTL-compatible function generator set at 1 Hz for the clock. If you use an on-board or external switch for the clock, make sure that it is debounced. Also, be sure to check your board's switches and 7-segment LEDs' configurations and pin numbers before making your pin assignments.

b) Create a new schematic named *bcd_ctr_display.bdf*.

c) Wire the input and output connectors to the circuit as shown in Figure S19-3.

d) Select **Project | Top-Level Entity**.

e) *Program the CPLD:* It is suggested that you use **PIN_50** for "**Enable**," **PIN_31** for "a2," **PIN_30** for "b2," **PIN_29** for "c2," **PIN_28** for "d2," **PIN_27** for "e2," **PIN_34** for "f2," **PIN_33** for "g2," **PIN_45** for "a," **PIN_44** for "b," **PIN_41** for "c," **PIN_37** for "d," **PIN_36** for "e," **PIN_48** for "f," and **PIN_40** for "g." Use **PIN_70** for "**Clear**." Also, use an external TTL compatible clock input at 1 Hz. If you choose to use an on-board or external switch to do your clocking, make sure it is debounced. To program your CPLD, refer to Appendix D, Exercise 3. A complete set of instructions will be found there. Briefly,

　1) *Recompile the project*: After making the pin assignments, compile your project. When the compilation is completed, the software will display an information box showing any warnings and any detected errors. If there are errors, fix them and recompile the project.

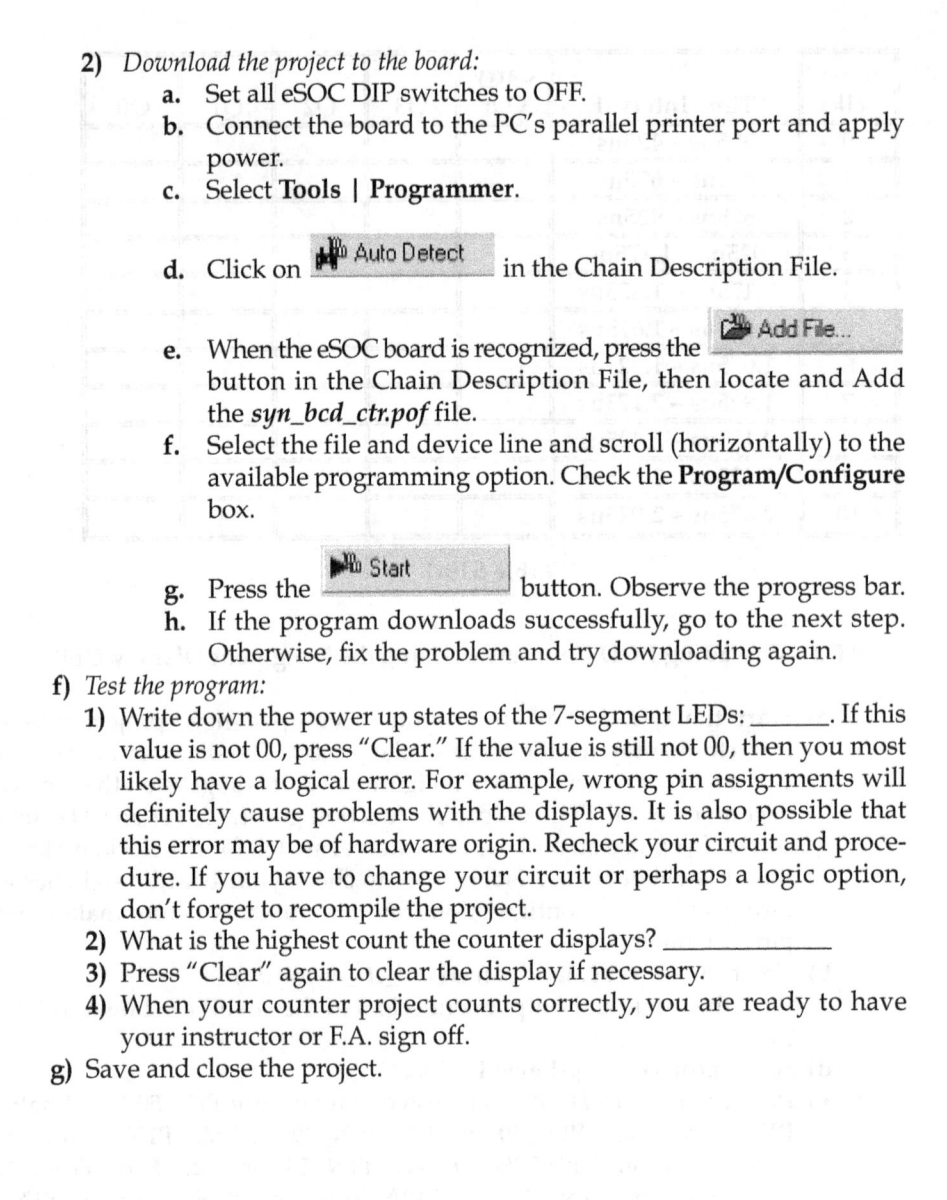

2) *Download the project to the board:*
   a. Set all eSOC DIP switches to OFF.
   b. Connect the board to the PC's parallel printer port and apply power.
   c. Select **Tools | Programmer**.
   d. Click on [Auto Detect] in the Chain Description File.
   e. When the eSOC board is recognized, press the [Add File...] button in the Chain Description File, then locate and Add the *syn_bcd_ctr.pof* file.
   f. Select the file and device line and scroll (horizontally) to the available programming option. Check the **Program/Configure** box.
   g. Press the [Start] button. Observe the progress bar.
   h. If the program downloads successfully, go to the next step. Otherwise, fix the problem and try downloading again.

f) *Test the program:*
   1) Write down the power up states of the 7-segment LEDs: _____. If this value is not 00, press "Clear." If the value is still not 00, then you most likely have a logical error. For example, wrong pin assignments will definitely cause problems with the displays. It is also possible that this error may be of hardware origin. Recheck your circuit and procedure. If you have to change your circuit or perhaps a logic option, don't forget to recompile the project.
   2) What is the highest count the counter displays? _____
   3) Press "Clear" again to clear the display if necessary.
   4) When your counter project counts correctly, you are ready to have your instructor or F.A. sign off.

g) Save and close the project.

PROCEDURE

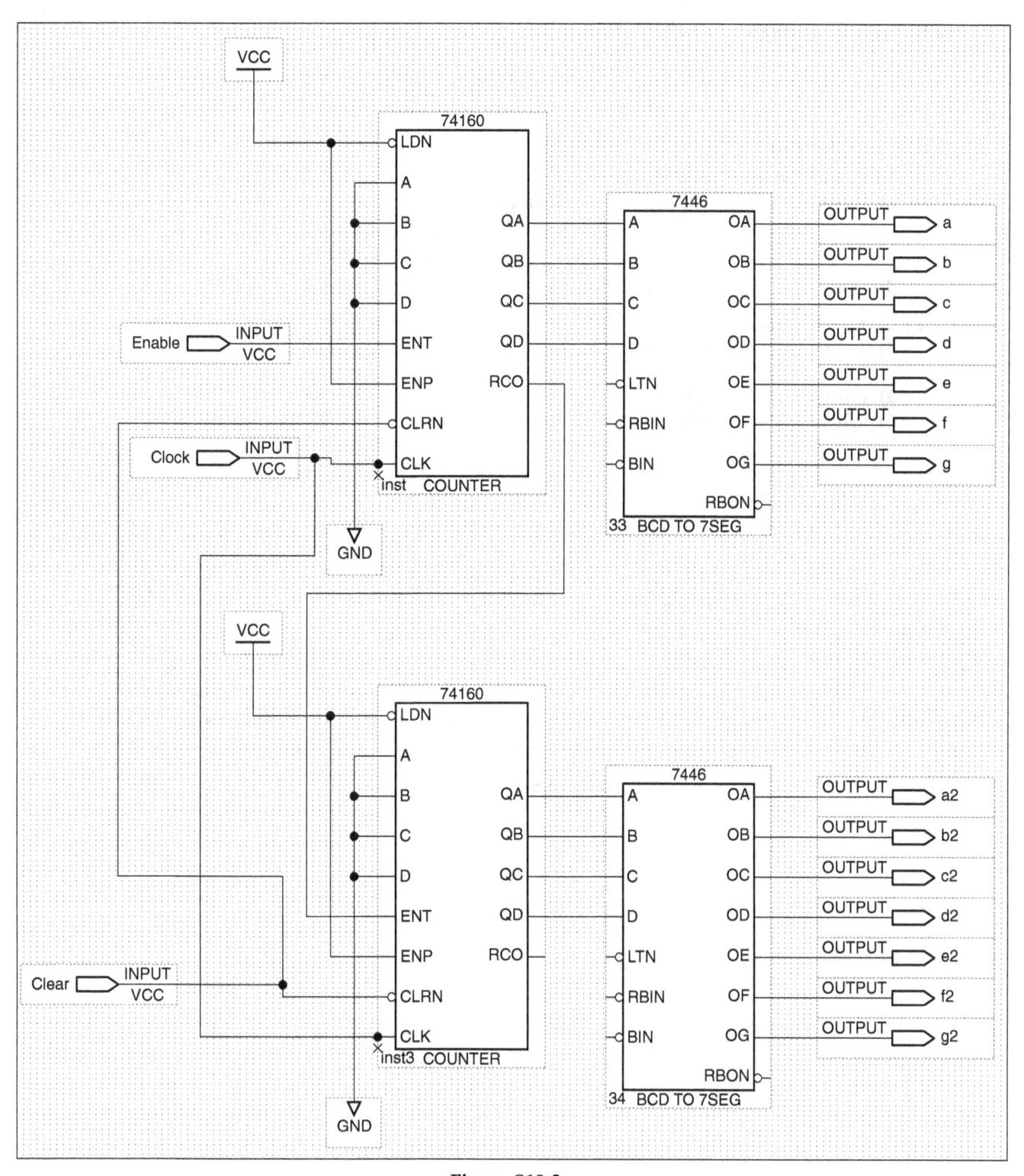

Figure S19-3

Data Collection Sheets for Lab S19   Name _____
                                      Date _____

Data collection for Part 1:
  o) *Record and analyze results:*
     3) What is the count sequence for this counter?
        _____
        _____

     4) Does the counter count up or down? _____

  q) *Test the program:*
     1) Toggle the Enable switch to HIGH. What value is displayed on the LEDs? _____
     2) Pulse the clock pushbutton through the counter's counting sequence.
     3) What is the counter's sequence?
        _____

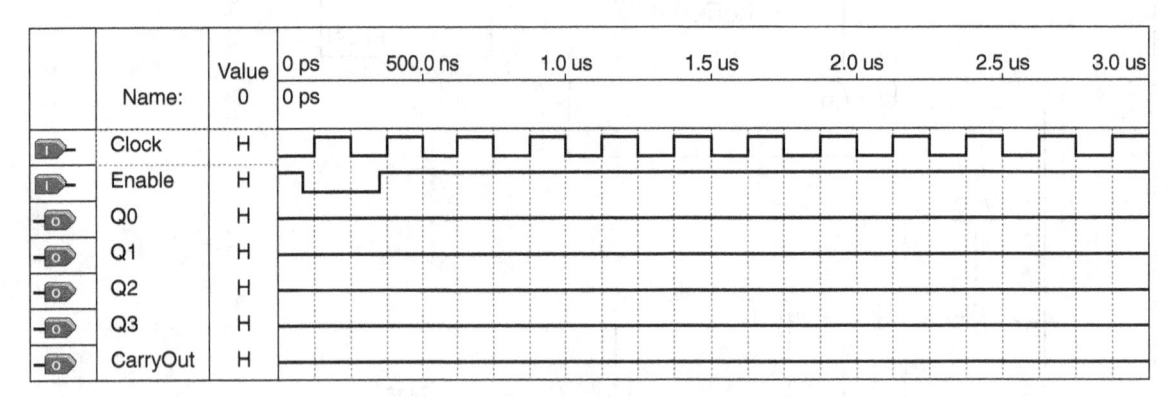

Name:	Value 0	0 ps	500.0 ns	1.0 us	1.5 us	2.0 us	2.5 us	3.0 us
Clock	H							
Enable	H							
Q0	H							
Q1	H							
Q2	H							
Q3	H							
CarryOut	H							

Figure S19-2

Demonstrated To
Instructor/FA _____ Date _____

Part 1 [ ]   Part 2 [ ]

DATA COLLECTION SHEET

Data Collection Sheets for Lab S19 Name _____

Date _____

clk	Time Interval	Carry Out	Q3	Q2	Q1	Q0
0	375ns – 425ns					
1	425ns – 675ns					
2	675ns – 925ns					
3	925ns – 1.175us					
4	1.175us – 1.425us					
5	1.425us – 1.675us					
6	1.675us – 1.925us					
7	1.925us – 2.175us					
8	2.175us – 2.425us					
9	2.425us – 2.675us					
10	2.675us – 2.925us					

Table S19-1

Data collection for Part 2:

    f) *Test the program:*
        1) Write down the power up states of the 7-segment LEDs: _____. If this value is not 00, press "Clear." If the value is still not 00, then you most likely have a logical error. For example, wrong pin assignments will definitely cause problems with the displays. It is also possible that this error may be of hardware origin. Recheck your circuit and procedure. If you have to change your circuit or perhaps a logic option, don't forget to recompile the project.
        2) What is the highest count the counter displays? _____

Demonstrated To
Instructor/FA _____ Date _____

Part 1 [ ] Part 2 [ ]

Experiment S20

Name _____

SHIFT REGISTER COUNTERS

OBJECTIVES

1. To practice using QUARTUS® II software to create schematics and waveforms.
2. To practice using QUARTUS® II software to compile and simulate digital circuits.
3. To investigate the operation of a ring counter.
4. To investigate the operation of a Johnson counter.

TEXT REFERENCES

Read section 7.20. Review Appendix D of this manual.

EQUIPMENT NEEDED

Components

Blank floppy, zip disk, or memory stick for storing projects;

QUARTUS® II software (Altera Corporation);

Desktop computer with minimum of INTEL Pentium PC @ 400MHz CPU running Microsoft Windows NT4SP1, 2000, or XP or better; Pentium III or 4 PC @ 400MHz running Red Hat Linux v7.3 or 8.0 or Red Hat Linux Enterprise 3; Sun Ultra running Solaris v8 or 9; HP9000 Series 700/800 running HP-UX v11.0 with ACE dated 11/1999 or later; or equivalent.

DISCUSSION

Shift register counters are flip-flop devices with feedback from the counter's last flip-flop output to the counter's first flip-flop input. This feedback could be from the un-inverted output to input (ring counter) or from the inverted output to input (Johnson counter).

In this experiment you will look at both the ring and Johnson counters. You will learn their characteristics and waveforms associated with each.

PROCEDURE

PART 1 – 4-bit Ring Counter

a) Start the QUARTUS® II program. Set up a new project and use **ring_ctr** for the name of the project. The diagram of the circuit you are to construct and its input waveforms are given in Figures S20-1 and S20-2.
b) Create a new graphic file for the schematic, *ring_ctr.bdf*.
c) Place the symbols on the schematic according to Figure S20-1.
d) Place the input and output connectors on the schematic. Label the inputs clk and clr; label the outputs Q0, Q1, Q2, and Q3.
e) Wire the input and output connectors to the circuit as shown in Figure S20-1. This configuration is for a 4-bit ring counter.
f) Select **Project | Top-Level Entity**.
g) Compile the project. If there are errors, fix them and save the file before recompiling the project. If there are no errors (or perhaps a warning concerning timing at most), then go on to the next step.
h) Create a new waveform *ring_ctr.vwf* and define the inputs for simulation. The input waveforms should look similar to those in Figure S20-2.
i) Select **Assignments | Settings** to access the Settings Window.
j) Click on *Simulator* at the left.
k) Choose *Functional* from the *Simulator Mode* pull-down menu.
l) Find and use for the *Simulator Input* file *ring_ctr.vwf* you just created above.
m) Select **Processing | Generate Functional Simulation Netlist**.
n) Simulate the project. If there are any errors and you change your circuit, remember to save and compile the circuit before re-simulating.
o) Record and analyze results:
1) What is the purpose of presetting the first flip-flop while keeping the other flip-flops cleared?

_____

2) Record the output waveforms on Figure S20-2 or print out the Simulation Waveforms report from QUARTUS® II and attach it to the data collection sheet at the end of this experiment.

PROCEDURE

3) You should complete this part of the experiment by converting the waveform information in Figure S20-2 to Table S20-1:
   a. Place the Time Bar in the 0 – 50ns time interval.
   b. Read the values for Q0, Q1, Q2, and Q3 in the Values column of the waveform diagram and place them in their corresponding columns in Table S20-1.
   c. Repeat steps 1 and 2 for each time interval.
4) The MOD number of the counter is _____.
5) What could you do to the circuit to change the count sequence to 1001, 1100, 0110, 0011, 1001, . . . ?

_____

_____

6) How long does the counter take in your simulation to complete one complete count cycle?

_____

7) Transfer the answers to the above questions, the waveform information, and state table to the data collection sheets at the end of this experiment.

p) Save and close the project.

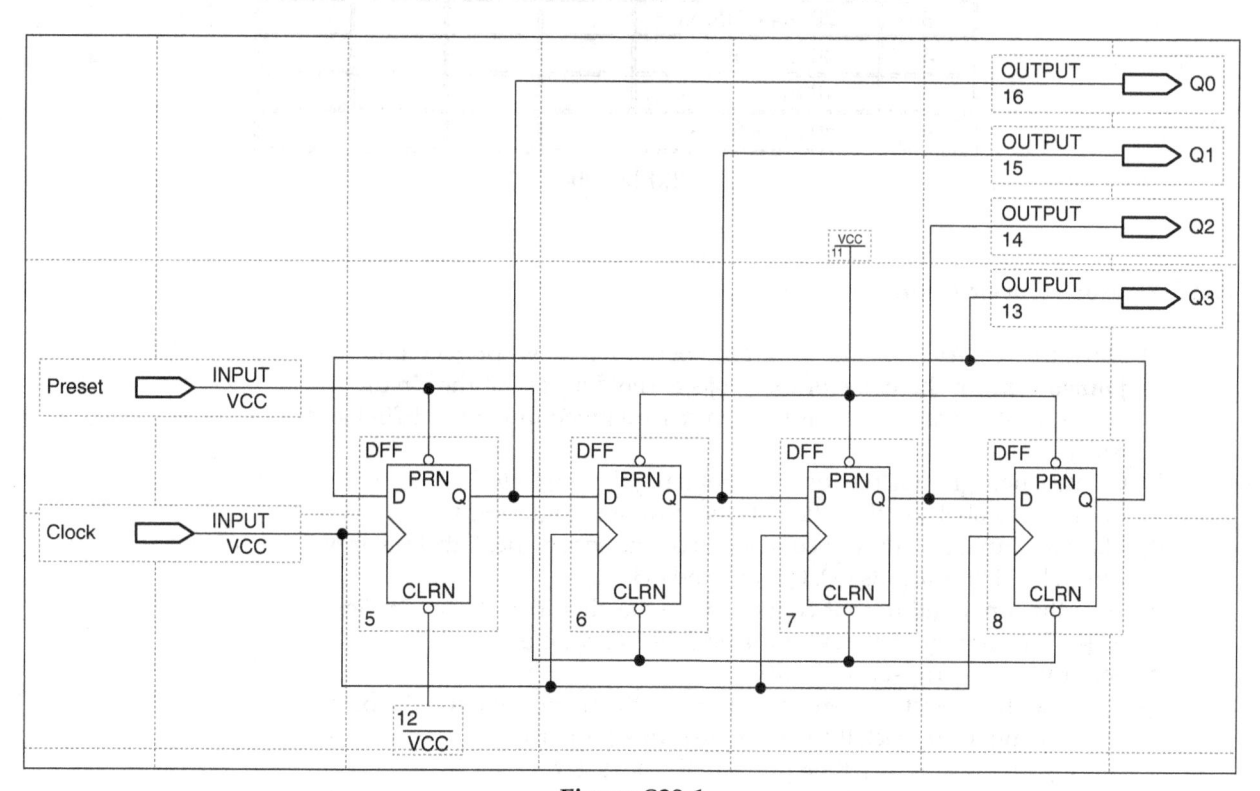

Figure S20-1

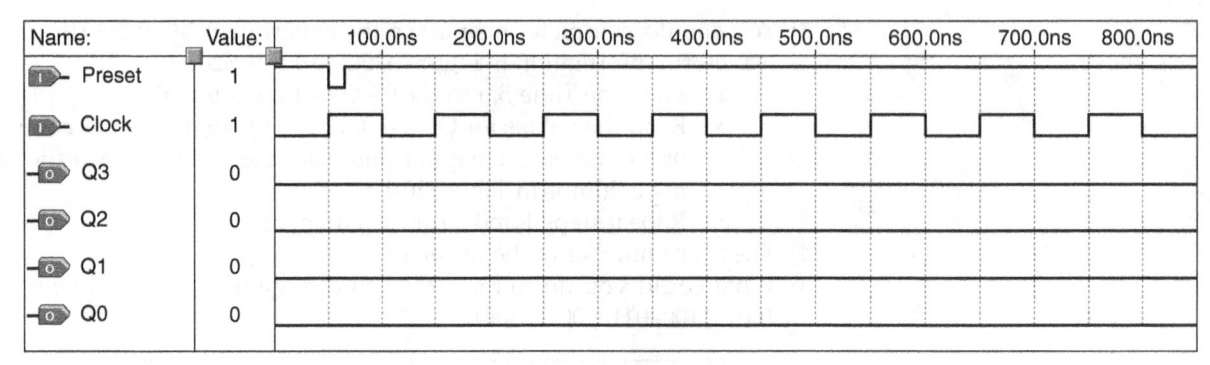

Name:	Value:	100.0ns	200.0ns	300.0ns	400.0ns	500.0ns	600.0ns	700.0ns	800.0ns
Preset	1								
Clock	1								
Q3	0								
Q2	0								
Q1	0								
Q0	0								

Figure S20-2

clk	Time Interval	Q0	Q1	Q2	Q3
0	0ns – 50ns				
1	50ns – 100ns				
2	100ns – 200ns				
3	200ns – 300ns				
4	300ns – 400ns				
5	400ns – 500ns				
6	500ns – 600ns				

Table S20-1

PART 2 – 4-bit Johnson Counter

a) Start the QUARTUS® II program. Set up a new project and use **johnson_ctr** for the name of the project. The diagram of the circuit you are to construct and its input waveforms are given in Figures S20-3 and S20-4.

b) Create a new graphic file for the schematic *johnson_ctr.bdf*.

c) Place the symbols on the schematic according to Figure S20-3.

d) Place the input and output connectors on the schematic. Label the input Clock; label the outputs Q0, Q1, Q2, and Q3.

e) Wire the input and output connectors to the circuit as shown in Figure S20-3. This configuration is for a 4-bit Johnson counter.

f) Select **Project | Top-Level Entity**.

g) Compile the project. If there are errors, fix them and save the file before recompiling the project. If there are no errors (or perhaps a warning concerning timing at most), then go on to the next step.

h) Create a new waveform *johnson_ctr.vwf* and define the inputs for simulation. The input waveforms should look similar to those in Figure S20-4.

i) Select **Assignments | Settings** to access the Settings Window.

j) Click on *Simulator* at the left.

k) Choose *Functional* from the *Simulator Mode* pull-down menu.

l) Find and use for the *Simulator Input* file *johnson_ctr.vwf* you just created above.

m) Select **Processing | Generate Functional Simulation Netlist.**

n) Simulate the project. If there are any errors and you change your circuit, remember to save and compile the circuit before re-simulating.

PROCEDURE

o) Record and analyze results:
  1) Record the output waveforms Q0, Q1, Q2, and Q3 on Figure S20-4 or print out the Simulation Waveforms report from QUARTUS® II and attach it to the data collection sheet at the end of this experiment.
  2) You should complete this part of the experiment by converting the waveform information in Figure S20-4 to Table S20-2:
     a. Place the Time Bar in the 0 – 50ns time interval.
     b. Read the values for Q0, Q1, Q2, and Q3 in the Values column of the waveform diagram and place them in their corresponding columns in Table S20-2.
     c. Repeat steps 1 and 2 for each time interval.
  3) What is the count sequence for this counter? _____
  4) The MOD number of the counter is _____.
  5) Name two major differences between the ring and Johnson counters: _____
  6) Transfer the answers to the above questions, the waveform information, and state table to the data collection sheets at the end of this experiment.

p) Close and save the project.

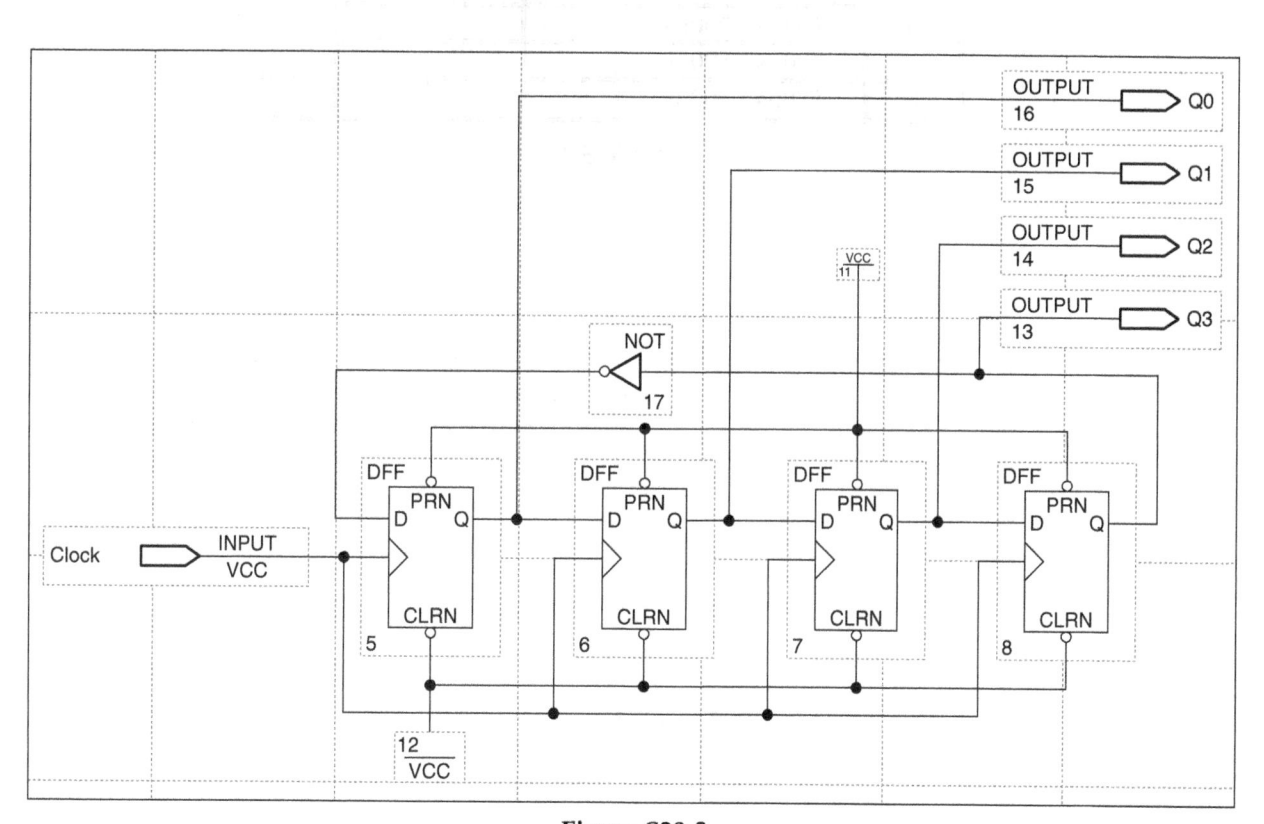

Figure S20-3

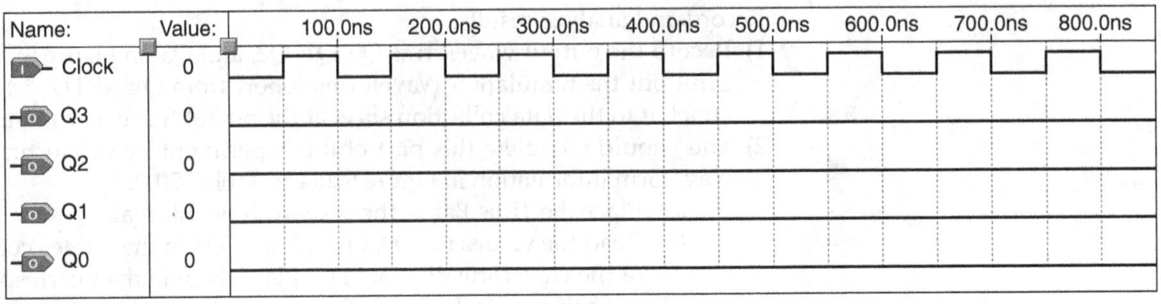

Name:	Value:	100.0ns	200.0ns	300.0ns	400.0ns	500.0ns	600.0ns	700.0ns	800.0ns
Clock	0								
Q3	0								
Q2	0								
Q1	0								
Q0	0								

Figure S20-4

clk	Time Interval	Q0	Q1	Q2	Q3
0	0ns – 100ns				
1	100ns – 200ns				
2	200ns – 300ns				
3	300ns – 400ns				
4	400ns – 500ns				
5	500ns – 600ns				
6	600ns – 700ns				
7	700ns – 800ns				
8	800ns – 900ns				

Table S20-2

DATA COLLECTION SHEET

Data Collection Sheets for Lab S20 Name _____
Date _____

Data collection for Part 1:

o) Record and analyze results:

1) What is the purpose of presetting the first flip-flop while keeping the other flip-flops cleared?
   _____
   _____

4) The MOD number of the counter is _____.

5) What could you do to the circuit to change the count sequence to 1001, 1100, 0110, 0011, 1001, . . . ?
   _____
   _____

6) How long does the counter take in your simulation to complete one complete count cycle?
   _____

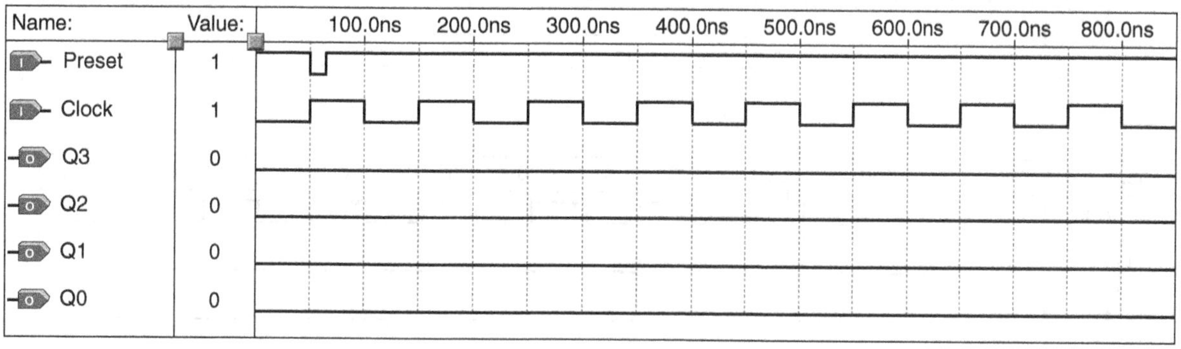

Name:	Value:	100.0ns	200.0ns	300.0ns	400.0ns	500.0ns	600.0ns	700.0ns	800.0ns
Preset	1								
Clock	1								
Q3	0								
Q2	0								
Q1	0								
Q0	0								

Figure S20-2

Demonstrated To
Instructor/FA _____ Date _____

Part 1 [ ]   Part 2 [ ]

Data Collection Sheets for Lab S20   Name _____
                                         Date _____

clk	Time Interval	Q0	Q1	Q2	Q3
0	0ns – 50ns				
1	50ns – 100ns				
2	100ns – 200ns				
3	200ns – 300ns				
4	300ns – 400ns				
5	400ns – 500ns				
6	500ns – 600ns				

Table S20-1

Data collection for Part 2:

o)

3) What is the count sequence for this counter?
   _____

4) The MOD number of the counter is _____.

5) Name two major differences between the ring and Johnson counters:
   _____
   _____

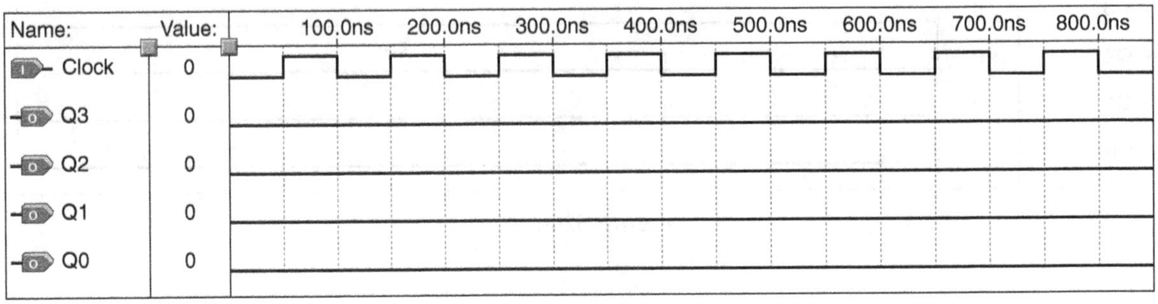

Figure S20-4

Demonstrated To
Instructor/FA _____ Date _____

Part 1 [ ]   Part 2 [ ]

DATA COLLECTION SHEET

Data Collection Sheets for Lab S20   Name _____
                                      Date _____

clk	Time Interval	Q0	Q1	Q2	Q3
0	0ns – 100ns				
1	100ns – 200ns				
2	200ns – 300ns				
3	300ns – 400ns				
4	400ns – 500ns				
5	500ns – 600ns				
6	600ns – 700ns				
7	700ns – 800ns				
8	800ns – 900ns				

Table S20-2

Demonstrated To
Instructor/FA _____ Date _____

Part 1 [ ]   Part 2 [ ]

Data Code-ion Sheet for Lab 520   Name:
                                    Date:

Time Interval	0s	Q1 (?)	Q2	Q3
0ns - 70ns				
100ns - 200ns				
200ns - 300ns				
300ns - 400ns				
400ns - 500ns				
500ns - 600ns				
600ns - 700ns				
700ns - 800ns				
800ns - 900ns				

Experiment S21

Name _____

IC REGISTERS

OBJECTIVES

1. To practice using QUARTUS® II software to create schematics and waveforms.
2. To practice using QUARTUS® II software to compile and simulate digital circuits.
3. To investigate the operation of the 74194A IC 4-bit shift register.
4. To investigate the operation of the 74194A IC as a ring counter.
5. To investigate parallel-to-serial data conversion using a 74194A IC.
6. To program the EPM7128SLC84 CPLD with a circuit having 74194A as a component.

TEXT REFERENCES

Read sections 7.15 through 7.19. Review Appendix D of this manual.

EQUIPMENT NEEDED

Components

Blank floppy, zip disk, or memory stick for storing projects;

QUARTUS® II software (Altera Corporation);

Desktop computer with minimum of INTEL Pentium PC @ 400MHz CPU running Microsoft Windows NT4SP1, 2000, or XP or better; Pentium III or 4 PC @ 400MHz running Red Hat Linux v7.3 or 8.0 or Red Hat Linux Enterprise 3; Sun Ultra running Solaris v8 or 9; HP9000 Series 700/800 running HP-UX v11.0 with ACE dated 11/1999 or later; or equivalent;

TTL compatible signal or function generator.

DISCUSSION

This experiment deals with the problem of converting between serial and parallel data formats. While much of the data in a digital system is processed in its native parallel format, data is often in serial format outside the system, so conversion from parallel-to-serial and serial-to-parallel formats is necessary.

One means of making these conversions is with registers. A typical IC register is a 74194A bidirectional universal shift register. This register can load data in parallel, load and shift right or left serially, and/or shift out in parallel. The 74194A must be programmed to make full use of its capabilities. Be sure to study the mode select table for the 74194A given in the data sheet for this device before beginning this experiment.

PROCEDURE

PART 1 – Serial Operation

a) Start the QUARTUS® II program. Set up a new project and use **serial_reg** for the name of the project. The diagram of the circuit you are to construct and its input waveforms are given in Figures S21-1 and S21-2.

b) Create a new graphic file for the schematic *serial_reg.bdf*.

c) Place the symbols on the schematic according to Figure S21-1.

d) Place the input and output connectors on the schematic. Label the inputs clk and clr; label the outputs QA, QB, QC, and QD.

e) Wire the input and output connectors to the circuit as shown in Figure S21-1.

f) Select **Project | Top-Level Entity**.

g) Compile the project. If there are errors, fix them and save the file before recompiling the project. If there are no errors (or perhaps a warning concerning timing at most), then go on to the next step.

h) Create a new waveform *serial_reg.vwf* and define the inputs for simulation. The input waveforms should look similar to those in Figure S21-2.

i) Select **Assignments | Settings** to access the Settings Window.

j) Click on *Simulator* at the left.

k) Choose *Functional* from the *Simulator Mode* pull-down menu.

l) Find and use for the *Simulator Input* file *serial_reg.vwf* you just created above.

m) Select **Processing | Generate Functional Simulation Netlist**.

n) Simulate the project. If there are any errors and you change your circuit, remember to save and compile the circuit before re-simulating.

o) Record and analyze results:

  1) Record the output waveforms on Figure S21-2 or print out the Simulation Waveforms report from QUARTUS® II and attach it to the data collection sheet at the end of this experiment.

PROCEDURE

2) You should complete this part of the experiment by converting the waveform information in Figure S21-2 to Table S21-1:
   a. Place the Time Bar in the 0 – 100ns time interval.
   b. Read the values for S0, S1, SLSI, SRSI, A, B, C, D, Clock, Q0, Q1, Q2, and Q3 in the Values column of the waveform diagram and place them in their corresponding columns in Table S21-1.
   c. Repeat steps 1 and 2 for each time interval.
3) What values of S0 and S1 are necessary for parallel loading? S0 = _____ S1 = ____
4) What values of S0 and S1 are necessary for shifting? S0 = _____ S1 = ____
5) Is parallel loading in the 74194A IC synchronous or asynchronous? _____
6) Transfer the answers to the above questions, the waveform information, and state table to the data collection sheets at the end of this experiment.

p) Close and save the project.

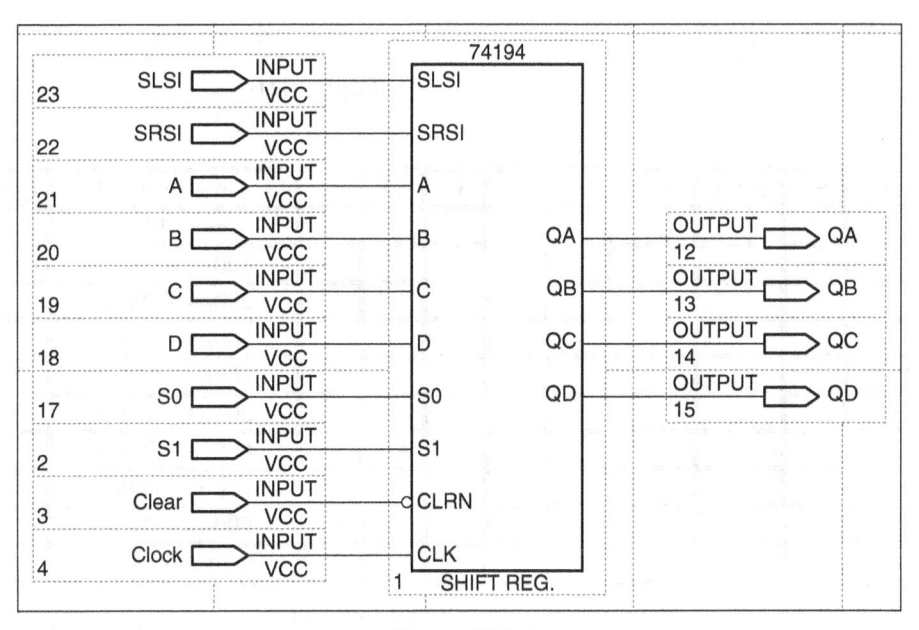

Figure S21-1

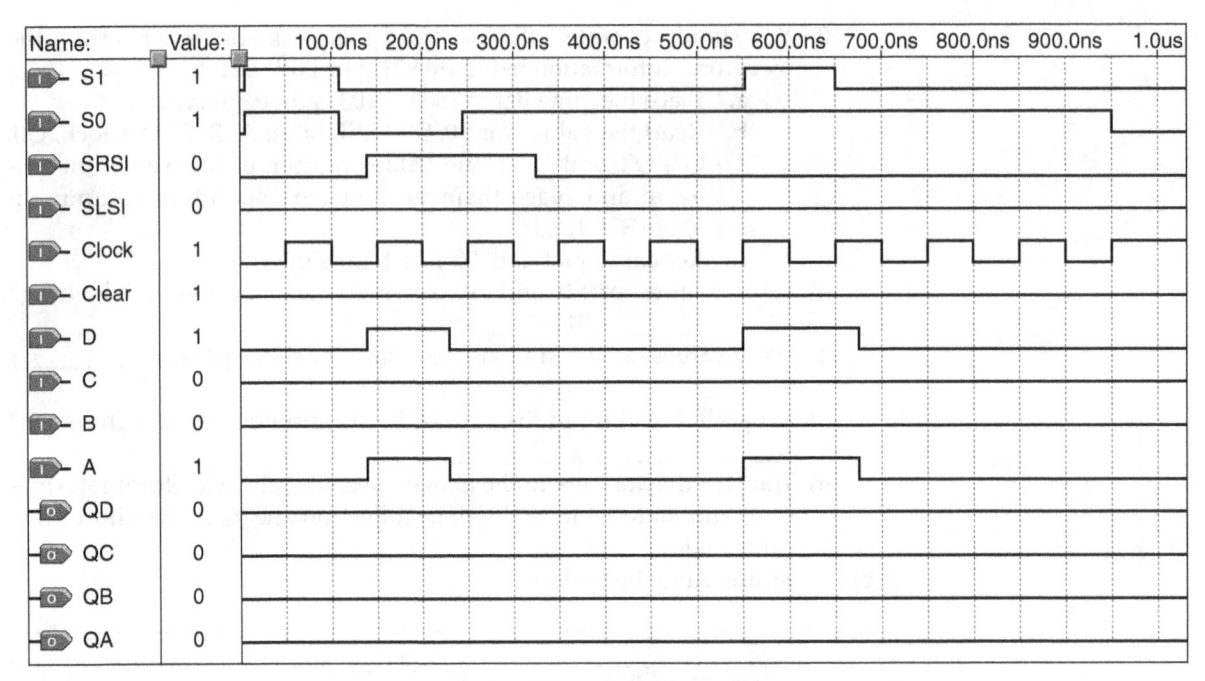

Name:	Value:	
S1	1	
S0	1	
SRSI	0	
SLSI	0	
Clock	1	
Clear	1	
D	1	
C	0	
B	0	
A	1	
QD	0	
QC	0	
QB	0	
QA	0	

Figure S21-2

S0	S1	SRSI	SLSI	A	B	C	D	clk	Time Interval	QA	QB	QC	QD
									0ns – 100ns				
									100ns – 200ns				
									200ns – 300ns				
									300ns – 400ns				
									400ns – 500ns				
									500ns – 600ns				
									600ns – 700ns				
									700ns – 800ns				
									800ns – 900ns				
									900ns – 1us				

Table S21-1

PART 2 – 4-bit Ring Counter

a) Start the QUARTUS® II program. Set up a new project and use **74194A_ring_ctr** for the name of the project. The diagram of the circuit you are to construct and its input waveforms are given in Figures S21-3 and S21-4.

b) Create a new graphic file for the schematic *74194A_ring_ctr.bdf*.

c) Place the symbols on the schematic according to Figure S21-3.

d) Place the input and output connectors on the schematic. Label the inputs Clock, Clear, S0, S1, SLSI, SRSI, A, B, C, and D; label the outputs QA, QB, QC, and QD.

e) Wire the input and output connectors to the circuit as shown in Figure S21-3. This configuration is for a 4-bit ring counter.

f) Select **Project | Top-Level Entity**.

PROCEDURE

g) Compile the project. If there are errors, fix them and save the file before recompiling the project. If there are no errors (or perhaps a warning concerning timing at most), then go on to the next step.
h) Create a new waveform *74194A_ring_ctr.vwf* and define the inputs for simulation. The input waveforms should look similar to those in Figure S21-4.
i) Select **Assignments | Settings** to access the Settings Window.
j) Click on *Simulator* at the left.
k) Choose *Functional* from the *Simulator Mode* pull-down menu.
l) Find and use for the *Simulator Input* file *74194A_ring_ctr.vwf* you just created above.
m) Select **Processing | Generate Functional Simulation Netlist**.
n) Simulate the project. If there are any errors and you change your circuit, remember to save and compile the circuit before re-simulating.
o) Record and analyze results:
   1) Record the output waveforms Q0, Q1, Q2, and Q3 on Figure S21-4 or print out the Simulation Waveforms report from QUARTUS® II and attach it to the data collection sheet at the end of this experiment.
   2) You should complete this part of the experiment by converting the waveform information in Figure S21-5 to Table S21-2:
      a. Place the Reference Cursor in the 0–50ns time interval.
      b. Read the values for Q0, Q1, Q2, and Q3 in the Values column of the waveform diagram and place them in their corresponding columns in Table S21-2.
      c. Repeat steps 1 and 2 for each time interval.
   3) What is the count sequence for this counter?
      _____
   4) The MOD number of the counter is _____.
   5) Name two major differences between the ring and Johnson counters:
      _____
      _____
   6) Transfer the answers to the above questions, the waveform information, and state table to the data collection sheets at the end of this experiment.
p) Close and save the project.

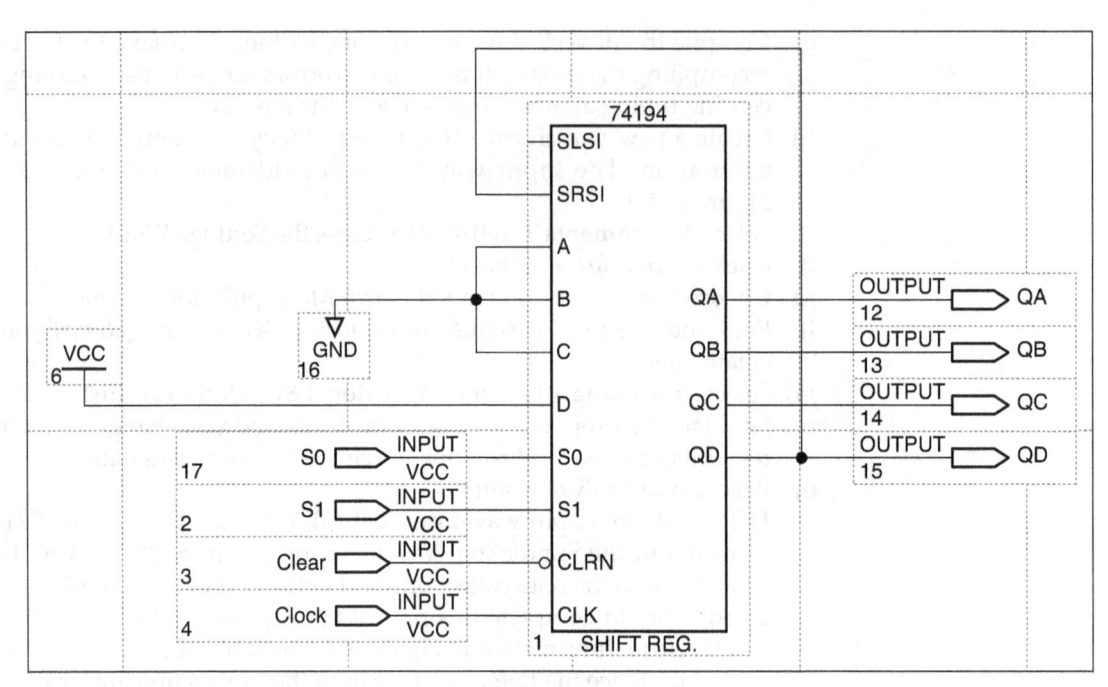

Figure S21-3

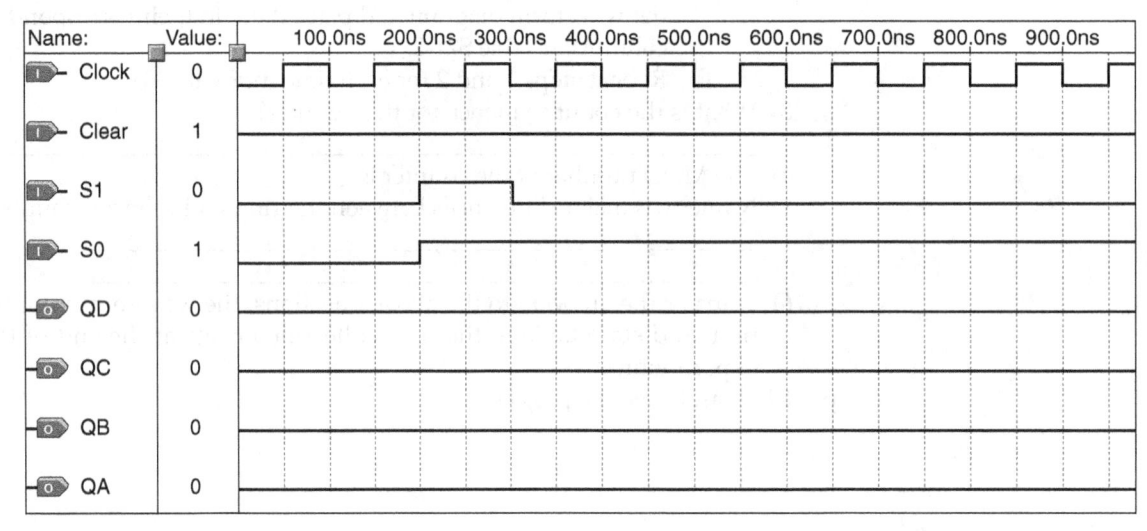

Figure S21-4

clk	Time Interval	Q0	Q1	Q2	Q3
0	0 – 100ns				
1	100ns – 200ns				
2	200ns – 300ns				
3	300ns – 400ns				
4	400ns – 500ns				
5	500ns – 600ns				
6	600ns – 700ns				
7	700ns – 800ns				
8	800ns – 900ns				

Table S21-2

PROCEDURE

PART 3 – Parallel-to-Serial Data Conversion

a) This part illustrates the process of converting parallel data to serial data. You must assume that the user can start the system any time (asynchronous) but that the data is to be shifted synchronously.

b) Start the QUARTUS® II program. Set up a new project and use **74194a_para_ser_shft** for the name of the project. The diagram of the circuit you are to construct and its input waveforms are given in Figures S21-5 and S21-6.

c) Create a new graphic file for the schematic *74194a_para_ser_shft.bdf*.

d) Place the symbols on the schematic according to Figure S21-5.

e) Place the input and output connectors on the schematic. Label the inputs CLOCK and START; label the outputs QA, QB, QC, and QD. There are three buried nodes, QY, QX, and QXn. START is an asynchronous input.

f) Wire the input and output connectors to the circuit as shown in Figure S21-5.

g) Select **Project | Top-Level Entity**.

h) Compile the project. If there are errors, fix them and save the file before recompiling the project. If there are no errors (or perhaps a warning concerning timing at most), then go on to the next step.

i) Create a new waveform *74194a_para_ser_shft.vwf* and define the inputs for simulation. The input waveforms should look similar to those in Figure S20-6.

j) Select **Assignments | Settings** to access the Settings Window.

k) Click on *Simulator* at the left.

l) Choose *Functional* from the *Simulator Mode* pull-down menu.

m) Find and use for the *Simulator Input* file *74194a_para_ser_shft.vwf* you just created above.

n) Select **Processing | Generate Functional Simulation Netlist**.

o) Simulate the project. If there are any errors and you change your circuit, remember to save and compile the circuit before re-simulating.

p) *Program the CPLD:* It is suggested that you use **PIN_70** for "**START**." Also, use an external TTL compatible clock input at 1 Hz. If you choose to use an on-board or external switch to do your clocking, make sure it is debounced. To program your CPLD, refer to Appendix D, Exercise 3. A complete set of instructions will be found there. Briefly,

   1) *Recompile the project*: After making the pin assignments, compile your project again. When the compilation is completed, the software will display an information box showing any warnings and any detected errors. If there are errors, fix them and recompile the project.

   2) *Download the project to the board:*
      a. Set all eSOC DIP switches to OFF.
      b. Connect the board to the PC's parallel printer port and apply power.
      c. Select **Tools | Programmer**.

d. Click on [Auto Detect] in the Chain Description File.

e. When the eSOC board is recognized, press the [Add File...] button in the Chain Description File, then locate and Add the *74194a_para_ser_shft.pof* file.

f. Select the file and device line and scroll (horizontally) to the available programming option. Check the **Program/Configure** box.

g. Press the [Start] button. Observe the progress bar.

h. If the program downloads successfully, go to the next step. Otherwise, fix the problem and try downloading again.

3) *Test the program:* After connections are made to your board, apply power and verify that the LEDs respond in a manner consistent with a MOD-256 counter.

a. Press the START pushbutton momentarily and then release it.

b. Observe the count's direction and sequence.

q) Record and analyze results:

1) Record the output waveforms QA, QB, QC, and QD on Figure S21-6 or print out the Simulation Waveforms report from QUARTUS® II and attach it to the data collection sheet at the end of this experiment.

2) Why is it necessary to synchronize the load and shift operations of the 74194A?

3) Explain how the flip-flops accomplish this synchronization in the circuit of Figure S21-5.

4) How would you modify the circuit of Figure S21-5 to change it into an 8-bit converter?

5) Transfer the answers to the above questions and the waveform information to the data collection sheets at the end of this experiment.

PROCEDURE

r) Save and close the project.

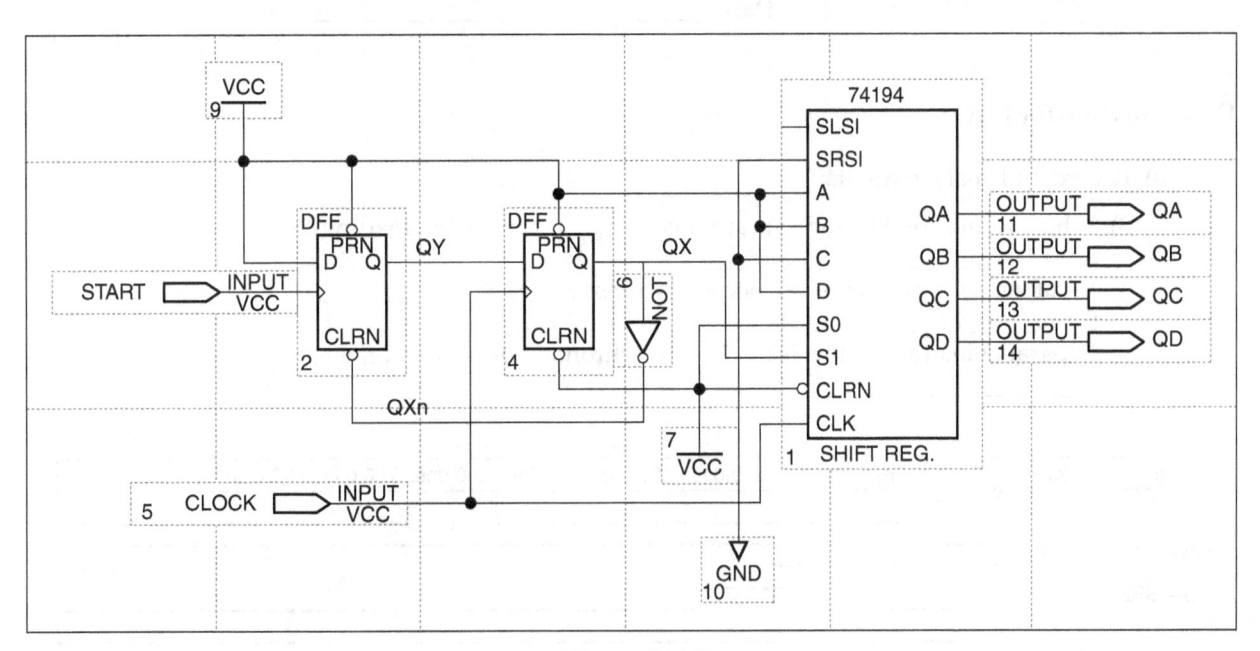

Figure S21-5

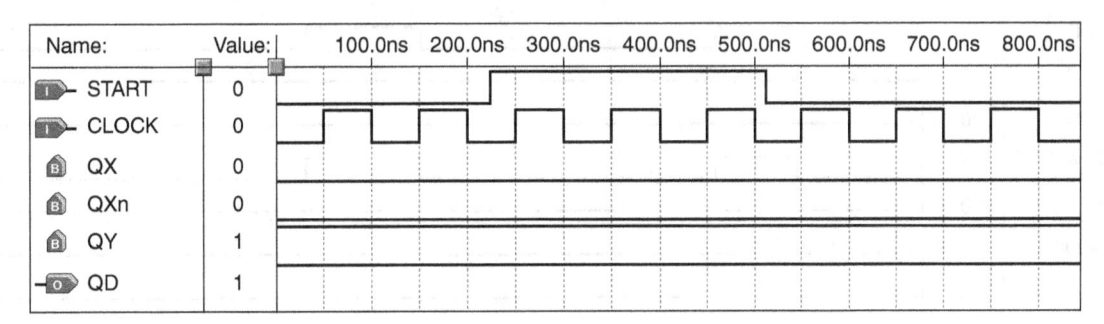

Figure S21-6

Data Collection Sheets for Lab S21   Name _____

                                          Date _____

Data collection for Part 1:

o) Record and analyze results:

3) What values of S0 and S1 are necessary for parallel loading?
   S0 = _____ S1 = _____

4) What values of S0 and S1 are necessary for shifting? S0 = _____
   S1 = _____

5) Is parallel loading in the 74194A IC synchronous or asynchronous?
   _____

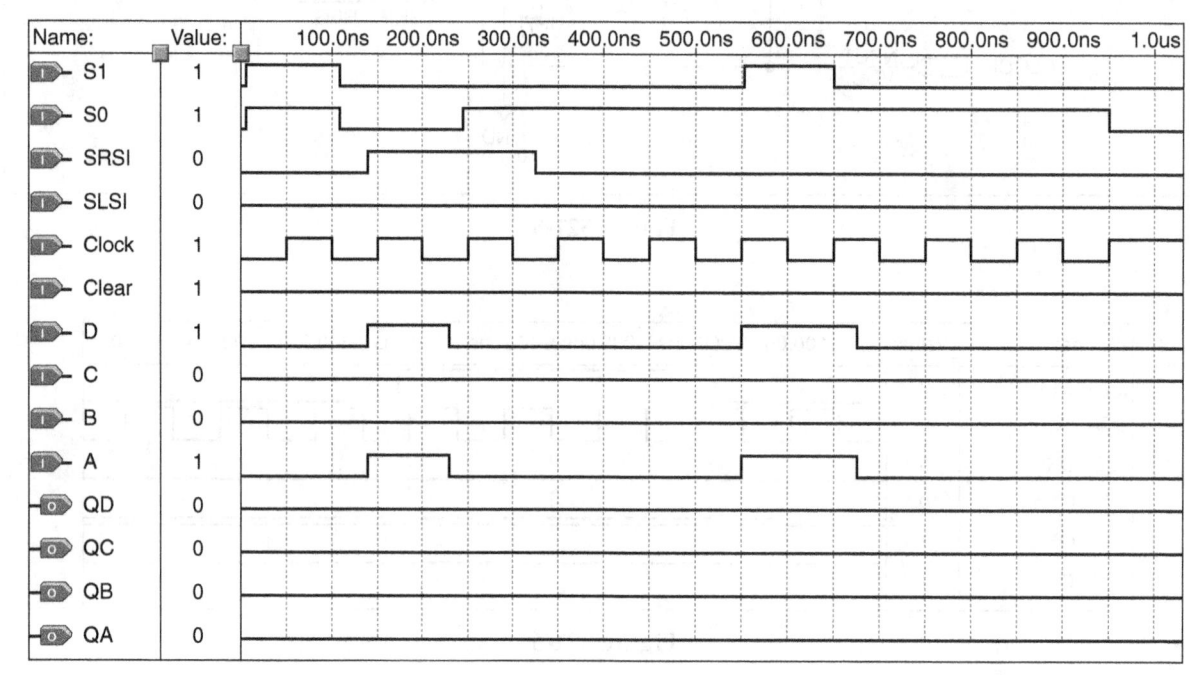

Figure S21-2

| Demonstrated To |
| Instructor/FA _____ Date _____ |
| |
| Part 1 [ ]   Part 2 [ ]   Part 3 [ ] |

DATA COLLECTION SHEET

Data Collection Sheets for Lab S21   Name _____

Date _____

S0	S1	SRSI	SLSI	A	B	C	D	clk	Time Interval	QA	QB	QC	QD
									0ns – 100ns				
									100ns – 200ns				
									200ns – 300ns				
									300ns – 400ns				
									400ns – 500ns				
									500ns – 600ns				
									600ns – 700ns				
									700ns – 800ns				
									800ns – 900ns				
									900ns – 1us				

Table S21-1

Data collection for Part 2:

 o) Record and analyze results:
  3) What is the count sequence for this counter?
   _____

  4)
  5) The MOD number of the counter is _____.
  6) Name two major differences between the ring and Johnson counters:
   _____
   _____

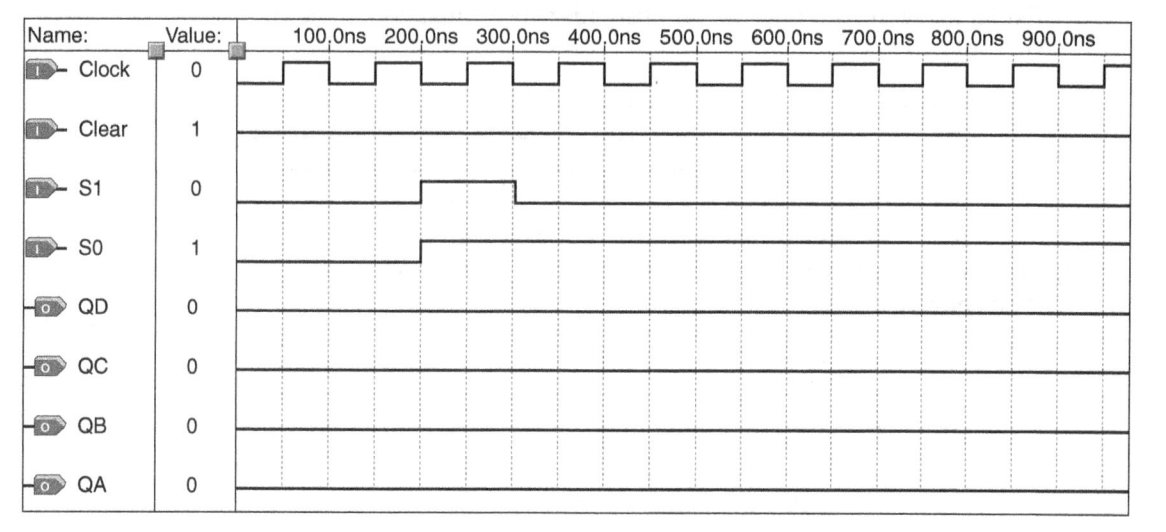

Figure S21-4

Demonstrated To
Instructor/FA _____ Date _____

Part 1 [ ]   Part 2 [ ]   Part 3 [ ]

Data Collection Sheets for Lab S21   Name _____

Date _____

clk	Time Interval	Q0	Q1	Q2	Q3
0	0 – 100ns				
1	100ns – 200ns				
2	200ns – 300ns				
3	300ns – 400ns				
4	400ns – 500ns				
5	500ns – 600ns				
6	600ns – 700ns				
7	700ns – 800ns				
8	800ns – 900ns				

Table S21-2

Data collection for Part 3:

    p) Record and analyze results:

        2) Why is it necessary to synchronize the load and shift operations of the 74194A?

        _____
        _____
        _____

        3) Explain how the flip-flops accomplish this synchronization in the circuit of Figure 21-5.

        _____
        _____
        _____
        _____
        _____

        4) How would you modify the circuit of Figure 21-5 to change it into an 9-bit converter.

        _____
        _____
        _____

Demonstrated To
Instructor/FA _____ Date _____

Part 1 [ ]   Part 2 [ ]   Part 3 [ ]

DATA COLLECTION SHEET

Data Collection Sheets for Lab S21   Name _____

Date _____

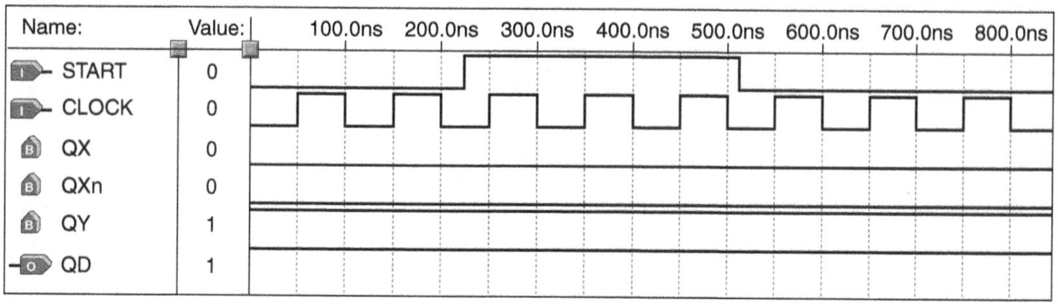

Name:	Value:	100.0ns	200.0ns	300.0ns	400.0ns	500.0ns	600.0ns	700.0ns	800.0ns
START	0								
CLOCK	0								
QX	0								
QXn	0								
QY	1								
QD	1								

Figure S21-6

Demonstrated To
Instructor/FA _____ Date _____

Part 1 [ ]   Part 2 [ ]   Part 3 [ ]

Experiment S22

Name _____

ONE-SHOTS, COUNTERS, AND REGISTERS WITH VHDL

OBJECTIVES

1. To practice using QUARTUS® II software to create schematics and waveforms.
2. To practice using QUARTUS® II software to compile and simulate digital circuits.
3. To implement one-shots with VHDL.
4. To implement counters with VHDL.
5. To implement shift registers with VHDL.
6. To implement a programmable timer using several devices created from VHDL in previous exercises.

TEXT REFERENCES

Read sections 7.11, 7.22, and 7.24. Review Appendix D of this manual.

EQUIPMENT NEEDED

Components

Blank floppy, zip disk, or memory stick for storing projects;
QUARTUS® II software (Altera Corporation);
Optional: DeVry University Board eSOC with EPM7128SLC84 CPLD (or Altera University Board with EPM7128SLC84 CPLD or any other equivalent board);
Desktop computer with minimum of INTEL Pentium PC @ 400MHz CPU running Microsoft Windows NT4SP1, 2000, or XP or better; Pentium III or 4 PC @ 400MHz running Red Hat Linux v7.3 or 8.0 or Red Hat Linux Enterprise 3; Sun Ultra running Solaris v8 or 9; HP9000 Series 700/800 running HP-UX v11.0 with ACE dated 11/1999 or later; or equivalent.

DISCUSSION

In this experiment you will investigate one-shots, counters, and registers implemented with VHDL. Recall that flip-flop devices are bistable, which means that they have two stable states. This class of devices includes registers and counters. One-shots on the other hand are monostable and therefore have only one stable state. You will start off by creating a VHDL version of the one-shot. Then you will implement a counter and finally a shift register with VHDL.

When designing these devices, you should be aware that you are not limited to the features normally included with certain ICs. For example, an IC synchronous up/down counter typically uses two clock inputs, one for counting up and the other for counting down. In VHDL you can use a single clock input. There are other such freedoms you can experience, such as the way you implement loading parallel data to change the starting count.

PROCEDURE

PART 1 – Implementing a Non-Retriggerable One-Shot with VHDL

a) Start the MAX+PLUS® II program. Set up a new project and use **oneshot** for the name of the project. The VHDL code for a non-retriggerable one-shot is given in Figure S22-1.
b) Create a new text file for the VHDL code *oneshot.bdf*.
c) Enter the program code given in Figure S22-1.
d) Select **Project | Top-Level Entity**.
e) Compile the project. If there are errors, fix them and save the file before recompiling the project. If there are no errors (or perhaps a warning concerning timing at most), then go on to the next step.
f) To include a circuit from a VHDL design in a schematic, the circuit must have a symbol. To create a symbol that can be imported into the Block Editor, select **File | Create/Update | Create Symbol Files for Current File**.
g) In response, QUARTUS® II generates a Block Symbol File *oneshot.bsf* in the **oneshot** project folder in the Project Navigator window.
h) To test our symbol, create a new project *oneshot_example* in the **oneshot** project folder.
i) Create a new schematic and save it as *oneshot_example.bdf*.
j) Double-click on the schematic window to access the **Symbol** window.
k) Use the Symbol browser to locate and select the *oneshot* symbol from the Project Folder. Place the symbol in the schematic. Note that the symbol is block-shaped.
l) Place input symbols and connect them to inputs "Trigger," "Reset," and "Delay[3..0]". Place an output symbol on the schematic and connect it to "q." Rename the connectors to agree with Figure S22-3.
m) Compile the project again.

PROCEDURE

n) Create a new waveform file *oneshot.vwf* and define the inputs for simulation. The input waveforms should look similar to those in Figure S22-2, including setting Delay[3..0] = [0101]. If necessary, review the manual methods of creating waveforms in Exercise 1 of Appendix D before trying to create the "Trigger" waveform.
o) Select **Assignments | Settings** to access the Settings Window.
p) Click on *Simulator* at the left.
q) Choose *Functional* from the *Simulator Mode* pull-down menu.
r) Find and use for the *Simulator Input* file *oneshot.vwf* you just created above.
s) Select **Processing | Generate Functional Simulation Netlist.**
t) Simulate the project. If there are any errors and you change your circuit, remember to save and compile the circuit before re-simulating.
u) Record and analyze results:
   1) Record the output waveform q on Figure S22-2 or print out the Simulation Waveforms report from QUARTUS® II and attach it to the data collection sheet at the end of this experiment.
   2) With the clock period used in the simulation, what is the maximum amount of delay possible? _____
   3) How would you modify this one-shot to get a maximum delay of 255 clock periods?
   _____
   4) Transfer the answers to the above questions and the waveform information to the data collection sheets at the end of this experiment.
v) Save and close the project.

```vhdl
LIBRARY ieee;
USE ieee.std_logic_1164.all;
entity OneShot is
        port( clock, trigger, reset : IN BIT;
                delay                           : IN INTEGER RANGE 0 TO 15;
                q                               : OUT BIT
                );
end OneShot;
ARCHITECTURE vhdl OF OneShot IS
BEGIN     -- rtl
        PROCESS(Clock, reset)
        VARIABLE count                          : INTEGER RANGE 0 TO 15;
        BEGIN
                IF reset = '0' THEN count := 0;
                ELSIF (clock'EVENT AND clock = '1')  THEN
                        IF trigger = '1' AND count = 0 THEN
                        count := delay;             -- load counter
                        ELSIF count = 0 THEN count := 0;
                        ELSE count := count - 1;
                        END IF;
                END IF;
                IF count /= 0 THEN q <= '1';
                ELSE q <= '0';
                END IF;
        END PROCESS;
END vhdl;
```

Figure S22-1

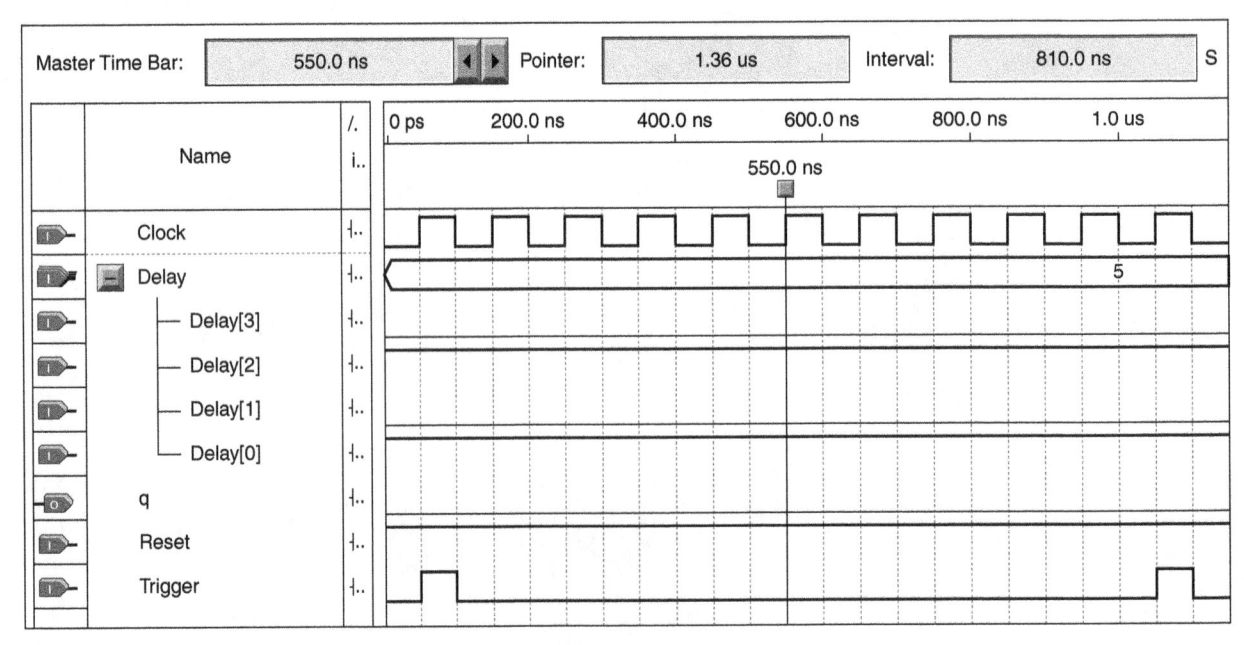

Figure S22-2

PROCEDURE

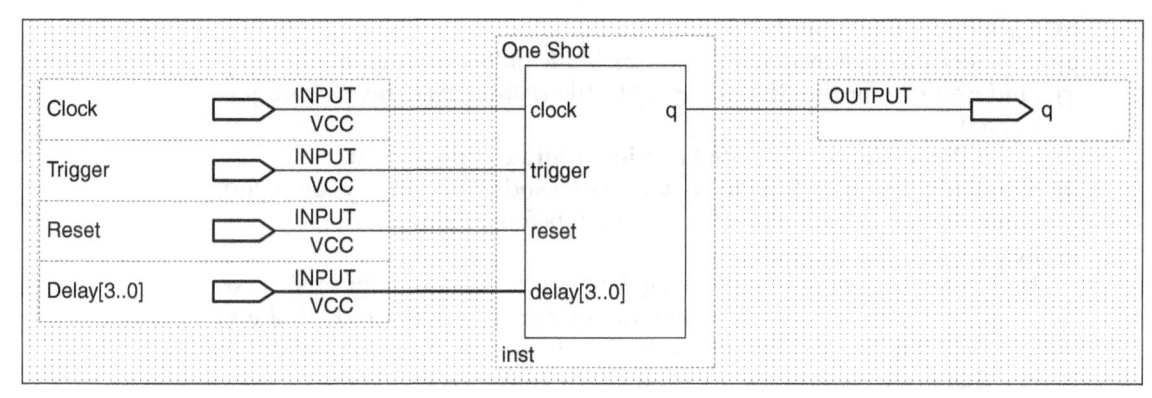

Figure S22-3

PART 2 – Implementing a Full-Featured Counter with VHDL

a) Start the QUARTUS® II program. Set up a new project and use **counter_vhd** for the name of the project. The VHDL code for a full-featured counter is given in Figure S22-4.
b) Create a new text file for the VHDL code *counter_vhd.bdf*.
c) Enter the text shown in Figure S22-4.
d) Select **Project | Top-Level Entity**.
e) Compile the project. If there are errors, fix them and save the file before recompiling the project. If there are no errors (or perhaps a warning concerning timing at most), then go on to the next step.
f) To include a circuit from a VHDL design in a schematic, the circuit must have a symbol. To create a symbol that can be imported into the Block Editor, select **File | Create/Update | Create Symbol Files for Current File**.
g) In response, QUARTUS® II generates a Block Symbol File, *counter_vhd.bsf* in the **counter_vhd** project folder in the Project Navigator window.
h) To test our symbol, create a new project *counter_vhd _example* in the **counter_vhd** project folder.
i) Create a new schematic and save it as *counter_vhd.example.bdf*.
j) Double-click on the schematic window to access the **Symbol** window.
k) Use the Symbol browser to locate and select the *counter_vhd* symbol from the Project Folder. Place the symbol in the schematic. Note that the symbol is block-shaped.
l) Place input symbols as shown in Figure S22-6 and connect them to inputs "load," "down," "cntenable," "clock," "clear," and "din[3..0]." Place output symbols for "term_ct" and "q[3..0]" on the schematic.
m) Compile the project again.
n) Create a new waveform file *counter_vhd.example.vwf* and define the inputs for simulation. The input waveforms should look similar to those in Figure S22-5. Make sure that din[3..0] is set to 1001. If necessary, review the manual methods of creating waveforms in Exercise 1 of Appendix D before trying to create the "load" waveform.
o) Select **Assignments | Settings** to access the Settings Window.

p) Click on *Simulator* at the left.
q) Choose *Functional* from the *Simulator Mode* pull-down menu.
r) Find and use for the *Simulator Input* file *counter_vhd.example* you just created.
s) Select **Processing | Generate Functional Simulation Netlist.**
t) Simulate the project. If there are any errors and you change your circuit, remember to save and compile the circuit before re-simulating.
u) Record and analyze results:
   1) Record the output waveforms q[3..0] and term_ct on Figure S22-5 or print out the Simulation Waveforms report from QUARTUS® II and attach it to the data collection sheet at the end of this experiment.
   2) What is the count sequence for this counter when din[3..0]=9?
   _____
   3) When the counter reaches its maximum count, to what value does it reset? _____
   4) Transfer the answers to the above questions, the waveform information, and state table to the data collection sheets at the end of this experiment.
v) Save and close the project.

PROCEDURE

```vhdl
LIBRARY ieee;
USE ieee.std_logic_1164.all;
ENTITY counter_vhd IS
        PORT( clock, clear, load, cntenable, down : IN BIT;
                din                             : IN INTEGER RANGE 0 TO 15;
                q                               : OUT INTEGER RANGE 0 TO 15;
                term_ct                         : OUT BIT);
END counter_vhd;
ARCHITECTURE vhdl OF counter_vhd IS
BEGIN      -- rtl
        PROCESS(clock, clear, down)
        VARIABLE count                  : INTEGER RANGE 0 TO 15;   --define a numeric signal
            BEGIN
                    IF clear = '1' THEN count := 0;   --asynchronous clear
                    ELSIF (clock = '1' AND clock'EVENT) THEN -- rising edge?
                        IF load = '1' THEN count := din; -- parallel load
                        ELSIF cntenable = '1' THEN    -- load counter
                            IF down = '0' THEN count := count + 1; -- increment ELSE count := count - 1; --decrement
                            END IF;
                        END IF;
                    END IF;
                IF (((count = 0) AND (down = '1')) OR
                ((count = 15) AND (down = '0'))) AND cntenable = '1'
                THEN term_ct <= '1';
                ELSE   term_ct <= '0';
                END IF;
                q <= count;   -- transfer register contents to outputs
        END PROCESS;
END vhdl;
```

Figure S22-4

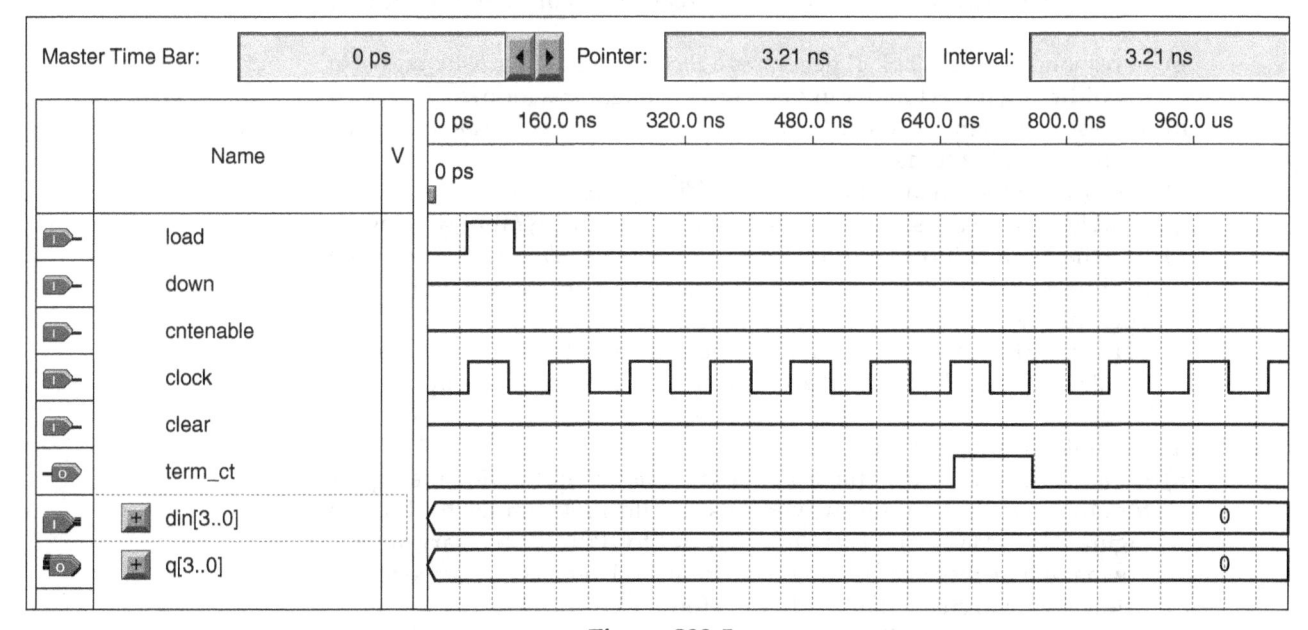

Figure S22-5

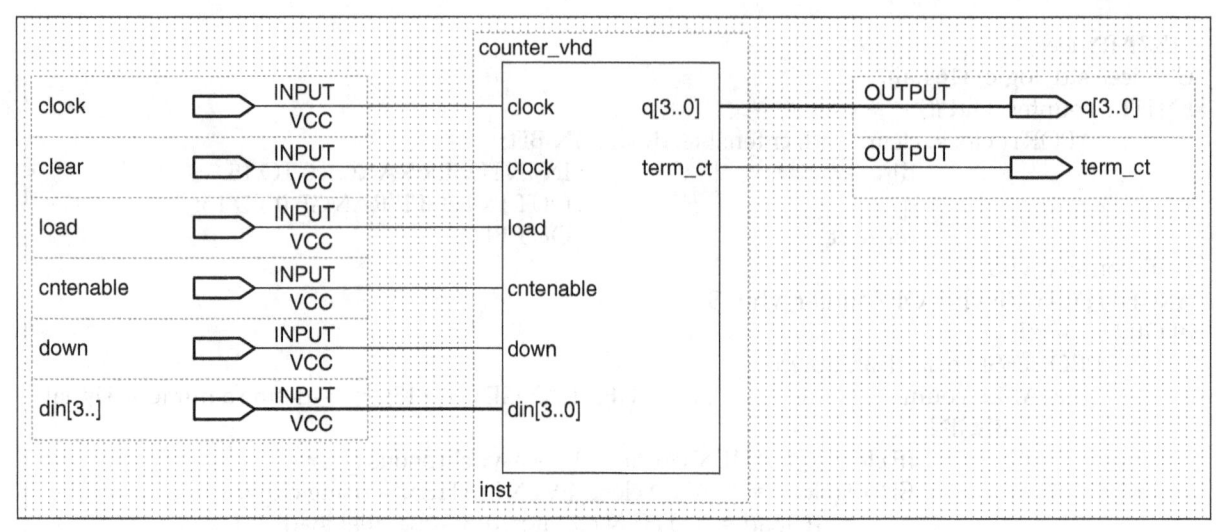

```
                            counter_vhd
clock   ─INPUT─┐            ┌clock    q[3..0]├──────OUTPUT──▶ q[3..0]
         VCC   │            │
clear   ─INPUT─┤            ┤clock    term_ct├──────OUTPUT──▶ term_ct
         VCC   │            │
load    ─INPUT─┤            ┤load
         VCC   │            │
cntenable─INPUT┤            ┤cntenable
         VCC   │            │
down    ─INPUT─┤            ┤down
         VCC   │            │
din[3..]─INPUT─┘            ┤din[3..0]
         VCC                │
                            inst
```

Figure S22-6

PART 3 – Implementing a Bidirectional Shift Register with VHDL

   a) Start the QUARTUS® II program. Set up a new project and use **bidir_reg** as the name of the project. The VHDL code for a bidirectional shift register is given in Figure S22-7.

   b) Create a new text file for the VHDL code: *bidir_reg.bdf*.

   c) Enter the text shown in Figure S22-7.

   d) Select **Project | Top-Level Entity**.

   e) Compile the project. If there are errors, fix them and save the file before recompiling the project. If there are no errors (or perhaps a warning concerning timing at most), then go on to the next step.

   f) To include a circuit from a VHDL design in a schematic, the circuit must have a symbol. To create a symbol that can be imported into the Block Editor, select **File | Create/Update | Create Symbol Files for Current File**.

   g) In response, QUARTUS® II generates a Block Symbol File, *bidir_reg.bsf* in the **bidir_reg** project folder in the Project Navigator window.

   h) To test our symbol, create a new project *bidir_reg _example* in the **bidir_reg** project folder.

   i) Create a new schematic and save it as *bidir_reg.example.bdf*.

   j) Double-click on the schematic window to access the **Symbol** window.

   k) Use the Symbol browser to locate and select the *bidir_reg* symbol from the Project Folder. Place the symbol in the schematic. Note that the symbol is block-shaped.

   l) Place input symbols and connect them to inputs "ser_in," "clock," "din[3..0]," "mode[1..0]." Place output symbols for "q[3..0]" on the schematic.

   m) Compile the project again.

   n) Create a new waveform file *bidir_reg.example.vwf* and define the inputs for simulation. The input waveforms should look similar to those in Figure S22-8. Make sure that din[3..0] is set to 1001. If necessary, review the manual methods of creating waveforms in Exercise 1 of Appendix D before trying to create the "load" waveform.

   o) Select **Assignments | Settings** to access the Settings Window.

PROCEDURE

p) Click on *Simulator* at the left.
q) Choose *Functional* from the *Simulator Mode* pull-down menu.
r) Find and use for the *Simulator Input* file *bidir_reg.example.example* you just created.
s) Select **Processing | Generate Functional Simulation Netlist.**
t) Simulate the project. If there are any errors and you change your circuit, remember to save and compile the circuit before re-simulating.
u) After collecting the data from the previous step, set mode = 1, ser_in to a clock signal with 8 times the period of the register's clock, and din = 0. Simulate the project with these parameters.
v) Record and analyze results:
   1) Record the output waveforms q[3..0] on Figure S22-8 or print out the Simulation Waveforms report from QUARTUS® II and attach it to the data collection sheet at the end of this experiment.
   2) What is the procedure for loading a 6 in parallel and then shifting the number out to the right replacing bits to the left with 0?
   _____
   _____

   3) In step **u)**, what counter can you recall has the same pattern of output? _____

   4) Transfer the answers to the above questions, the waveform information, and state table to the data collection sheets at the end of this experiment.
w) Save and close the project.

```
LIBRARY ieee;
USE ieee.std_logic_1164.all;
ENTITY bidir_reg IS
        PORT( clock, ser_in      : IN BIT;
                    din          : IN BIT_VECTOR (3 DOWNTO 0);
                    mode         : IN INTEGER RANGE 0 TO 3;
                    q            : OUT BIT_VECTOR (3 DOWNTO 0)
                    );
END bidir_reg;
ARCHITECTURE vhdl OF bidir_reg IS
SIGNAL ff           : BIT_VECTOR (3 DOWNTO 0);
BEGIN   -- rtl
     PROCESS (clock)             -- respond to clock
     BEGIN
               IF (clock = '1' AND clock'EVENT) THEN
               CASE mode IS
               WHEN 0 => ff    <= ff;         -- hold shift
               WHEN 1 => ff(2 DOWNTO 0) <= ff (3 DOWNTO 1); -- shift right
                              ff(3)      <= ser_in;
               WHEN 2 => ff(3 DOWNTO 1) <= ff (2 DOWNTO 0); -- shift left
                              ff(0)      <= ser_in;
               WHEN OTHERS =>      ff <= din; -- parallel load
               END CASE;
          END IF;
          END PROCESS;
       q <= ff;   --update outputs
END vhdl;
```

Figure S22-7

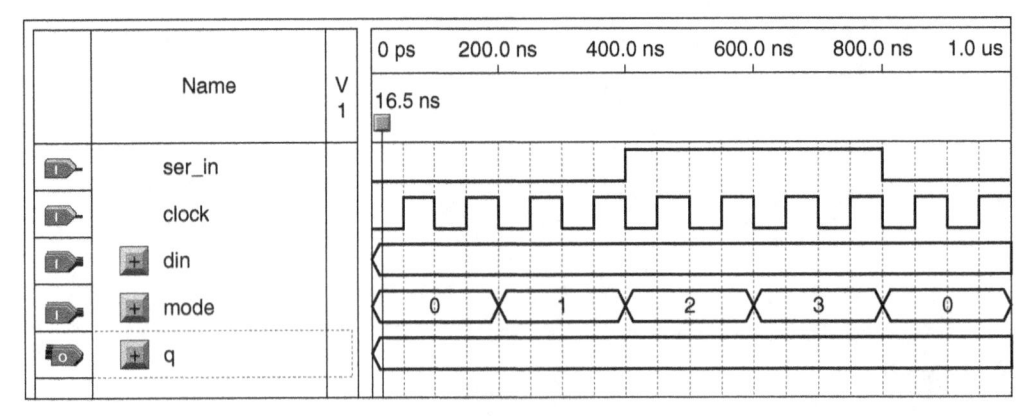

Figure S22-8

PROCEDURE

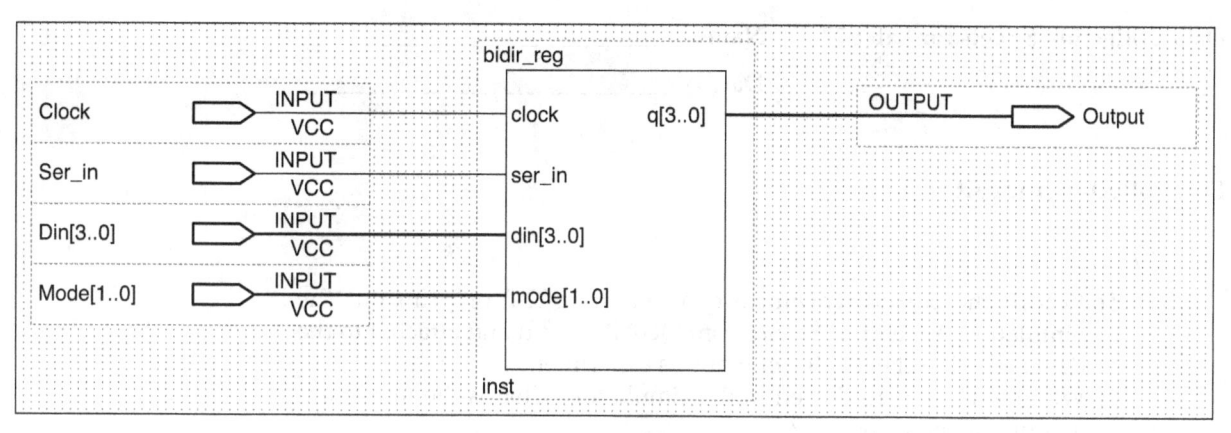

Figure S22-9

PART 4 – Designing Using VHDL

a) Refer to Experiment 33 in the TTL portion of the lab manual. Figure 33-2 is a schematic for a programmable timer. In the space below, draw a diagram for the same timer using VHDL components you have covered so far in the laboratory.

b) When you have completed the design, your instructor may either require you to implement this timer using your eSOC board or simulate the timer.

542 Exper. S22

Data Collection Sheets for Lab S22   Name _____

Date _____

Data collection for Part 1:

u)

1) Record the output waveform q on Figure S22-2 or print out the Simulation Waveforms report from QUARTUS® II and attach it to the data collection sheet at the end of this experiment.
2) With the clock period used in the simulation, what is the maximum amount of delay possible? _____
3) How would you modify this one-shot to get a maximum delay of 255 clock periods?
   _____
   _____

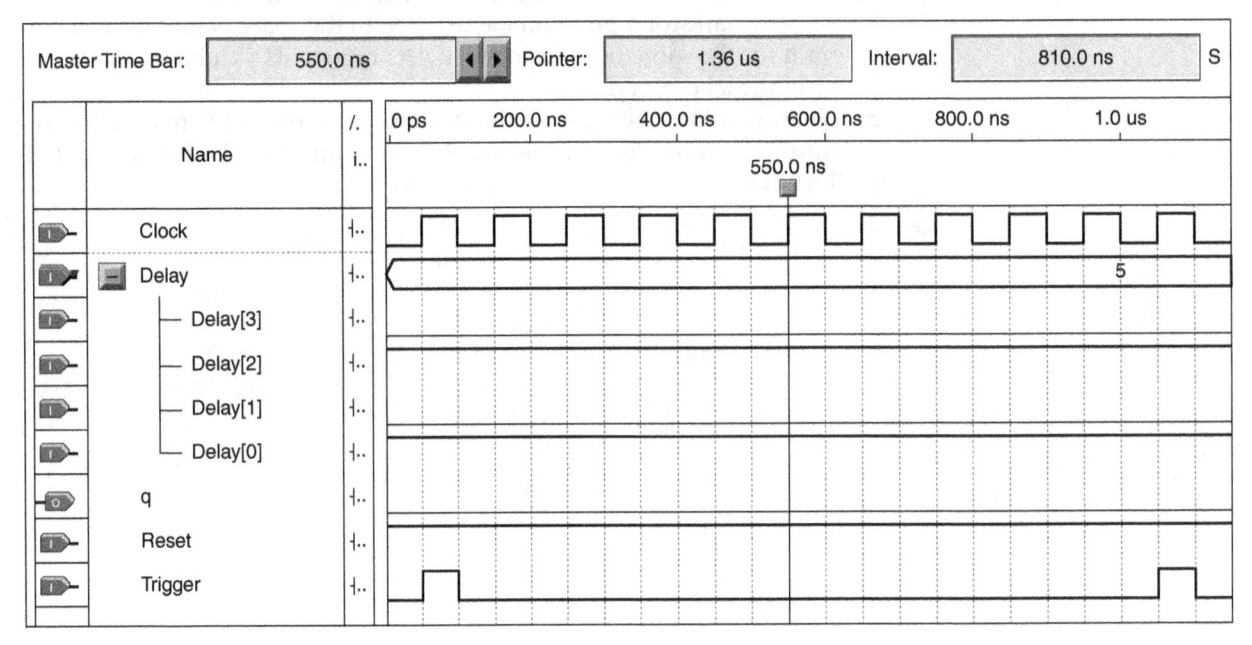

| Master Time Bar: | 550.0 ns | ◀▶ | Pointer: | 1.36 us | Interval: | 810.0 ns | S |

Name		0 ps	200.0 ns	400.0 ns	600.0 ns	800.0 ns	1.0 us
Clock							
Delay					5		
— Delay[3]							
— Delay[2]							
— Delay[1]							
— Delay[0]							
q							
Reset							
Trigger							

Figure S22-2

Demonstrated To
Instructor/FA _____ Date _____

Part 1 [ ]   Part 2 [ ]   Part 3 [ ]   Part 4 [ ]

DATA COLLECTION SHEET

Data Collection Sheets for Lab S22   Name _____

Date _____

Data collection for Part 2:

u) Record and analyze results:

2) What is the count sequence for this counter when din[3..0]=9? _____

3) When the counter reaches its maximum count, to what value does it reset? _____

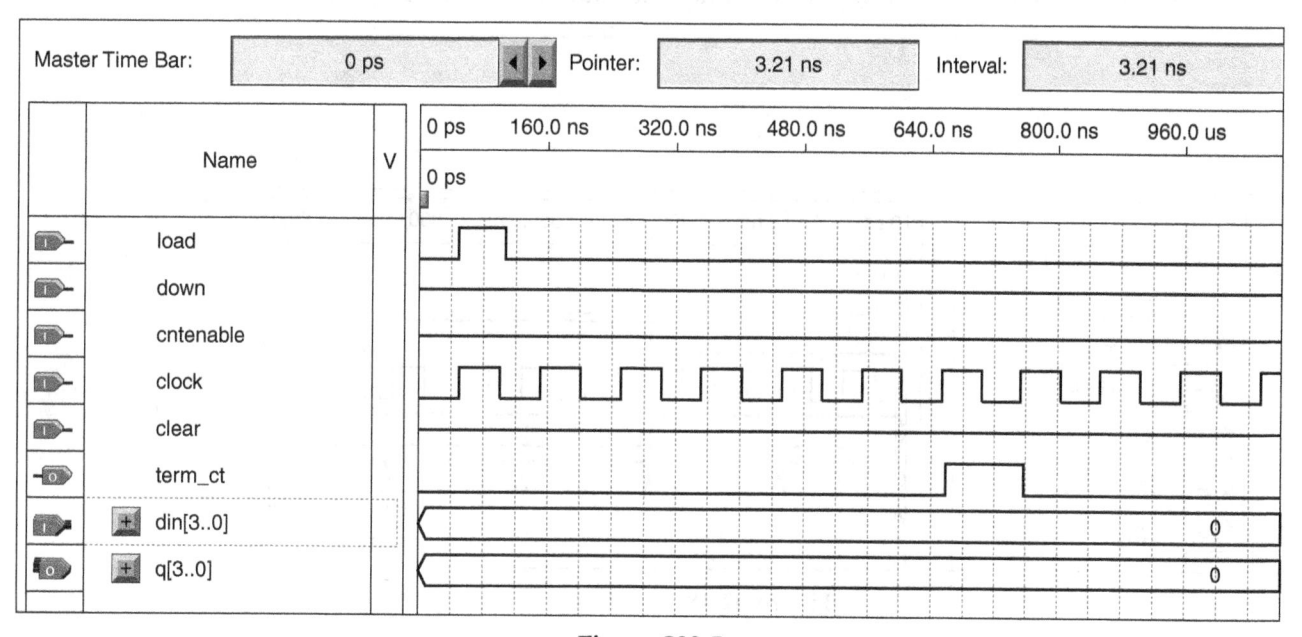

| Master Time Bar: | 0 ps | Pointer: | 3.21 ns | Interval: | 3.21 ns |

| Name | V | 0 ps | 160.0 ns | 320.0 ns | 480.0 ns | 640.0 ns | 800.0 ns | 960.0 us |

- load
- down
- cntenable
- clock
- clear
- term_ct
- din[3..0]   0
- q[3..0]   0

Figure S22-5

Demonstrated To
Instructor/FA _____ Date _____

Part 1 [ ]   Part 2 [ ]   Part 3 [ ]   Part 4 [ ]

Data Collection Sheets for Lab S22   Name _____
                                         Date _____

Data collection for Part 3:

  v) Record and analyze results:

  2) What is the procedure for loading a 6 in parallel and then shifting the number out to the right replacing bits to the left with 0?
  _____
  _____

  In step **u)**, what counter can you recall has the same pattern of output?
  _____

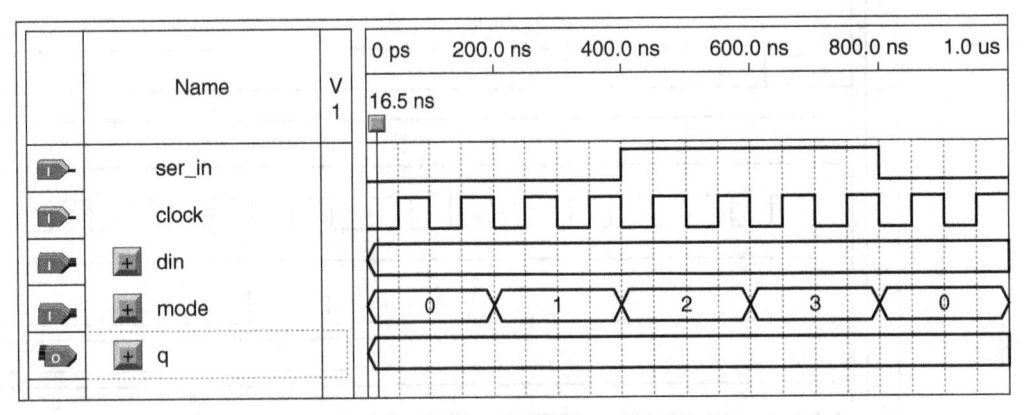

Figure S22-8

Data collection for Part 4:

  a) Draw your diagram here:

Demonstrated To
Instructor/FA _____ Date _____

Part 1 [ ]   Part 2 [ ]   Part 3 [ ]   Part 4 [ ]

Experiment S23

Name _____

IC DECODERS AND ENCODERS

OBJECTIVES

1. To practice using QUARTUS® II software to create schematics and waveforms.
2. To practice using QUARTUS® II software to compile and simulate digital circuits.
3. To investigate the operation of the 74138 IC octal decoder.
4. To investigate the operation of the 7442 IC BCD decoder.
5. To investigate the operation of the 74147 IC encoder.

TEXT REFERENCES

Read sections 9.1 and 9.4. Review Appendix D of this manual.

EQUIPMENT NEEDED

Components

Blank floppy, zip disk, or memory stick for storing projects;

QUARTUS® II software (Altera Corporation);

Desktop computer with minimum of INTEL Pentium PC @ 400MHz CPU running Microsoft Windows NT4SP1, 2000, or XP or better; Pentium III or 4 PC @ 400MHz running Red Hat Linux v7.3 or 8.0 or Red Hat Linux Enterprise 3; Sun Ultra running Solaris v8 or 9; HP9000 Series 700/800 running HP-UX v11.0 with ACE dated 11/1999 or later; or equivalent.

DISCUSSION

A decoder is a logic device that accepts a binary input and outputs one output that corresponds to that binary input. If a decoder has N inputs representing a binary number, then there will be 2^N possible outputs of which only one will be selected. A special case is the BCD decoder. Since BCD uses only 10 of the 16 possible binary 4-bit combinations, a BCD decoder will have only 10 outputs instead of 16.

PROCEDURE

PART 1 – The Operation of the 74LS138 IC Octal Decoder

a) Start the QUARTUS® II program. Set up a new project and use **octal_decode** for the name of the project. The diagram of the circuit you are to construct and its input waveforms are given in Figures S23-1 and S23-2.
b) Create a new graphic file for the schematic *octal_decode.bdf*.
c) Place the symbols on the schematic according to Figure S23-1.
d) Place the input and output connectors on the schematic. Label the inputs A, B, C, and G1; label the outputs O0n, O1n . . . O7n.
e) Wire the input and output connectors to the circuit as shown in Figure S23-1.
f) Select **Project | Top-Level Entity**.
g) Compile the project. If there are errors, fix them and save the file before recompiling the project. If there are no errors (or perhaps a warning concerning timing at most), then go on to the next step.
h) Compile the project again.
i) Create a new waveform file *octal_decode.vwf* and define the inputs for simulation. The input waveforms should look similar to those in Figure S23-2. If necessary, review the manual methods of creating waveforms in Exercise 1 of Appendix D before trying to create the "G1" waveform.
j) Select **Assignments | Settings** to access the Settings Window.
k) Click on *Simulator* at the left.
l) Choose *Functional* from the *Simulator Mode* pull-down menu.
m) Find and use for the *Simulator Input* file *octal_decode.vwf* you just created above.
n) Select **Processing | Generate Functional Simulation Netlist**.
o) Simulate the project. If there are any errors and you change your circuit, remember to save and compile the circuit before re-simulating.
p) Record and analyze results:
  1) Record the output waveforms O0n, O1n, . . ., O7n on Figure S23-2 or print out the Simulation Waveforms report from QUARTUS® II and attach it to the data collection sheet at the end of this experiment.

PROCEDURE

2) You should complete this part of the experiment by converting the waveform information in Figure S23-2 to Table S23-1:
   a. Place the Time Bar in the 0 – 50ns time interval.
   b. Read the values for outputs O0n, O1n, . . ., O7n, and A, B, C, and G1 in the Values column of the waveform diagram and place them in their corresponding columns in Table S23-1.
   c. Repeat steps 1 and 2 for each time interval.
3) What values of A, B, and C are required to enable O6n? A = _____ B = _____ C = _____
4) If G1 is LOW, what are the output values for O0n – O7n?
   _____
5) Is it possible for two outputs to be enabled simultaneously?
   _____
6) Transfer the answers to the above questions, the waveform information, and state table to the data collection sheets at the end of this experiment.
q) Save and close the project.

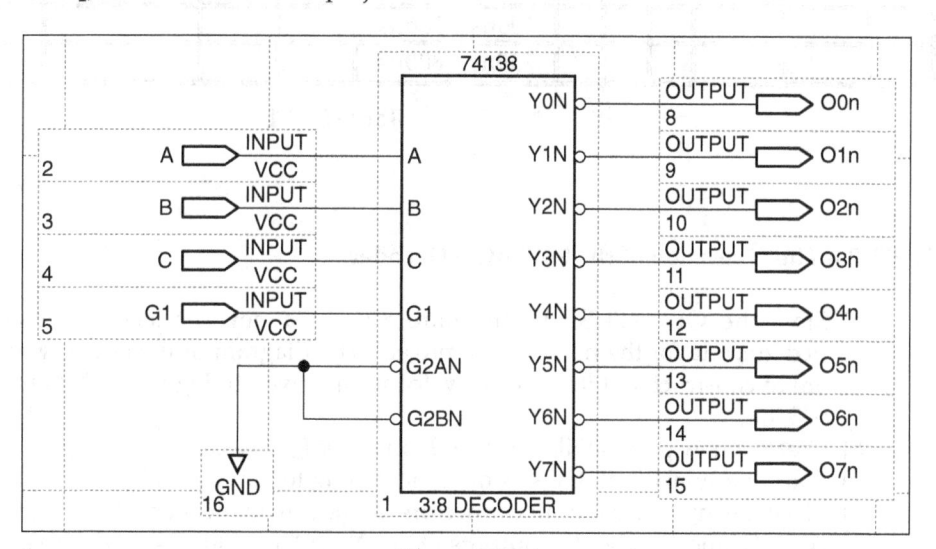

Figure S23-1

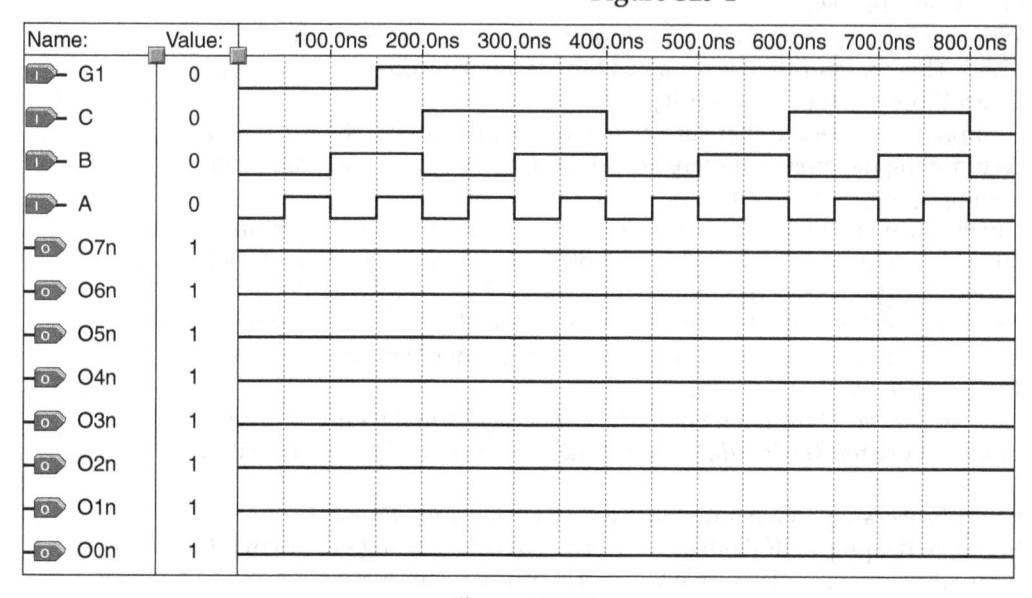

Figure S23-2

G1	G2An	G2Bn	A	B	C	Time Interval	O0n	O1n	O2n	O3n	O4n	O5n	O6n	O7n
						0ns – 50ns								
						50ns – 100ns								
						100ns – 150ns								
						150ns – 200ns								
						200ns – 250ns								
						250ns – 300ns								
						300ns – 350ns								
						350ns – 400ns								
						400ns – 450ns								
						450ns – 500ns								
						500ns – 550ns								
						550ns – 600ns								
						600ns – 650ns								
						650ns – 700ns								
						700ns – 750ns								
						750ns – 800ns								

Table S23-1

PART 2 – The Operation of the 7442 BCD Decoder

a) Start the QUARTUS® II program. Set up a new project and use **bcd_decode** for the name of the project. The diagram of the circuit you are to construct and its input waveforms are given in Figures S23-3 and S23-4.

b) Create a new graphic file for the schematic *bcd_decode.bdf*.

c) Place the symbols on the schematic according to Figure S23-3.

d) Place the input and output connectors on the schematic. Label the inputs A, B, C, and D; label the outputs O0n, O1n, O2n, O3n, O4n, O5n, O6n, O7n, O8n, and O9n.

e) Wire the input and output connectors to the circuit as shown in Figure S23-3. This configuration is for a BCD-to-Decimal decoder.

f) Select **Project | Top-Level Entity**.

g) Compile the project. If there are errors, fix them and save the file before recompiling the project. If there are no errors (or perhaps a warning concerning timing at most), then go on to the next step.

h) Create a new waveform file *bcd_decode.vwf* and define the inputs for simulation. The input waveforms should look similar to those in Figure S23-4. If necessary, review the manual methods of creating waveforms in Exercise 1 of Appendix D before trying to create the "G1" waveform.

i) Select **Assignments | Settings** to access the Settings Window.

j) Click on *Simulator* at the left.

k) Choose *Functional* from the *Simulator Mode* pull-down menu.

l) Find and use for the *Simulator Input* file *bcd_decode.vwf* you just created above.

m) Select **Processing | Generate Functional Simulation Netlist**.

n) Simulate the project. If there are any errors and you change your circuit, remember to save and compile the circuit before re-simulating.

PROCEDURE

o) Record and analyze results:
  1) Record the output waveforms O0n, O1n, O2n, O3n, O4n, O5n, O6n, O7n, O8n, and O9n on Figure S23-4 or print out the Simulation Waveforms report from QUARTUS® II and attach it to the data collection sheet at the end of this experiment.
  2) What are the output values if DCBA = 1010? _____
  3) Can you suggest a way to convert the 7442 to an octal decoder with enable?
  _____
  _____
  4) Transfer the answers to the above questions, the waveform information, and state table to the data collection sheets at the end of this experiment.

p) Save and close the project.

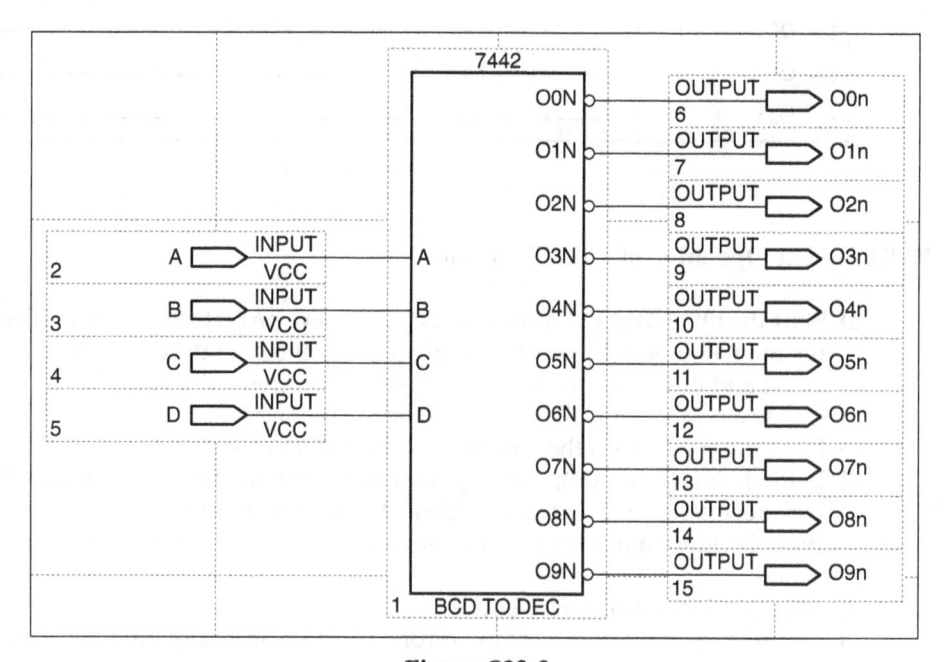

Figure S23-3

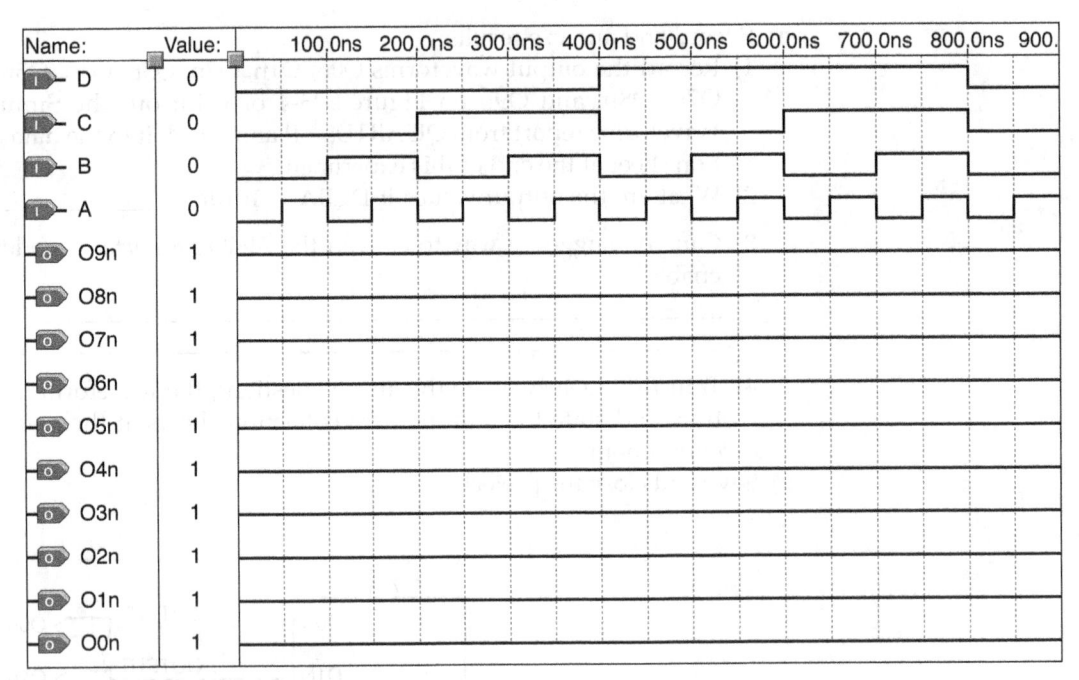

Name:	Value:	100.0ns 200.0ns 300.0ns 400.0ns 500.0ns 600.0ns 700.0ns 800.0ns 900
D	0	
C	0	
B	0	
A	0	
O9n	1	
O8n	1	
O7n	1	
O6n	1	
O5n	1	
O4n	1	
O3n	1	
O2n	1	
O1n	1	
O0n	1	

Figure S23-4

PART 3 – The Operation of a 74147 Encoder

a) Start the QUARTUS® II program. Set up a new project and use **encoder** for the name of the project. The diagram of the circuit you are to construct and its input waveforms are given in Figures S23-5 and S23-6.

b) Create a new graphic file for the schematic *encoder.bdf*.

c) Place the symbols on the schematic according to Figure S23-5.

d) Place the input and output connectors on the schematic. Label the input waveforms 1n – 9n ; label the outputs An, Bn, Cn, and Dn.

e) Wire the input and output connectors to the circuit as shown in Figure S23-5.

f) Select **Project | Top-Level Entity**.

g) Compile the project. If there are errors, fix them and save the file before recompiling the project. If there are no errors (or perhaps a warning concerning timing at most), then go on to the next step.

h) Create a new waveform file *encoder.vwf* and define the inputs for simulation. The input waveforms should look similar to those in Figure S23-6. If necessary, review the manual methods of creating waveforms in Exercise 1 of Appendix D before trying to create the "G1" waveform.

i) Select **Assignments | Settings** to access the Settings Window.

j) Click on *Simulator* at the left.

k) Choose *Functional* from the *Simulator Mode* pull-down menu.

l) Find and use for the *Simulator Input* file *encoder.vwf* you just created above.

m) Select **Processing | Generate Functional Simulation Netlist.**

n) Simulate the project. If there are any errors and you change your circuit, remember to save and compile the circuit before re-simulating.

PROCEDURE

o) Record and analyze results:
1) Record the output waveforms An, Bn, Cn, and Dn on Figure S23-6 or print out the Simulation Waveforms report from QUARTUS® II and attach it to the data collection sheet at the end of this experiment.
2) Transfer the waveform information to the data collection sheets at the end of this experiment.

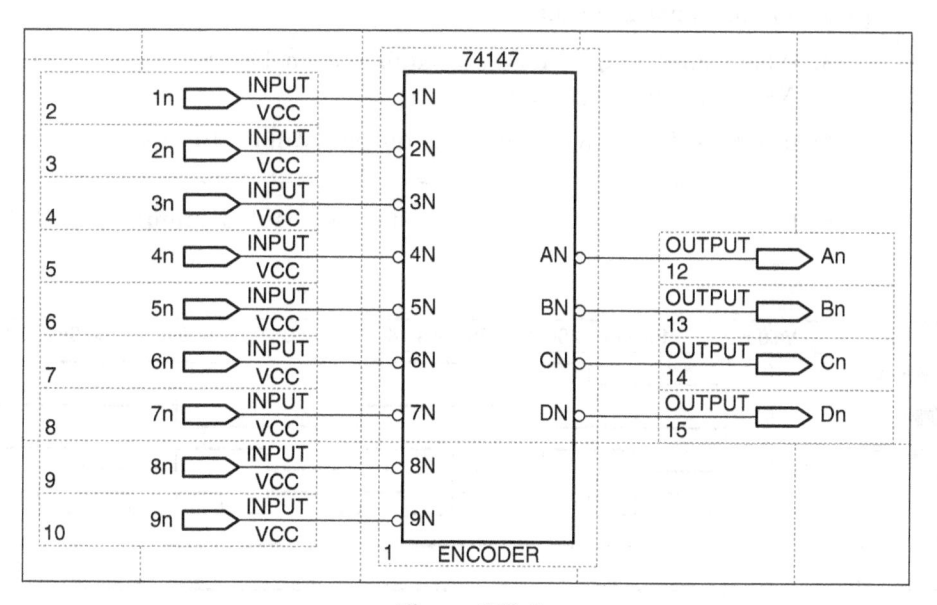

Figure S23-5

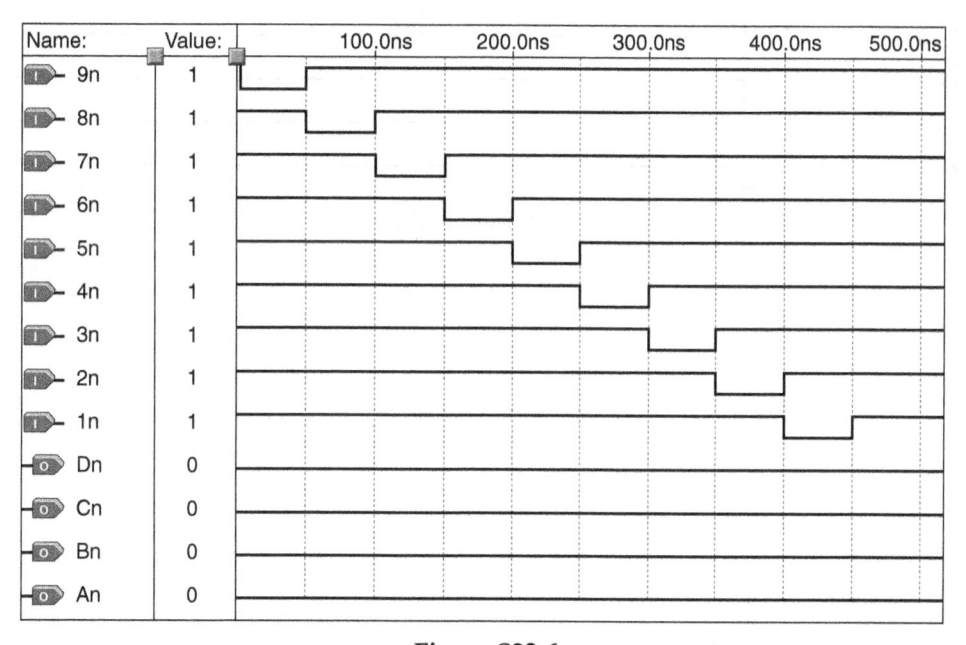

Figure S23-6

Data Collection Sheets for Lab S23   Name _____

                                          Date _____

Data collection for Part 1:

    p) Record and analyze results:

        3) What values of A, B, and C are required to enable O6n?
            A = _____ B = _____ C = _____

        4) If G1 is LOW, what are the output values for O0n – O7n?
           _____

        5) Is it possible for two outputs to be enabled simultaneously?
           _____

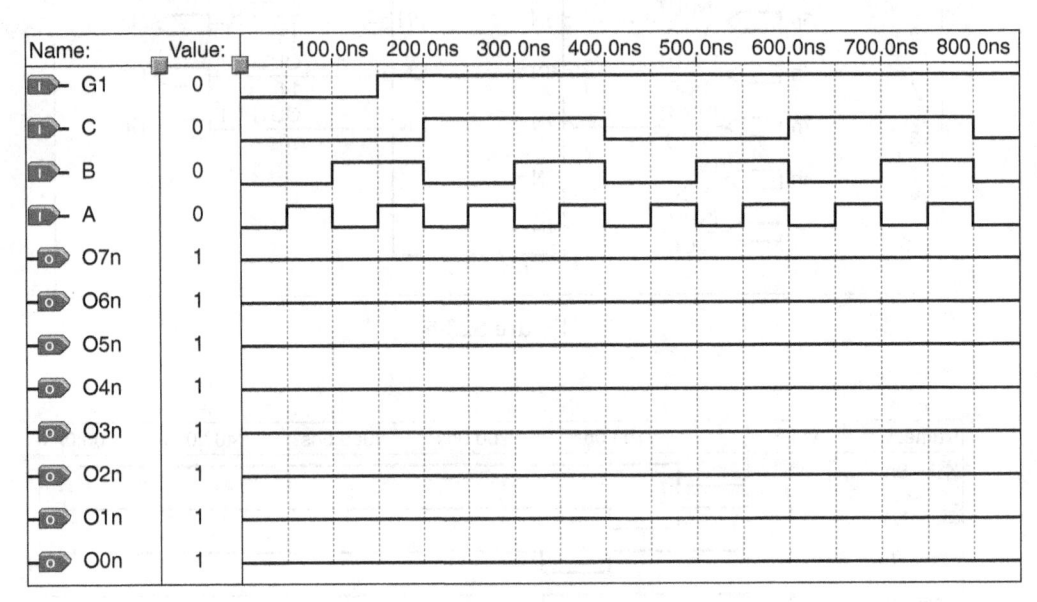

Name:	Value:
G1	0
C	0
B	0
A	0
O7n	1
O6n	1
O5n	1
O4n	1
O3n	1
O2n	1
O1n	1
O0n	1

Figure S23-2

Demonstrated To
Instructor/FA _____ Date _____
Part 1 [ ]   Part 2 [ ]   Part 3 [ ]

DATA COLLECTION SHEET

Data Collection Sheets for Lab S23 Name _____

Date _____

G1	G2An	G2Bn	A	B	C	Time Interval	O0n	O1n	O2n	O3n	O4n	O5n	O6n	O7n
						0ns – 50ns								
						50ns – 100ns								
						100ns – 150ns								
						150ns – 200ns								
						200ns – 250ns								
						250ns – 300ns								
						300ns – 350ns								
						350ns – 400ns								
						400ns – 450ns								
						450ns – 500ns								
						500ns – 550ns								
						550ns – 600ns								
						600ns – 650ns								
						650ns – 700ns								
						700ns – 750ns								
						750ns – 800ns								

Table S23-1

Data collection for Part 2:

o) Record and analyze results:

2) What are the output values if DCBA = 1010? _____

3) Can you suggest a way to convert the 7442 to an octal decoder with enable?

_____
_____
_____

Demonstrated To
Instructor/FA _____ Date _____
Part 1 [ ]   Part 2 [ ]   Part 3 [ ]

Data Collection Sheets for Lab S23   Name _____

Date _____

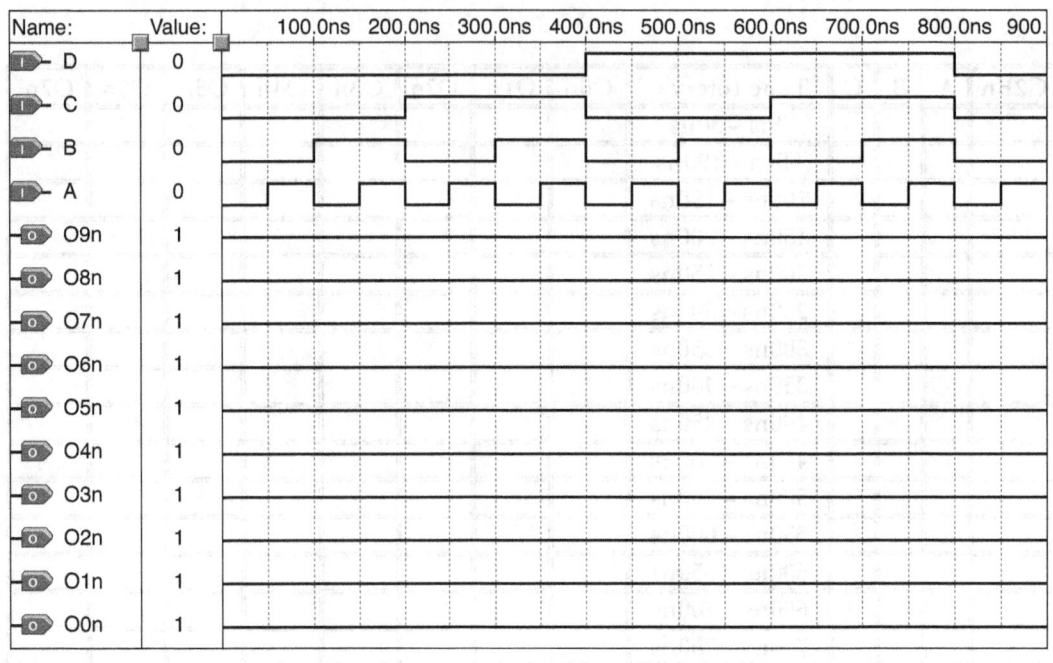

Name:	Value:
D	0
C	0
B	0
A	0
O9n	1
O8n	1
O7n	1
O6n	1
O5n	1
O4n	1
O3n	1
O2n	1
O1n	1
O0n	1

Figure S23-4

Data collection for Part 3:

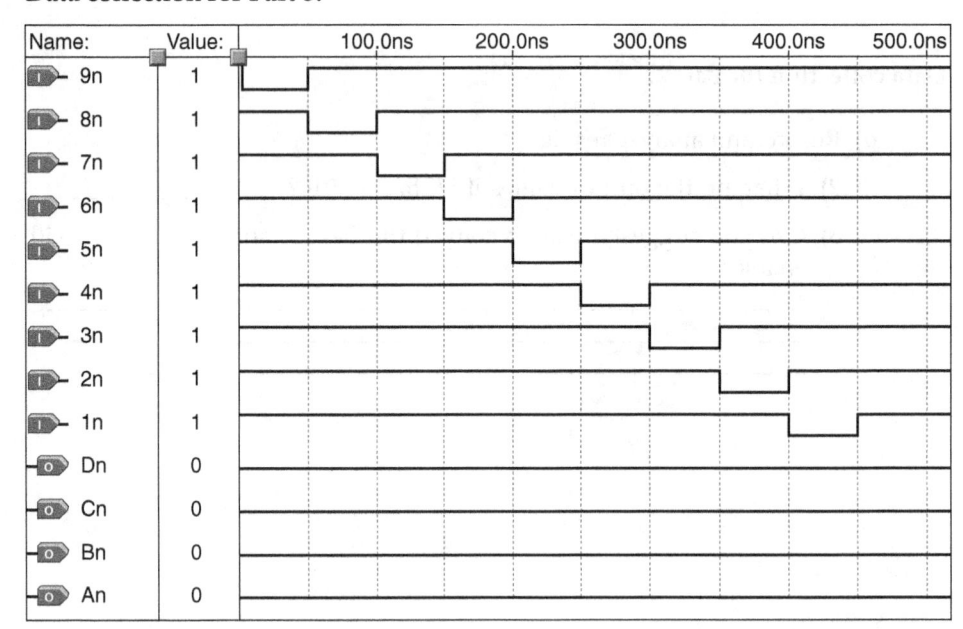

Name:	Value:
9n	1
8n	1
7n	1
6n	1
5n	1
4n	1
3n	1
2n	1
1n	1
Dn	0
Cn	0
Bn	0
An	0

Figure S23-6

Demonstrated To
Instructor/FA _____ Date _____

Part 1 [ ]   Part 2 [ ]   Part 3 [ ]

Experiment S24

Name _____

IC MULTIPLEXERS AND DEMULTIPLEXERS

OBJECTIVES

1. To practice using QUARTUS® II software to create schematics and waveforms.
2. To practice using QUARTUS® II software to compile and simulate digital circuits.
3. To investigate the operation of the 74151 IC multiplexer.
4. To investigate the operation of the 74138 IC as a demultiplexer.

TEXT REFERENCES

Read sections 9.6 through 9.8. Review Appendix D of this manual.

EQUIPMENT NEEDED

Components

Blank floppy, zip disk, or memory stick for storing projects;

QUARTUS® II software (Altera Corporation);

Optional: DeVry University Board eSOC with EPM7128SLC84 CPLD (or Altera University Board with EPM7128SLC84 CPLD or any other equivalent board);

Desktop computer with minimum of INTEL Pentium PC @ 400MHz CPU running Microsoft Windows NT4SP1, 2000, or XP or better; Pentium III or 4 PC @ 400MHz running Red Hat Linux v7.3 or 8.0 or Red Hat Linux Enterprise 3; Sun Ultra running Solaris v8 or 9; HP9000 Series 700/800 running HP-UX v11.0 with ACE dated 11/1999 or later; or equivalent.

DISCUSSION

Switches are often used to select data from several numbered input sources in electronic systems. Digital systems use electronic circuits called multiplexers to simulate the switches. A multiplexer consists of several inputs, one output, and a number of SELECT inputs. When a binary code is applied to the SELECT inputs, the data with the input number represented by the code will be routed to the output. In this exercise you will investigate the operation of the 74151 multiplexer. This device has eight input lines with a complementary output (referred to as 8-line-to-1 line), and an enable. You will then use this multiplexer in a frequency selector.

A device that performs the opposite of multiplexing is a demultiplexer. The demultiplexer receives its data on a single input line and distributes it over several output lines. Each output is selected by the SELECT inputs, and each gets a "slice" of the data present on the line. In this exercise you will investigate the demultiplexer function of the 74LS138, whose decoder function was investigated in Supplementary Experiment 23. This device has a single data input and eight outputs (referred to as 1-line-to-8-line).

PROCEDURE

PART 1 – The Operation of the 74LS151 IC Multiplexer

a) Start the QUARTUS® II program. Set up a new project and use **mux** for the name of the project. The diagram of the circuit you are to construct and its input waveforms are given in Figures S24-1 and S24-2.
b) Create a new graphic file for the schematic *mux.bdf*.
c) Place the symbols on the schematic according to Figure S24-1.
d) Place the input and output connectors on the schematic. Label the inputs A, B, C, GN, and D0, D1... D7; label the outputs Y and WN.
e) Wire the input and output connectors to the circuit as shown in Figure S24-1.
f) Select **Project | Top-Level Entity**.
g) Compile the project. If there are errors, fix them and save the file before recompiling the project. If there are no errors (or perhaps a warning concerning timing at most), then go on to the next step.
h) Create a new waveform file *mux.vwf* and define the inputs for simulation. Group A, B, and C as SELECT. The input waveforms should look similar to those in Figure S24-2:
   1) For D0 use a clock signal with 100ns period.
   2) For D1 use a clock signal with 200ns period.
   3) For D2 use a clock period with 300ns period.
   4) For D3 use a clock signal with 400ns period.
   5) For D4 use a clock signal with 500ns period.
   6) For D5 use a clock period with 600ns period.
   7) For D6 use a clock signal with 700ns period.
   8) For D7 use a clock signal with 800ns period.
   9) For SELECT use a counter signal that increments by 1 every 2 μs.
   10) Set GN to LOW.
   11) Set END TIME to 16us.

PROCEDURE

    i) Select **Assignments | Settings** to access the Settings Window.
    j) Click on *Simulator* at the left.
    k) Choose *Functional* from the *Simulator Mode* pull-down menu.
    l) Find and use for the *Simulator Input* file *mux.vwf* you just created.
    m) Select **Processing | Generate Functional Simulation Netlist**.
    n) Simulate the project. If there are any errors and you change your circuit, remember to save and compile the circuit before re-simulating.
    o) Record and analyze results:
        1) Print out the Simulation Waveforms report from QUARTUS® II and attach it to the data collection sheet at the end of this experiment.
        2) Observe that Y is a composite made from "slices" of each input in the order of their particular SELECT numbers.
        3) For each count of SELECT, measure the period of Y for the corresponding "slice" and identify the input (D0, D1...) the signal came from. Place the input number in the Y column of Table S24-1 in the appropriate row.
        4) What values of A, B, and C are required to select from D3?
            A = _____ B = _____ C = _____ .
        5) If GN is HIGH, what are the output values for Y and WN?

        _____

        6) Is it possible for two inputs to be selected simultaneously?
        _____

        7) Transfer the answers to the above questions, the waveform information, and state table to the data collection sheets at the end of this experiment.
    p) Save and close the project.

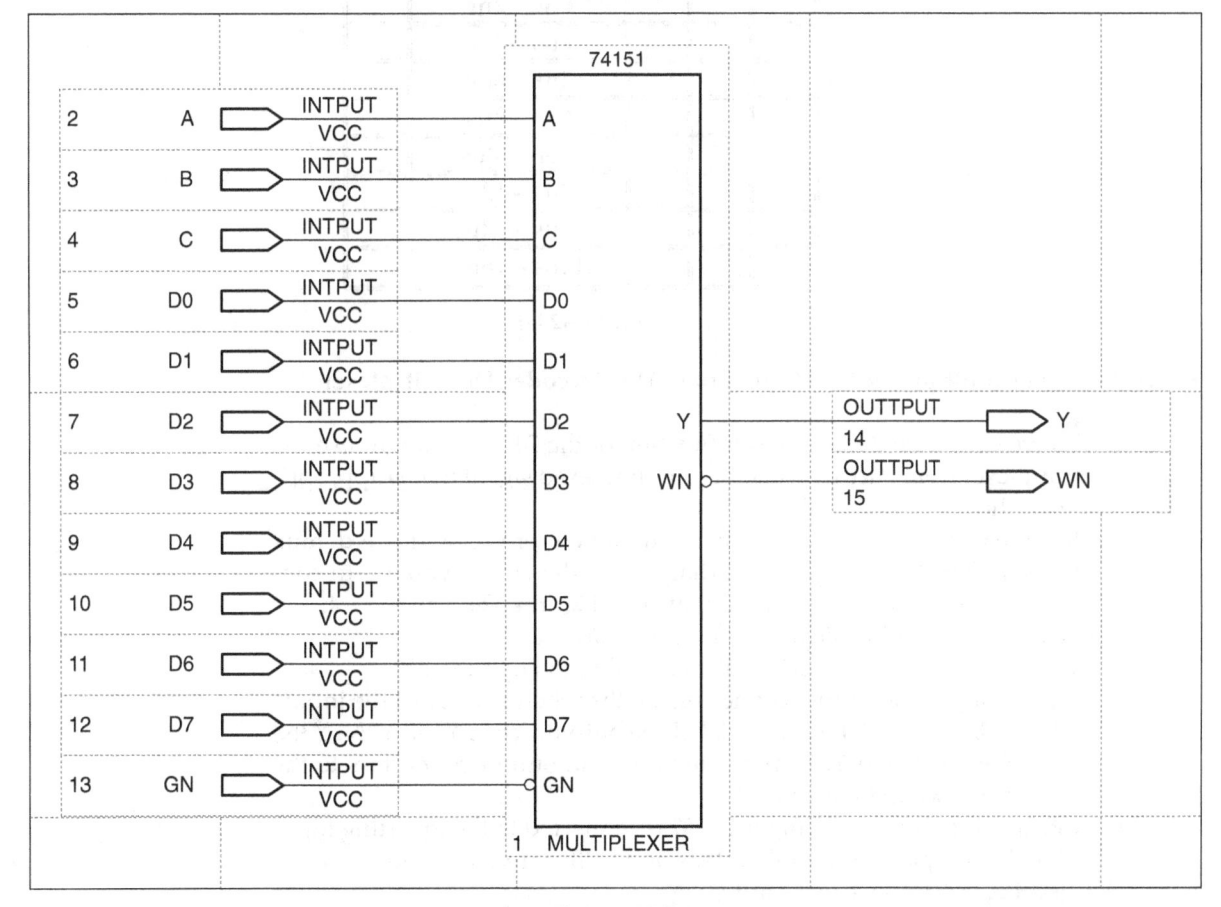

Figure S24-1

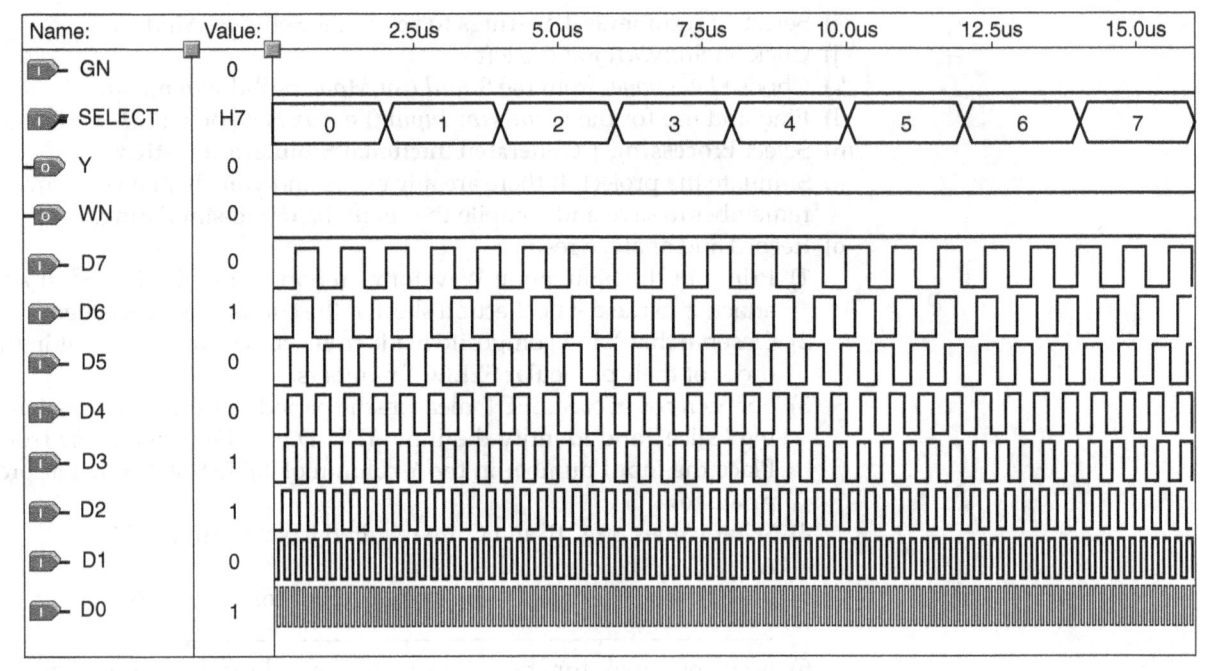

Figure S24-2

SELECT			Time Interval	Y
A	B	C		
			0ns – 2ns	
			2us – 4us	
			4us – 6us	
			6us – 8us	
			8us – 10us	
			10us – 12us	
			12us – 14us	
			14us – 16us	

Table S24-1

PART 2 – The Demultiplexer Function of the 74138 Decoder/Demultiplexer

a) To investigate the demultiplexer function of the 74138, you will use the multiplexer from Part 1 to generate a multiplexed signal that is appropriate for the 74138 to demultiplex.
b) Start the QUARTUS® II program. Set up a new project and use **demux** for the name of the project. The diagram of the circuit you are to construct and its input waveforms are given in Figures S24-3 and S24-4.
c) Create a new graphic file for the schematic *demux.bdf*.
d) Place the symbols on the schematic according to Figure S24-3.
e) Place the input and output connectors on the schematic. Label the inputs A, B, C, GN, and D0, D1,..., D7; label the outputs Y0N, Y1N, Y2N, Y3N, Y4N, Y5N, Y6N, and Y7N. Wire the input and output connectors to the circuit as shown in Figure S24-3.
f) Note that the Y output from the 74151 is connected to the inverting input of the demultiplexer G1 while G2AN and G2BN are grounded.
g) Select **Project | Top-Level Entity.**

PROCEDURE

h) Compile the project. If there are errors, fix them and save the file before recompiling the project. If there are no errors (or perhaps a warning concerning timing at most), then go on to the next step.

i) Create a new waveform file and define the inputs for simulation. Use the same periods as Part 1. The waveforms should look similar to those in Figure S24-4.

j) Select **Assignments | Settings** to access the Settings Window.

k) Click on *Simulator* at the left.

l) Choose *Functional* from the *Simulator Mode* pull-down menu.

m) Find and use for the *Simulator Input* file *demux.vwf* you just created above.

n) Select **Processing | Generate Functional Simulation Netlist.**

o) Record and analyze results:
   1) Print out the Simulation Waveforms report from QUARTUS® II and attach it to the data collection sheet at the end of this experiment.
   2) Observe that each output of the 74138 has a frequency that is dependent on the input time slice selected.
   3) For each output of the 74138, measure the period of its waveform and identify the original 74151 input (D0, D1, ...) it represents. Place the input number in the appropriate Y column of Table S24-2 and in the appropriate row.
   4) Suppose separate counters were used to generate the SELECT for the multiplexer and the demultiplexer. Explain what would happen if the counters did not have the same count sequence:

   _____
   _____

   5) Why is G1 referred to as the 74138's inverting input?

   _____
   _____

   6) Transfer the answers to the above questions, the waveform information, and state table to the data collection sheets at the end of this experiment.

p) Save and close the project.

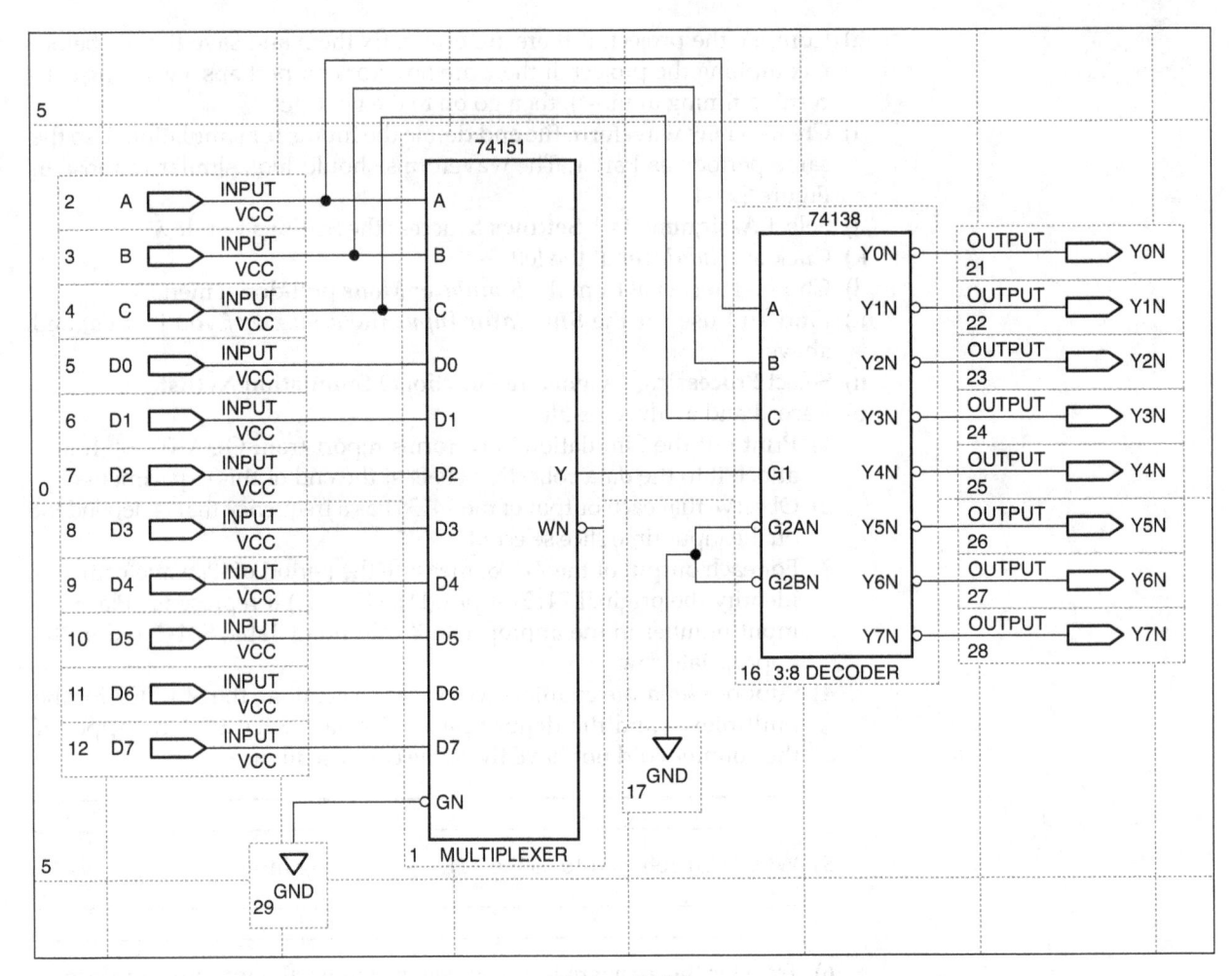

Figure S24-3

PROCEDURE

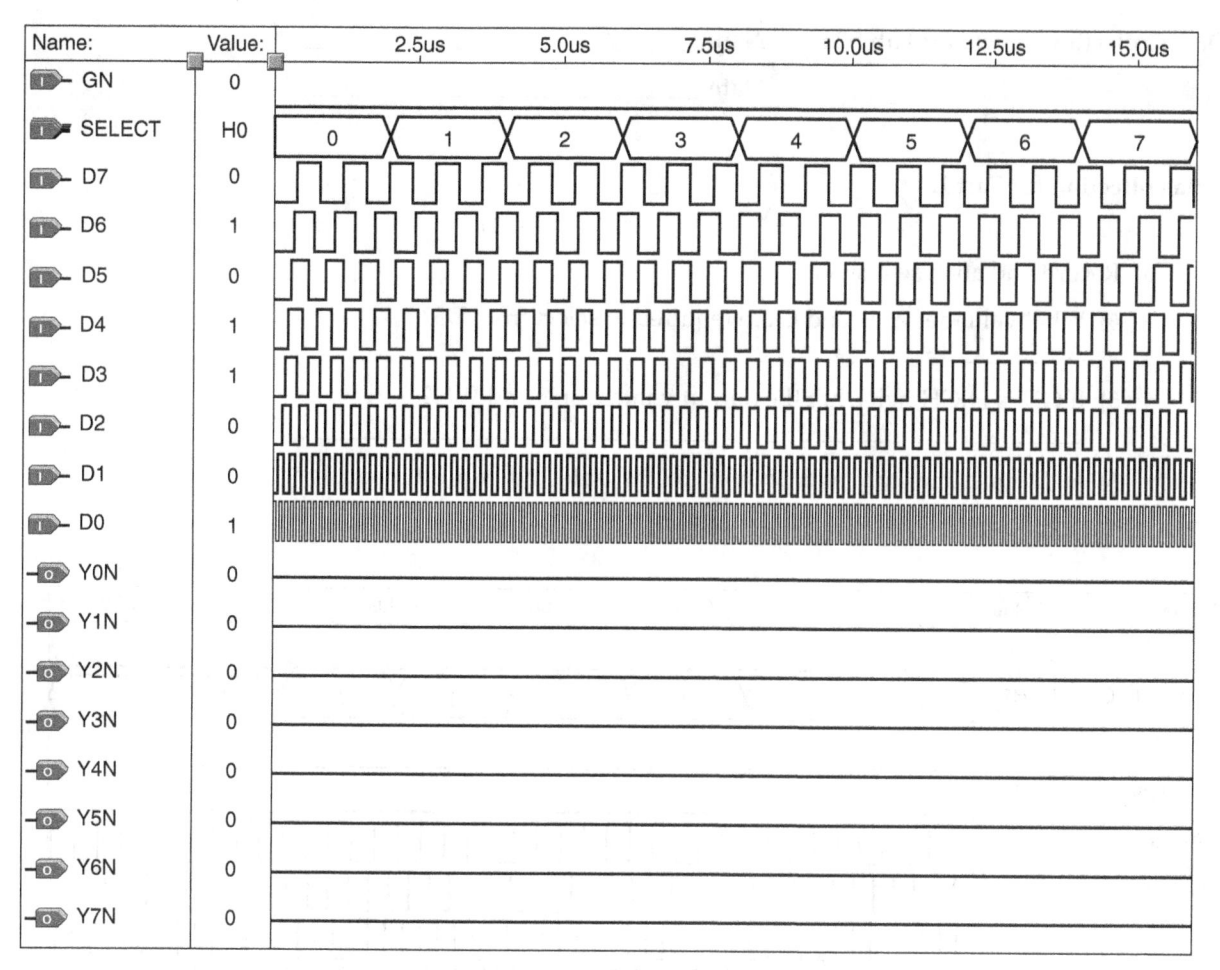

Figure S24-4

| SELECT | | | Time Interval | Y0N | Y1N | Y2N | Y3N | Y4N | Y5N | Y6N | Y7N |
A	B	C									
			0ns-2ns								
			2us–4us								
			4us–6us								
			6us–8us								
			8us-10us								
			10us-12us								
			12us–14us								
			14us–16us								

Table S24-2

Data Collection Sheets for Lab S24   Name _____

Date _____

Data collection for Part 1:

o) Record and analyze results:

4) What values of A, B, and C are required to select from D3?
   A = _____   B = _____   C = _____ .

5) If GN is HIGH, what are the output values for Y and WN?
   _____

6) Is it possible for two inputs to be selected simultaneously?
   _____

Name:	Value:
GN	0
SELECT	H7
Y	0
WN	0
D7	0
D6	1
D5	0
D4	0
D3	1
D2	1
D1	0
D0	1

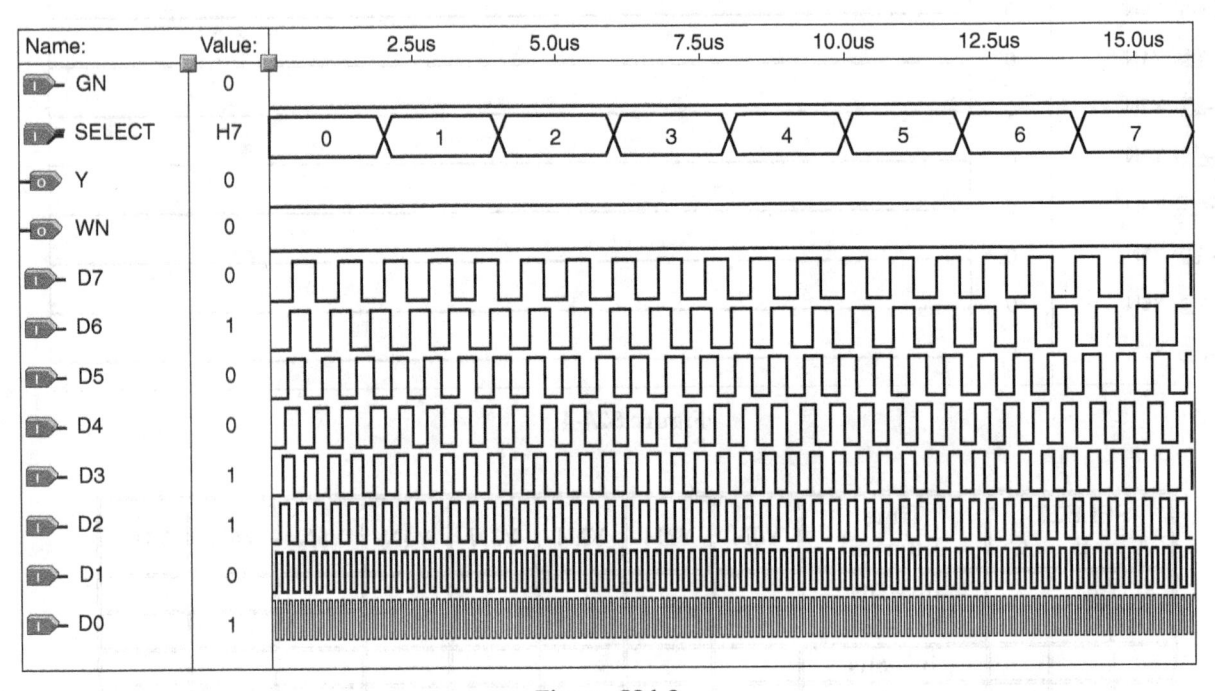

Figure S24-2

Demonstrated To
Instructor/FA _____ Date _____

Part 1 [ ]   Part 2 [ ]

DATA COLLECTION SHEET

Data Collection Sheets for Lab S24   Name _____

Date _____

SELECT A	SELECT B	SELECT C	Time Interval	Y
			0ns – 2ns	
			2us – 4us	
			4us – 6us	
			6us – 8us	
			8us – 10us	
			10us – 12us	
			12us – 14us	
			14us – 16us	

Table S24-1

Data collection for Part 2:

o) Record and analyze results:

4) Suppose separate counters were used to generate the SELECT for the multiplexer and the demultiplexer. Explain what would happen if the counters did not have the same count sequence:

_____
_____
_____

5) Why is G1 referred to as the 74138's inverting input?

_____
_____
_____

Demonstrated To
Instructor/FA _____ Date _____

Part 1 [ ]   Part 2 [ ]

Data Collection Sheets for Lab S24 **Name** ______________________________

Date ______________________________

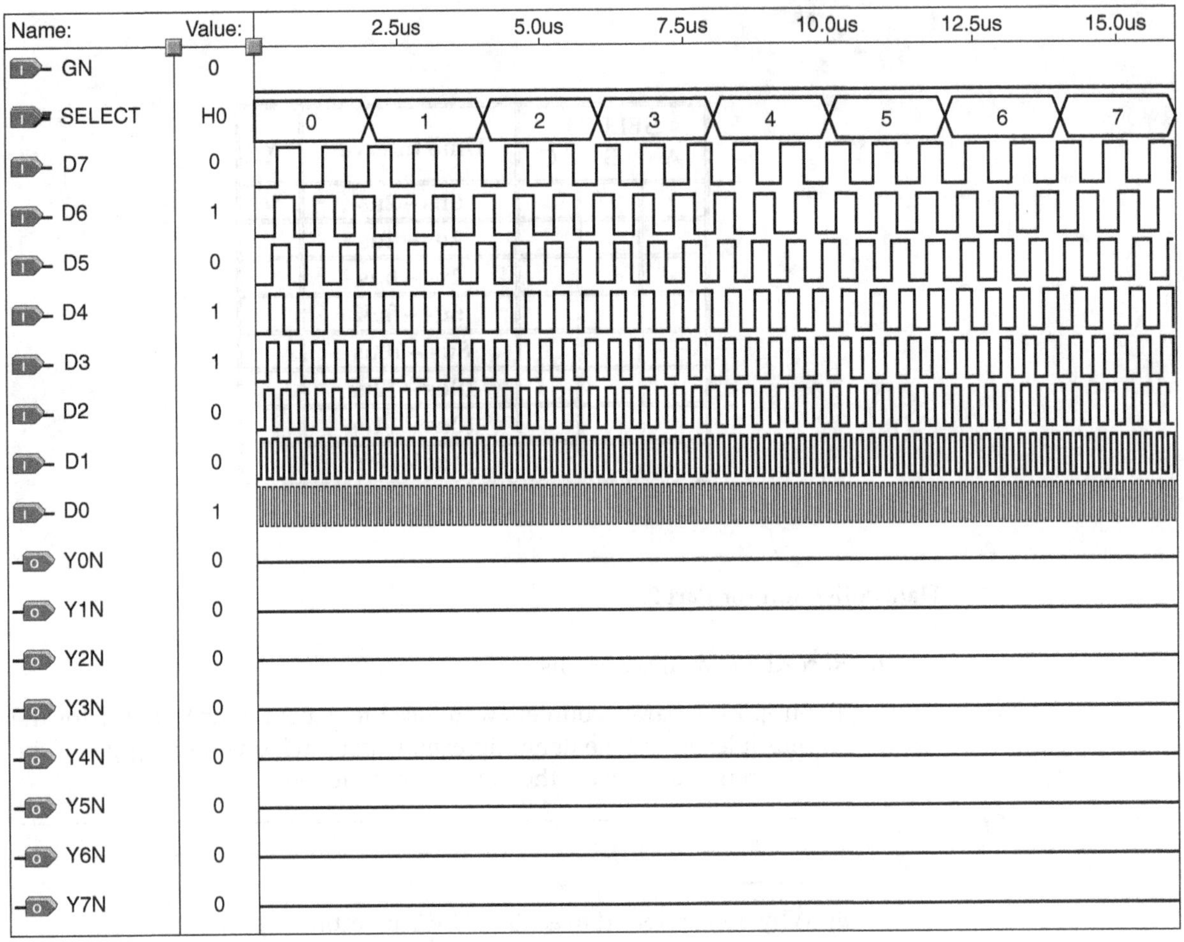
Name:
Value:
2.5us
5.0us
7.5us
10.0us
12.5us
15.0us
GN 0
SELECT H0
D7 0
D6 1
D5 0
D4 1
D3 1
D2 0
D1 0
D0 1
Y0N 0
Y1N 0
Y2N 0
Y3N 0
Y4N 0
Y5N 0
Y6N 0
Y7N 0
0
1
2
3
4
5
6
7

Figure S24-4

SELECT			Time Interval	Y0N	Y1N	Y2N	Y3N	Y4N	Y5N	Y6N	Y7N
A	B	C									
			0ns-2ns								
			2us–4us								
			4us–6us								
			6us–8us								
			8us-10us								
			10us-12us								
			12us–14us								
			14us–16us								

Table S24-2

Demonstrated To
Instructor/FA ______________________________ Date ______________________________

Part 1 [ ] Part 2 [ ]

Experiment S25

Name _____

VHDL STATE MACHINES

OBJECTIVES

1. To practice using QUARTUS® II software to create schematics and waveforms.
2. To practice using QUARTUS® II software to compile and simulate digital circuits.
3. To investigate the application of state machines.
4. To program the EPM7128SLC84 CPLD.

TEXT REFERENCES

Read section 7.14. Review Appendix D of this lab manual.

EQUIPMENT NEEDED

Components

Blank floppy, zip disk, or memory stick for storing projects;
QUARTUS® II software (Altera Corporation);
DeVry University Board eSOC with EPM7128SLC84 CPLD (or Altera University Board with EPM7128SLC84 CPLD or any other equivalent board);
Desktop computer with minimum of INTEL Pentium PC @ 400MHz CPU running Microsoft Windows NT4SP1, 2000, or XP or better; Pentium III or 4 PC @ 400MHz running Red Hat Linux v7.3 or 8.0 or Red Hat Linux Enterprise 3; Sun Ultra running Solaris v8 or 9; HP9000 Series 700/800 running HP-UX v11.0 with ACE dated 11/1999 or later; or equivalent.

DISCUSSION

A **finite state machine** (FSM) is a behavioral model, which consists of states, transitions, and actions. A state stores past information. It indicates the input changes from start to now. A transition is a state change. It is described by a condition that causes the transition. An action is what is to be carried out at a given instant in time.

There are two basic types of finite state machines. A **Moore** machine is an FSM whose outputs are determined by the current state alone (no dependency on the input). The state diagram for a Moore machine will include an output signal for each state. Compare this with a **Mealy** machine, which maps transitions in the machine to outputs.

Most electronic systems are designed as clocked sequential systems. Clocked sequential systems are a type of Moore machine where the state changes only when the global clock signal changes. Usually the current state is stored in flip-flops, and global clock signal is connected to the "clock" input of the flip-flops.

A **Mealy** machine is an FSM whose outputs are determined by the current state and the input. The state diagram will include an output signal for each transition edge. For example, in going from state 1 to state 2 on input '0,' the output might be '1' (its edge would be labeled 0/1). Compare this to the Moore FSM above.

In this experiment, you are asked to program a simple state machine, a model for a washing machine, in VHDL. You will then simulate the machine and download its code to the eSOC board.

Machine Description

The machine has four states: idle, fill, agitate, and spin. However, since the machine essentially repeats these states (the machine executes each state during a wash sequence and then again during a rinse sequence), it is simpler to differentiate between the wash and the rinse states. So, the states are idle, fill1, agitate1, spin1, fill2, agitate2, and spin2. The transitions are

1. idle -> fill1
2. fill1 -> agitate1 (wash)
3. agitate1 (wash) -> spin1
4. spin1 -> fill2
5. fill2 -> agitate2 (rinse)
6. agitate2 (rinse) -> spin2
7. spin2 -> idle

The first transition is caused by depressing a **Start** switch. The second transition is caused by a switch indicating full when ON.
The third transition is caused by a timer "timing out" (**Time Out 1**).
The fourth transition is caused by the timer timing out (**Time Out 2**).
The fifth transition is caused by a switch indicating full when ON.
The sixth transition is caused by a timer timing out (**Time Out 3**).
Finally the seventh transition is caused by the timer timing out (**Time Out 4**).

To simplify things, you can assume that the agitate cycles (wash and rinse) are the same and that all timer settings are the same, say one minute. Because one minute is a long time in simulation, the actual time should be scaled down to perhaps microseconds for simulation and seconds for the actual finished product. You could modify this machine to enter the conditions from either switches or an external sequencer with timers.

PROCEDURE

PART 1 – Construct the Transition Table and State Diagram

a) A transition table for the machine is given in Table S25-1. The top row gives the names of the current states while the left column gives the names of the conditions for the next step.

b) Using the machine description, fill in the next states into the appropriate cells.

Current State/ Condition	Idle	Fill	Agitate	Spin
Start				
Full 1				
Time Out 1 (1 min)				
Time Out 2 (1 min)				
Full 2				
Time Out 3 (1 min)				
Dry				

Table S25-1

c) In the space provided below, draw a state diagram for the machine:

PART 2 – Program the Machine in VHDL

a) Start the QUARTUS® II program. Set up a new project and use **state_washer** for the name of the project.

b) Create a new text file *state_washer.vhd* for writing the washing machine's VHDL description.

c) Type in the VHDL program from the listing in Figure S25-1.

```vhdl
ENTITY   state_washer IS
PORT    (       clock, start, full1, full2, timeout1, timeout2, timeout3, dry : IN  BIT;
                drivers   :OUT    bit_vector (3 DOWNTO 0)); --water, ag_mode, sp_mode
END      state_washer;

ARCHITECTURE vhdl OF state_washer IS
TYPE state_machine IS (idle, fill, agitate, spin, fill2, agitate2, spin2);
BEGIN
        PROCESS (clock, start)
        VARIABLE cycle : state_machine;
        BEGIN
                IF (clock'EVENT AND clock = '1') THEN
                        CASE cycle IS
                                WHEN idle =>
                                        IF (start = '1') THEN       cycle := fill;
                                        ELSE                        cycle := idle;
                                        END IF;
                                WHEN fill =>
                                        IF (full1 = '1') THEN       cycle := agitate;
                                        ELSE                        cycle := fill;
                                        END IF;
                                WHEN agitate =>
                                        IF (timeout1 = '1') THEN    cycle := spin;
                                        ELSE                        cycle := agitate;
                                        END IF;
                                WHEN spin =>
                                        IF (timeout2 = '1') THEN    cycle := fill2;
                                        ELSE                        cycle := spin;
                                        END IF;
                                WHEN fill2 =>
                                        IF (full2 = '1') THEN       cycle := agitate2;
                                        ELSE                        cycle := fill2;
                                        END IF;
                                WHEN agitate2 =>
                                        IF (timeout3 = '1') THEN    cycle := spin2;
                                        ELSE                        cycle := agitate2;
                                        END IF;
                                WHEN spin2 =>
                                        IF (dry = '1') THEN         cycle := idle;
                                        ELSE                        cycle := spin2;
                                        END IF;
                        END CASE;
                END IF;
                CASE cycle IS           --      ("water, ag_mode, sp_mode")
                        WHEN idle           => drivers <= "0000";
                        WHEN spin           => drivers <= "0001";
                        WHEN agitate        => drivers <= "0010";
                        WHEN fill           => drivers <= "0100";
                        WHEN spin2          => drivers <= "1001";
                        WHEN agitate2       => drivers <= "1010";
                        WHEN fill2          => drivers <= "1100";
                        WHEN OTHERS         => drivers <= "0000";
                END CASE;
        END PROCESS;
END vhdl;
```

Figure S25-1

PROCEDURE

d) Compile the project. If there are errors, fix them and save the file before recompiling the project. If there are no errors (or perhaps a warning concerning timing at most), then go on to the next step.
e) Create a block symbol for the machine. You will use this as part of an example application to illustrate its functions. See Figure S25-2 for an example of the symbol.

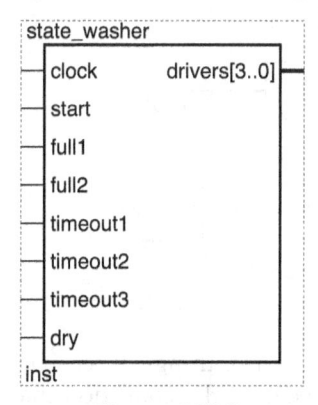

Figure S25-2

f) Save and close the project.

PART 3 – Test the FSM with an Example

a) Examine the block diagram of Figure S25-3, which illustrates the example project that will be used to test the FSM.

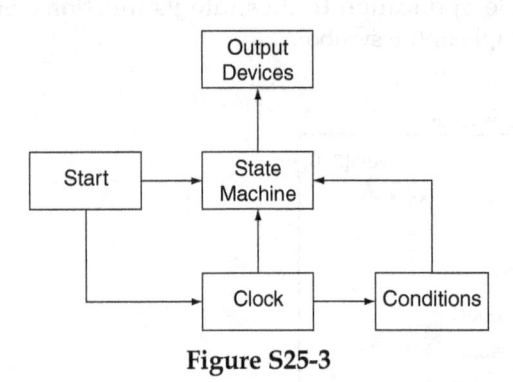

Figure S25-3

b) The diagram shows the FSM at the center of things. It receives a "start" and sets the current state. When it receives the condition that says the current state's task has been accomplished from "Condition" it advances to the next state and continues to repeat this sequence state followed by a condition until the idle state is reached.
c) A start button will be used to start the clock and the machine. This is an asynchronous start.
d) The block labeled "Conditions" is in turn started by the clock. Its function will be to inform the FSM that a particular task has been done. You will use a ring counter to simulate this.
e) The clock is used to time events.
f) The outputs from the FSM will drive the output devices which, in this example, are two 7-segment displays, which display codes that represent the current state of the FSM. Table S25-2 shows the states and their respective codes.

Current State	Codes
Idle/Dry	00
Spin # 1	01
Agitate #1	02
Fill 1	04
Spin # 2	11
Agitate #2	12
Fill 2	14

Table S25-2

a) *Build the condition sequencer*. Create a new design file *six_bit_ring_ctr.bdf* for the condition sequencer. Place the symbols onto the design file as shown in Figure S25-4.
b) There are two inputs: **seq_clr,** which will be used to set the first D flip-flop and clear the others with a negative-going pulse, and the other is **seq_clk,** the clock for the sequencer.

PROCEDURE

c) There are six outputs, one for each condition in the order shown: **full1**, **finish_agitate1**, **finish_spin1**, **full2**, **finish_agitate2**, and **dry**.
d) Compile the design. Once the sequencer works, create a block symbol for it. An example block symbol for the sequencer is shown in Figure S25-5.
e) Close the design file.

    a. *Build the clock divider.* Create a new design file *clock_divider.vhd* for the clock divider. Type in the VHDL code in Figure S25-6.
    b. There is one input: **clock_4MHz**, which is from the eSOC board master clock.
    c. There are seven outputs, ranging from 100 kHz down to 1 Hz. You will use the **clock_1Hz** to time the example project when it is downloaded to the eSOC board. For functional simulation, you can use one of the faster outputs.
    d. Compile the design. Once the clock divider works, create a block symbol for it. An example block symbol for the sequencer is shown in Figure S25-7.
    e. Save and close the project.

    a. *Build the example.* Create a new project and name it **state_washer_example**.
    b. Create a schematic design file, *state_washer_example.bdf*.
    c. Place the symbols onto the schematic as shown in Figure S25-8. Remember that the design files for each of the symbols you created must be a part of the project.
    d. Note the connection between the clock divider and the NAND gate is taken at the 1 MHz output for simulation purposes.
    e. Compile the design.
    f. Select **Processing | Create Functional Simulation Netlist**.
    g. Create a new vector waveform file and name it *state_machine_example.vwf*. An example is shown in Figure S25-9.
    h. Set the clock to 4 MHz and the grid size to 500 ns.
    i. Make sure the Start input is about 2 microseconds. Even though an actual Start pulse will be random in both time and duration, the design requires several clock cycles of Start for the rest of the circuit to begin functioning. There is ample room here for improvement.
    j. Simulate the design.

    a. *Program the eSOC board.* Once the example circuit behaves correctly, disconnect the clock_1MHz output of the clock divider and connect it to the clock_1Hz output.
    b. Assign pins of the eSOC board to the inputs and outputs.
    c. Make sure that all of the eSOC DIP switches are OFF.
    d. Select **Tools | Programmer** to program and download the design to the eSOC board.
    e. Turn the necessary DIP switches to ON.
    f. Apply power.
    g. Apply that Start pulse. You will probably need to press and hold the switch for a second or two.
    h. Record the sequence displayed by the 7-segment LEDs in Table S25-3.
    i. Save and close the project.

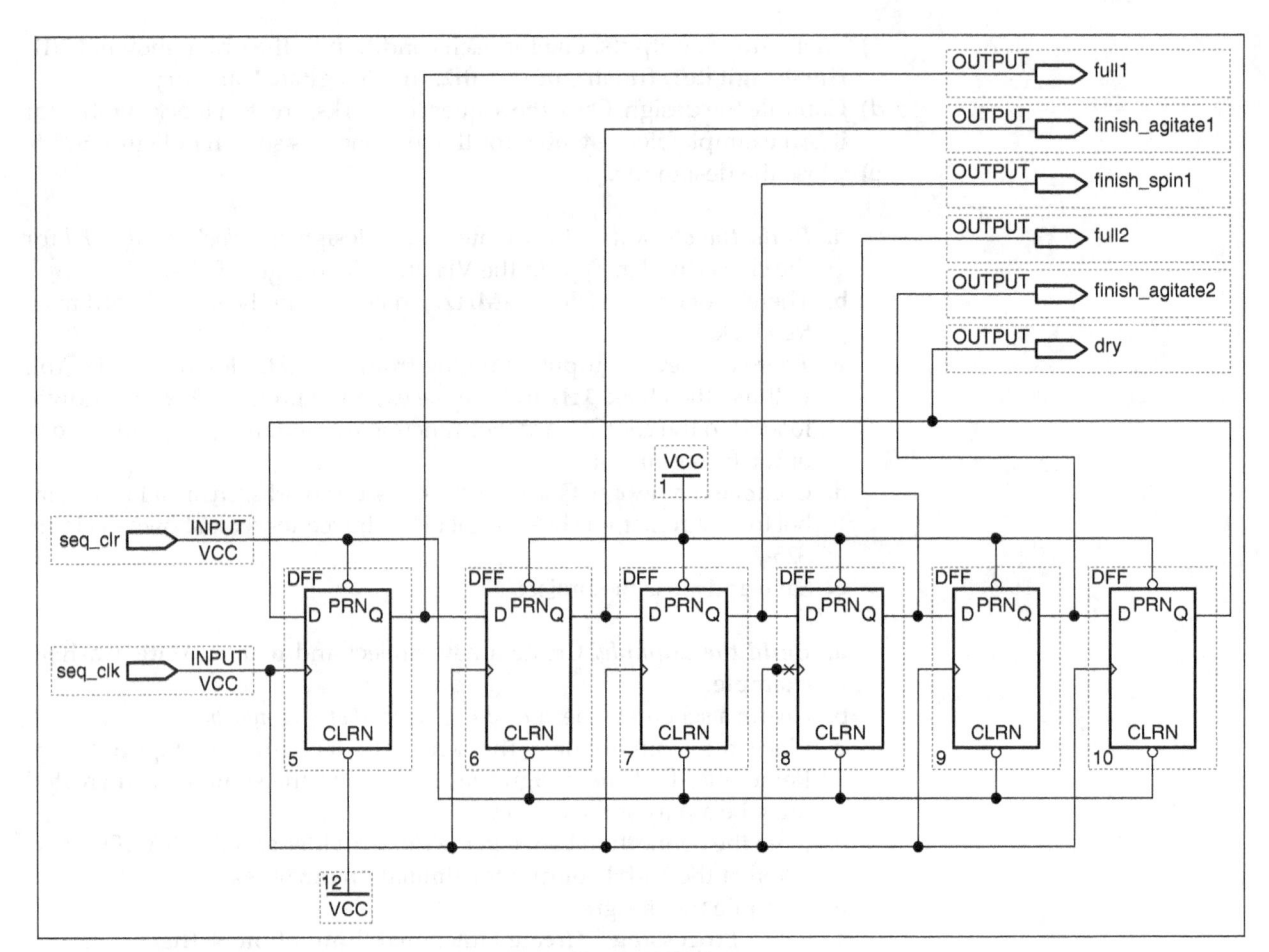

Figure S25-4

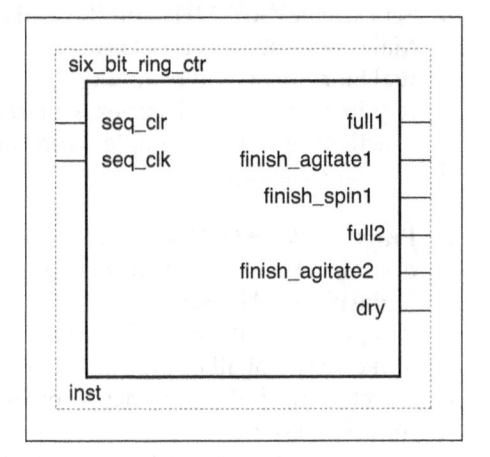

Figure S25-5

PROCEDURE

```vhdl
LIBRARY IEEE;
USE IEEE.STD_LOGIC_1164.all;
USE IEEE.STD_LOGIC_ARITH.all;
USE IEEE.STD_LOGIC_UNSIGNED.all;

ENTITY clock_divider IS

    PORT
    (
        clock_4MHz              : IN    STD_LOGIC;
        clock_1MHz              : OUT   STD_LOGIC;
        clock_100kHz            : OUT   STD_LOGIC;
        clock_10kHz             : OUT   STD_LOGIC;
        clock_1kHz              : OUT   STD_LOGIC;
        clock_100Hz             : OUT   STD_LOGIC;
        clock_10Hz              : OUT   STD_LOGIC;
        clock_1Hz               : OUT   STD_LOGIC);

END clock_divider;

ARCHITECTURE a OF clock_divider IS
    SIGNAL count_1MHz: STD_LOGIC_VECTOR(4 DOWNTO 0);
    SIGNAL count_100kHz, count_10kHz, count_1kHz : STD_LOGIC_VECTOR(2 DOWNTO 0);
    SIGNAL count_100Hz, count_10Hz, count_1Hz : STD_LOGIC_VECTOR(2 DOWNTO 0);
    SIGNAL clock_1MHz_int, clock_100kHz_int, clock_10kHz_int, clock_1kHz_int: STD_LOGIC;
    SIGNAL clock_1MHz_int, clock_100kHz_int, clock_10kHz_int, clock_1kHz_int: STD_LOGIC;
    SIGNAL clock_100Hz_int, clock_10Hz_int, clock_1Hz_int : STD_LOGIC;
BEGIN
    PROCESS
    BEGIN
-- Divide by 4
        WAIT UNTIL clock_4MHz'EVENT and clock_4MHz = '1';
            IF count_1MHz < 3 THEN
                count_1MHz <= count_1MHz + 1;
            ELSE
                count_1MHz <= "00000";
            END IF;
            IF count_1MHz < 2 THEN
                clock_1MHz_int <= '0';
            ELSE
                clock_1MHz_int <= '1';
            END IF;
        -- Sync all clock prescalar outputs back to master clock signal
            clock_1MHz <= clock_1MHz_int;
            clock_100kHz <= clock_100kHz_int;
            clock_10kHz <= clock_10kHz_int;
            clock_1kHz <= clock_1kHz_int;
            clock_100Hz <= clock_100Hz_int;
            clock_10Hz <= clock_10Hz_int;
            clock_1Hz <= clock_1Hz_int;
    END PROCESS;

    -- Divide by 10
    PROCESS
    BEGIN
        WAIT UNTIL clock_1MHz_int'EVENT and clock_1MHz_int = '1';
            IF count_100kHz /= 4 THEN
                count_100kHz <= count_100kHz + 1;
            ELSE
                count_100kHz <= "000";
                clock_100kHz_int <= NOT clock_100kHz_int;
            END IF;
    END PROCESS;

        -- Divide by 10
```

```vhdl
            PROCESS
            BEGIN
                    WAIT UNTIL clock_100kHz_int'EVENT and clock_100kHz_int = '1';
                            IF count_10kHz /= 4 THEN
                                    count_10kHz <= count_10kHz + 1;
                            ELSE
                                    count_10kHz <= "000";
                                    clock_10kHz_int <= NOT clock_10kHz_int;
                            END IF;
            END PROCESS;

                    -- Divide by 10
            PROCESS
            BEGIN
                    WAIT UNTIL clock_10kHz_int'EVENT and clock_10kHz_int = '1';
                            IF count_1kHz /= 4 THEN
                                    count_1kHz <= count_1kHz + 1;
                            ELSE
                                    count_1kHz <= "000";
                                    clock_1kHz_int <= NOT clock_1kHz_int;
                            END IF;
            END PROCESS;

                    -- Divide by 10
            PROCESS
            BEGIN
                    WAIT UNTIL clock_1kHz_int'EVENT and clock_1kHz_int = '1';
                            IF count_100Hz /= 4 THEN
                                    count_100Hz <= count_100Hz + 1;
                            ELSE
                                    count_100Hz <= "000";
                                    clock_100Hz_int <= NOT clock_100Hz_int;
                            END IF;
            END PROCESS;

                    -- Divide by 10
            PROCESS
            BEGIN
                    WAIT UNTIL clock_100Hz_int'EVENT and clock_100Hz_int = '1';
                            IF count_10Hz /= 4 THEN
                                    count_10Hz <= count_10Hz + 1;
                            ELSE
                                    count_10Hz <= "000";
                                    clock_10Hz_int <= NOT clock_10Hz_int;
                            END IF;
            END PROCESS;

                    -- Divide by 10
            PROCESS
            BEGIN
                    WAIT UNTIL clock_10Hz_int'EVENT and clock_10Hz_int = '1';
                            IF count_1Hz /= 4 THEN
                                    count_1Hz <= count_1Hz + 1;
                            ELSE
                                    count_1Hz <= "000";
                                    clock_1Hz_int <= NOT clock_1Hz_int;
                            END IF;
            END PROCESS;
END a;
```

Figure S25-6

PROCEDURE

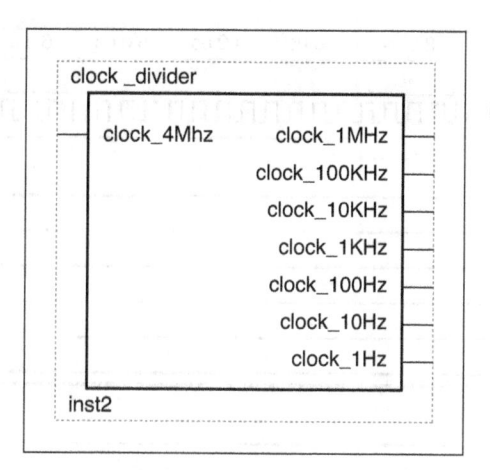

Figure S25-7

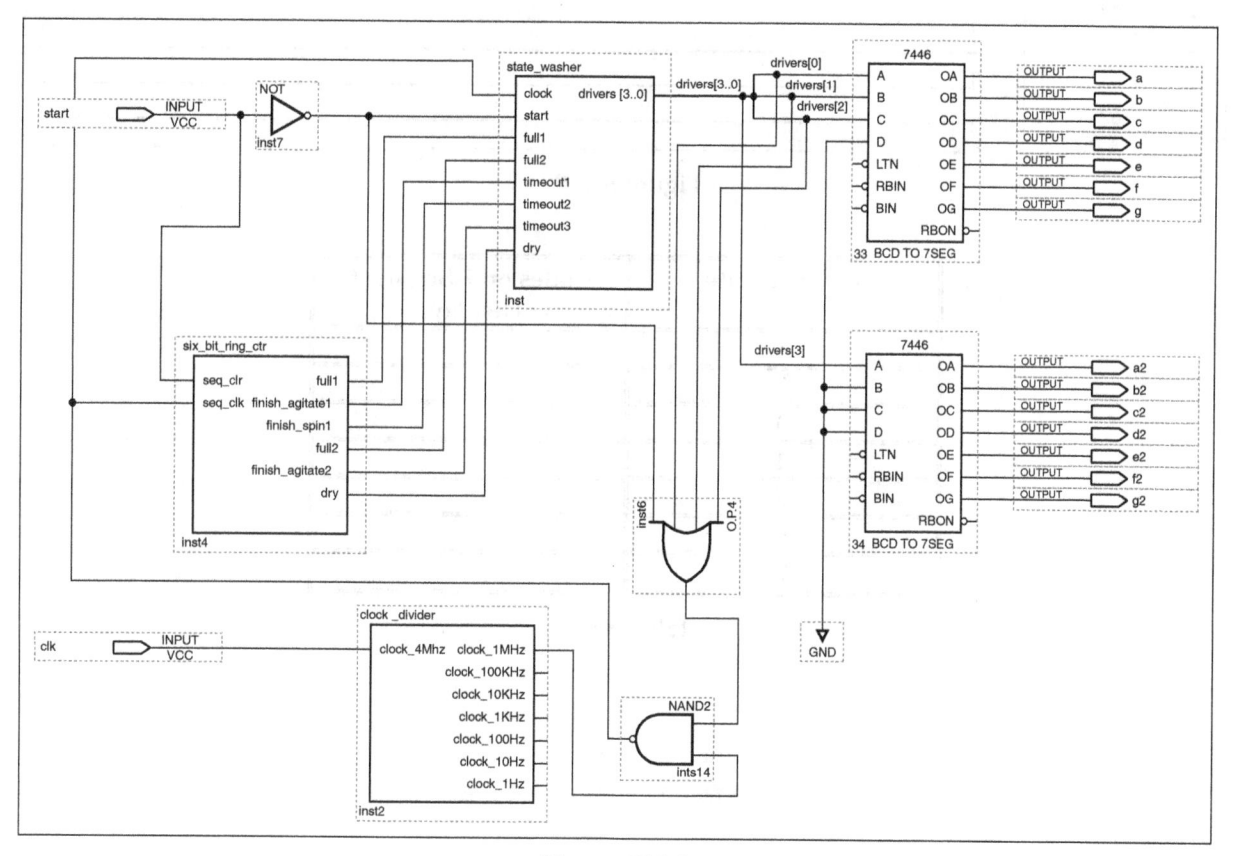

Figure S25-8

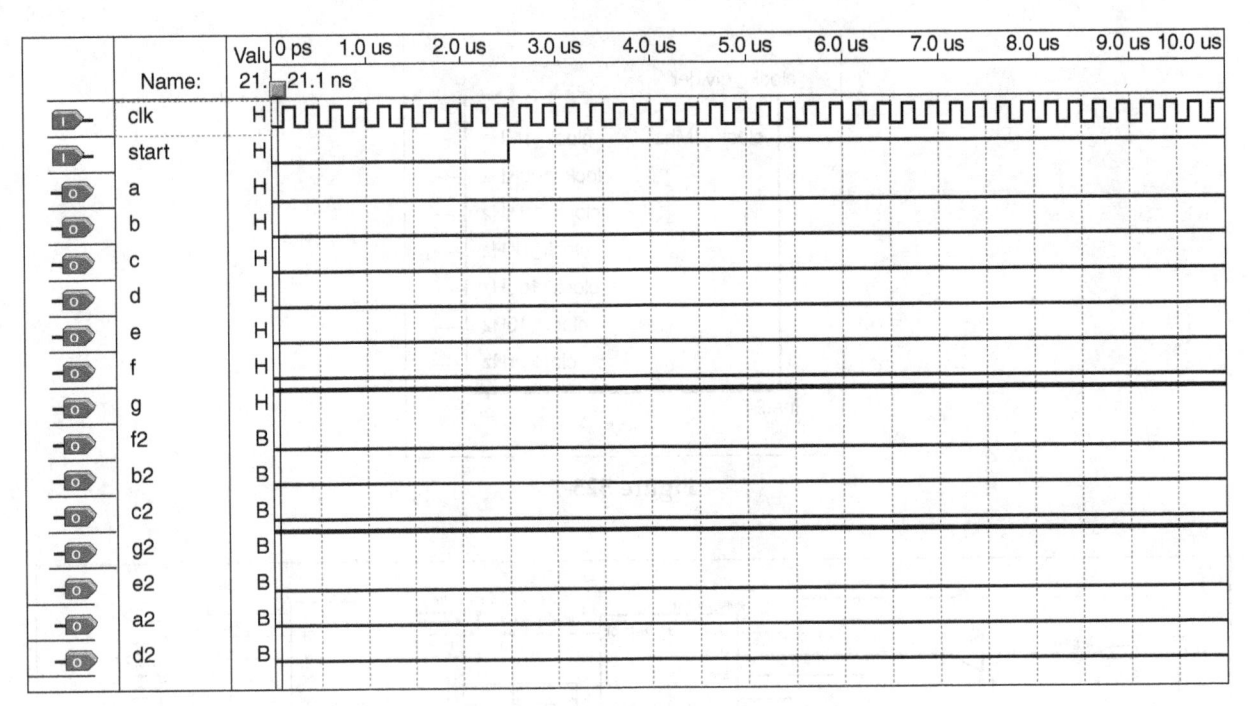

Name:	Valu 21.	0 ps	1.0 us	2.0 us	3.0 us	4.0 us	5.0 us	6.0 us	7.0 us	8.0 us	9.0 us	10.0 us
clk	H											
start	H											
a	H											
b	H											
c	H											
d	H											
e	H											
f	H											
g	H											
f2	B											
b2	B											
c2	B											
g2	B											
e2	B											
a2	B											
d2	B											

Figure S25-9

Current State	Codes on 7-Segment Displays
1	
2	
3	
4	
5	
6	
7	

Table S25-3

DATA COLLECTION SHEET

Data Collection Sheets for Lab S25   Name _____

Date _____

Data collection for Part 1:

Current State/Condition	Idle	Fill	Agitate	Spin
Start				
Full 1				
Time Out 1 (1 min)				
Time Out 2 (1 min)				
Full 2				
Time Out 3 (1 min)				
Dry				

Table S25-1

Data collection for Part 3:

Current State	Codes on 7-Segment Displays
1	
2	
3	
4	
5	
6	
7	

Table S25-3

Demonstrated To
Instructor/FA _____ Date _____

Part 1 [ ]   Part 2 [ ]   Part 3 [ ]

Project SP 1

Name _____

IMPLEMENTING A SIMPLE FREQUENCY COUNTER

OBJECTIVES

1. To create a simple frequency counter in QUARTUS® II.
2. Program the EPM7128SLC84 CPLD.

TEXT REFERENCES

Read sections 7.4 through 7.6, 7.24, and 10.5. Review Appendix D of this manual.

EQUIPMENT NEEDED

Components

Blank floppy, zip disk, or memory stick for storing projects;

QUARTUS® II software (Altera Corporation);

DeVry University Board eSOC with EPM7128SLC84 CPLD (or Altera University Board with EPM7128SLC84 CPLD or any other equivalent board);

Desktop computer with minimum of INTEL Pentium PC @ 400MHz CPU running Microsoft Windows NT4SP1, 2000, or XP or better; Pentium III or 4 PC @ 400MHz running Red Hat Linux v7.3 or 8.0 or Red Hat Linux Enterprise 3; Sun Ultra running Solaris v8 or 9; HP9000 Series 700/800 running HP-UX v11.0 with ACE dated 11/1999 or later; or equivalent;

Signal generator with TTL compatible output.

DISCUSSION

A frequency counter measures and displays the frequency of a signal. Its basic premise is to use BCD counting blocks to count the individual pulses for a precisely fixed period known as the sampling time. In general, the sampling time is accomplished by using a clock divider to divide a reference frequency, usually produced by a crystal oscillator, to a 1 Hz signal (1 second period). Other time intervals such as 0.001 seconds, 0.01 seconds, 0.1 seconds or even 10 seconds are useful, in order to change the range of frequencies that can be measured.

The frequency counter project depicted here is a variation of the counter suggested by the text. It is more like the counter in Experiment 24 of this lab manual, which will measure frequencies between 1 – 99 Hz. This version will better fit the DeVry eSOC (or Altera UP1 or UP2) board and all but the frequency to be measured is implemented internally. It consists of the following units:

- Clock divider
- Control
- BCD counter/display units

PROCEDURE

Step 1 – Project

a) Set up a new project and use **freq_ctr** for the name of the project and
b) Create a schematic file *freq_ctr.bdf*.

Step 2 – Clock Divider

a) The clock divider is a VHDL routine that takes the DeVry eSOC board's 4 MHz reference frequency and divides it into frequencies ranging from 1 MHz to 1 Hz in decades.
b) Type in the VHDL program from the listing in Figure SP1-1.
c) Compile and simulate the clock divider.
d) Once the clock divider compiles and simulates correctly, create a symbol for the routine and place it on your project schematic.

Step 3 – Retriggerable Edge-Triggered OS

a) The retriggerable edge-triggered oneshot delays the "clear" pulse to the BCD counter by up to 15 clock cycles. Typically, the delay is set to 3.
b) Type in the VHDL program from the listing in Figure SP1-2.
c) Compile and simulate the oneshot.
d) Once the oneshot compiles and simulates correctly, create a symbol for the routine and place it on your project schematic.

PROCEDURE

Step 4 – BCD Counter/Display

a) The BCD Counter/Display in part 2 of Experiment S19 can be used in this project.
b) Load the *bcd_ctr_display.bdf* file from your storage disk, save it in the project director, and add it to the project.
c) Compile the counter/display unit.
d) Once the counter/display unit compiles correctly, create a symbol for the routine and place it on your project schematic.

Step 5 – Counter Control

a) Examine Figure SP1–3, which is a schematic of the counter control circuit. Note the two flip-flops and 3-input NAND gate that make up the counter control along with the oneshot. Note that sample_pulses, X, and ctr_clock are internal signals used for troubleshooting.
b) The relationships between the control signals is illustrated in Figure SP1–4.
c) Place the symbols onto the schematic. Do not place any symbols that are used for troubleshooting.
d) You are now ready to wire the counter.

> Note: The switches for delay[3..0] are not necessary. You can hard wire the inputs to any value from 0 -15.

Step 6 – Wire the Counter

a) Using Figure SP1-5 as your guide, wire all of your components together.
b) Place input and output symbols as shown.
c) Save the project.

Step 7 – Compile and Simulate the Project

a) Compile the project.
b) Create a vector waveform file *freq_ctr.vwf* similar to the one shown in Figure SP1-4.
c) Simulate the project. Since this requires a large amount of your computer's resources, depending on the model, clock speed, and amount of available memory, the simulation time can exceed five minutes or more.

Step 8 – Assign Pins to the Project

a) Access the Assignment Editor, select pins category, and carefully assign pins to the circuit's inputs and outputs. Refer to Appendix D.
b) Choose a pin that is free for I/O to assign to *f_input.* If you choose a pin that is being used by a toggle switch, remember that you must leave its associated DIP switch OFF all of the time the board is powered up.

c) Make sure the power to the eSOC board and signal source is OFF before proceeding.
d) Set the signal source up to output a square wave TTL compatible signal @ 51 Hz.
e) Find the pin's location on the eSOC header and insert the "hot" wire from your TTL compatible signal source.
f) Connect the source's ground to the ground block at the far right end of the board (when looking at the header with the board right-side up.)

Step 9 – Download the Project to the eSOC Board

a) Select the programmer tool and set up the programmer to download your project.
b) Connect the eSOC board to the PC's LPT1 port.
c) Turn all DIP switches to OFF.
d) When the program is downloaded, turn ON only the DIP switches that are necessary
e) Apply power and observe the reading on the two 7-segment LEDs. It should read 51 ±1 Hz. If it does, adjust the signal source frequency between 1-99 Hz. The readout should follow the changes.

Step 10 – Demonstrate your project to your instructor or lab assistant and secure his or her signature below.

```vhdl
LIBRARY IEEE;
USE IEEE.STD_LOGIC_1164.all;
USE IEEE.STD_LOGIC_ARITH.all;
USE IEEE.STD_LOGIC_UNSIGNED.all;

ENTITY clock_divider IS
    PORT
    (
        clock_4Mhz      : IN    STD_LOGIC;
        clock_1MHz      : OUT   STD_LOGIC;
        clock_100KHz    : OUT   STD_LOGIC;
        clock_10KHz     : OUT   STD_LOGIC;
        clock_1KHz      : OUT   STD_LOGIC;
        clock_100Hz     : OUT   STD_LOGIC;
        clock_10Hz      : OUT   STD_LOGIC;
        clock_1Hz       : OUT   STD_LOGIC);
END clock_divider;

ARCHITECTURE a OF clock_divider IS
    SIGNAL  count_1Mhz: STD_LOGIC_VECTOR(4 DOWNTO 0);
    SIGNAL  count_100Khz, count_10Khz, count_1Khz : STD_LOGIC_VECTOR(2 DOWNTO 0);
    SIGNAL  count_100hz, count_10hz, count_1hz : STD_LOGIC_VECTOR(2 DOWNTO 0);
    SIGNAL  clock_1Mhz_int, clock_100Khz_int, clock_10Khz_int, clock_1Khz_int : STD_LOGIC;
    SIGNAL  clock_100hz_int, clock_10Hz_int, clock_1Hz_int : STD_LOGIC;
```

PROCEDURE

```vhdl
BEGIN
PROCESS
BEGIN
        -- Divide by 4
WAIT UNTIL clock_4Mhz'EVENT and clock_4Mhz = '1';
                IF count_1Mhz < 3 THEN
                    count_1Mhz <= count_1Mhz + 1;
                ELSE
                    count_1Mhz <= "00000";
                END IF;
                IF count_1Mhz < 2 THEN
                    clock_1Mhz_int <= '0';
                ELSE
                    clock_1Mhz_int <= '1';
                END IF;
        -- Sync all clock prescalar outputs back to master clock signal
                clock_1Mhz <= clock_1Mhz_int;
                clock_100Khz <= clock_100Khz_int;
                clock_10Khz <= clock_10Khz_int;
                clock_1Khz <= clock_1Khz_int;
                clock_100hz <= clock_100hz_int;
                clock_10hz <= clock_10hz_int;
                clock_1hz <= clock_1hz_int;
END PROCESS;
        -- Divide by 10
PROCESS
BEGIN
WAIT UNTIL clock_1Mhz_int'EVENT and clock_1Mhz_int = '1';
                IF count_100Khz /= 4 THEN
                    count_100Khz <= count_100Khz + 1;
                ELSE
                    count_100khz <= "000";
                    clock_100Khz_int <= NOT clock_100Khz_int;
                END IF;
        END PROCESS;
            -- Divide by 10
PROCESS
BEGIN
WAIT UNTIL clock_100Khz_int'EVENT and clock_100Khz_int = '1';
                IF count_10Khz /= 4 THEN
                    count_10Khz <= count_10Khz + 1;
                ELSE
                    count_10khz <= "000";
                    clock_10Khz_int <= NOT clock_10Khz_int;
                END IF;
        END PROCESS;
```

```vhdl
                    -- Divide by 10
PROCESS
BEGIN
WAIT UNTIL clock_10Khz_int'EVENT and clock_10Khz_int = '1';
            IF count_1Khz /= 4 THEN
                count_1Khz <= count_1Khz + 1;
            ELSE
                count_1khz <= "000";
                clock_1Khz_int <= NOT clock_1Khz_int;
            END IF;
    END PROCESS;

                -- Divide by 10
PROCESS
BEGIN
WAIT UNTIL clock_1Khz_int'EVENT and clock_1Khz_int = '1';
            IF count_100hz /= 4 THEN
                count_100hz <= count_100hz + 1;
            ELSE
                count_100hz <= "000";
                clock_100hz_int <= NOT clock_100hz_int;
            END IF;
    END PROCESS;

                -- Divide by 10
PROCESS
BEGIN
WAIT UNTIL clock_100hz_int'EVENT and clock_100hz_int = '1';
            IF count_10hz /= 4 THEN
                count_10hz <= count_10hz + 1;
            ELSE
                count_10hz <= "000";
                clock_10hz_int <= NOT clock_10hz_int;
            END IF;
    END PROCESS;

                -- Divide by 10
PROCESS
BEGIN
WAIT UNTIL clock_10hz_int'EVENT and clock_10hz_int = '1';
            IF count_1hz /= 4 THEN
                count_1hz <= count_1hz + 1;
            ELSE
                count_1hz <= "000";
                clock_1hz_int <= NOT clock_1hz_int;
            END IF;
    END PROCESS;
END a;
```

Figure SP1–1

PROCEDURE

```vhdl
LIBRARY ieee;
USE ieee.std_logic_1164.all;
entity RETOneShot is              -- Retriggerable Edge-Triggered OS
        port( clock, trigger, reset    : IN BIT;
          delay                        : IN INTEGER RANGE 0 TO 15;
          q                            : OUT BIT
                              );
end RETOneShot;
ARCHITECTURE vhdl OF RETOneShot IS
BEGIN
        PROCESS(Clock, reset)
        VARIABLE count              : INTEGER RANGE 0 TO 15;
        VARIABLE trigger_last       : BIT;
        BEGIN
            IF reset = '0' THEN count := 0;
            ELSIF (clock'EVENT AND clock = '1') THEN
                IF trigger = '1' AND trigger_last = '0' THEN
                    count := delay;            -- load counter
                    trigger_last := '1';       -- store edge
                ELSIF count = 0 THEN count := 0;
                ELSE count := count - 1;
                END IF;
                If trigger = '0' THEN trigger_last := '0';
                END IF;
            END IF;
            IF count /= 0 THEN q <= '1';
            ELSE q <= '0';
            END IF;
        END PROCESS;
END vhdl;
```

Figure SP1–2

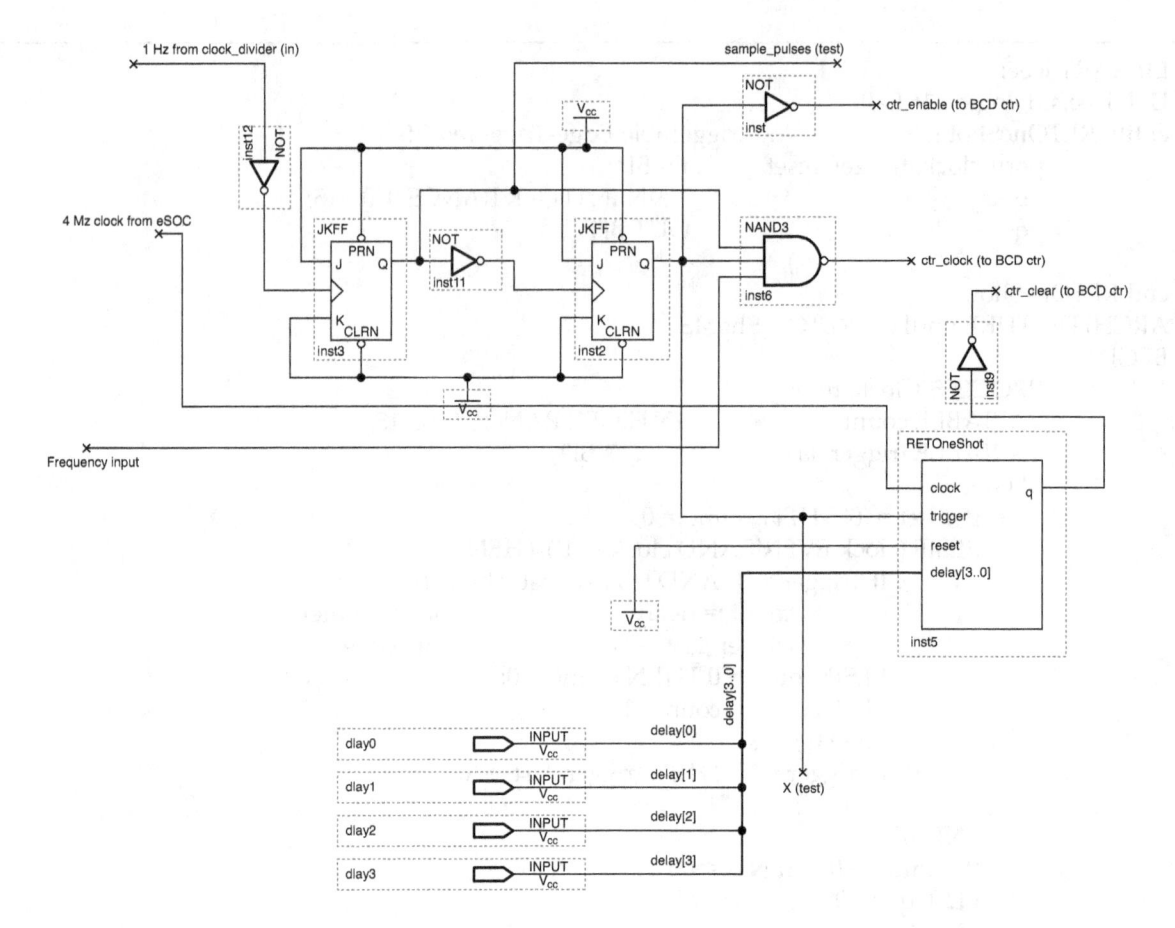

Figure SP1–3

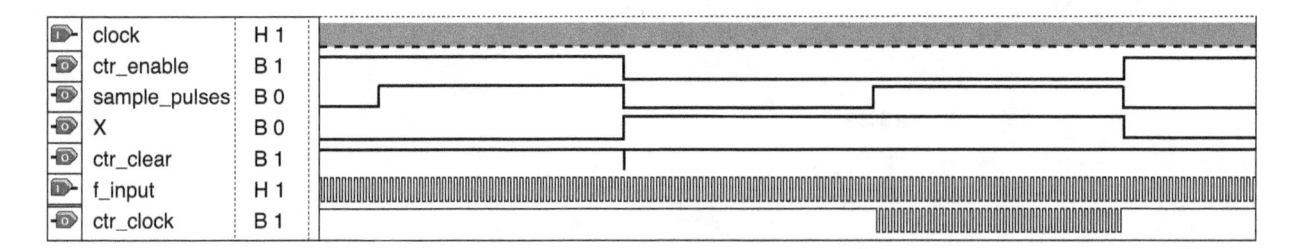

Figure SP1–4

PROCEDURE

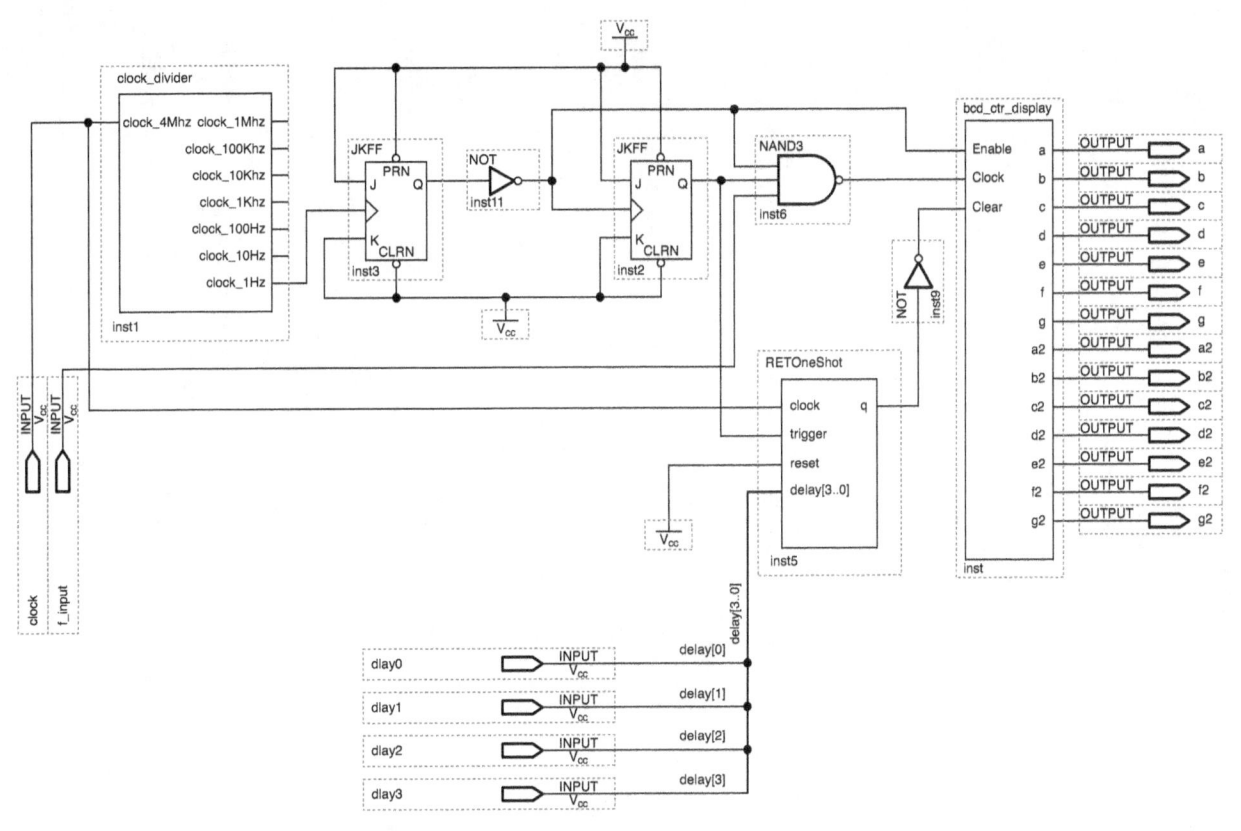

Figure SP1–5

Step 11 – Two Ways of Extending the Project

a) The frequency counter can be extended to count frequencies up to 9999 by adding two more BCD counter/Display units.
b) The clock divider outputs can be multiplexed to provide smaller sample intervals and therefore provide a way to measure higher frequencies.

Appendix A

WIRING AND TROUBLESHOOTING DIGITAL CIRCUITS

OBJECTIVES

1. To discuss general wiring procedures for digital circuits.
2. To introduce the student to formalized troubleshooting procedures.
3. To list some of the common faults found in digital systems.

DISCUSSION

The experiments in this manual are designed to give you hands-on experience with digital circuits. More than that, they provide you with an opportunity to develop sound breadboarding and troubleshooting skills that will be invaluable to you whether you eventually become an engineer or a technician. This appendix will present some very basic information and suggestions concerning each area. It is not meant to replace any laboratory standards. However, much of the information given here can be used as reference material that can be, and should be, reviewed from time to time.

Most of the experiments contain operational testing of various ICs. At times, this may appear to be a tedious undertaking on your part. Don't fall into the trap of treating this sort of experimentation mechanically, taking for granted that an IC will operate just as it did in the classroom lecture. In the classroom, you are working with the ideal. In the lab, you will occasionally work with ICs that are less than ideal. In fact, they may not work at all, or at least not in the manner they were designed to work. If you keep in mind that lab experimentation is not only to verify principles but also to learn to recognize common problems associated with the circuits, you will get more out of the experiments. As you will learn, verification of a circuit's operation is one of the first steps taken in troubleshooting.

A.1 Prototype Circuit Wiring

It is assumed that you will be wiring circuits using a prototype circuit board. Such boards come in different sizes, but most have the following features:

a) Two horizontal rows of holes, one at the top and one at the bottom. The contacts underneath the holes on each of these rows are connected together to form a bus. They are not directly connected to the other holes on the board.

b) At least two sections of holes, with each section arranged so that the holes are in vertical groups called circuit blocks. Each circuit block is isolated from all others. This permits several wires to be joined at common junctions. The two sections are separated by a horizontal gap. This gap separates the sections electrically as well as physically. Thus, a vertical circuit block in the top section of the board is not connected to the block directly below it in the bottom section. ICs will straddle this gap so that each IC pin will be inserted into its own block. Connections to each pin will be brought to its block.

Installing ICs: ICs should be installed or mounted on the board to permit wires going from the top section of the board to the bottom to go between the ICs. It is not advisable to pass wires over ICs, although sometimes it is hard to avoid. Strapping ICs to the board in this manner will present problems if the IC has to be removed. The consequences of this are obvious.

As you mount an IC, check to make sure that none of its pins are being tucked beneath it. If it is necessary to remove an IC, always use an IC puller. *Never* remove an IC with your fingers or with a pair of pliers. The first causes a definite safety hazard, while the second will often result in eventual damage to the IC.

Wiring the circuit: Wires should be dressed so that 3/8" insulation is stripped from each end and the length of wire is no more than needed to make a neat connection between circuit blocks. If the wires are too long, some circuits will malfunction, especially flip-flops and flip-flop devices such as counters. You may have to rearrange the ICs on the board to solve this problem, if it occurs. Another way to solve the problem is by inserting a 2 k-ohm resistor in series with the wire at the input end of the wire.

Have a lab partner call out each connection to be made. Route the wires along the circuit board neatly, bending them smoothly wherever necessary. Avoid bending the wire sharply, since this will increase the likelihood of fracture beneath the insulation, resulting in an open circuit or an intermittent open. Minimize the number of crossovers, that is, wires routed over other wires.

The overall appearance should be neat, not like a bowl of spaghetti. If you have made all of your connections as outlined above, it may not be picture perfect, but the neatness will pay off in reduced troubleshooting time and easier IC replacement.

A.2 Testing the Circuit

Circuit testing is also known as troubleshooting. You are probably accustomed to discrete circuit (e.g., a transistor amplifier) troubleshooting methods. Since each circuit element of a discrete circuit is accessible to the troubleshooter, faulty circuit elements can be isolated by making basic measurements such as voltage, resistance, capacitance, and inductance, using conventional test equipment. Modern digital circuits and systems, on the other hand, consist mainly of digital ICs. The ICs components are not accessible to the troubleshooter, so the troubleshooter must rely on knowledge of the ICs operation(s) in order to isolate the IC as being faulty. The experiments in this manual are designed to give you the necessary experience to test for and recognize proper operation of ICs.

A digital IC is considered defective or faulty if its outputs do not respond correctly, according to its truth table, for each set of input conditions and for each of its various operating modes. A similar statement can be made for digital circuits

Appendix A

and systems. Once it has been verified that a circuit or system is not responding correctly, a *fault is said to exist*, and further troubleshooting is indicated. The next troubleshooting step is to *isolate the cause of the fault*, which may be in one or more smaller circuits or subsystems. By progressively isolating smaller circuits, and perhaps smaller subsystems, the troubleshooter will eventually isolate the defective components, which may be one or more ICs and/or discrete components. After *replacing the defective components*, the circuit or system is tested for proper operation once more. Once proper operation is established for all operating modes, the troubleshooter's task is completed.

Now that the student is acquainted with the nature of digital troubleshooting, a procedure for fault isolation is presented. The student should, when applying the procedure in the lab, perform each step in the order given. After sufficient experience with digital circuits is attained, common sense and intuition may lead the student directly to the faulty device and thereby reduce the amount of troubleshooting time.

Step 1: Perform a visual inspection of the system or circuit. Look for loose or damaged connecting wires, cables, and printed circuit (PC) boards, evidence of burning or extreme overheating, missing components, and blown fuses. If the circuit or system is mounted on prototype boards, look for wiring errors, damaged boards, and digital ICs improperly inserted. Also check for incorrect circuit design.

Step 2: Check all power source levels, and confirm that power is actually being applied to the circuit or system.

Step 3: Study all relevant documentation on the circuit or system, such as block diagrams, schematics, and operating instructions. Learn how the circuit or system operates normally.

Step 4: Verify all operating modes of the circuit by running tests.

Step 5: Record results of the tests run in step 4. Test results often show patterns that may lead to the faulty device. Repeat steps 4 and 5 at least once before proceeding to step 6.

Step 6: If the circuit passes all tests, end the procedure. If the circuit fails at least one test, continue to the next step.

Step 7: Analyze the test results recorded above and select a possible location for the fault.

Step 8: Check all signals and static logic levels at this location, and record them. If nothing appears abnormal, return to step 7.

Step 9: Analyze the test results recorded above, and select a possible faulty device.

Step 10: Check the device for proper functioning. If it is a discrete component, take basic Ohm's Law measurements and/or use a device tester to determine if the device is faulty. If the device is an IC, check the IC for proper functioning. This includes checking inputs and outputs for stuck-HIGH and stuck-LOW conditions and other types of digital IC faults (see *Common Digital IC Faults*). If the device passes all tests, then return to step 9.

Step 11: Repair or replace the faulty device and return to step 4.

A.3 Common Causes of Faults in Digital Systems

In this section, several common causes of faults in digital systems are listed along with symptoms given for each cause and steps that may be taken to correct or minimize its effects on the system.

Defective components: Components normally fail because of age, because the maximum voltage or current rating of the device was exceeded due to improper design or because of the breakdown of another component, improper connections, or excessive ambient temperature. In the case of digital ICs, overheating caused by improper connections (especially prototype circuits), overvoltage, or ambient temperature may result in the IC operating only sporadically. After cooling down, the IC will usually operate normally.

IC loading problems: Exceeding the fan-out of a TTL logic output may result in the output voltage dropping below $V_{OH}(min)$ or rising above $V_{OL}(max)$. To verify this condition, the output voltage should be checked for a level of 0 V–0.8 V for a LOW and 2 V–5 V for a HIGH. If not, the excessive fan-out is causing a problem.

CMOS and MOS logic outputs will not be affected significantly by exceeding the fan-out limits. However, any transitions at the output will show an increase in risetime and falltime. This is because each CMOS or MOS input loads the output capacitively (5 pf each input).

Some common symptoms to look for are flip-flops, counters, and flip-flop registers that do not respond to the signal at the clock input. Measure t_R and t_F of the clock signal, and compare the measurements to the minimum required by the flip-flop, counter, or register for proper triggering.

To correct the problem caused by excessive fan-out, a buffer should be used or the fan-out reduced by load splitting. Another solution is to insert a pulse-shaping circuit such as a Schmitt trigger between the overloaded output and the clock input.

Improper signal characteristics: A digital IC may function improperly if logic signals not meeting its requirements are applied to its inputs. Minimum requirements are given for amplitude, pulse duration, and transition times. A signal that fails to meet any one of these requirements can cause the IC to function incorrectly.

Common symptoms brought on by improper signal characteristics include flip-flops, counters, and flip-flop registers that respond incorrectly to signals at clock, clear, and preset inputs.

The characteristic(s) causing the problem must be determined and brought back into specification.

Power supply—Improper levels: Since all IC logic devices use voltage to represent logic levels, trouble with the output level of the supply can cause ICs to function improperly. A common cause of improper power supply levels is overload. This can be particularly true in prototype systems and circuits.

Symptoms of this type of trouble include the condition where logic HIGH at circuit outputs is less than V_{OH}. Disconnecting a few ICs from the power supply will usually cause the level of V_{CC} to rise if this is the case.

Using a larger power supply or redesigning the existing one for higher current output will solve the problem.

Power supply—Poor regulation: Poor regulation in a power supply will cause V_{CC} to fluctuate when large numbers of logic circuits are switching states. These fluctuations act like noise pulses and can cause false triggering of logic devices. This problem is especially significant in TTL circuits.

A symptom caused by this problem is flip-flops, counters, and registers triggering when they are not supposed to, and triggering instead at the time other devices in the system are changing states. To verify that poor regulation is the problem, V_{CC} should be examined with an oscilloscope. If spikes or pulses are riding on the V_{CC} level causing V_{CC} to drop by more than 0.2 V, then the power supply has poor regulation.

There are two ways to correct this problem: (1) improve the power supply regulation by either replacing or redesigning the current one, or (2) use RF decoupling capacitors (refer to text, Chapter 8).

Appendix A

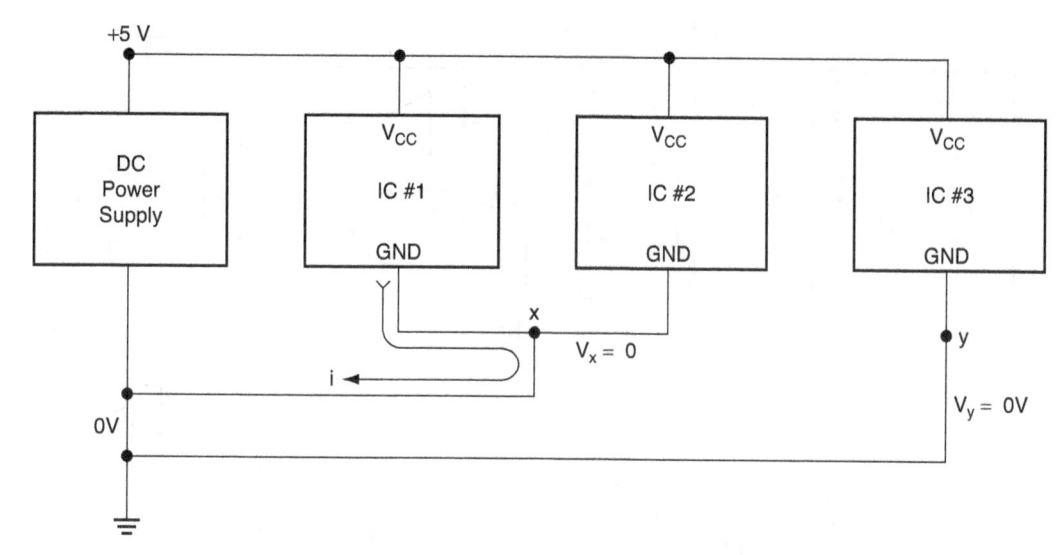

Figure A-1

Grounding problems: Poorly designed ground return circuits can cause the voltage at IC ground pins to be nonzero. This is because currents flowing through the ground system can cause resistive and inductive voltage drops (see Figure A-1). To avoid this problem, all ground wires should have low resistance and inductance, and each IC ground pin should be connected to the power supply separately. PC board ground returns should be large conductive traces.

Noise problems: Circuit noise can be externally or internally generated. Internally generated noise was discussed earlier. Externally generated noise can cause sporadic triggering of logic circuits. Common sources are electromechanical devices (e.g., motors and relays that produce electromagnetic radiation) and electronic power control circuitry using SCRs and TRIACs. This type of problem can be minimized by using special AC power line filtering devices to prevent noise from entering through the AC lines and grounded shields or conducting planes to short radiated noise signals to ground.

A.4 Common Digital IC Faults

Digital IC faults are classified as either internal or external faults. We begin our discussion with internal faults.

Internal digital IC faults: There are four types of internal failures:

1) inputs or outputs shorted to ground or V_{CC}
2) inputs or outputs open
3) shorts between pins (not to ground or V_{CC})
4) internal circuitry failure

These failures are corrected by replacing the faulty IC. A discussion on each type of failure follows.

Short to ground or V_{CC}: This failure causes the inputs or outputs to be either permanently HIGH or permanently LOW (referred to as stuck-HIGH or stuck-LOW). Figure A-2(a) shows a NAND gate with a stuck-LOW input and a stuck-HIGH output. The stuck-HIGH condition may be the result of an internal short in input A, an internal short at output X, or both.

Figure A-2

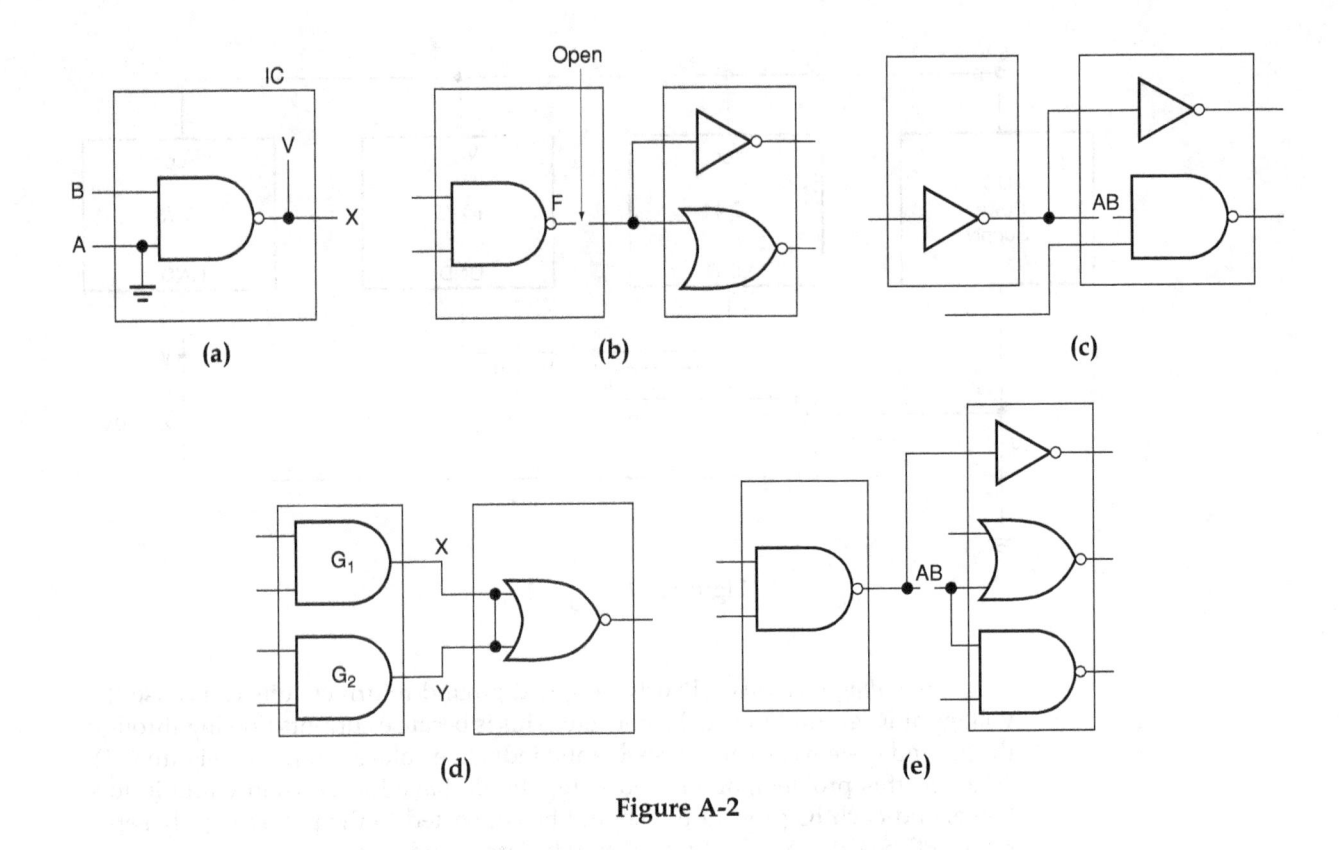

Connections to output X are also forced HIGH; connections to input A are forced LOW. Shorts of this type in emitter-coupled logic (ECL) devices result in neither a HIGH nor a LOW.

In troubleshooting this type of failure, the student should be aware that signals may not change beyond the point where the short is located.

Open inputs or outputs: An open output will result in an open input for all inputs driven by the output. Open inputs in TTL logic devices generally act as HIGHs, causing inputs that are tied to open outputs to resemble a stuck-HIGH input, though not always. Open CMOS inputs do not generally act as a HIGH or a LOW. This being the case, inputs tied to an open TTL or CMOS output will resemble a stuck-LOW or stuck-HIGH input or may even oscillate between HIGH and LOW. Open inputs for ECL devices, with inputs pulled down by a resistor, are LOW.

Figure A-2(b) illustrates an open output in a NAND gate. Figure A-2(c) shows an open input. In the latter diagram, the student should note that all signals before point A are unaffected.

Short between two pins: Figure A-2(d) shows two input pins shorted together. This means that the outputs of the two driver gates are also shorted. This condition will cause a fault in TTL and CMOS devices only if the two driver outputs try to go to opposite levels, say, X to HIGH and Y to LOW. In this case, if the device is TTL, X will be stuck-LOW. In other words, if one output is LOW, both will be LOW. However, if the device is CMOS, this condition typically produces an intermediate level (see Figure A-3). There is obviously no fault that occurs when both outputs are supposed to be at the same level. ECL device outputs can normally be connected together, so no logic faults will occur unless the driver gates are damaged by excessive currents.

Appendix A

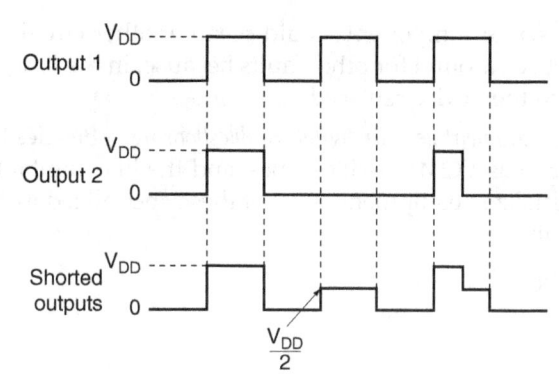

Figure A-3

Internal circuitry failure: Failure in the circuits within a digital IC can cause its inputs and outputs to be stuck-HIGH or stuck-LOW.

External digital IC faults: In addition to the four types of internal failure, there are four types of external failures that can occur:

1) line shorted to ground or power supply
2) open signal line
3) short between signal lines
4) failure of a discrete component

To discover these faults, look for poor soldering joints, solder bridges, open wires or traces, or test components such as capacitors and resistors for opens, shorts, and/or values that are out of tolerance.

Line shorted to ground or power supply: This type of failure will appear like an internal short and can't be distinguished from it. Perform a careful visual inspection to isolate this fault.

Open signal line: Figure A-2(e) shows an open signal line that results in an open input only for points beyond point B. All inputs before A are unaffected by the open line. Signal tracing and/or continuity checks are useful techniques for discovering this type of fault.

Short between two signal lines: This type of fault cannot be distinguished from an internal short. Often, poor soldering on PC boards results in solder bridges across the signal lines. On prototype boards, look for bare connecting wires (poorly dressed) too close together. In either case, a visual inspection is necessary to locate this fault.

Shorted signal lines in TTL will appear different from shorts in CMOS circuits. In TTL, if one signal is trying to go HIGH and the other is going LOW, the level at the short will be about 0 V. This is because resistance at TTL outputs is lower in the LOW state than in the HIGH state. For CMOS and MOS devices, the level at the short will be about midway between 0 V and 5 V for this same situation, because their output resistance is about the same in both states. See Figure A-3 for an example of how waveforms would look for shorted signal lines in CMOS and MOS circuits. Note the 2.5 V levels. These levels would not normally appear on the waveforms.

Failure of discrete components: While most digital components are ICs, there is still circuitry that requires discrete components such as resistors, capacitors, transistors, and diodes. These components can be tested either completely out of the circuit or by unsoldering one or more of their leads and checking them with an appropriate test instrument such as an ohmmeter, capacitance checker, or transistor checker. Faulty discrete components could mean another circuit caused the failure. Be sure and check around for other faults because, in the long run, this avoids repeated failures in the device replaced.

Common test equipment used in digital troubleshooting: Besides the usual analog test equipment, such as VOMs, oscilloscopes, and the like, digital troubleshooting requires some specialized equipment. A list of these specialized instruments would include the following:

1) logic probe
2) logic pulser
3) current tracer
4) logic analyzer

Of the four, the logic probe is the most useful in general troubleshooting. The pulser is useful when it is necessary to trigger gates, flip-flops, counters, or other types of circuits to check for proper operation. Both the logic probe and pulser are described in Experiment 3.

The current tracer is a more specialized test probe used in locating shorts in digital circuits. Whenever a short circuit is suspected, the current tracer can assist the troubleshooter in pinpointing the exact location of the short.

The logic analyzer is a complex instrument used to compare many different logic signals at one time. However, it is expensive and is used mostly in complex systems to solve the more difficult problems that occur in digital systems. The operation of a typical logic analyzer, the Tektronix Model 7D01, is discussed in Appendix B of this manual.

The experiments in this manual provide opportunities to gain experience with the logic probe, logic pulser, and the logic analyzer. It is recommended that you become acquainted with the logic probe you will be using by reading the user manual that should accompany the probe. If your laboratory has a logic analyzer, it would also be to your advantage to learn as much as possible about the instrument before you attempt to use it. If there is an operator's manual for the analyzer, get it and read it.

CONCLUDING REMARKS

The material in this appendix will be of more use to you if you review it from time to time. There is too much information relating to troubleshooting to include all of it in a short appendix. Your learning resource center or library may have some videotapes, journals, or books on the subject.

Appendix B

LOGIC ANALYZERS

A logic analyzer is a device that can store and display several channels of digital data. Although it resembles a multitrace oscilloscope, the logic analyzer differs from the oscilloscope in several ways:

- The logic analyzer displays only digital data; it cannot display both analog and digital signals like the oscilloscope can.
- The logic analyzer stores its data in memory first, then causes this data to be continually displayed on its CRT screen. Thus, the data that the analyzer displays has already occurred and is not real-time like that displayed by the oscilloscope.
- The logic analyzer is capable of displaying several channels of data simultaneously. A multitrace oscilloscope either chops or alternates between traces (channels) and thus cannot display the channels simultaneously.
- The logic analyzer can display its data in tabular format, using binary, octal, or hexadecimal characters, as well as in timing diagram format. The oscilloscope uses the timing diagram format only.

PART 1—LOGIC ANALYZER FUNDAMENTALS

In this section, we will describe the principles of operation of a typical logic analyzer. The model to be described is the Tektronix 7D01/7D01F. You should consult the operator's manual for your particular logic analyzer and/or ask your instructor for assistance in applying the information covered here to your analyzer.

B.1 Basic Block Diagram

Figure B-1 shows a block diagram for a simple logic analyzer. The probes are connected to a bus that carries the signals you wish to examine. The examination begins when the control block causes the trigger circuitry to issue a write enable to memory. The data on the bus is then clocked into memory, where it is stored. When enough data is stored, the write enable to memory is removed and replaced with a read enable. This permits the data that was stored to be read and displayed on the multitrace oscilloscope.

The logic analyzer described here could be modified to display tabular data. What is needed is a computer to control the probes and display. The probes would be connected to the computer via a peripheral interface adapter (PIA), and the computer would be programmed to clock in the data, store it, and then display the data in the format desired.

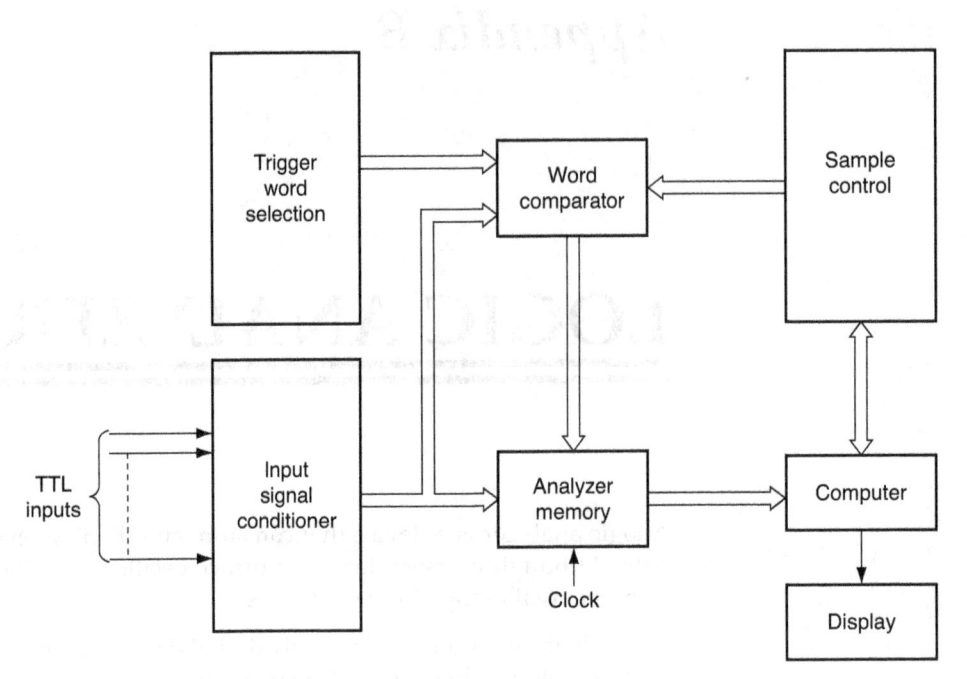

Figure B-1

B.2 Operating Principles

At the heart of the logic analyzer is its capability to store data in its memory. The memory is organized as serial shift registers, one per channel. Each channel's register size depends on the number of channels being used. For example, the Tektronix 7D01 has three possibilities:

No. data channels	No. bits/channel
4 (Ch. 0–3)	1024
8 (Ch. 0–7)	508
16 (Ch. 0–15)	254

In each case, the memory capacity is 4064 bits (about 4K).

Figure B-2 shows the basic structure for a single data channel. This structure is the same for all of the other channels (up to 16 channels). The input signal that is to be examined is connected to the probe for that channel and delivered to a comparator. Regardless of whether the signal is digital or analog, the comparator compares the input to a reference (threshold) voltage, converting the input signal to a pure digital signal. For TTL logic, the threshold voltage is set to +1.4 volts.

Figures B-3 and B-4 show two cases of input signals. In Figure B-3, the input signal is digital, and the output of the comparator is the same as its input. Figure B-4 shows that in the case of a nondigital input, the comparator output will be a digitized version of the input.

Appendix B

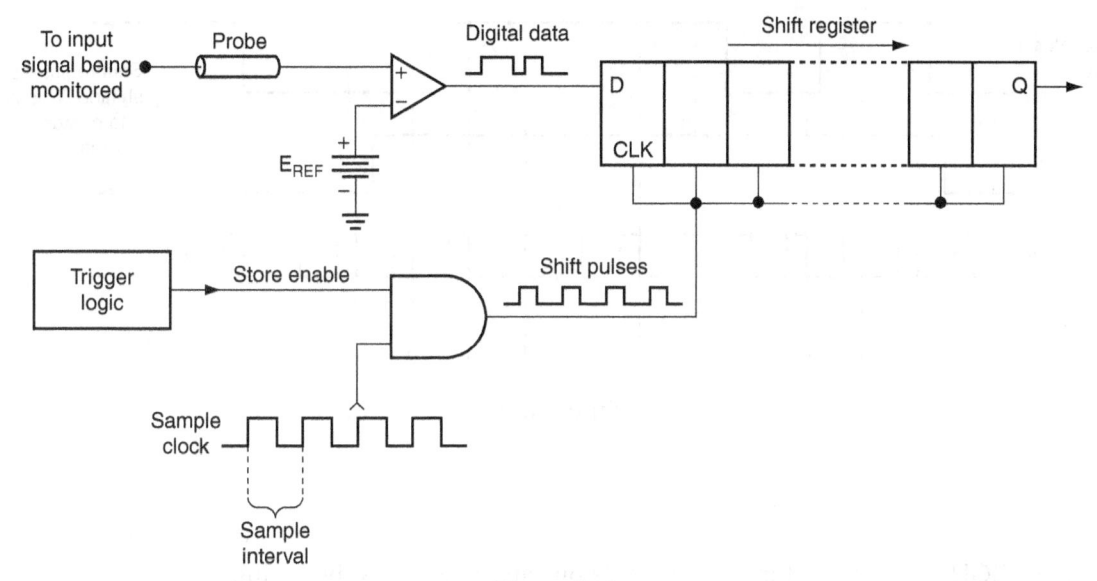

Figure B-2

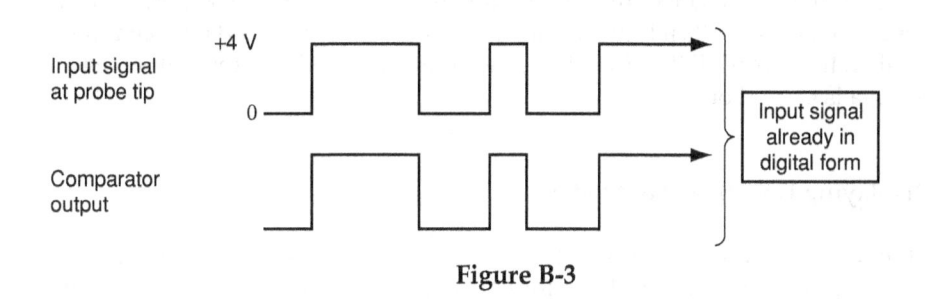

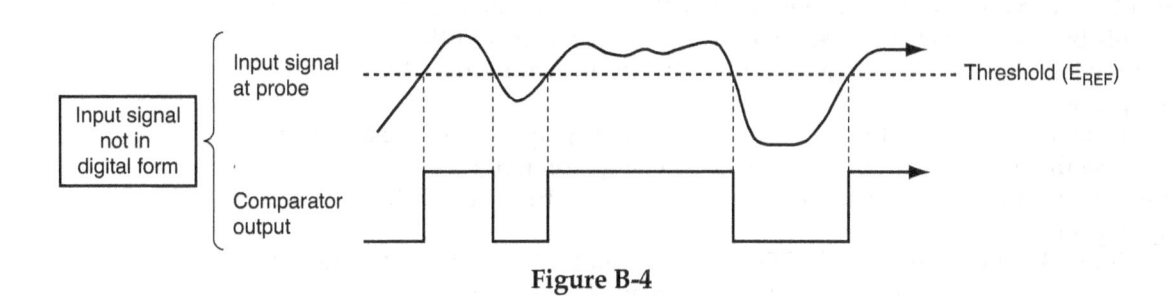

Figure B-3

Figure B-4

B.3 Storing Data (Store Mode)

After the comparator converts the input data to digital, the data is sent to the input of a shift register, where it is clocked into the shift register by the SAMPLE CLOCK for as long as the STORE ENABLE is HIGH. The shift register stores the data until it is needed for display. Figure B-5 shows the timing relationship between the output of the comparator, the STORE ENABLE signal, SAMPLE CLOCK, and the data shifted into the shift register. Note that while the STORE

Figure B-5

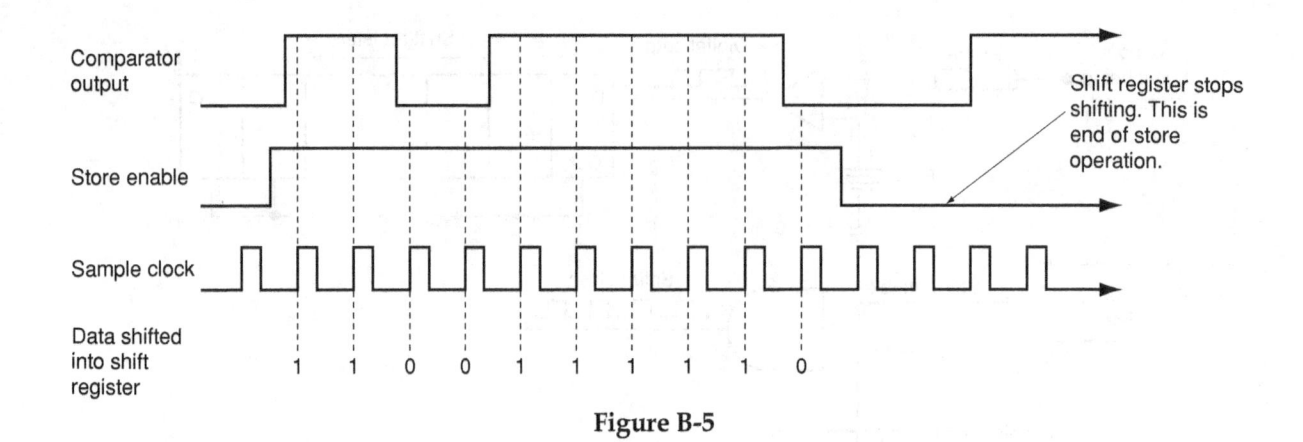

ENABLE is HIGH, the level of the comparator's output at each positive-going transition of the SAMPLE CLOCK is shifted into the shift register from left to right, and that when the STORE ENABLE goes LOW, the shifting stops. Even though the data at the comparator output is changing, the data currently in the shift register does not change. In other words, the SAMPLE CLOCK periodically samples the data and shifts it into the register for storage. The period between samples is called the SAMPLE INTERVAL. As new data is shifted into the shift register, the old data is shifted out.

B.4 Displaying Data (Display Mode)

After the logic analyzer has completed the STORE operation, its memory contains a digital representation of the input signals that are connected to its inputs. As soon as the STORE ENABLE goes LOW, the logic analyzer goes into its DISPLAY mode of operation. In this mode, the logic analyzer displays the contents of each channel's shift register on the CRT as a serial digital waveform. It does this by applying clock pulses to each shift register to shift the stored data out of the last flip-flop of the register and to the vertical amplifier of the CRT to be displayed.

As mentioned earlier, the 7D01 can be used to display 4, 8, or 16 channels. It is possible that the actual number of channels being monitored is fewer than the number of channels being displayed. If this is the case, the unused channels are simply ignored.

Figure B-6 illustrates how the 7D01 can be used to display four channels of stored data on the CRT screen. Note that the channels are numbered from 0 to 3 starting at the top of the display. Each waveform represents the complete contents of the channel memory register. For the 7D01, this means that each waveform represents the 1016 bits that were shifted into the register during the STORE mode, one bit per SAMPLE INTERVAL.

If the logical analyzer were to be used in the 8-channel display mode, there would be eight channels, numbered 0 through 7, displayed on the screen. Each channel waveform would represent 508 bits of stored data on the screen. In the 16-channel display mode, there would be 16 waveforms displayed on the screen, each one representing 254 bits of data. Thus, as mentioned earlier, the total number of displayed bits is constant at about 4K, but the number per channel depends on how many channels are used.

Appendix B 601

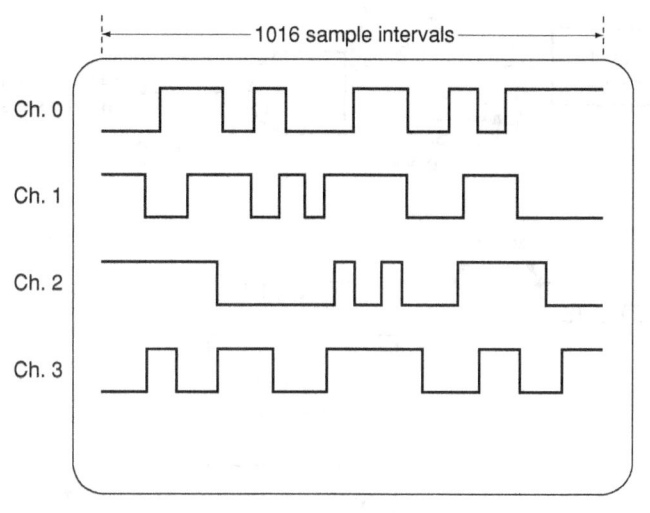

Figure B-6

EXAMPLE 1:
The data displayed in Figure B-6 was obtained using a SAMPLE INTERVAL of 5 microseconds during the STORE operation. What is the total time duration of each waveform?

SOLUTION:
Each waveform consists of 1016 bits, which were shifted into the memory register at the rate of one every 5 microseconds. Thus, the waveform duration is 1016 × 5 microseconds = 5080 microseconds.

EXAMPLE 2:
Repeat Example 1 for a 16-channel display.

SOLUTION:
For a 16-channel display, each waveform consists of 254 sample intervals or 254 × 5 microseconds = 1270 microseconds.

B.5 Types of Sampling

The 7D01 logic analyzer permits input signals to be sampled either *synchronously*, using the clock signal from the system under test, or *asynchronously*, using the 7D01's internal clock (refer to Figure B-7).

B.5.A Synchronous

Most digital systems, and all digital computers, operate from a master clock signal that synchronizes all of the system operations. The logic states of all the system signals can change only on the appropriate clock edge. The logic analyzer can use synchronous sampling to examine the system's data signals at the precise time that system clock edges occur. Figure B-8 illustrates the use of the negative-going edge of the system clock. The 7D01 permits either clock edge to be selected. Note that the stored data represents data on the input signal line when the negative clock edges occur. Any glitches that occur in the intervals between clock edges are ignored.

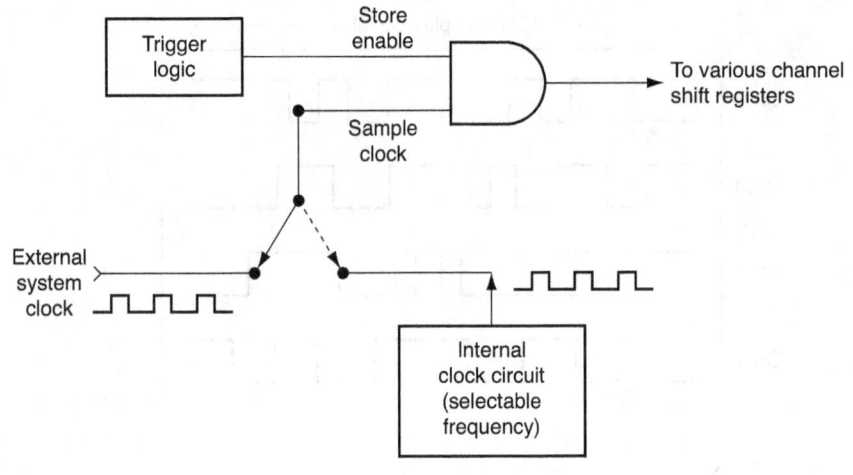

Figure B-7

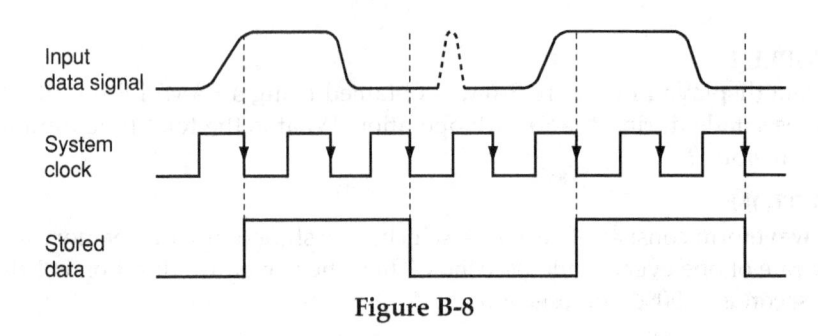

Figure B-8

B.5.B Asynchronous

Asynchronous sampling uses the 7D01's internal clock, which allows selectable sampling intervals from 10 nanoseconds to 5 milliseconds. The smaller sampling intervals permit you to acquire more information about the data signals over a shorter time duration. In other words, for a given number of samples, the use of a shorter sampling interval provides a more detailed look at the data signals. Figure B-9 shows the data signal used in Figure B-8 being sampled at a much higher rate (shorter interval between samples) using the 7D01's internal (asynchronous) clock.

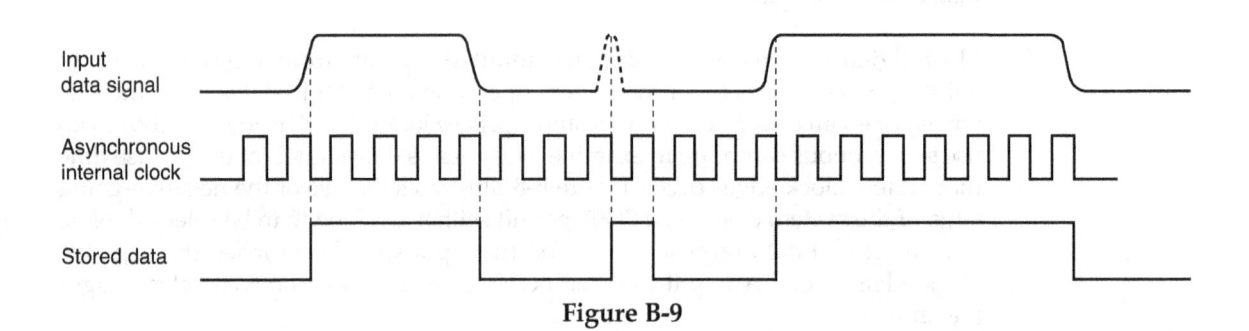

Figure B-9

Appendix B

Note that the shorter sample interval will store a more accurate picture of the input data signal, including glitches or erroneous pulses that occur in the intervals between system clock edges.

In troubleshooting, the synchronous sampling mode is often used to locate the general problem area; then, the asynchronous mode is used to take a closer look at the suspected signals for detection of glitches or erroneous pulses.

B.6 Triggering

In our previous discussion, we stated that the TRIGGER LOGIC circuit (see Figure B-2) generates the STORE ENABLE to control the store and display operations. If the STORE ENABLE is HIGH, the input data is sampled and stored; if the STORE ENABLE is LOW, the stored data is displayed on the CRT.

B.6.A Data Storage Relative to Trigger Event

The TRIGGER LOGIC controls STORE ENABLE in accordance with the occurrence of a selected TRIGGER EVENT. (We will describe the different TRIGGER EVENTS later.) When the TRIGGER EVENT occurs, the TRIGGER LOGIC decides when to make the STORE ENABLE go LOW to end the STORE operation and begin the DISPLAY operation. The 7D01 permits three different possibilities:

1) *Pre-trigger:* In this mode, STORE ENABLE will go LOW right after the TRIGGER EVENT occurs. Thus, the data that is stored in the different channel shift registers represents data that occurred prior to the TRIGGER EVENT. When the data is displayed on the CRT, the display will show those portions of the input waveforms that occurred before the TRIGGER EVENT.

Figure B-10 illustrates how the pre-trigger display appears for the 7D01 logic analyzer. The intensified dots on the display show when the TRIGGER EVENT occurred relative to the various channel waveforms. The greater part of the waveforms occurs prior to the TRIGGER EVENT. The 7D01 also displays a small portion of the waveforms that occur after the TRIGGER EVENT.

Intensified dots show when trigger event occurred

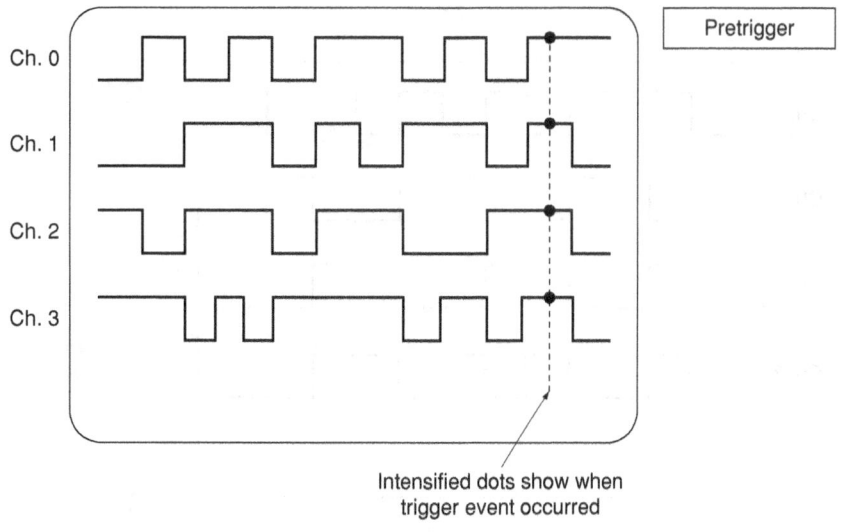

Figure B-10

2) *Post-trigger:* In this mode, the logic analyzer stores and displays data that occurs after the TRIGGER EVENT. Figure B-11 shows how the post-trigger appears for the 7D01 and shows that the greater part of the displayed waveforms occurs after the TRIGGER EVENT. The 7D01 also displays a small portion of the waveforms that occur before the TRIGGER EVENT.

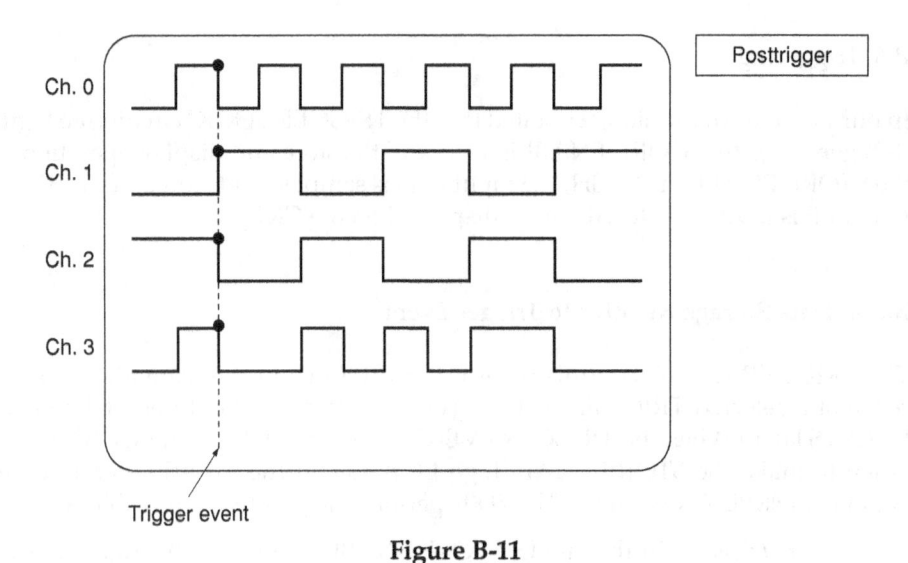

Figure B-11

3) *Center-trigger:* In this mode, the logic analyzer stores and displays data that occur both before and after the TRIGGER EVENT. This is illustrated in Figure B-12, which shows that the TRIGGER EVENT, indicated by the intensified dots, is in the center of the displayed waveforms.

Each of these three trigger modes is useful in different situations, depending on what portion of the waveforms you are interested in examining relative to a particular TRIGGER EVENT. We will now describe the possible TRIGGER EVENT sources that are used in the 7D01.

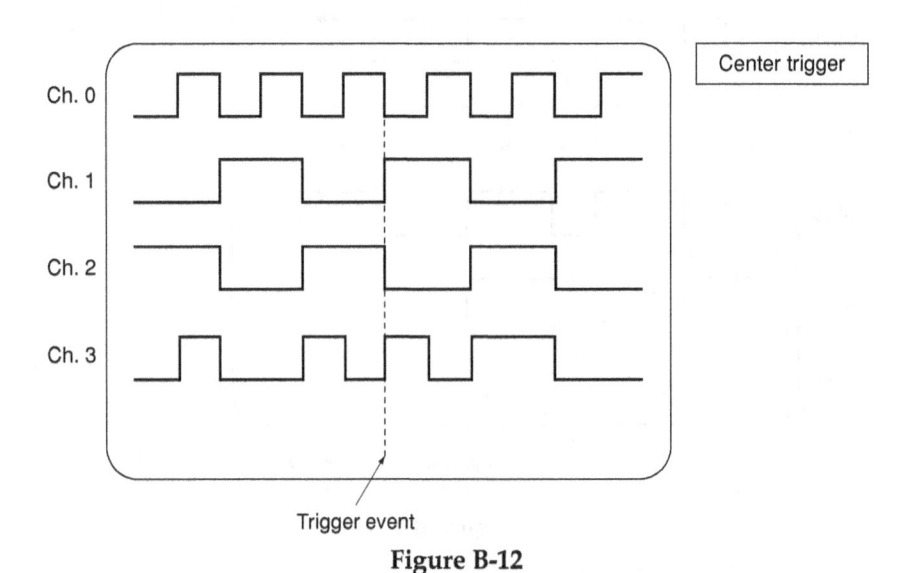

Figure B-12

B.6.B Sources of Trigger Events

The 7D01 allows you to select one of four different sources to produce the TRIGGER EVENT. They are described below:

1) *Manual.* The TRIGGER EVENT occurs when the operator presses the manual trigger button. This is used for the initial setting-up and positioning of the display or for getting a display of stored data when the expected TRIGGER EVENT has failed to occur.

2) *Channel 0 (Ch. 0).* The TRIGGER EVENT is the positive-going transition of the signal on Channel 0.

3) *External.* The TRIGGER EVENT occurs on the selected transition of the signal applied to the external trigger input. This signal is not one of the signals being displayed.

4) *Word Recognizer (WR).* When the WR is selected as the source of the TRIGGER EVENT, the WR unit generates a TRIGGER EVENT pulse when the logic levels present at the 16-channel inputs match a specific 16-bit word. The operator can select any 16-bit word using the WR channel switches on the 7D01 front panel.

B.7 Examining and Interpreting the Display

The logic analyzer has no time scale for measuring time intervals like that of an oscilloscope. Time is measured in units of one SAMPLE INTERVAL and is referenced to the TRIGGER EVENT. The examples in Figures B-13–B-15 show how this is done for the 7D01 4-channel display; the same idea may be extended to the 8- and 16-channel displays.

In Figure B-13, the post-trigger mode is used. The intensified dots on the left indicate when the TRIGGER EVENT occurred. The other set of dots is called the CURSOR and can be positioned anywhere along the waveforms using the CURSOR controls on the 7D01 front panel. The position of the CURSOR relative to the TRIGGER EVENT is displayed on the upper right-hand portion of the CRT screen. Here it is given as TRIG +45, which means that the CURSOR is 45 SAMPLE

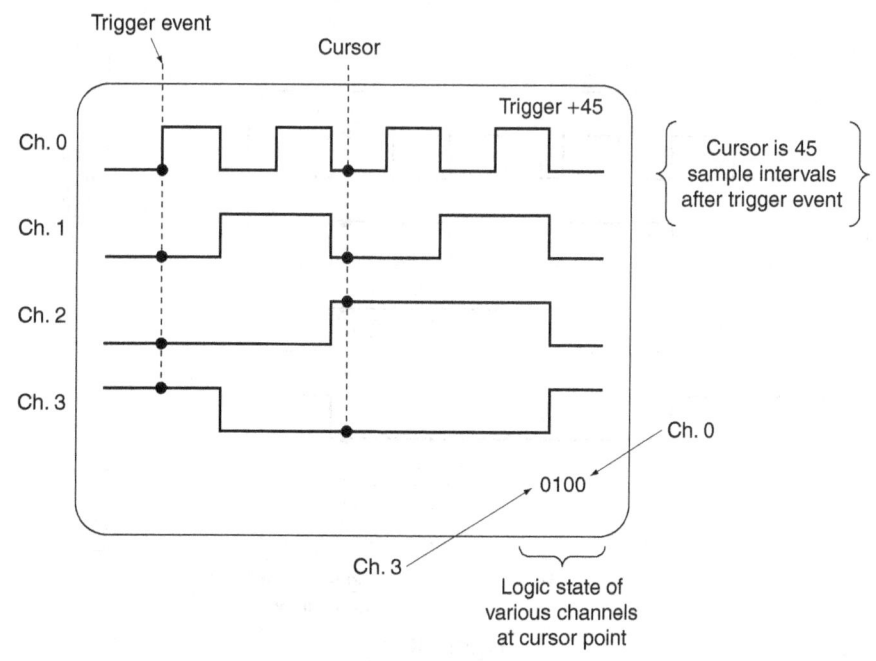

Figure B-13

INTERVALS after the TRIGGER EVENT. The data levels that are present on the various channel waveforms at the CURSOR point are displayed on the bottom of the CRT screen, with Channel 0 being the rightmost bit. Here it is given as 0100, which can be verified by looking at the waveform levels at the CURSOR position.

Figure B-14 illustrates the pre-trigger mode. Here the CURSOR is at −30 SAMPLE INTERVALS before the TRIGGER. The data levels at the CURSOR point are given as 0101.

Figure B-14

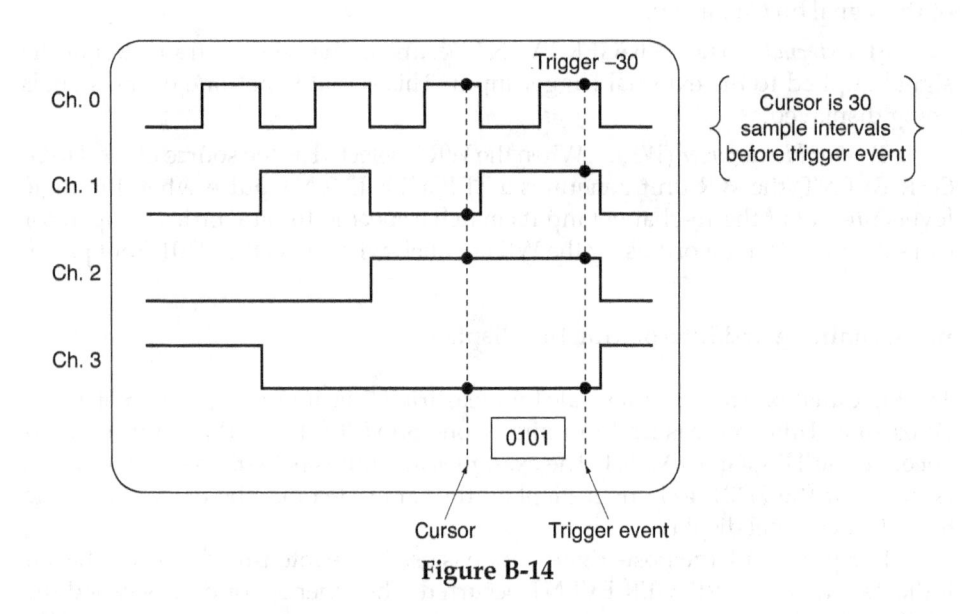

Figure B-15 shows the CURSOR positioned right at the TRIGGER point, indicated by TRIG +0. The data levels at this point are 0111.

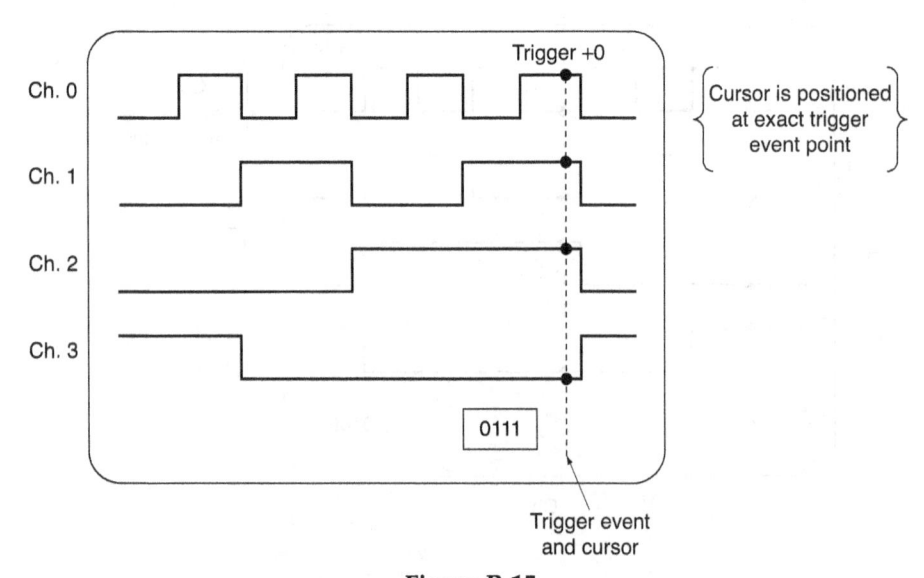

Figure B-15

Appendix B

EXAMPLE:
The waveforms in Figure B-15 were obtained using a SAMPLE INTERVAL of 50 nanoseconds. What is the time duration between the CURSOR and the TRIGGER EVENT?

SOLUTION:
The cursor is 30 SAMPLE INTERVALS before the TRIGGER or 30×50 n sec = 1500 n sec.

PART 2—USING THE STATE TABLE MODE

In Part 1, you saw how the 7D01 logic analyzer was used to sample, store, and display the logic levels present on up to 16 different signals simultaneously. The mode of display used was the TIMING DIAGRAM mode. Although this mode is very useful in many applications, it is not suitable for handling the large amounts of data present in the microprocessor and memory systems.

In Part 2, we will describe the STATE TABLE mode of operation. In this mode, all of the data in the logic analyzer's memory is displayed on the CRT in *tabular* form. Recall that the logic analyzer memory consists of shift registers, one per channel. These registers store the logic levels that were present on the data lines at each SAMPLE CLOCK edge during the STORE cycle.

The STATE TABLE mode is just another way the logic analyzer can display its memory data. The STORE cycle and TRIGGER operation are exactly the same as described in Part 1. Once the STORE cycle ends and the DISPLAY cycle begins, the logic analyzer will display the data in a table instead of as timing diagrams.

B.8 State Table Format

Figure B-16 shows a typical display in the TIMING DIAGRAM mode for four channels. The intensified dots on the left end of each waveform indicate the TRIGGER POINT. These dots also represent the CURSOR position, because the CURSOR is positioned at TRIG +0 as indicated.

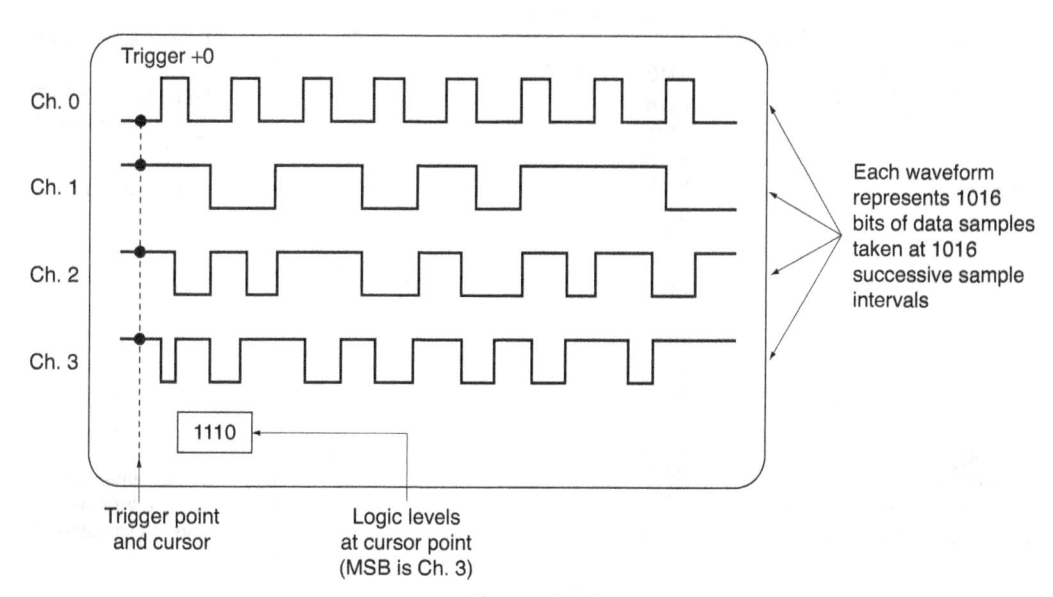

Figure B-16

You may recall that each channel waveform represents 1016 bits of data. These data correspond to samples that were taken at 1016 successive SAMPLE INTERVALS during the STORE cycle. The binary word shown at the lower part of the display tells us what the channel logic levels are at the CURSOR POINT, starting with channel 3 as the MSB. Since the CURSOR has been positioned at the TRIGGER POINT, this binary word also indicates the channel logic levels at the TRIGGER POINT.

Of course, the operator can move the CURSOR along the waveforms in steps of one SAMPLE INTERVAL. As this is done, the binary word will change to indicate the change in channel logic levels. If you were to record the binary words each time you moved the CURSOR one step, you would get a table of 4-bit words corresponding to the channel logic levels at the various sample intervals after the TRIGGER POINT. This is precisely what the logic analyzer does if you use it in the STATE TABLE mode.

B.8.A Binary Table

Figure B-17 shows what the display will look like when the operator selects the BINARY STATE TABLE display instead of the TIMING DIAGRAM display. The display is a table of *eighteen* 4-bit words. The top word in the table is *always* the CURSOR WORD. That is, it represents the logic levels present on the waveforms at the CURSOR POINT. The position of the CURSOR relative to the TRIGGER POINT is given as TRIG +0 in this example.

The next 16 entries in the table show the logic levels present for the next 16 SAMPLE INTERVALS past the CURSOR POINT. Thus, the second table entry would be the data at TRIG +1, the third entry would be at TRIG +2, and so on, until the 17th entry, which is at TRIG +16.

The 18th word (at the bottom of the display) is the TRIGGER WORD representing the logic levels that are present at the TRIGGER POINT. This TRIGGER WORD is *not* part of the data table; it is simply there as a reminder for the operator.

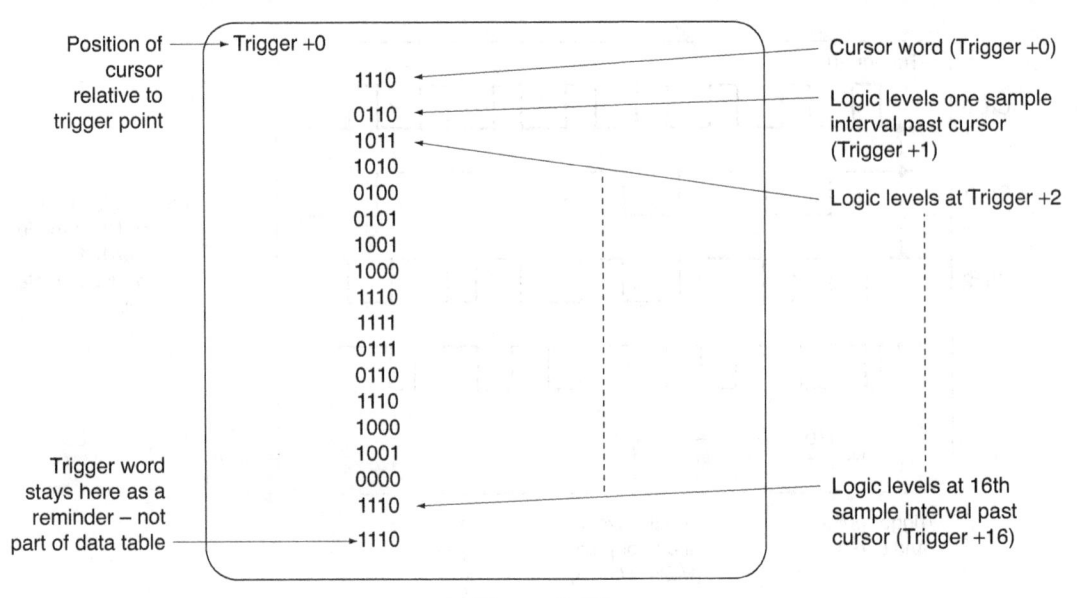

```
Position of ──┐  ┌── Trigger +0                           ── Cursor word (Trigger +0)
cursor                                                        
relative to       1110  ◄──                               ── Logic levels one sample
trigger point     0110  ◄──                                  interval past cursor
                  1011  ◄──                                  (Trigger +1)
                  1010
                  0100                                    ── Logic levels at Trigger +2
                  0101
                  1001
                  1000
                  1110
                  1111
                  0111
                  0110
                  1110
                  1000
                  1001
Trigger word      0000
stays here as a   1110  ◄──                               ── Logic levels at 16th
reminder – not                                               sample interval past
part of data table ──► 1110                                  cursor (Trigger +16)
```

Figure B-17

Appendix B

B.8.B Blinking Trigger Word

Whenever the TRIGGER WORD appears as one of the 17 entries in the displayed data table, it will be blinking. For the example in Figure B-17, the TRIGGER WORD is the top entry in the table, since the CURSOR POINT is TRIG +0. Thus, this 4-bit word will be blinking on the display.

B.8.C 8- or 16-Channel Operation

The same binary table format can be used for 8- or 16-channel operation. The only difference is in the number of bits per word. We are using the 4-channel case here simply for convenience.

B.8.D Examining the Rest of the Data

The display contains only 17 data words, corresponding to the CURSOR WORD and the 16 following SAMPLE INTERVALS. The logic analyzer memory, however, holds a total of 1016 data words (508 for 8-channel operation, 254 for 16-channel operation). The operator can change the displayed data table by using the CURSOR position control. This is illustrated in the two tables of Figure B-18.

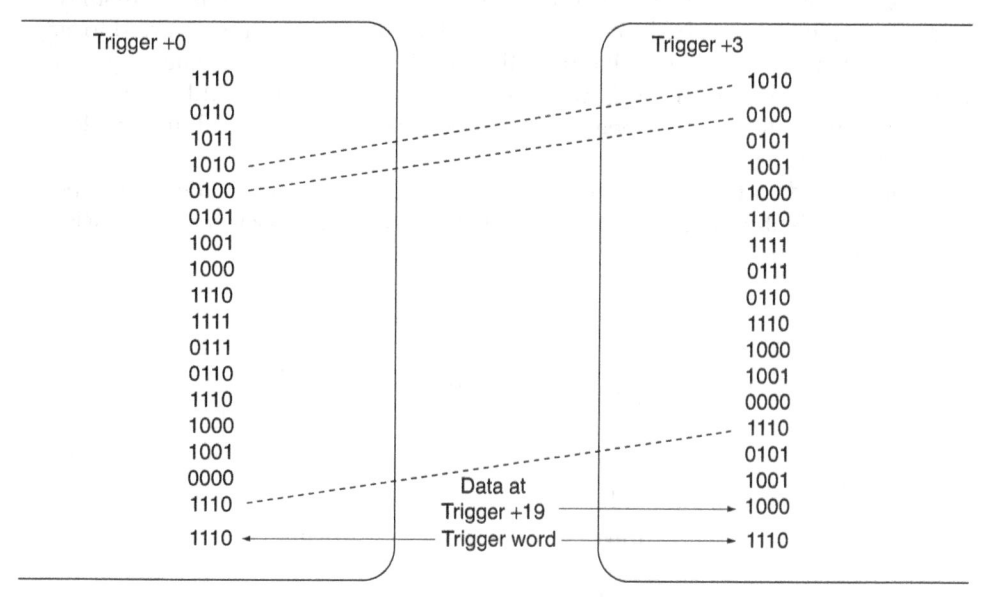

Trigger +0	Trigger +3
1110	1010
0110	0100
1011	0101
1010	1001
0100	1000
0101	1110
1001	1111
1000	0111
1110	0110
1111	1110
0111	1000
0110	1001
1110	0000
1000	1110
1001	0101
0000	1001
1110	1000 ← Data at Trigger +19
1110 ← Trigger word	1110

Figure B-18

The table on the left is the same as that in Figure B-17, where the CURSOR has been set at TRIG +0. The table on the right corresponds to a CURSOR position of TRIG +3. This means that the top word in this table is at TRIG +19. Thus, this new table contains three new entries at the bottom, which were not contained in the original table. Of course, the new table has lost three of the entries from the top of the original table.

The operator can move the CURSOR to other positions relative to the TRIGGER POINT by using the logic analyzer CURSOR controls. In this manner, the operator can examine all of the data in the logic analyzer memory.

Note that the TRIG WORD at the bottom of both table displays is the same. This will not change with the CURSOR position because it is not part of the data tables. It is there to remind the operator what the TRIG WORD is.

B.8.E Octal and Hex Tables

In many applications, it is more convenient to have the data tables displayed in octal or hexadecimal rather than in binary. The operator can select either octal or hex tables, and the logic analyzer will convert the binary data to the selected format.

PART 3—USING THE REFERENCE TABLE

The REF TABLE is used whenever the operator wishes to save the complete data table taken during a STORE operation so that it can be used as a reference to which subsequent new data can be compared. The following steps will illustrate the basic use of the REF TABLE.

B.8.F Transferring 7D01 Data to the REF TABLE

After the logic analyzer has executed a STORE cycle, it will automatically display the data table in the manner described earlier. This data table will be referred to as the 7D01 data table to distinguish it from the REF TABLE. The 7D01 data table can be transferred to the REF TABLE memory by actuating the 7D01 REF control on the front panel. When this is done, there will be *two* tables displayed on the CRT, as shown in Figure B-19.

The 7D01 data table is on the left, and the REF TABLE is on the right. The two tables are identical because the 7D01 data table was just transferred to the REF TABLE memory.

7D01	Trigger +0	Ref	Trigger +0	
	1110		1110	
Cursor word →	0110		0110	← Cursor word
	1011		1011	
	1010		1010	
	0100		0100	
	0101		0101	
	1001		1001	
	1000		1000	
	1110		1110	
	1111		1111	
	0111		0111	
	0110		0110	
	1110		1110	
	1000		1000	
	1001		1001	
	0000		0000	
	1110		1110	
Trigger word →	1110		1110	← Trigger word

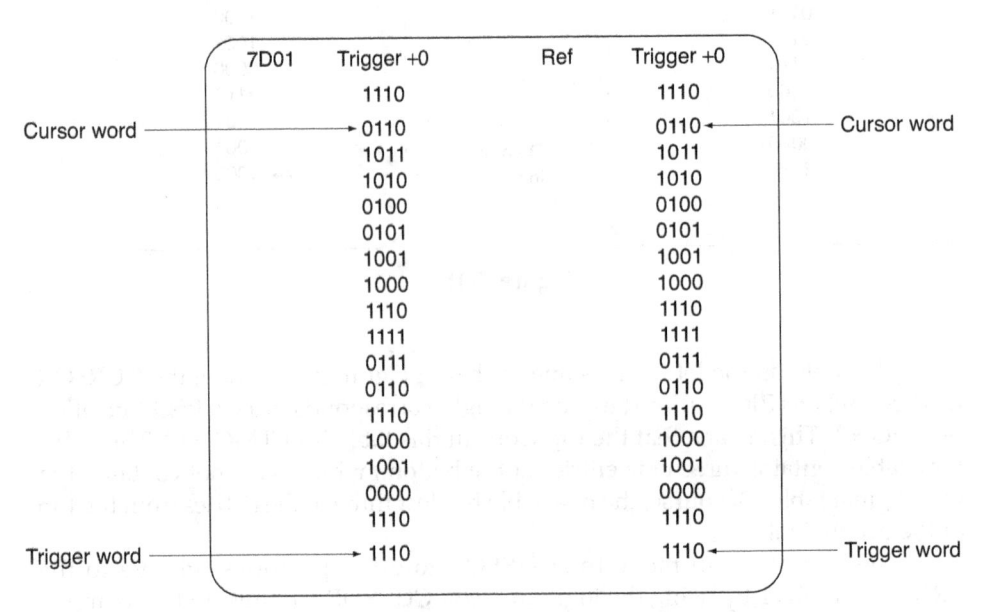

Figure B-19

Appendix B

B.8.G Executing a New STORE Cycle

Once the data has been transferred to the REF TABLE, the operator can have the logic analyzer execute a STORE cycle to obtain a new set of data *without* affecting the REF TABLE. In other words, the REF TABLE will retain its current data while the logic analyzer fills the 7D01 data memory with new data. When the STORE cycle is completed, the logic analyzer will again display both tables. The 7D01 table will have the new set of data while the REF TABLE will have the old set of data.

B.8.H Comparing the 7D01 Data Table and the REF TABLE

The operator can now compare the new data table with the REF TABLE to see if and where any differences occur. This is extremely valuable in testing and troubleshooting situations where data from a known good circuit can be placed in the REF TABLE. The 7D01 helps the operator to see any differences in the two tables by intensifying any part of the 7D01 data table that is different from the REF TABLE. This is illustrated in Figure B-20.

Note that two bits in the 7D01 data table are intensified because they are different from the corresponding bits in the REF TABLE. The operator can use this information to help troubleshoot the circuit from which the new data was taken.

7D01 Trigger +0	Ref Trigger +0
1110	1110
0110	0110
1011	1011
1010	1010
010①	0100
1001	1001
1000	1000
1110	1110
⓪111	1111
0111	0111
0110	0110
1110	1110
1000	1000
1001	1001
0000	0000
1110	1110
1110	1110

Figure B-20

Appendix C

MANUFACTURERS' DATA SHEETS

LINEAR INTEGRATED CIRCUITS

TYPES LM124, LM224, LM324 QUADRUPLE OPERATIONAL AMPLIFIERS

BULLETIN NO. DL-S 12248, SEPTEMBER 1975 – REVISED OCTOBER 1979

- Wide Range of Supply Voltages
  Single Supply ... 3 V to 30 V
  or Dual Supplies
- Low Supply Current Drain
  Independent of Supply Voltage
  ... 0.8 mA Typ
- Common-Mode Input Voltage Range Includes Ground Allowing Direct Sensing near Ground
- Low Input Bias and Offset Parameters
  Input Offset Voltage ... 2 mV Typ
  Input Offset Current ... 3 nA Typ (LM124)
  Input Bias Current ... 45 nA Typ
- Differential Input Voltage Range Equal to Maximum-Rated Supply Voltage ... ±32 V
- Open-Loop Differential Voltage Amplification ... 100 V/mV Typ
- Internal Frequency Compensation

schematic (each amplifier)

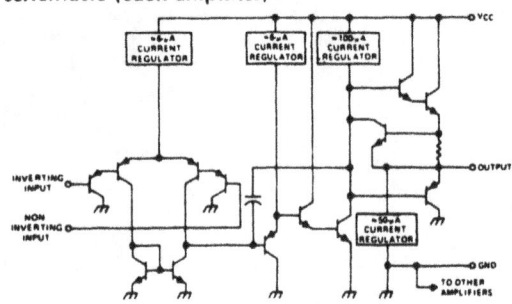

J OR N DUAL-IN-LINE OR W FLAT PACKAGE (TOP VIEW)

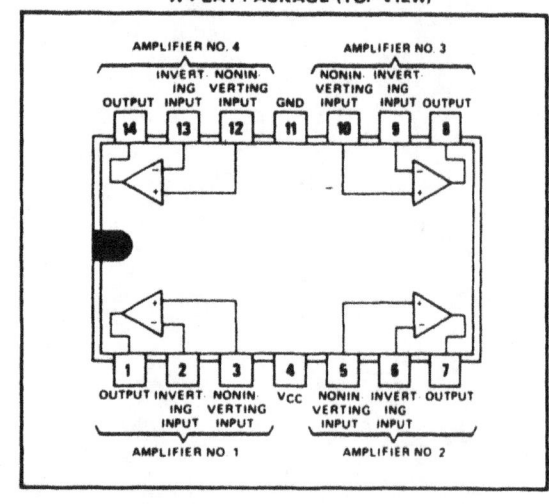

description

These devices consist of four independent, high-gain, frequency-compensated operational amplifiers that were designed specifically to operate from a single supply over a wide range of voltages. Operation from split supplies is also possible so long as the difference between the two supplies is 3 volts to 30 volts and Pin 4 is at least 1.5 volts more positive than the input common-mode voltage. The low supply current drain is independent of the magnitude of the supply voltage.

Applications include transducer amplifiers, d-c amplification blocks, and all the conventional operational amplifier circuits that now can be more easily implemented in single-supply-voltage systems. For example, the LM124 can be operated directly off of the standard five-volt supply that is used in digital systems and will easily provide the required interface electronics without requiring additional ± 15-volt supplies.

absolute maximum ratings over operating free-air temperature range (unless otherwise noted)

Supply voltage, V_{CC} (see Note 1)	32 V
Differential input voltage (see Note 2)	±32 V
Input voltage range (either input)	−0.3 V to 32 V
Duration of output short-circuit (one amplifier) to ground at (or below) 25°C free-air temperature ($V_{CC} \leq 15$ V) (see Note 3)	unlimited
Continuous total dissipation at (or below) 25°C free-air temperature (see Note 4)	900 mW
Operating free-air temperature range: LM124	−55°C to 125°C
LM224	−25°C to 85°C
LM324	0°C to 70°C
Storage temperature range	−65°C to 150°C
Lead temperature 1/16 inch (1,6 mm) from case for 60 seconds: J or W package	300°C
Lead temperature 1/16 inch (1,6 mm) from case for 10 seconds: N package	260°C

NOTES: 1. All voltage values, except differential voltages, are with respect to the network ground terminal.
2. Differential voltages are at the noninverting input terminal with respect to the inverting input terminal.
3. Short circuits from outputs to V_{CC} can cause excessive heating and eventual destruction.
4. For operation above 25°C free-air temperature, refer to Dissipation Derating Table. In the J package, LM124 chips are alloy-mounted; LM224 and LM324 chips are glass-mounted.

Copyright © 1979 by Texas Instruments Incorporated

TEXAS INSTRUMENTS
INCORPORATED
POST OFFICE BOX 225012 • DALLAS, TEXAS 75265

Appendix C 615

TYPES LM124, LM224, LM324
QUADRUPLE OPERATIONAL AMPLIFIERS

electrical characteristics at specified free-air temperature, V_{CC} = 5 V (unless otherwise noted)

PARAMETER		TEST CONDITIONS†		LM124, LM224 MIN	TYP	MAX	LM324 MIN	TYP	MAX	UNIT
V_{IO}	Input offset voltage	V_O = 1.4 V, V_{CC} = 5 V to 30 V	25°C		2	5		2	7	mV
			Full range			7			9	
I_{IO}	Input offset current	V_O = 1.4 V	25°C		3	30		5	50	nA
			Full range			100			150	
I_{IB}	Input bias current	V_O = 1.4 V, See Note 5	25°C		−45	−150		−45	−250	nA
			Full range			−300			−500	
V_{ICR}	Common-mode input voltage range	V_{CC} = 30 V	25°C	0 to V_{CC}−1.5			0 to V_{CC}−1.5			V
			Full range	0 to V_{CC}−2			0 to V_{CC}−2			
V_{OH}	High-level output voltage	V_{CC} = 30 V, R_L = 2 kΩ	Full range	26			26			V
		V_{CC} = 30 V, $R_L \geq$ 10 kΩ	Full range	27	28		27	28		
V_{OL}	Low-level output voltage	$R_L \leq$ 10 kΩ	Full range		5	20		5	20	mV
A_{VD}	Large-signal differential voltage amplification	V_{CC} = 15 V, V_O = 1 V to 11 V, $R_L \geq$ 2 kΩ	25°C	50	100		25	100		V/mV
			Full range	25			15			
CMRR	Common-mode rejection ratio	$R_S \leq$ 10 kΩ	25°C	70	85		65	85		dB
k_{SVR}*	Supply voltage rejection ratio	$R_S \leq$ 10 kΩ	25°C	65	100		65	100		dB
V_{o1}/V_{o2}	Channel separation	f = 1 kHz to 20 kHz	25°C		120			120		dB
I_O	Output current	V_{CC} = 15 V, V_{ID} = 1 V, V_O = 0 V	25°C	−20	−40		−20	−40		mA
			Full range	−10	−20		−10	−20		
		V_{CC} = 15 V, V_{ID} = −1 V, V_O = 5 V	25°C	10	20		10	20		
			Full range	5	8		5	8		
		V_{ID} = −1 V, V_O = 200 mV	25°C	12	50		12	50		µA
I_{CC}	Supply current (four amplifiers)	No load, No signal	25°C		0.8			0.8		mA
			Full range			1.2			1.2	

*$k_{SVR} = \Delta V_{CC}/\Delta V_{IO}$

† All characteristics are specified under open-loop conditions. Full range is −55°C to 125°C for LM124, −25°C to 85°C for LM224, and 0°C to 70°C for LM324.

NOTE 5: The direction of the bias current is out of the device due to the P-N-P input stage. This current is essentially constant, regardless of the state of the output, so no loading change is presented to the input lines.

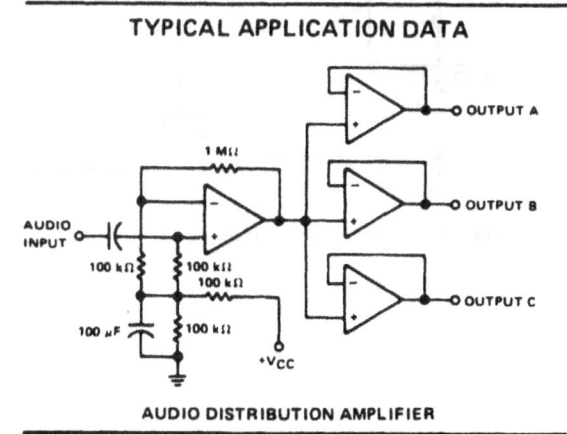

TYPICAL APPLICATION DATA

AUDIO DISTRIBUTION AMPLIFIER

THERMAL INFORMATION

DISSIPATION DERATING TABLE

PACKAGE	POWER RATING	DERATING FACTOR	ABOVE T_A
J (Alloy-Mounted Chip)	900 mW	11.0 mW/°C	68°C
J (Glass-Mounted Chip)	900 mW	8.2 mW/°C	40°C
N	900 mW	9.2 mW/°C	52°C
W	900 mW	8.0 mW/°C	37°C

Also see Dissipation Derating Curves, Section 2.

TEXAS INSTRUMENTS
INCORPORATED
POST OFFICE BOX 225012 • DALLAS, TEXAS 75265

SE555, SE555C, SA555, NE555 PRECISION TIMERS

D1669, SEPTEMBER 1973 – REVISED OCTOBER 1988

- Timing from Microseconds to Hours
- Astable or Monostable Operation
- Adjustable Duty Cycle
- TTL-Compatible Output Can Sink or Source Up to 200 mA
- Functionally Interchangeable with the Signetics SE555, SE555C, SA555, NE555; Have Same Pinout

SE555C FROM TI IS NOT RECOMMENDED FOR NEW DESIGNS

SE555, SE555C . . . JG PACKAGE
SA555, NE555 . . . D, JG, OR P PACKAGE
(TOP VIEW)

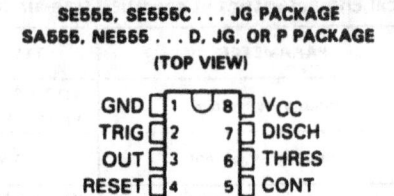

```
GND  [1    8] VCC
TRIG [2    7] DISCH
OUT  [3    6] THRES
RESET[4    5] CONT
```

SE555, SE555C . . . FK PACKAGE
(TOP VIEW)

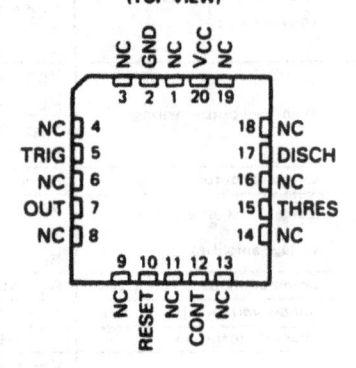

Pins (top, left-right): NC, GND, NC, VCC, NC (3 2 1 20 19)
Left side: NC 4, TRIG 5, NC 6, OUT 7, NC 8
Right side: 18 NC, 17 DISCH, 16 NC, 15 THRES, 14 NC
Bottom: 9 10 11 12 13 — NC, RESET, NC, CONT, NC

NC – No internal connection

description

These devices are monolithic timing circuits capable of producing accurate time delays or oscillation. In the time-delay or monostable mode of operation, the timed interval is controlled by a single external resistor and capacitor network. In the astable mode of operation, the frequency and duty cycle may be independently controlled with two external resistors and a single external capacitor.

The threshold and trigger levels are normally two-thirds and one-third, respectively, of V_{CC}. These levels can be altered by use of the control voltage terminal. When the trigger input falls below the trigger level, the flip-flop is set and the output goes high. If the trigger input is above the trigger level and the threshold input is above the threshold level, the flip-flop is reset and the output is low. The reset input can override all other inputs and can be used to initiate a new timing cycle. When the reset input goes low, the flip-flop is reset and the output goes low. Whenever the output is low, a low-impedance path is provided between the discharge terminal and ground.

The output circuit is capable of sinking or sourcing current up to 200 mA. Operation is specified for supplies of 5 to 15 V. With a 5-V supply, output levels are compatible with TTL inputs.

The SE555 and SE555C are characterized for operation over the full military range of −55 °C to 125 °C. The SA555 is characterized for operation from −40 °C to 85 °C, and the NE555 is characterized for operation from 0 °C to 70 °C.

functional block diagram

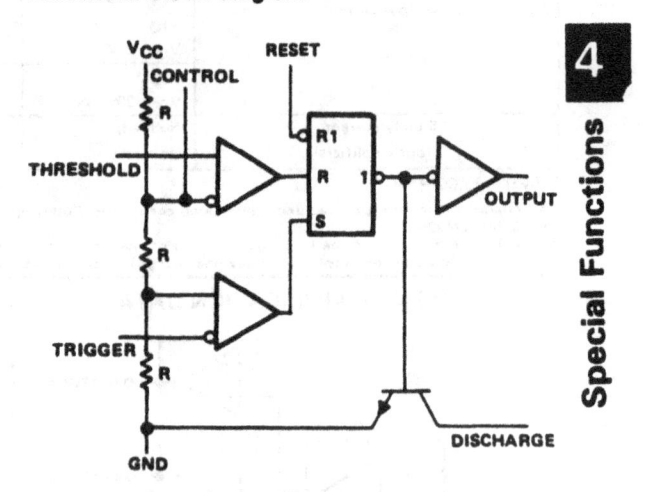

Reset can override Trigger, which can override Threshold.

PRODUCTION DATA documents contain information current as of publication date. Products conform to specifications per the terms of Texas Instruments standard warranty. Production processing does not necessarily include testing of all parameters.

Copyright © 1983, Texas Instruments Incorporated

TEXAS INSTRUMENTS
POST OFFICE BOX 655012 • DALLAS, TEXAS 75265

SE555, SE555C, SA555, NE555 PRECISION TIMERS

electrical characteristics at 25 °C free-air temperature, V_{CC} = 5 V to 15 V (unless otherwise noted)

PARAMETER	TEST CONDITIONS		SE555 MIN	SE555 TYP	SE555 MAX	SE555C, SA555, NE555 MIN	SE555C, SA555, NE555 TYP	SE555C, SA555, NE555 MAX	UNIT
Threshold voltage level	V_{CC} = 15 V		9.4	10	10.6	8.8	10	11.2	V
	V_{CC} = 5 V		2.7	3.3	4	2.4	3.3	4.2	
Threshold current (see Note 2)				30	250		30	250	nA
Trigger voltage level	V_{CC} = 15 V		4.8	5	5.2	4.5	5	5.6	V
	V_{CC} = 5 V		1.45	1.67	1.9	1.1	1.67	2.2	
Trigger current	Trigger at 0 V			0.5	0.9		0.5	2	µA
Reset voltage level			0.3	0.7	1	0.3	0.7	1	V
Reset current	Reset at V_{CC}			0.1	0.4		0.1	0.4	mA
	Reset at 0 V			−0.4	−1		−0.4	−1.5	
Discharge switch off-state current				20	100		20	100	nA
Control voltage (open circuit)	V_{CC} = 15 V		9.6	10	10.4	9	10	11	V
	V_{CC} = 5 V		2.9	3.3	3.8	2.6	3.3	4	
Low-level output voltage	V_{CC} = 15 V	I_{OL} = 10 mA		0.1	0.15		0.1	0.25	V
		I_{OL} = 50 mA		0.4	0.5		0.4	0.75	
		I_{OL} = 100 mA		2	2.2		2	2.5	
		I_{OL} = 200 mA		2.5			2.5		
	V_{CC} = 5 V	I_{OL} = 5 mA		0.1	0.2		0.1	0.35	
		I_{OL} = 8 mA		0.15	0.25		0.15	0.4	
High-level output voltage	V_{CC} = 15 V	I_{OH} = −100 mA	13	13.3		12.75	13.3		V
		I_{OH} = −200 mA		12.5			12.5		
	V_{CC} = 5 V	I_{OH} = −100 mA	3	3.3		2.75	3.3		
Supply current	Output low, No load	V_{CC} = 15 V		10	12		10	15	mA
		V_{CC} = 5 V		3	5		3	6	
	Output high, No load	V_{CC} = 15 V		9	10		9	13	
		V_{CC} = 5 V		2	4		2	5	

NOTE 2: This parameter influences the maximum value of the timing resistors R_A and R_B in the circuit of Figure 12. For example, when V_{CC} = 5 V, the maximum value is R = R_A + R_B ≈ 3.4 MΩ, and for V_{CC} = 15 V, the maximum value is 10 MΩ.

operating characteristics, V_{CC} = 5 V and 15 V

PARAMETER	TEST CONDITIONS†		SE555 MIN	SE555 TYP	SE555 MAX	SE555C, SA555, NE555 MIN	SE555C, SA555, NE555 TYP	SE555C, SA555, NE555 MAX	UNIT
Initial error of timing interval‡	Each timer, monostable§	T_A = 25 °C		0.5	1.5		1	3	%
	Each timer, astable¶			1.5			2.25		
Temperature coefficient of timing interval	Each timer, monostable§	T_A = MIN to MAX		30	100		50		ppm/°C
	Each timer, astable¶			90			150		
Supply voltage sensitivity of timing interval	Each timer, monostable§	T_A = 25 °C		0.05	0.2		0.1	0.5	%/V
	Each timer, astable¶			0.15			0.3		
Output pulse rise time	C_L = 15 pF, T_A = 25 °C			100	200		100	300	ns
Output pulse fall time				100	200		100	300	

†For conditions shown as MIN or MAX, use the appropriate value specified under recommended operating conditions.
‡Timing interval error is defined as the difference between the measured value and the average value of a random sample from each process run.
§Values specified are for a device in a monostable circuit similar to Figure 9, with component values as follow: R_A = 2 kΩ to 100 kΩ, C = 0.1 µF.
¶Values specified are for a device in an astable circuit similar to Figure 12, with component values as follow: R_A = 1 kΩ to 100 kΩ, C = 0.1 µF.

TEXAS INSTRUMENTS
POST OFFICE BOX 655012 • DALLAS, TEXAS 75265

SE555, SE555C, SA555, NE555
PRECISION TIMERS

TYPICAL APPLICATION DATA

sequential timer

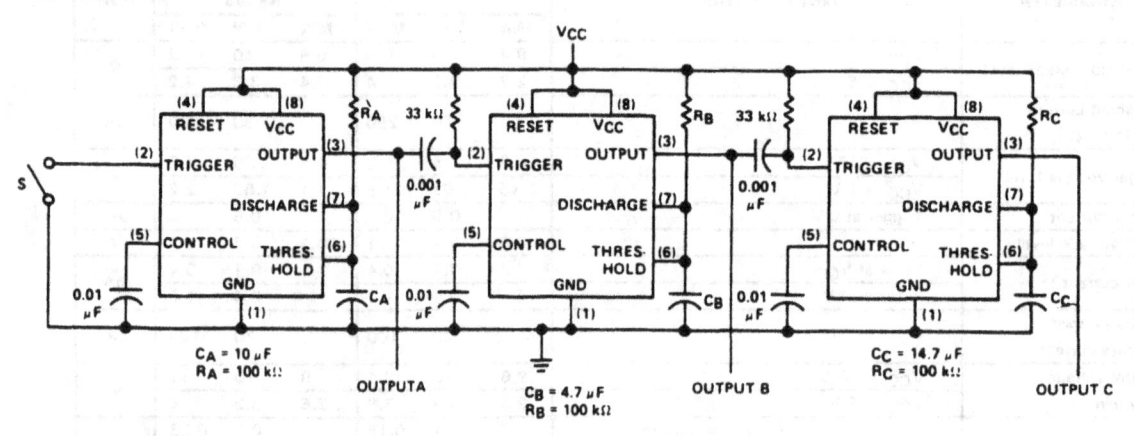

$C_A = 10\ \mu F$
$R_A = 100\ k\Omega$

$C_B = 4.7\ \mu F$
$R_B = 100\ k\Omega$

$C_C = 14.7\ \mu F$
$R_C = 100\ k\Omega$

S closes momentarily at t = 0.

FIGURE 22. SEQUENTIAL TIMER CIRCUIT

Many applications, such as computers, require signals for initializing conditions during start-up. Other applications, such as test equipment, require activation of test signals in sequence. These timing circuits may be connected to provide such sequential control. The timers may be used in various combinations of astable or monostable circuit connections, with or without modulation, for extremely flexible waveform control. Figure 22 illustrates a sequencer circuit with possible applications in many systems, and Figure 23 shows the output waveforms.

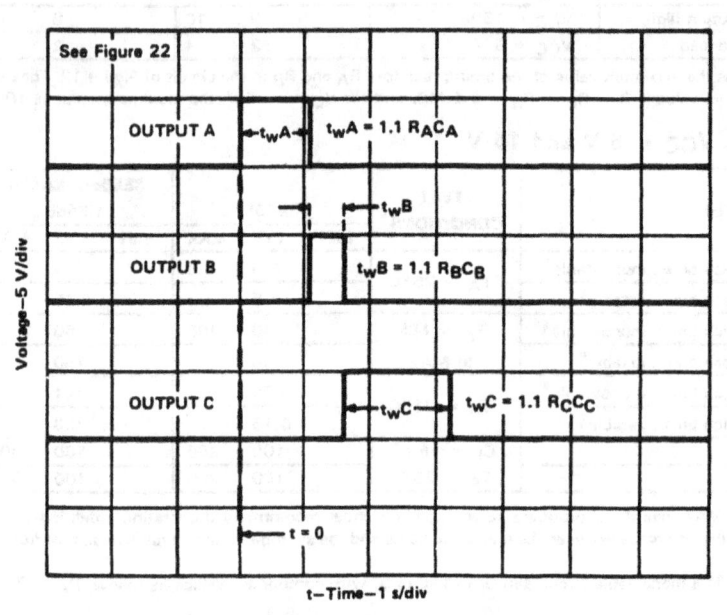

$t_wA = 1.1\ R_A C_A$

$t_wB = 1.1\ R_B C_B$

$t_wC = 1.1\ R_C C_C$

FIGURE 23. SEQUENTIAL TIMER WAVEFORMS

TEXAS INSTRUMENTS
POST OFFICE BOX 655012 • DALLAS, TEXAS 75265

Appendix C

MOTOROLA

MC1408 MC1508

Specifications and Applications Information

EIGHT-BIT MULTIPLYING DIGITAL-TO-ANALOG CONVERTER

... designed for use where the output current is a linear product of an eight-bit digital word and an analog input voltage.

- Eight-Bit Accuracy Available in Both Temperature Ranges
  Relative Accuracy: ±0.19% Error maximum
  (MC1408L8, MC1408P8, MC1508L8)
- Seven and Six-Bit Accuracy Available with MC1408 Designated by 7 or 6 Suffix after Package Suffix
- Fast Settling Time – 300 ns typical
- Noninverting Digital Inputs are MTTL and CMOS Compatible
- Output Voltage Swing – +0.4 V to –5.0 V
- High-Speed Multiplying Input
  Slew Rate 4.0 mA/µs
- Standard Supply Voltages: +5.0 V and –5.0 V to –15 V

EIGHT-BIT MULTIPLYING DIGITAL-TO-ANALOG CONVERTER

SILICON MONOLITHIC INTEGRATED CIRCUIT

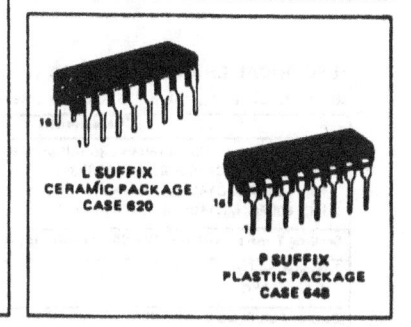

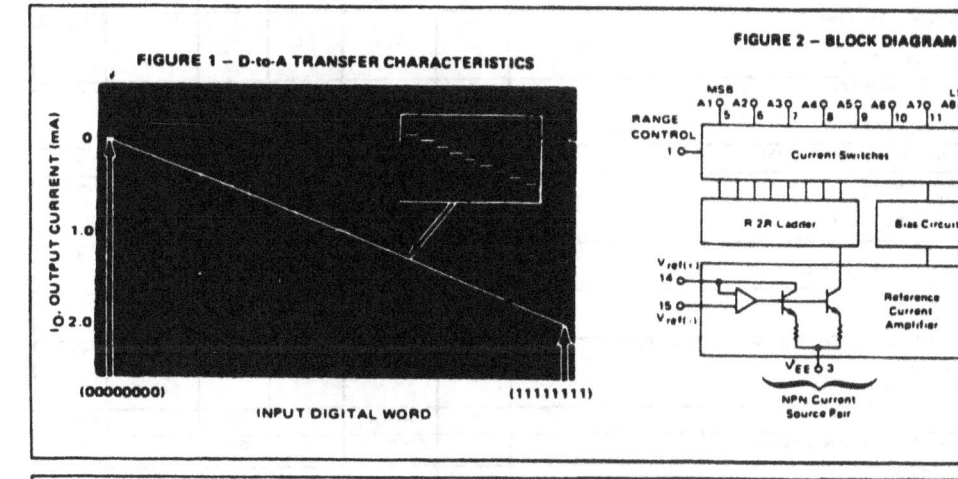

L SUFFIX
CERAMIC PACKAGE
CASE 620

P SUFFIX
PLASTIC PACKAGE
CASE 648

FIGURE 1 – D-to-A TRANSFER CHARACTERISTICS

FIGURE 2 – BLOCK DIAGRAM

TYPICAL APPLICATIONS

- Tracking A-to-D Converters
- Successive Approximation A-to-D Converters
- 2 1/2 Digit Panel Meters and DVM's
- Waveform Synthesis
- Sample and Hold
- Peak Detector
- Programmable Gain and Attenuation
- CRT Character Generation
- Audio Digitizing and Decoding
- Programmable Power Supplies
- Analog-Digital Multiplication
- Digital-Digital Multiplication
- Analog-Digital Division
- Digital Addition and Subtraction
- Speech Compression and Expansion
- Stepping Motor Drive

MC1408, MC1508

MAXIMUM RATINGS (T_A = +25°C unless otherwise noted.)

Rating	Symbol	Value	Unit
Power Supply Voltage	V_{CC}	+5.5	Vdc
	V_{EE}	-16.5	
Digital Input Voltage	V_5 thru V_{12}	0 to +5.5	Vdc
Applied Output Voltage	V_O	+0.5, -5.2	Vdc
Reference Current	I_{14}	5.0	mA
Reference Amplifier Inputs	V_{14}, V_{15}	V_{CC}, V_{EE}	Vdc
Operating Temperature Range			°C
MC1508	T_A	-55 to +125	
MC1408 Series		0 to +75	
Storage Temperature Range	T_{stg}	-65 to +150	°C

ELECTRICAL CHARACTERISTICS (V_{CC} = +5.0 Vdc, V_{EE} = -15 Vdc, $\frac{V_{ref}}{R14}$ = 2.0 mA, MC1508L8 T_A = -55°C to +125°C. MC1408L Series T_A = 0 to +75°C unless otherwise noted. All digital inputs at high logic level.)

Characteristic	Figure	Symbol	Min	Typ	Max	Unit
Relative Accuracy (Error relative to full scale I_O)	4	E_r				%
MC1508L8, MC1408L8, MC1408P8			–	–	±0.19	
MC1408P7, MC1408L7, See Note 1			–	–	±0.39	
MC1408P6, MC1408L6, See Note 1			–	–	±0.78	
Settling Time to within ±1/2 LSB (includes t_{PLH}) (T_A = +25°C) See Note 2	5	t_s	–	300	–	ns
Propagation Delay Time T_A = +25°C	5	t_{PLH}, t_{PHL}	–	30	100	ns
Output Full Scale Current Drift		TCI_O	–	-20	–	PPM/°C
Digital Input Logic Levels (MSB)	3					Vdc
High Level, Logic "1"		V_{IH}	2.0	–	–	
Low Level, Logic "0"		V_{IL}	–	–	0.8	
Digital Input Current (MSB)	3					mA
High Level, V_{IH} = 5.0 V		I_{IH}	–	0	0.04	
Low Level, V_{IL} = 0.8 V		I_{IL}	–	-0.4	-0.8	
Reference Input Bias Current (Pin 15)	3	I_{15}	–	-1.0	-5.0	µA
Output Current Range	3	I_{OR}				mA
V_{EE} = -5.0 V			0	2.0	2.1	
V_{EE} = -15 V, T_A = 25°C			0	2.0	4.2	
Output Current V_{ref} = 2.000 V, R14 = 1000 Ω	3	I_O	1.9	1.99	2.1	mA
Output Current (All bits low)	3	$I_{O(min)}$	–	0	4.0	µA
Output Voltage Compliance ($E_r \leq 0.19\%$ at T_A = +25°C)	3	V_O				Vdc
Pin 1 grounded			–	–	-0.55, +0.4	
Pin 1 open, V_{EE} below -10 V					-5.0, +0.4	
Reference Current Slew Rate	6	SR I_{ref}		4.0		mA/µs
Output Current Power Supply Sensitivity		PSRR(-)	–	0.5	2.7	µA/V
Power Supply Current	3					mA
(All bits low)		I_{CC}	–	+13.5	+22	
		I_{EE}		-7.5	-13	
Power Supply Voltage Range	3	V_{CCR}	+4.5	+5.0	+5.5	Vdc
(T_A = +25°C)		V_{EER}	-4.5	-15	-16.5	
Power Dissipation	3	P_D				mW
All bits low						
V_{EE} = -5.0 Vdc				105	170	
V_{EE} = -15 Vdc			–	190	305	
All bits high						
V_{EE} = -5.0 Vdc			–	90	–	
V_{EE} = -15 Vdc			–	160	–	

Note 1. All current switches are tested to guarantee at least 50% of rated output current.
Note 2. All bits switched.

Appendix C

intel

2114
1024 X 4 BIT STATIC RAM

	2114-2	2114-3	2114	2114L3	2114L
Max. Access Time (ns)	200	300	450	300	450
Max. Power Dissipation (mw)	710mw	710mw	710mw	370mw	370mw

- High Density 18 Pin Package
- Identical Cycle and Access Times
- Single +5V Supply
- No Clock or Timing Strobe Required
- Completely Static Memory
- Directly TTL Compatible: All Inputs and Outputs
- Common Data Input and Output Using Three-State Outputs
- Pin-Out Compatible with 3605 and 3625 Bipolar PROMs

The Intel® 2114 is a 4096-bit static Random Access Memory organized as 1024 words by 4-bits using N-channel Silicon-Gate MOS technology. It uses fully DC stable (static) circuitry throughout — in both the array and the decoding — and therefore requires no clocks or refreshing to operate. Data access is particularly simple since address setup times are not required. The data is read out nondestructively and has the same polarity as the input data. Common input/output pins are provided.

The 2114 is designed for memory applications where high performance, low cost, large bit storage, and simple interfacing are important design objectives. The 2114 is placed in an 18-pin package for the highest possible density.

It is directly TTL compatible in all respects: inputs, outputs, and a single +5V supply. A separate Chip Select ($\overline{CS}$) lead allows easy selection of an individual package when outputs are or-tied.

The 2114 is fabricated with Intel's N-channel Silicon-Gate technology — a technology providing excellent protection against contamination permitting the use of low cost plastic packaging.

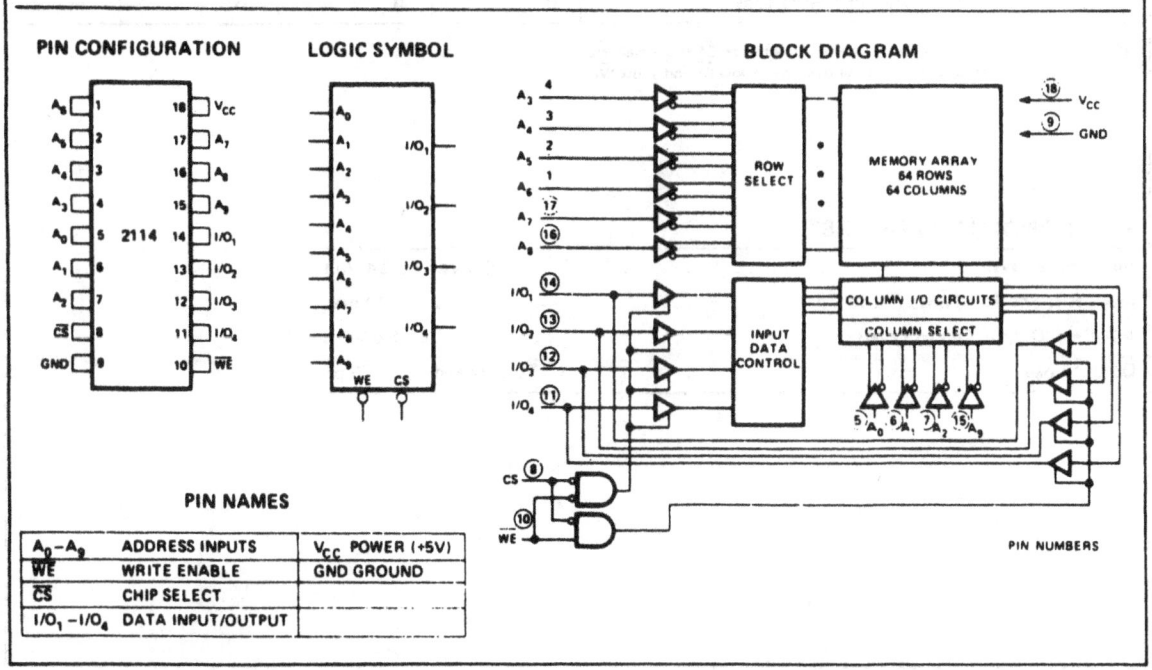

PIN CONFIGURATION   LOGIC SYMBOL   BLOCK DIAGRAM

PIN NAMES

A_0–A_9	ADDRESS INPUTS	V_{CC}	POWER (+5V)
$\overline{WE}$	WRITE ENABLE	GND	GROUND
$\overline{CS}$	CHIP SELECT		
I/O_1–I/O_4	DATA INPUT/OUTPUT		

(Courtesy of Intel Corporation)

2114 FAMILY

A.C. CHARACTERISTICS $T_A = 0°C$ to $70°C$, $V_{CC} = 5V \pm 5\%$, unless otherwise noted.

READ CYCLE [1]

SYMBOL	PARAMETER	2114-2 Min.	2114-2 Max.	2114-3, 2114L3 Min.	2114-3, 2114L3 Max.	2114, 2114L Min.	2114, 2114L Max.	UNIT
t_{RC}	Read Cycle Time	200		300		450		ns
t_A	Access Time		200		300		450	ns
t_{CO}	Chip Selection to Output Valid		70		100		100	ns
t_{CX}	Chip Selection to Output Active	0		0		0		ns
t_{OTD}	Output 3-state from Deselection	0	40	0	80	0	100	ns
t_{OHA}	Output Hold from Address Change	10		10		10		ns

WRITE CYCLE [2]

SYMBOL	PARAMETER	2114-2 Min.	2114-2 Max.	2114-3, 2114L3 Min.	2114-3, 2114L3 Max.	2114, 2114L Min.	2114, 2114L Max.	UNIT
t_{WC}	Write Cycle Time	200		300		450		ns
t_W	Write Time	100		150		200		ns
t_{WR}	Write Release Time	20		0		0		ns
t_{OTW}	Output 3-state from Write	0	40	0	80	0	100	ns
t_{DW}	Data to Write Time Overlap	100		150		200		ns
t_{DH}	Data Hold From Write Time	0		0		0		ns

NOTES 1. A Read occurs during the overlap of a low $\overline{CS}$ and a high $\overline{WE}$.
2. A Write occurs during the overlap of a low $\overline{CS}$ and a low $\overline{WE}$.

A.C. CONDITIONS OF TEST

Input Pulse Levels	0.8 Volt to 2.4 Volt
Input Rise and Fall Times	10 nsec
Input and Output Timing Levels	1.5 Volts
Output Load	1 TTL Gate and C_L = 50 pF

Appendix C

National Semiconductor

CD4016BM/CD4016BC Quad Bilateral Switch

General Description

The CD4016BM/CD4016BC is a quad bilateral switch intended for the transmission or multiplexing of analog or digital signals. It is pin-for-pin compatible with CD4066BM/CD4066BC.

Features

- Wide supply voltage range 3V to 15V
- Wide range of digital and analog switching ±7.5 V_{PEAK}
- "ON" resistance for 15V operation 400Ω (typ.)
- Matched "ON" resistance over 15V signal input $\Delta R_{ON} = 10\Omega$ (typ.)
- High degree of linearity 0.4% distortion (typ.)
  @ f_{IS} = 1 kHz, V_{IS} = 5 V_{p-p},
  $V_{DD} - V_{SS}$ = 10V, R_L = 10 kΩ
- Extremely low "OFF" switch leakage 0.1 nA (typ.)
  @ $V_{DD} - V_{SS}$ = 10V
  T_A = 25°C
- Extremely high control input impedance $10^{12}\Omega$ (typ.)
- Low crosstalk between switches −50 dB (typ.)
  @ f_{IS} = 0.9 MHz, R_L = 1 kΩ
- Frequency response, switch "ON" 40 MHz (typ.)

Applications

- Analog signal switching/multiplexing
  - Signal gating
  - Squelch control
  - Chopper
  - Modulator/Demodulator
  - Commutating switch
- Digital signal switching/multiplexing
- CMOS logic implementation
- Analog-to-digital/digital-to-analog conversion
- Digital control of frequency, impedance, phase, and analog-signal gain

Schematic and Connection Diagrams

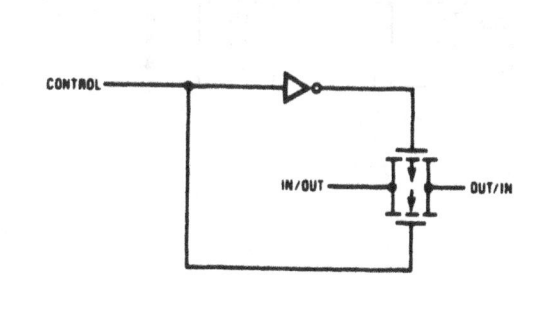

Dual-In-Line Package

Pin assignments (TOP VIEW):
1. IN/OUT — SWA
2. OUT/IN
3. OUT/IN — SWD
4. IN/OUT
5. CONTROL B — SWB
6. CONTROL C
7. V_{SS} — SWC
8. IN/OUT
9. OUT/IN
10. OUT/IN
11. IN/OUT
12. CONTROL D
13. CONTROL A
14. V_{DD}

TL/F/5661-1

Order Number CD4016BMJ or CD4016BCJ
NS Package J14A

Order Number CD4016BMN or CD4016BCN
NS Package N14A

AC Electrical Characteristics

$T_A = 25°C$, $C_L = 50\,pF$, $R_L = 200\,K$, $t_r = t_f = 20\,ns$, unless otherwise specified.

Symbol	Parameter	Conditions	Min	Typ	Max	Units
Clocked Operation						
t_{PHL}, t_{PLH}	Propagation Delay Time	$V_{DD} = 5V$		230	350	ns
		$V_{DD} = 10V$		80	160	ns
		$V_{DD} = 15V$		60	120	ns
t_{THL}, t_{TLH}	Transition Time	$V_{DD} = 5V$		100	200	ns
		$V_{DD} = 10V$		50	100	ns
		$V_{DD} = 15V$		40	80	ns
t_{WL}, t_{WM}	Minimum Clock Pulse-Width	$V_{DD} = 5V$		160	250	ns
		$V_{DD} = 10V$		60	110	ns
		$V_{DD} = 15V$		50	85	ns
t_{rCL}, t_{fCL}	Clock Rise and Fall Time	$V_{DD} = 5V$			15	µS
		$V_{DD} = 10V$			15	µS
		$V_{DD} = 15V$			15	µS
t_{SU}	Minimum Data Set-Up Time	$V_{DD} = 5V$		50	100	ns
		$V_{DD} = 10V$		20	40	ns
		$V_{DD} = 15V$		15	30	ns
f_{CL}	Maximum Clock Frequency	$V_{DD} = 5V$	2	3.5		MHz
		$V_{DD} = 10V$	4.5	8		MHz
		$V_{DD} = 15V$	6	11		MHz
C_{IN}	Input Capacitance	Clock Input		7.5	10	pF
		Other Inputs		5	7.5	pF
Reset Operation						
$t_{PHL(R)}$	Propagation Delay Time	$V_{DD} = 5V$		200	400	ns
		$V_{DD} = 10V$		100	200	ns
		$V_{DD} = 15V$		80	160	ns
$t_{WH(R)}$	Minimum Reset Pulse Width	$V_{DD} = 5V$		135	250	ns
		$V_{DD} = 10V$		40	80	ns
		$V_{DD} = 15V$		30	60	ns

CD4015BM/CD4015BC

Appendix C

CD4016BM/CD4016BC

Absolute Maximum Ratings
(Notes 1 and 2)

V_{DD} Supply Voltage	−0.5V to +18V
V_{IN} Input Voltage	−0.5V to V_{DD} + 0.5V
T_S Storage Temperature Range	−65°C to +150°C
P_D Package Dissipation	500 mW
Lead Temperature (Soldering, 10 seconds)	260°C

Recommended Operating Conditions (Note 2)

V_{DD} Supply Voltage	3V to 15V
V_{IN} Input Voltage	0V to V_{DD}
T_A Operating Temperature Range	
CD4016BM	−55°C to +125°C
CD4016BC	−40°C to +85°C

DC Electrical Characteristics CD4016BM (Note 2)

Symbol	Parameter	Conditions	−55°C Min	−55°C Max	25°C Min	25°C Typ	25°C Max	125°C Min	125°C Max	Units
I_{DD}	Quiescent Device Current	$V_{DD}=5V$, $V_{IN}=V_{DD}$ or V_{SS}		0.25		0.01	0.25		7.5	μA
		$V_{DD}=10V$, $V_{IN}=V_{DD}$ or V_{SS}		0.5		0.01	0.5		15	μA
		$V_{DD}=15V$, $V_{IN}=V_{DD}$ or V_{SS}		1.0		0.01	1.0		30	μA

Signal Inputs and Outputs

Symbol	Parameter	Conditions	−55°C Min	−55°C Max	25°C Min	25°C Typ	25°C Max	125°C Min	125°C Max	Units
R_{ON}	"ON" Resistance	$R_L=10\ k\Omega$ to $\frac{V_{DD}-V_{SS}}{2}$ $V_C=V_{DD}$, $V_{IS}=V_{SS}$ or V_{DD}								
		$V_{DD}=10V$		600		250	660		960	Ω
		$V_{DD}=15V$		360		200	400		600	Ω
		$R_L=10\ k\Omega$ to $\frac{V_{DD}-V_{SS}}{2}$ $V_C=V_{DD}$								
		$V_{DD}=10V$, $V_{IS}=4.75$ to $5.25V$		1870		850	2000		2600	Ω
		$V_{DD}=15V$, $V_{IS}=7.25$ to $7.75V$		775		400	850		1230	Ω
ΔR_{ON}	Δ"ON" Resistance Between any 2 of 4 Switches (In Same Package)	$R_L=10\ k\Omega$ to $\frac{V_{DD}-V_{SS}}{2}$ $V_C=V_{DD}$, $V_{IS}=V_{SS}$ to V_{DD}								
		$V_{DD}=10V$				15				Ω
		$V_{DD}=15V$				10				Ω
I_{IS}	Input or Output Leakage Switch "OFF"	$V_C=0$, $V_{DD}=15V$ $V_{IS}=15V$ and 0V, $V_{OS}=0V$ and 15V		±50		±0.1	±50		±500	nA

Control Inputs

Symbol	Parameter	Conditions	−55°C Min	−55°C Max	25°C Min	25°C Typ	25°C Max	125°C Min	125°C Max	Units
V_{ILC}	Low Level Input Voltage	$V_{IS}=V_{SS}$ and V_{DD} $V_{OS}=V_{DD}$ and V_{SS} $I_{IS}=\pm 10\ \mu A$								
		$V_{DD}=5V$		0.9			0.7		0.5	V
		$V_{DD}=10V$		0.9			0.7		0.5	V
		$V_{DD}=15V$		0.9			0.7		0.5	V
V_{IHC}	High Level Input Voltage	$V_{DD}=5V$	3.5		3.5			3.5		V
		$V_{DD}=10V$ (see Note 6 and	7.0		7.0			7.0		V
		$V_{DD}=15V$ Figure 8)	11.0		11.0			11.0		V
I_{IN}	Input Current	$V_{DD}-V_{SS}=15V$ $V_{DD}\geq V_{IS}\geq V_{SS}$ $V_{DD}\geq V_C\geq V_{SS}$		±0.1		±10⁻⁵	±0.1		±1.0	μA

DC Electrical Characteristics CD4016BC (Note 2) (Continued)

Symbol	Parameter	Conditions	−40°C Min	−40°C Max	25°C Min	25°C Typ	25°C Max	85°C Min	85°C Max	Units
I_{DD}	Quiescent Device Current	$V_{DD} = 5V$, $V_{IN} = V_{DD}$ or V_{SS}		1.0		0.01	1.0		7.5	µA
		$V_{DD} = 10V$, $V_{IN} = V_{DD}$ or V_{SS}		2.0		0.01	2.0		15	µA
		$V_{DD} = 15V$, $V_{IN} = V_{DD}$ or V_{SS}		4.0		0.01	4.0		30	µA
Signal Inputs and Outputs										
R_{ON}	"ON" Resistance	$R_L = 10\ k\Omega$ to $\frac{V_{DD} - V_{SS}}{2}$ $V_C = V_{DD}$, $V_{IS} = V_{SS}$ or V_{DD} $V_{DD} = 10V$ $V_{DD} = 15V$		610 370		275 200	660 400		840 520	Ω Ω
		$R_L = 10\ k\Omega$ to $\frac{V_{DD} - V_{SS}}{2}$ $V_C = V_{DD}$ $V_{DD} = 10V$, $V_{IS} = 4.75$ to $5.25V$ $V_{DD} = 15V$, $V_{IS} = 7.25$ to $7.75V$		1900 790		850 400	2000 850		2380 1080	Ω Ω
ΔR_{ON}	Δ"ON" Resistance Between any 2 of 4 Switches (In Same Package)	$R_L = 10\ k\Omega$ to $\frac{V_{DD} - V_{SS}}{2}$ $V_C = V_{DD}$, $V_{IS} = V_{SS}$ to V_{DD} $V_{DD} = 10V$ $V_{DD} = 15V$				15 10				Ω Ω
I_{IS}	Input or Output Leakage Switch "OFF"	$V_C = 0$, $V_{DD} = 15V$ $V_{IS} = 0V$ or $15V$, $V_{OS} = 15V$ or $0V$		±50		±0.1	±50		±200	nA
Control Inputs										
V_{ILC}	Low Level Input Voltage	$V_{IS} = V_{SS}$ and V_{DD} $V_{OS} = V_{DD}$ and V_{SS} $I_{IS} = \pm 10\ \mu A$ $V_{DD} = 5V$ $V_{DD} = 10V$ $V_{DD} = 15V$		0.9 0.9 0.9			0.7 0.7 0.7		0.4 0.4 0.4	V V V
V_{IHC}	High Level Input Voltage	$V_{DD} = 5V$ $V_{DD} = 10V$ (see Note 6 and $V_{DD} = 15V$ Figure 8)	3.5 7.0 11.0		3.5 7.0 11.0			3.5 7.0 11.0		V V V
I_{IN}	Input Current	$V_{CC} - V_{SS} = 15V$ $V_{DD} \geq V_{IS} \geq V_{SS}$ $V_{DD} \geq V_C \geq V_{SS}$		±0.3		±10⁻⁵	±0.3		±1.0	µA

AC Electrical Characteristics $T_A = 25°C$, $t_r = t_f = 20$ ns and $V_{SS} = 0V$ unless otherwise specified

Symbol	Parameter	Conditions	Min	Typ	Max	Units
t_{PHL}, t_{PLH}	Propagation Delay Time Signal Input to Signal Output	$V_C = V_{DD}$, $C_L = 50$ pF, (Figure 1) $R_L = 200k$ $V_{DD} = 5V$ $V_{DD} = 10V$ $V_{DD} = 15V$		58 27 20	100 50 40	ns ns ns
t_{PZH}, t_{PZL}	Propagation Delay Time Control Input to Signal Output High Impedance to Logical Level	$R_L = 1.0\ k\Omega$, $C_L = 50$ pF, (Figures 2 and 3) $V_{DD} = 5V$ $V_{DD} = 10V$ $V_{DD} = 15V$		20 18 17	50 40 35	ns ns ns
t_{PHZ}, t_{PLZ}	Propagation Delay Time Control Input to Signal Output Logical Level to High Impedance	$R_L = 1.0\ k\Omega$, $C_L = 50$ pF, (Figures 2 and 3) $V_{DD} = 5V$ $V_{DD} = 10V$ $V_{DD} = 15V$		15 11 10	40 25 22	ns ns ns
	Sine Wave Distortion	$V_C = V_{DD} = 5V$, $V_{SS} = -5$ $R_L = 10\ k\Omega$, $V_{IS} = 5\ V_{p-p}$, $f = 1$ kHz, (Figure 4)		0.4		%

Appendix C

AC Electrical Characteristics (Continued)

$T_A = 25°C$, $t_r = t_f = 20$ ns and $V_{SS} = 0V$ unless otherwise specified

Symbol	Parameter	Conditions	Min	Typ	Max	Units
	Frequency Response — Switch "ON" (Frequency at −3 dB)	$V_C = V_{DD} = 5V$, $V_{SS} = -5V$, $R_L = 1$ kΩ, $V_{IS} = 5$ V_{P-P}, $20 \log_{10} V_{OS}/V_{OS}$ (1 kHz) −dB, (Figure 4)		40		MHz
	Feedthrough — Switch "OFF" (Frequency at −50 dB)	$V_{DD} = 5V$, $V_C = V_{SS} = -5V$, $R_L = 1$ kΩ, $V_{IS} = 5$ V_{P-P}, $20 \log_{10} (V_{OS}/V_{IS}) = -50$ dB, (Figure 4)		1.25		MHz
	Crosstalk Between Any Two Switches (Frequency at −50 dB)	$V_{DD} = V_{C(A)} = 5V$; $V_{SS} = V_{C(B)} = -5V$, $R_L = 1$ kΩ $V_{IS(A)} = 5$ V_{P-P}, $20 \log_{10} (V_{OS(B)}/V_{OS(A)}) = -50$ dB, (Figure 5)		0.9		MHz
	Crosstalk; Control Input to Signal Output	$V_{DD} = 10V$, $R_L = 10$ kΩ $R_{IN} = 1$ kΩ, $V_{CC} = 10V$ Square Wave, $C_L = 50$ pF (Figure 6)		150		mV$_{P-P}$
	Maximum Control Input	$R_L = 1$ kΩ, $C_L = 50$ pF, (Figure 7) $V_{OS(f)} = ½ V_{OS}$(1 kHz)				
		$V_{DD} = 5V$		6.5		MHz
		$V_{DD} = 10V$		8.0		MHz
		$V_{DD} = 15V$		9.0		MHz
C_{IS}	Signal Input Capacitance			4		pF
C_{OS}	Signal Output Capacitance	$V_{DD} = 10V$		4		pF
C_{IOS}	Feedthrough Capacitance	$V_C = 0V$		0.2		pF
C_{IN}	Control Input Capacitance			5	7.5	pF

Note 1: "Absolute Maximum Ratings" are those values beyond which the safety of the device cannot be guaranteed. They are not meant to imply that the devices should be operated at these limits. The tables of "Recommended Operating Conditions" and "Electrical Characteristics" provide conditions for actual device operation.

Note 2: $V_{SS} = 0V$ unless otherwise specified.

Note 3: These devices should not be connected to circuits with the power "ON".

Note 4: In all cases, there is approximately 5 pF of probe and jig capacitance on the output; however, this capacitance is included in C_L wherever it is specified.

Note 5: V_{IS} is the voltage at the in/out pin and V_{OS} is the voltage at the out/in pin. V_C is the voltage at the control input.

Note 6: If the switch input is held at V_{DD}, V_{IHC} is the control input level that will cause the switch output to meet the standard "B" series V_{OH} and I_{OH} output levels. If the analog switch input is connected to V_{SS}, V_{IHC} is the control input level — which allows the switch to sink standard "B" series $|I_{OH}|$, high level current, and still maintain a $V_{OL} \leq$ "B" series. These currents are shown in Figure 8.

AC Test Circuits and Switching Time Waveforms

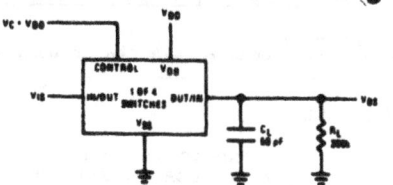

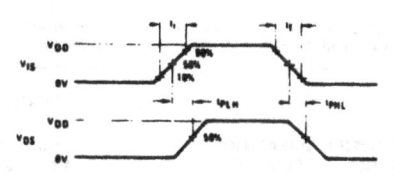

Figure 1. t_{PLH}, t_{PLH} Propagation Delay Time Signal Input to Signal Output

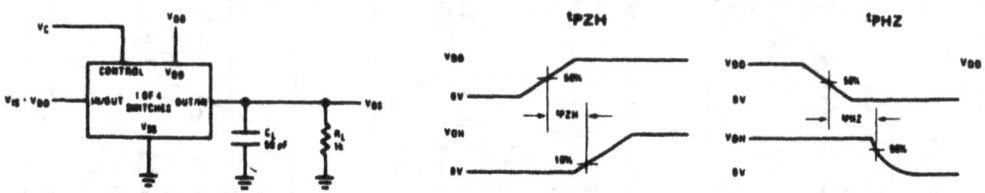

FIGURE 2. t_{PZH}, t_{PHZ} Propagation Delay Time Control to Signal Output

4023B
TRIPLE 3-INPUT NAND GATE

DESCRIPTION — This CMOS logic element provides a 3-input positive NAND function. The outputs are fully buffered for highest noise immunity and pattern insensitivity of output impedance.

LOGIC AND CONNECTION DIAGRAM
DIP (TOP VIEW)

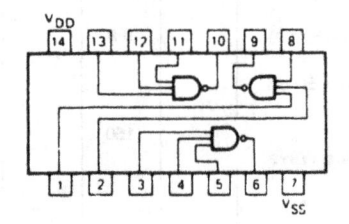

NOTE:
The Flatpak version has the same pinouts (Connection Diagram) as the Dual In-line Package.

DC CHARACTERISTICS: V_{DD} as shown, V_{SS} = 0 V (See Note 1)

SYMBOL	PARAMETER		V_{DD} = 5 V MIN	TYP	MAX	V_{DD} = 10 V MIN	TYP	MAX	V_{DD} = 15 V MIN	TYP	MAX	UNITS	TEMP	TEST CONDITIONS
I_{DD}	Quiescent Power Supply Current	XC			1			2			4	µA	MIN, 25°C	All inputs at 0 V or V_{DD}
					7.5			15			30		MAX	
		XM			0.25			0.5			1	µA	MIN, 25°C	
					7.5			15			30		MAX	

AC CHARACTERISTICS: V_{DD} as shown, V_{SS} = 0 V, T_A = 25°C (See Note 2)

SYMBOL	PARAMETER	V_{DD} = 5 V MIN	TYP	MAX	V_{DD} = 10 V MIN	TYP	MAX	V_{DD} = 15V MIN	TYP	MAX	UNITS	TEST CONDITIONS
t_{PLH}	Propagation Delay		45	110		25	60		19	48	ns	C_L = 50 pF, R_L = 200 kΩ
t_{PHL}			51	110		25	60		12	48	ns	
t_{TLH}	Output Transition Time		45	135		18	70		17	45	ns	Input Transition Times ≤ 20 ns
t_{THL}			45	135		18	70		12	45	ns	

NOTES:
1. Additional DC Characteristics are listed in this section under 4000B Series CMOS Family Characteristics.
2. Propagation Delays and Output Transition Times are graphically described in this section under 4000B Series CMOS Family Characteristics.

TYPICAL ELECTRICAL CHARACTERISTICS

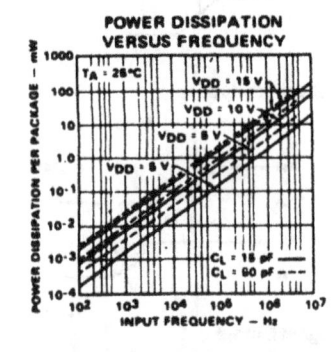

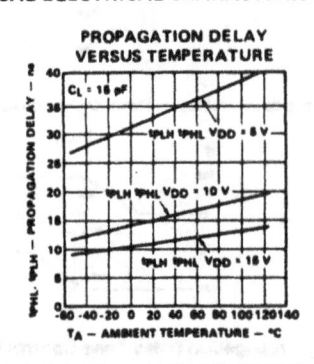

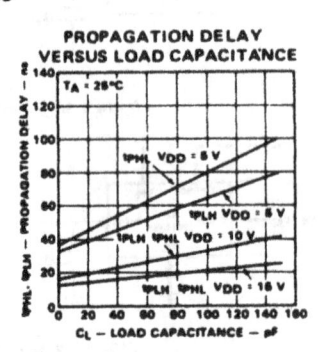

POWER DISSIPATION VERSUS FREQUENCY

PROPAGATION DELAY VERSUS TEMPERATURE

PROPAGATION DELAY VERSUS LOAD CAPACITANCE

(Courtesy of Fairchild – A Schlumberger Company)

Appendix C

SN5490A, SN5492A, SN5493A, SN54LS90, SN54LS92, SN54LS93, SN7490A, SN7492A, SN7493A, SN74LS90, SN74LS92, SN74LS93
DECADE, DIVIDE-BY-TWELVE AND BINARY COUNTERS

MARCH 1974 – REVISED MARCH 1988

- '90A, 'LS90 . . . Decade Counters
- '92A, 'LS92 . . . Divide By-Twelve Counters
- '93A, 'LS93 . . . 4-Bit Binary Counters

TYPES	TYPICAL POWER DISSIPATION
'90A	145 mW
'92A, '93A	130 mW
'LS90, 'LS92, 'LS93	45 mW

SN5490A, SN54LS90 . . . J OR W PACKAGE
SN7490A . . . N PACKAGE
SN74LS90 . . . D OR N PACKAGE
(TOP VIEW)

```
CKB   [ 1    14 ] CKA
R0(1) [ 2    13 ] NC
R0(2) [ 3    12 ] QA
NC    [ 4    11 ] QD
VCC   [ 5    10 ] GND
R9(1) [ 6     9 ] QB
R9(2) [ 7     8 ] QC
```

SN5492A, SN54LS92 . . . J OR W PACKAGE
SN7492A . . . N PACKAGE
SN74LS92 . . . D OR N PACKAGE
(TOP VIEW)

```
CKB   [ 1    14 ] CKA
NC    [ 2    13 ] NC
NC    [ 3    12 ] QA
NC    [ 4    11 ] QB
VCC   [ 5    10 ] GND
R0(1) [ 6     9 ] QC
R0(2) [ 7     8 ] QD
```

SN5493A, SN54LS93 . . . J OR W PACKAGE
SN7493 . . . N PACKAGE
SN74LS93 . . . D OR N PACKAGE
(TOP VIEW)

```
CKB   [ 1    14 ] CKA
R0(1) [ 2    13 ] NC
R0(2) [ 3    12 ] QA
NC    [ 4    11 ] QD
VCC   [ 5    10 ] GND
NC    [ 6     9 ] QB
NC    [ 7     8 ] QC
```

NC – No internal connection

description

Each of these monolithic counters contains four master-slave flip-flops and additional gating to provide a divide-by-two counter and a three-stage binary counter for which the count cycle length is divide-by-five for the '90A and 'LS90, divide-by-six for the '92A and 'LS92, and the divide-by-eight for the '93A and 'LS93.

All of these counters have a gated zero reset and the '90A and 'LS90 also have gated set-to-nine inputs for use in BCD nine's complement applications.

To use their maximum count length (decade, divide-by-twelve, or four-bit binary) of these counters, the CKB input is connected to the Q_A output. The input count pulses are applied to CKA input and the outputs are as described in the appropriate function table. A symmetrical divide-by-ten count can be obtained from the '90A or 'LS90 counters by connecting the Q_D output to the CKA input and applying the input count to the CKB input which gives a divide-by-ten square wave at output Q_A.

PRODUCTION DATA documents contain information current as of publication date. Products conform to specifications per the terms of Texas Instruments standard warranty. Production processing does not necessarily include testing of all parameters.

TEXAS INSTRUMENTS
POST OFFICE BOX 655012 • DALLAS TEXAS 75265

TTL Devices

SN5490A, '92A, '93A, SN54LS90, 'LS92, 'LS93, SN7490A, '92A, '93A, SN74LS90, 'LS92, 'LS93
DECADE, DIVIDE-BY-TWELVE, AND BINARY COUNTERS

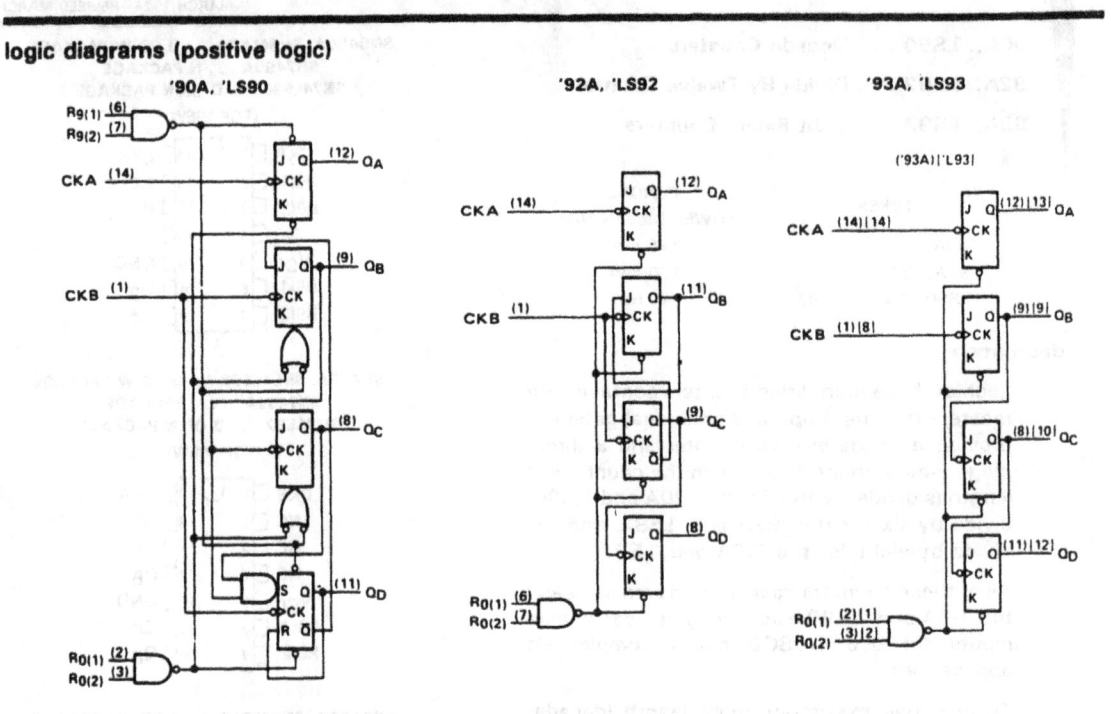

logic diagrams (positive logic)

'90A, 'LS90

'92A, 'LS92

'93A, 'LS93 ('93A)['L93]

The J and K inputs shown without connection are for reference only and are functionally at a high level.
Pin numbers shown in ( ) are for the 'LS93 and '93A and pin numbers shown in [ ] are for the 54L93.

schematics of inputs and outputs

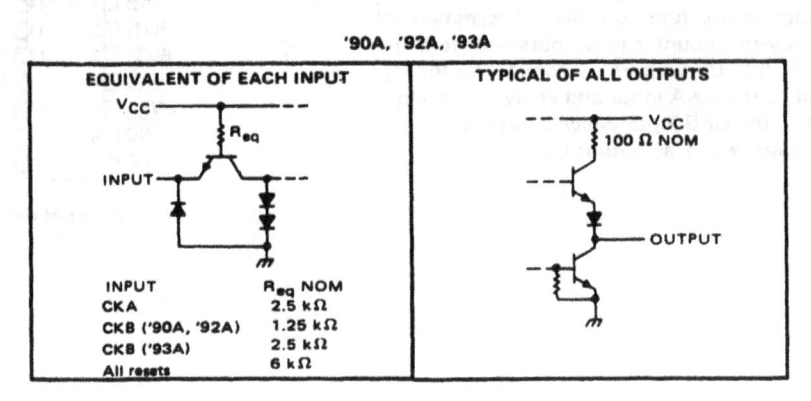

'90A, '92A, '93A

EQUIVALENT OF EACH INPUT	TYPICAL OF ALL OUTPUTS

INPUT	R_{eq} NOM
CKA	2.5 kΩ
CKB ('90A, '92A)	1.25 kΩ
CKB ('93A)	2.5 kΩ
All resets	6 kΩ

Output: 100 Ω NOM

Texas Instruments
POST OFFICE BOX 655012 • DALLAS, TEXAS 75265

Appendix C

intel

2732A
32K (4K x 8) PRODUCTION AND UV ERASABLE PROMS

- 200 ns (2732A-2) Maximum Access Time ... HMOS*-E Technology
- Compatible with High-Speed Microcontrollers and Microprocessors ... Zero WAIT State
- Two Line Control
- 10% V_{CC} Tolerance Available
- Low Current Requirement
  - 100 mA Active
  - 35 mA Standby
- int$_e$ligent Identifier™ Mode
  - Automatic Programming Operation
- Industry Standard Pinout ... JEDEC Approved 24 Pin Ceramic and Plastic Package

(See Packaging Spec. Order #221369)

The Intel 2732A is a 5V-only, 32,768-bit ultraviolet erasable (cerdip) Electrically Programmable Read-Only Memory (EPROM). The standard 2732A access time is 250 ns with speed selection (2732A-2) available at 200 ns. The access time is compatible with high performance microprocessors such as the 8 MHz iAPX 186. In these systems, the 2732A allows the microprocessor to operate without the addition of WAIT states.

The 2732A is currently available in two different package types. Cerdip packages provide flexibility in prototyping and R & D environments where reprogrammability is required. Plastic DIP EPROMs provide optimum cost effectiveness in production environments. Inventoried in the unprogrammed state, the P2732A is programmed quickly and efficiently when the need to change code arises. Costs incurred for new ROM masks or obsoleted ROM inventories are avoided. The tight package dimensional controls, inherent non-erasability, and high reliability of the P2732A make it the ideal component for these production applications.

An important 2732A feature is Output Enable ($\overline{OE}$) which is separate from the Chip Enable ($\overline{CE}$) control. The $\overline{OE}$ control eliminates bus contention in microprocessor systems. The $\overline{CE}$ is used by the 2732A to place it in a standby mode ($\overline{CE}$ = V_{IH}) which reduces power consumption without increasing access time. The standby mode reduces the current requirement by 65%; the maximum active current is reduced from 100 mA to a standby current of 35 mA.

*HMOS is a patented process of Intel Corporation.

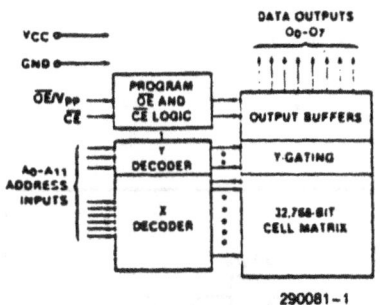

Figure 1. Block Diagram

Pin Names

A_0–A_{11}	Addresses
$\overline{CE}$	Chip Enable
$\overline{OE}/V_{pp}$	Output Enable/V_{pp}
O_0–O_7	Outputs

Figure 2. Cerdip/Plastic DIP Pin Configuration

NOTE:
Intel "Universal Site" compatible EPROM configurations are shown in the blocks adjacent to the 2732A pins.

intel 2732A

EXTENDED TEMPERATURE (EXPRESS) EPROMs

The Intel EXPRESS EPROM family is a series of electrically programmable read only memories which have received additional processing to enhance product characteristics. EXPRESS processing is available for several densities of EPROM, allowing the choice of appropriate memory size to match system applications. EXPRESS EPROM products are available with 168 ±8 hour, 125°C dynamic burn-in using Intel's standard bias configuration. This process exceeds or meets most industry specifications of burn-in. The standard EXPRESS EPROM operating temperature range is 0°C to 70°C. Extended operating temperature range (−40°C to +85°C) EXPRESS products are available. Like all Intel EPROMs, the EXPRESS EPROM family is inspected to 0.1% electrical AQL. This may allow the user to reduce or eliminate incoming inspection testing.

READ OPERATION

D.C. CHARACTERISTICS

Electrical Parameters of EXPRESS EPROM products are identical to standard EPROM parameters except for:

Symbol	Parameter	TD2732A LD2732A Min	TD2732A LD2732A Max	Test Conditions
I_{SB}	V_{CC} Standby Current (mA)		45	$\overline{CE} = V_{IH}$, $\overline{OE} = V_{IL}$
I_{CC_1}[1]	V_{CC} Active Current (mA)		150	$\overline{OE} = \overline{CE} = V_{IL}$
	V_{CC} Active Current at High Temperature (mA)		125	$\overline{OE} = \overline{CE} = V_{IL}$, $V_{PP} = V_{CC}$, $T_{Ambient} = 85°C$

NOTE:
1. Maximum current value is with outputs O_0 to O_7 unloaded.

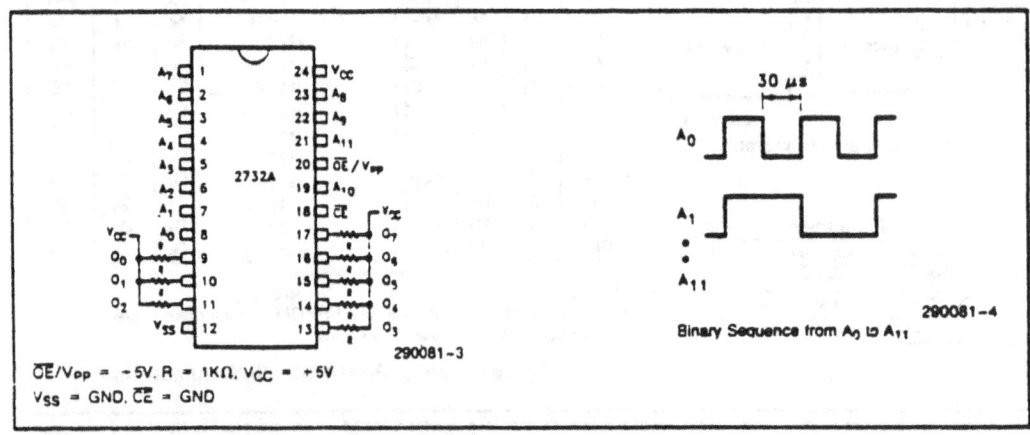

EXPRESS EPROM PRODUCT FAMILY

PRODUCT DEFINITONS

Type	Operating Temperature	Burn-in 125°C (hr)
Q	0°C to +70°C	168 ±8
T	−40°C to −85°C	None
L	−40°C to −85°C	168 ±8

EXPRESS OPTIONS

2732A Versions

Packaging Options		
Speed Versions	Cerdip	Plastic
−2	Q	
STD	Q, T, L	
−3	Q	
−4	Q, T, L	
−20	Q	
−25	Q, T, L	
−30	Q	
−45	Q, T, L	

$\overline{OE}/V_{PP} = -5V$, $R = 1K\Omega$, $V_{CC} = +5V$
$V_{SS} = GND$, $\overline{CE} = GND$

Binary Sequence from A_0 to A_{11}

Burn-In Bias and Timing Diagrams

intel 2732A

ABSOLUTE MAXIMUM RATINGS*

Operating Temp. During Read .... 0°C to +70°C
Temperature Under Bias .... −10°C to +80°C
Storage Temperature .... −65°C to +125°C
All Input or Output Voltages with
  Respect to Ground .... −0.3V to +6V
Voltage on A9 with Respect
  to Ground .... −0.3V to +13.5V
V_{PP} Supply Voltage with Respect to Ground
  During Programming .... −0.3V to +22V
V_{CC} Supply Voltage with
  Respect to Ground .... −0.3V to +7.0V

*Notice: Stresses above those listed under "Absolute Maximum Ratings" may cause permanent damage to the device. This is a stress rating only and functional operation of the device at these or any other conditions above those indicated in the operational sections of this specification is not implied. Exposure to absolute maximum rating conditions for extended periods may affect device reliability.

READ OPERATION

D.C. CHARACTERISTICS $0°C \leq T_A \leq +70°C$

Symbol	Parameter	Limits Min	Limits Typ(3)	Limits Max	Units	Conditions
I_{LI}	Input Load Current			10	µA	V_{IN} = 5.5V
I_{LO}	Output Leakage Current			10	µA	V_{OUT} = 5.5V
I_{SB}(2)	V_{CC} Current (Standby)			35	mA	$\overline{CE} = V_{IH}, \overline{OE} = V_{IL}$
I_{CC1}(2)	V_{CC} Current (Active)			100	mA	$\overline{OE} = \overline{CE} = V_{IL}$
V_{IL}	Input Low Voltage	−0.1		0.8	V	
V_{IH}	Input High Voltage	2.0		V_{CC} + 1	V	
V_{OL}	Output Low Voltage			0.45	V	I_{OL} = 2.1 mA
V_{OH}	Output High Voltage	2.4			V	I_{OH} = −400 µA

A.C. CHARACTERISTICS $0°C \leq T_A \leq 70°C$

Versions	V_{CC} ±5%	2732A-2 / P2732A-2		2732A / P2732A		2732A-3 / P2732A-3		2732A-4 / P2732A-4		Units	Test Conditions
	V_{CC} ±10%	2732A-20		2732A-25		2732A-30		2732A-45			
Symbol	Parameter	Min	Max	Min	Max	Min	Max	Min	Max		
t_{ACC}	Address to Output Delay		200		250		300		450	ns	$\overline{CE} = \overline{OE} = V_{IL}$
t_{CE}	$\overline{CE}$ to Output Delay		200		250		300		450	ns	$\overline{OE} = V_{IL}$
t_{OE}	$\overline{OE}/V_{PP}$ to Output Delay		70		100		150		150	ns	$\overline{CE} = V_{IL}$
t_{DF}(4)	$\overline{OE}/V_{PP}$ High to Output Float	0	60	0	60	0	130	0	130	ns	$\overline{CE} = V_{IL}$
t_{OH}	Output Hold from Addresses, $\overline{CE}$ or $\overline{OE}/V_{PP}$, Whichever Occurred First	0		0		0		0		ns	$\overline{CE} = \overline{OE} = V_{IL}$

NOTES:
1. V_{CC} must be applied simultaneously or before $\overline{OE}/V_{PP}$ and removed simultaneously or after $\overline{OE}/V_{PP}$.
2. The maximum current value is with outputs O_0 to O_7 unloaded.
3. Typical values are for T_A = 25°C and nominal supply voltages.
4. This parameter is only sampled and is not 100% tested. Output Float is defined as the point where data is no longer driven—see timing diagram.

intel 2732A

CAPACITANCE [2] $T_A = 25°C$, $f = 1$ MHz

Symbol	Parameter	Typ	Max	Unit	Conditions
C_{IN1}	Input Capacitance Except $\overline{OE}/V_{PP}$	4	6	pF	$V_{IN} = 0V$
C_{IN2}	$\overline{OE}/V_{PP}$ Input Capacitance		20	pF	$V_{IN} = 0V$
C_{OUT}	Output Capacitance	8	12	pF	$V_{OUT} = 0V$

A.C. TESTING INPUT/OUTPUT WAVEFORM

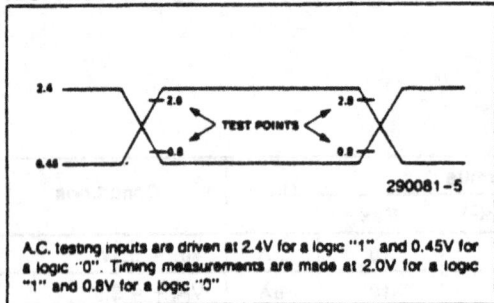

A.C. testing inputs are driven at 2.4V for a logic "1" and 0.45V for a logic "0". Timing measurements are made at 2.0V for a logic "1" and 0.8V for a logic "0".

A.C. TESTING LOAD CIRCUIT

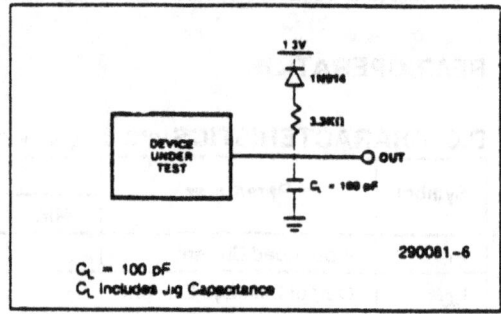

$C_L = 100$ pF
C_L Includes Jig Capacitance

A.C. WAVEFORMS

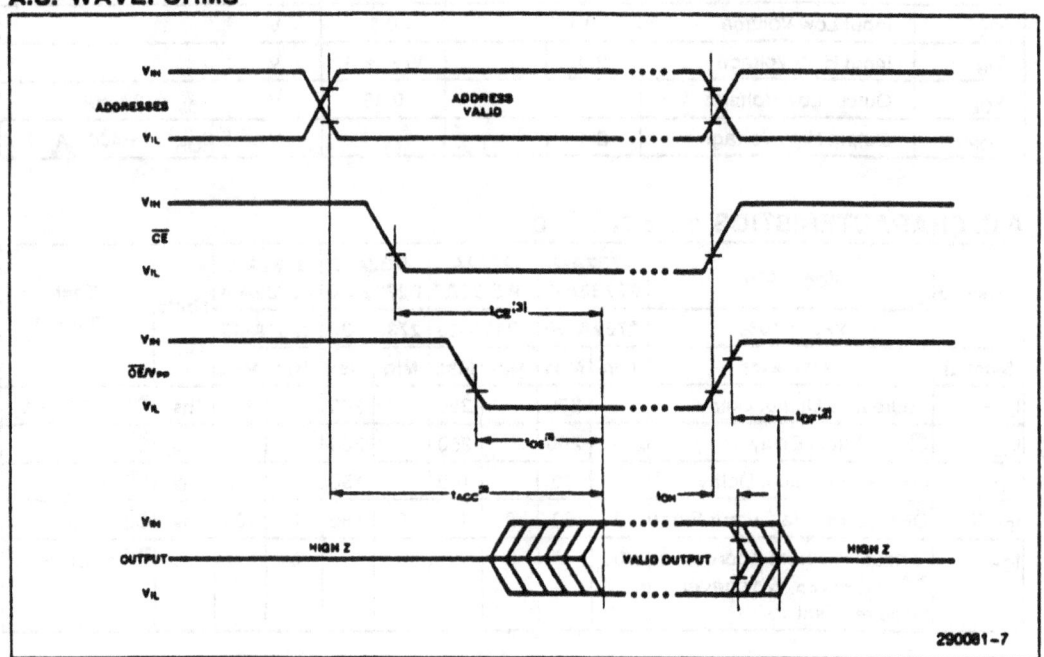

NOTES:
1. Typical values are for $T_A = 25°C$ and nominal supply voltages.
2. This parameter is only sampled and is not 100% tested. Output float is defined as the point where data is no longer driven—see timing diagram.
3. $\overline{OE}/V_{PP}$ may be delayed up to $t_{ACC} - t_{OE}$ after the falling edge of $\overline{CE}$ without impacting t_{CE}.

intel 2732A

PLASTIC EPROM APPLICATIONS

Intel's P2732A is the result of a multi-year effort to make EPROMs more cost effective for production applications. The benefits of a plastic package enable the P2732A to be used for high volume production with lower profile boards and easier production assembly (no cover over UV transparent windows).

The reliability of plastic EPROMs is equivalent to traditional CERDIP packaging. The plastic is rugged and durable making it optimal for auto insertion and auto handling equipment. Design and testing ensures device programmability, data integrity, and impermeability to moisture.

Intel's Plastic EPROMs are designed for total compatibility with their CERDIP packaged predecessors. This encompasses quality, reliability, and programming. All Intel Plastic EPROMs have passed Intel's strict process and product reliability qualifications.

DEVICE OPERATION

The modes of operation of the 2732A are listed in Table 1. A single 5V power supply is required in the read mode. All inputs are TTL levels except for $\overline{OE}/V_{PP}$ during programming and 12V on A_9 for the int$_e$ligent Identifier™ mode. In the program mode the $\overline{OE}/V_{PP}$ input is pulsed from a TTL level to 21V.

Table 1. Mode Selection

Mode \ Pins	$\overline{CE}$	$\overline{OE}/V_{PP}$	A_9	A_0	V_{CC}	Outputs
Read/Program Verify	V_{IL}	V_{IL}	X	X	V_{CC}	D_{OUT}
Output Disable	V_{IL}	V_{IH}	X	X	V_{CC}	High Z
Standby	V_{IH}	X	X	X	V_{CC}	High Z
Program	V_{IL}	V_{PP}	X	X	V_{CC}	D_{IN}
Program Inhibit	V_{IH}	V_{PP}	X	X	V_{CC}	High Z
Int$_e$ligent Identifier[3]						
—Manufacturer	V_{IL}	V_{IL}	V_H	V_{IL}	V_{CC}	89H
—Device	V_{IL}	V_{IL}	V_H	V_{IH}	V_{CC}	01H

NOTES:
1. X can be V_{IH} or V_{IL}.
2. V_H = 12V ±0.5V.
3. A_1–A_8, A_{10}, A_{11} = V_{IL}.

Read Mode

The 2732A has two control functions, both of which must be logically active in order to obtain data at the outputs. Chip Enable ($\overline{CE}$) is the power control and should be used for device selection. Output Enable ($\overline{OE}/V_{PP}$) is the output control and should be used to gate data from the output pins, independent of device selection. Assuming that addresses are stable, address access time (t_{ACC}) is equal to the delay from $\overline{CE}$ to output (t_{CE}). Data is available at the outputs after the falling edge of $\overline{OE}/V_{PP}$, assuming that $\overline{CE}$ has been low and addresses have been stable for at least $t_{ACC} - t_{OE}$.

Standby Mode

EPROMs can be placed in a standby mode which reduces the maximum active current of the device by applying a TTL-high signal to the $\overline{CE}$ input. When in standby mode, the outputs are in a high impedance state, independent of the $\overline{OE}/V_{PP}$ input.

Two Line Output Control

Because EPROMs are usually used in larger memory arrays, Intel has provided two control lines which accommodate this multiple memory connection. The two control lines allow for:

a) The lowest possible memory power dissipation, and

b) complete assurance that output bus contention will not occur.

To use these two control lines most efficiently, $\overline{CE}$ should be decoded and used as the primary device selecting function, while $\overline{OE}/V_{PP}$ should be made a common connection to all devices in the array and connected to the $\overline{READ}$ line from the system control bus. This assures that all deselected memory devices are in their low power standby mode and that the output pins are active only when data is desired from a particular memory device.

SYSTEM CONSIDERATION

The power switching characteristics of EPROMs require careful decoupling of the devices. The supply current, I_{CC}, has three segments that are of interest to the system designer—the standby current level, the active current level, and the transient current peaks that are produced by the falling and rising edges of Chip Enable. The magnitude of these transient current peaks is dependent on the output capacitive and inductive loading of the device. The associated transient voltage peaks can be suppressed by complying with Intel's two-line control and by use of properly selected decoupling capacitors. It is recommended that a 0.1 μF ceramic capacitor be used on every device between V_{CC} and GND. This should be a high frequency capacitor of low inherent inductance and should be placed as close to the device as possible. In addition, a 4.7 μF bulk electrolytic capacitor should be used between V_{CC} and GND for

intel 2732A

every eight devices. The bulk capacitor should be located near where the power supply is connected to the array. The purpose of the bulk capacitor is to overcome the voltage droop caused by the inductive effects of PC board traces.

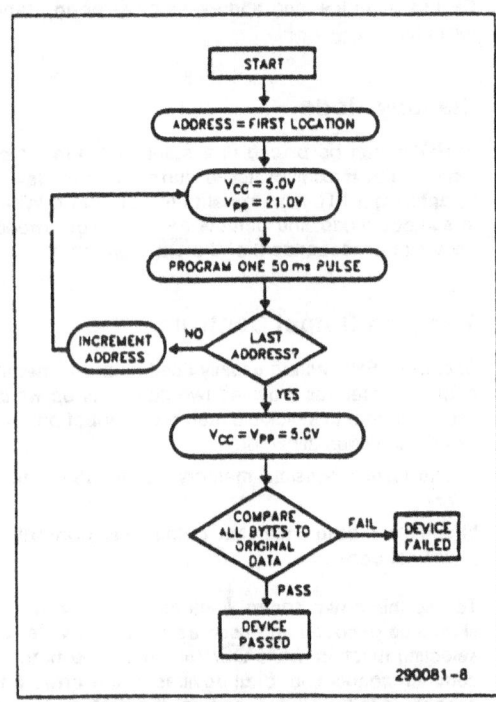

Figure 3. Standard Programming Flowchart

PROGRAMMING MODES

CAUTION: Exceeding 22V on $\overline{OE}/V_{PP}$ will permanently damage the device.

Initially, and after each erasure (cerdip EPROMs), all bits of the EPROM are in the "1" state. Data is introduced by selectively programming "0s" into the bit locations. Although only "0s" will be programmed, both "1s" and "0s" can be present in the data word. The only way to change a "0" to a "1" in cerdip EPROMs is by ultraviolet light erasure.

The device is in the programming mode when the $\overline{OE}/V_{PP}$ input is at 21V. It is required that a 0.1 μF capacitor be placed across $\overline{OE}/V_{PP}$ and ground to suppress spurious voltage transients which may damage the device. The data to be programmed is applied 8 bits in parallel to the data output pins. The levels required for the address and data inputs are TTL.

When the address and data are stable, a 20 ms (50 ms typical) active low, TTL program pulse is applied to the $\overline{CE}$ input. A program pulse must be applied at each address location to be programmed (see Figure 3). Any location can be programmed at any time—either individually, sequentially, or at random. The program pulse has a maximum width of 55 ms. The EPROM must not be programmed with a DC signal applied to the $\overline{CE}$ input.

Programming of multiple 2732As in parallel with the same data can be easily accomplished due to the simplicity of the programming requirements. Like inputs of the paralleled 2732As may be connected together when they are programmed with the same data. A low level TTL pulse applied to the $\overline{CE}$ input programs the paralleled 2732As.

Program Inhibit

Programming of multiple EPROMs in parallel with different data is easily accomplished by using the Program Inhibit mode. A high level $\overline{CE}$ input inhibits the other EPROMs from being programmed. Except for $\overline{CE}$, all like inputs (including $\overline{OE}/V_{PP}$) of the parallel EPROMs may be common. A TTL low level pulse applied to the $\overline{CE}$ input with $\overline{OE}/V_{PP}$ at 21V will program that selected device.

Program Verify

A verify (Read) should be performed on the programmed bits to determine that they have been correctly programmed. The verify is performed with $\overline{OE}/V_{PP}$ and $\overline{CE}$ at V_{IL}. Data should be verified t_{DV} after the falling edge of $\overline{CE}$.

int_eligent Identifier™ Mode

The int_eligent Identifier Mode allows the reading out of a binary code from an EPROM that will identify its manufacturer and type. This mode is intended for use by programming equipment for the purpose of automatically matching the device to be programmed with its corresponding programming algorithm. This mode is functional in the 25°C ±5°C ambient temperature range that is required when programming the device.

To activate this mode, the programming equipment must force 11.5V to 12.5V on address line A9 of the EPROM. Two identifier bytes may then be sequenced from the device outputs by toggling address line A0 from V_{IL} to V_{IH}. All other address lines must be held at V_{IL} during the int_eligent Identifier Mode.

Byte 0 (A0 = V_{IL}) represents the manufacturer code and byte 1 (A0 = V_{IH}) the device identifier code. These two identifier bytes are given in Table 1.

intel 2732A

INTEL EPROM PROGRAMMING SUPPORT TOOLS

Intel offers a full line of EPROM Programmers providing state-of-the-art programming for all Intel programmable devices. The modular architecture of Intel's EPROM programmers allows you to add new support as it becomes available, with very low cost add-ons. For example, even the earliest users of the iUP-FAST 27/K module may take advantage of Intel's new Quick-Pulse Programming™ Algorithm, the fastest in the industry.

Intel EPROM programmers may be controlled from a host computer using Intel's PROM Programming software (iPPS). iPPS makes programming easy for a growing list of industry standard hosts, including the IBM PC, XT, AT; and PCDOS compatibles, Intellec Development Systems, Intel's iPDS Personal Development System, and the Intel Network Development System (iNDS-II). Stand-alone operation is also available, including device previewing, editing, programming, and download of programming data from any source over an RS232C port.

For further details consult the EPROM Programming section of the Development Systems Handbook.

ERASURE CHARACTERISTICS (FOR CERDIP EPROMS)

The erasure characteristics are such that erasure begins to occur upon exposure to light with wavelengths shorter than aproximately 4000 Angstroms (Å). It should be noted that sunlight and certain types of fluorescent lamps have wavelengths in the 3000–4000Å range. Data shows that constant exposure to room level fluorescent lighting could erase the EPROM in approximately 3 years, while it would take approximately 1 week to cause erasure when exposed to direct sunlight. If the device is to be exposed to these types of lighting conditions for extended periods of time, opaque labels should be placed over the window to prevent unintentional erasure.

The recommended erasure procedure is exposure to shortwave ultraviolet light which has a wavelength of 2537 Angstroms (Å). The integrated dose (i.e., UV intensity $\times$ exposure time) for erasure should be a minimum of 15 Wsec/cm^2. The erasure time with this dosage is approximately 15 to 20 minutes using an ultraviolet lamp with a 12000 μW/cm^2 power rating. The EPROM should be placed within 1 inch of the lamp tubes during erasure. The maximum integrated dose an EPROM can be exposed to without damage is 7258 Wsec/cm^2 (1 week @ 12000 μW/cm^2). Exposure of the device to high intensity UV light for longer periods may cause permanent damage.

PROGRAMMING

D.C. PROGRAMMING CHARACTERISTICS

$T_A = 25°C \pm 5°C$, $V_{CC} = 5V \pm 5\%$, $V_{PP} = 21V \pm 0.5V$

Symbol	Parameter	Min	Typ(3)	Max	Units	Test Conditions (Note 1)
I_{LI}	Input Current (All Inputs)			10	μA	$V_{IN} = V_{IL}$ or V_{IH}
V_{IL}	Input Low Level (All Inputs)	-0.1		0.8	V	
V_{IH}	Input High Level (All Inputs Except $\overline{OE}/V_{PP}$)	2.0		V_{CC} + 1	V	
V_{OL}	Output Low Voltage During Verify			0.45	V	I_{OL} = 2.1 mA
V_{OH}	Output High Voltage During Verify	2.4			V	I_{OH} = -400 μA
I_{CC_2}(4)	V_{CC} Supply Current (Program and Verify)		85	100	mA	
I_{PP_2}(4)	V_{PP} Supply Current (Program)			30	mA	$\overline{CE} = V_{IL}$, $\overline{OE}/V_{PP} = V_{PP}$
V_{ID}	A$_9$ intelligent Identifier Voltage	11.5		12.5	V	

intel 2732A

A.C. PROGRAMMING CHARACTERISTICS

$T_A = 25°C \pm 5°C$, $V_{CC} = 5V \pm 5\%$, $V_{PP} = 21V \pm 0.5V$

Symbol	Parameter	Limits Min	Limits Typ(3)	Limits Max	Units	Test Conditions* (Note 1)
t_{AS}	Address Setup Time	2			µs	
t_{OES}	$\overline{OE}/V_{PP}$ Setup Time	2			µs	
t_{DS}	Data Setup Time	2			µs	
t_{AH}	Address Hold Time	0			µs	
t_{DH}	Data Hold Time	2			µs	
t_{DFP}	$\overline{OE}/V_{PP}$ High to Output Not Driven	0		130	ns	(Note 2)
t_{PW}	$\overline{CE}$ Pulse Width During Programming	20	50	55	ms	
t_{OEH}	$\overline{OE}/V_{PP}$ Hold Time	2			µs	
t_{DV}	Data Valid from $\overline{CE}$			1	µs	$\overline{CE} = V_{IL}$, $\overline{OE}/V_{PP} = V_{IL}$
t_{VR}	V_{PP} Recovery Time	2			µs	
t_{PRT}	$\overline{OE}/V_{PP}$ Pulse Rise Time During Programming	50			ns	

NOTES:
1. V_{CC} must be applied simultaneously or before $\overline{OE}/V_{PP}$ and removed simultaneously or after $\overline{OE}/V_{PP}$.
2. This parameter is only sampled and is not 100% tested. Output Float is defined as the point where data is no longer driven—see timing diagram.
3. Typical values are for $T_A = 25°C$ and nominal supply voltages.
4. The maximum current value is with outputs O_0 to O_7 unloaded.

*A.C. TEST CONDITIONS

Input Rise and Fall Time (10% to 90%) ..... ≤20 ns
Input Pulse Levels .... 0.45V to 2.4V
Input Timing Reference Level .... 0.8V and 2.0V
Output Timing Reference Level ...... 0.8V and 2.0V

Appendix C

intel 2732A

PROGRAMMING WAVEFORMS

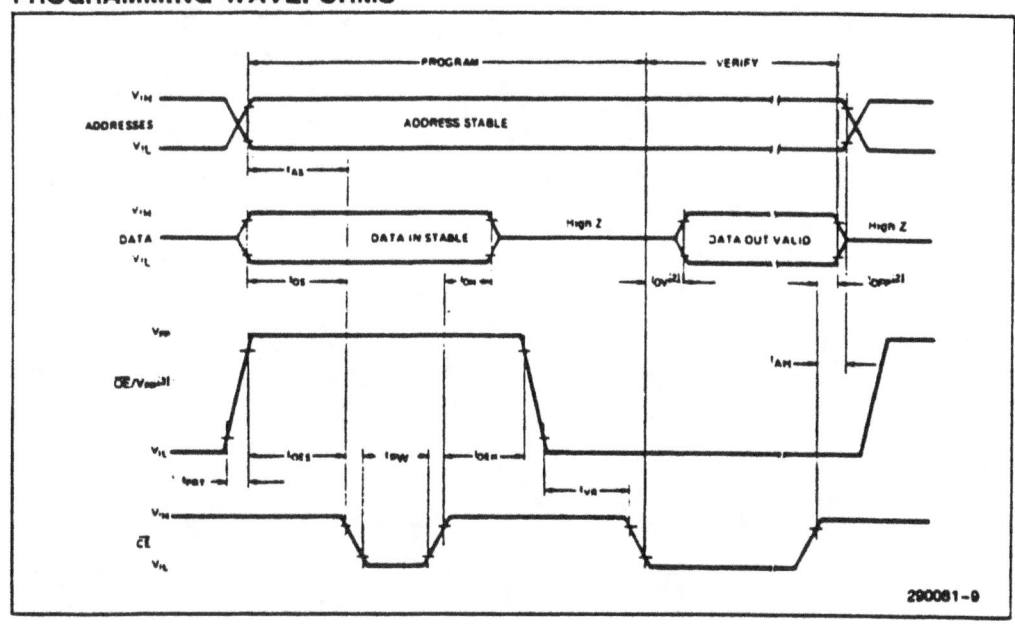

NOTES:
1. The input timing reference level is 0.8V for a V_{IL} and 2V for a V_{IH}.
2. t_{OV} and t_{DFP} are characteristics of the device but must be accommodated by the programmer.
3. When programming the 2732A, a 0.1μF capacitor is required across $\overline{OE}/V_{PP}$ and ground to suppress spurious voltage transients which can damage the device.

Appendix D

USING QUARTUS® II AND PROGRAMMING PLD BOARDS WITH QUARTUS® II PROJECTS

What is presented in the Supplemental Experiment set assumes that you have a DeVry eSOC board that is to be programmed using QUARTUS® II software. There are other boards available such as the Altera University Project Board (UP-1 and UP-2) and the RSR PLDT-2 board. While the QUARTUS® II software works identically for each board, the actual physical implementation will be somewhat different. The difference is mainly in the on-board devices that are available and their connection to the EPM7128SLC84 CPLD. These connections limit the number of pins and their pin numbers for use by your projects.

This appendix covers the use of QUARTUS II® software and the main differences in devices (switches and LEDs, mainly) and their pin assignments for the three boards mentioned above:

1. DeVry eSOC Board
2. Altera UP-1 Board
3. RSR PLDT-2 Board

This appendix also gives the general programming and downloading procedure that will be referred to throughout the Supplementary Experiments. If you are going to use any other board that is QUARTUS® II compatible, make sure that you keep information on devices and pin assignments that are available to the user (you) handy.

PART 1 - TUTORIAL FOR QUARTUS® SOFTWARE

Introduction. This appendix assumes that the student is using a computer on which QUARTUS® II is installed. Although QUARTUS® II operates essentially the same on all computers supported, there are some differences. Also, students using operating systems other than Microsoft Windows may experience some slight differences from this appendix. For example, directories (or folders) may vary from those shown in the examples. Your instructor will give you instructions on how to locate and place the files used in each experiment. These variations are minor and will not affect the student's ability to follow the instructions given here.

QUARTUS II® Files and Folders. A logic circuit designed with QUARTUS® II is called a *project*. Only one project at a time can be active in QUARTUS® II and all information for that project is kept in a single folder (directory). This folder is created early in the design procedure. To hold the design files for this tutorial, we will use a folder named *TocciLM*. The location and name of the directory is not really important, so the reader may use any valid folder name. In many cases your instructor will require or suggest a particular folder name.

Upon starting the QUARTUS® II software, you should see a display similar to the one in Figure D-1. This display consists of several windows that provide access to all features of QUARTUS® II, which the user selects with the computer mouse.

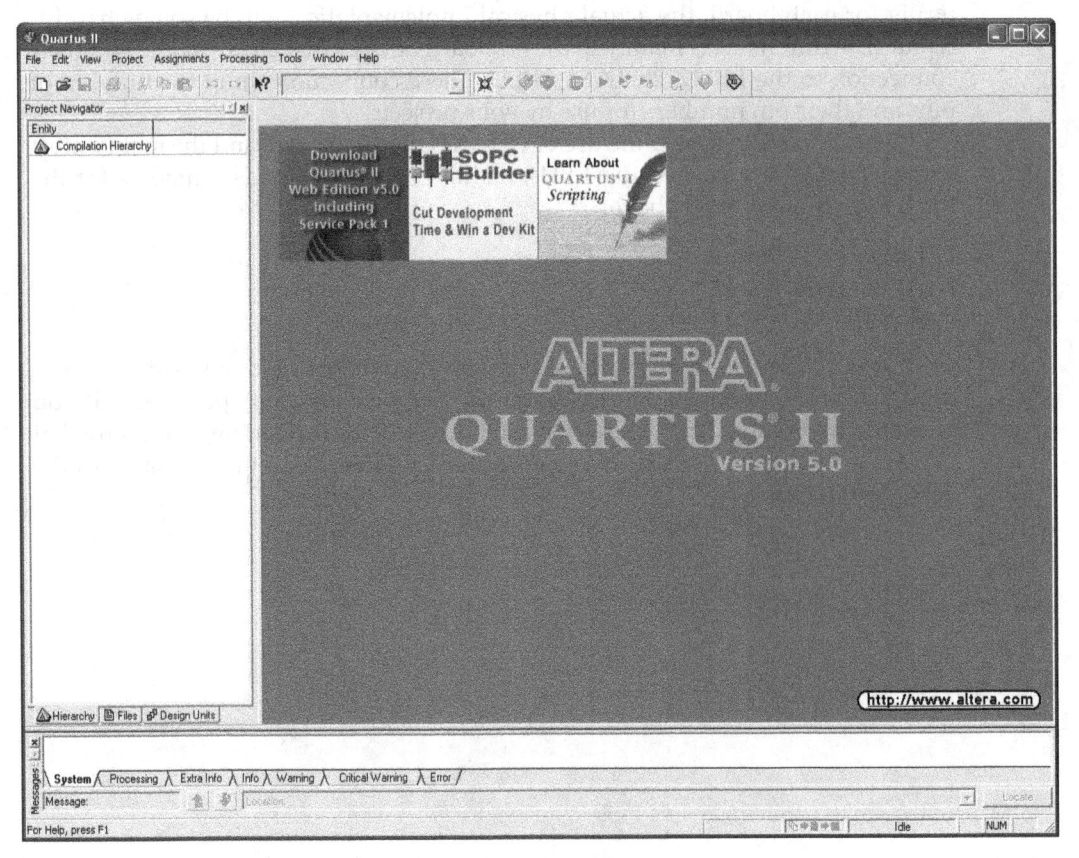

Figure D-1

Appendix D

Most of the commands provided by QUARTUS® II can be accessed by using the main menu bar located below the title bar. Clicking the mouse on the menu named File opens the menu shown in Figure D-2. Clicking the mouse button on the File menu item Exit exits from QUARTUS® II. To access some commands, it may be necessary to go through several menus in sequence. To indicate this, the convention **Menu1 | Menu2 | . . . | Command** will be used. For example, to access the command in **Menu1 | Menu2 | Command**, the student will first click the mouse on Menu1, select (click) Menu2 from the Menu1 list, and finally select (click) the command from the Menu2 list.

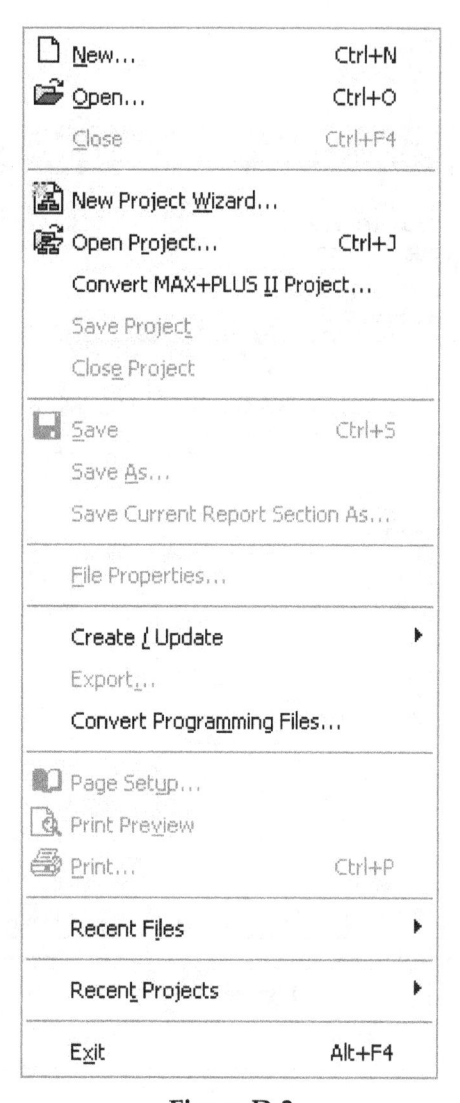

Figure D-2

Starting a New Project. To create a new design we first have to define a new *design project*. To make this task easier, QUARTUS® II provides support in the form of a *New Project Wizard*.

a) Select **File | New Project Wizard** to access the features of this wizard. You will get an introduction window, which requires no input.
b) Click **Next** in the introduction window to get the *Directory, Name, Top-Level Entity* window shown in Figure D-3.

c) Set the working directory to be C:/*altera*/*TocciLM*/*example*.
d) Name the project, which may optionally be the same as the name of the directory. The name used here is *example*. Observe that QUARTUS® II automatically fills the next entry with the same name. [*Note:* This entry can be changed but it is suggested that the default entry be used.]
e) Click **Next**. A pop-up box, like the one shown in Figure D-4 will appear, indicating that the desired directory needs to be created.
f) Click **Yes**. This will bring up the *Add Files* window, shown in Figure D-5, in which the student can specify which existing files (if any) should be included in the project.

Figure D-3

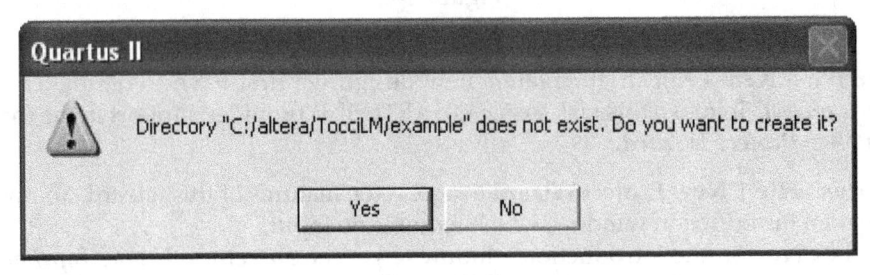

Figure D-4

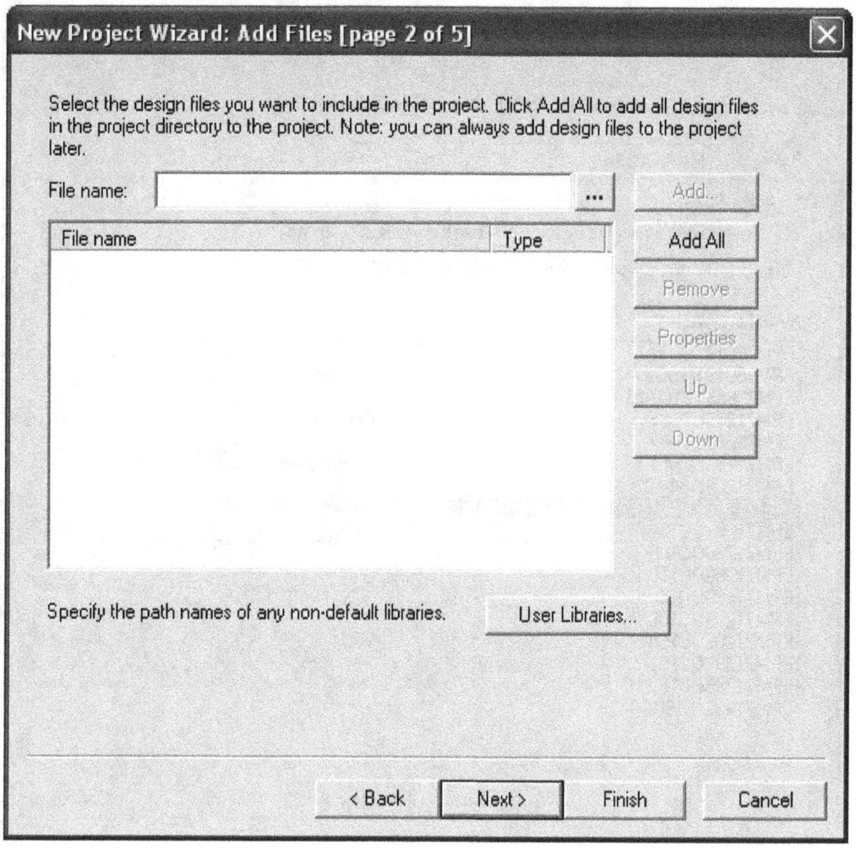
New Project Wizard: Add Files [page 2 of 5]
Select the design files you want to include in the project. Click Add All to add all design files in the project directory to the project. Note: you can always add design files to the project later.
File name:
...
Add...
File name
Type
Add All
Remove
Properties
Up
Down
Specify the path names of any non-default libraries.
User Libraries...
< Back
Next >
Finish
Cancel

Figure D-5

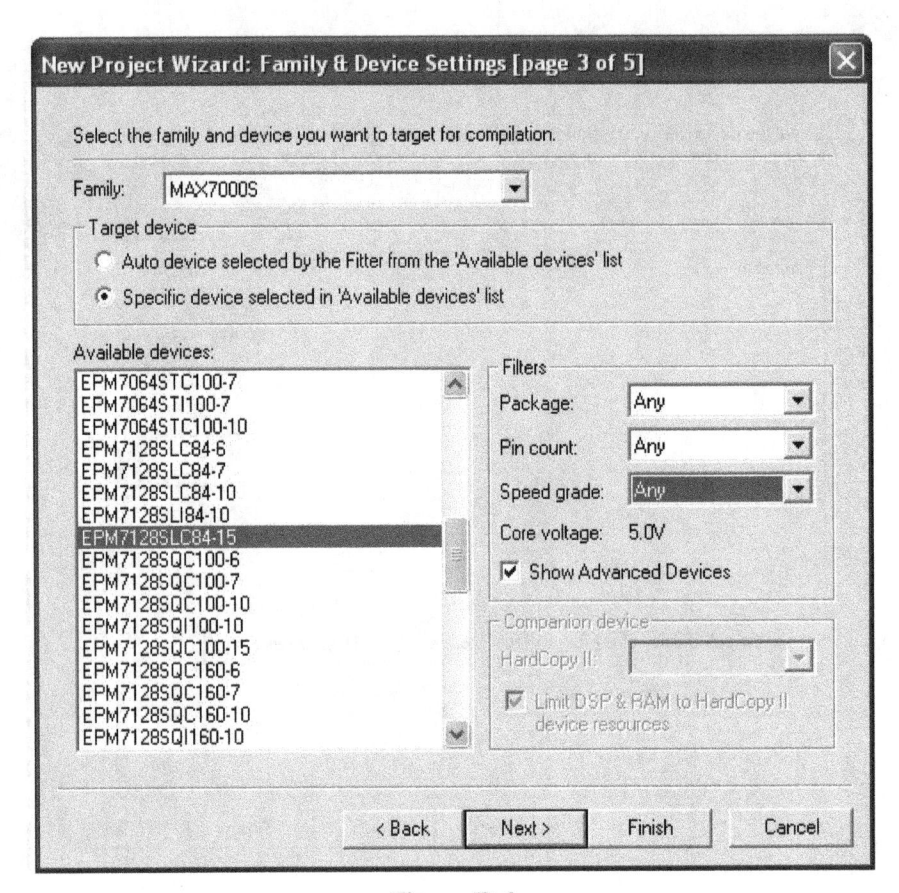

Figure D-6

g) There are no existing files, so click **Next**. A window like the one in Figure D-6 appears, which allows the student to specify the device to be programmed *Family and Device Settings*. Since this lab manual assumes the use of the DeVry eSOC board, the family should be MAX7000S and the target device EPM7128SLC84-15. Your instructor will provide you with the necessary information if your board is not the DeVry eSOC board or your target device is different.

h) When the information has been entered, click **Next**. The *EDA Tool Settings* window like the one in Figure D-7 appears, which allows the student to specify third-party CAD (computer aided design) or EDA (electronic design automation) tools (i.e., tools developed by companies other than Altera) that should be used.

i) There are no third-party CAD or EDA tools used in this lab manual, so click **Next**. The *Summary* window like the one in Figure D-8 is displayed, which will display a summary of the project's settings.

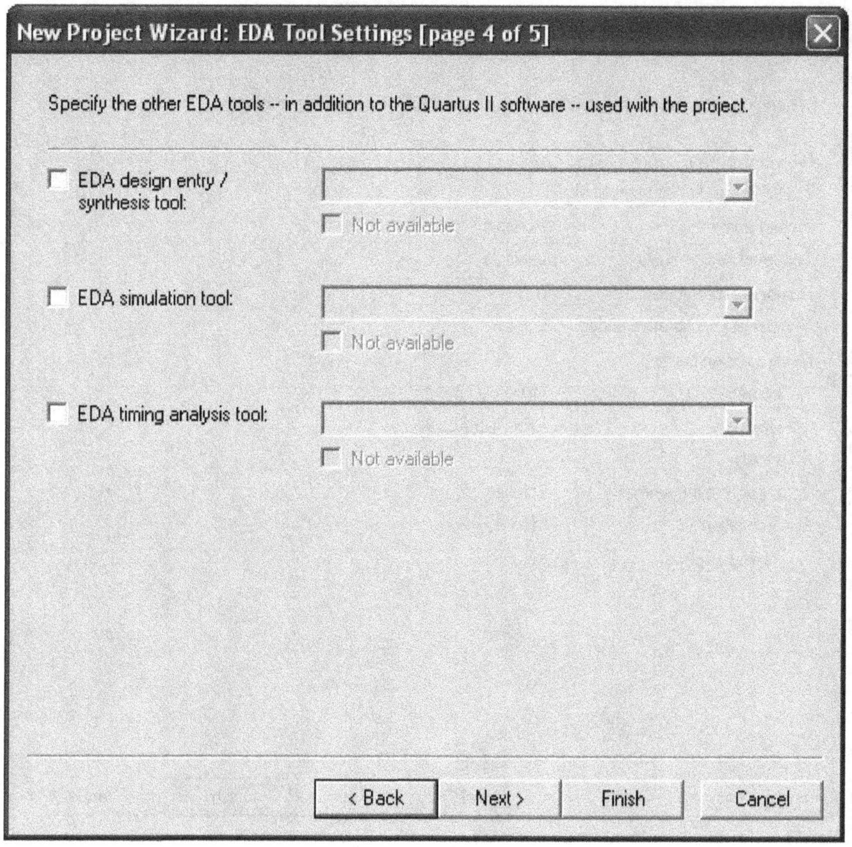
New Project Wizard: EDA Tool Settings [page 4 of 5]
Specify the other EDA tools -- in addition to the Quartus II software -- used with the project.
EDA design entry / synthesis tool:
Not available
EDA simulation tool:
Not available
EDA timing analysis tool:
Not available
< Back
Next >
Finish
Cancel

Figure D-7

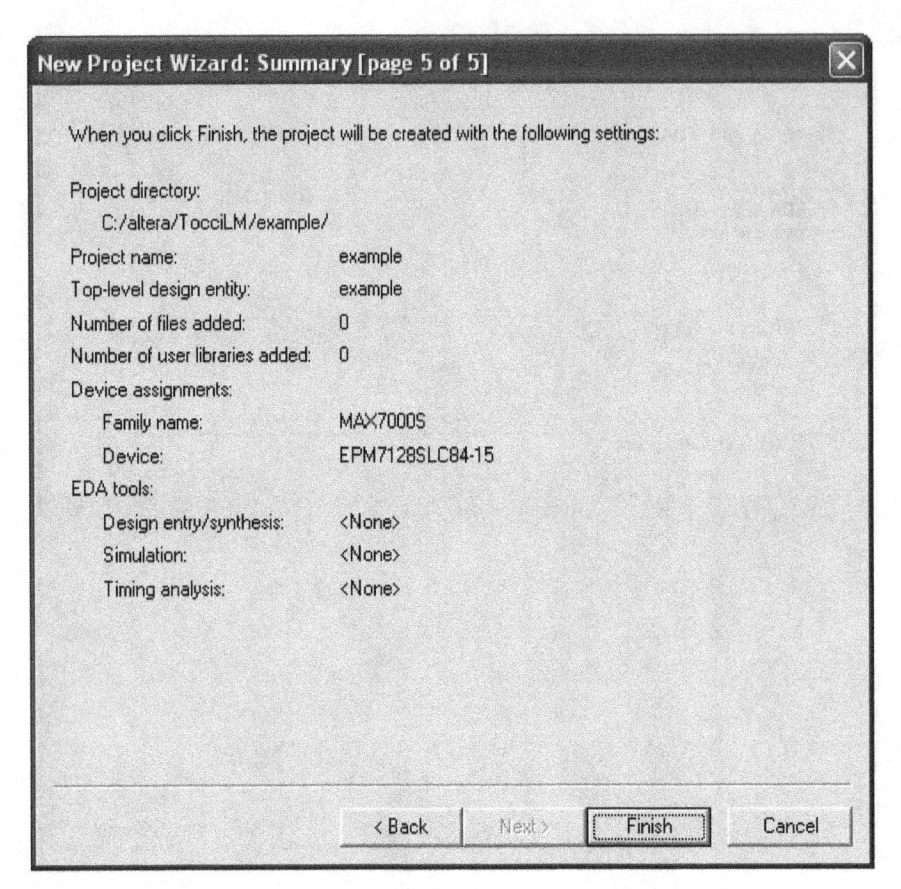

Figure D-8

j) The creation of the *example* project is now complete, so press **Finish**.

Creating a Design. The QUARTUS® II software has two tools that this lab manual will employ to create a design: the **Block Editor** for creating schematic designs and the **Text Editor** to create an HDL design using the AHDL, Verilog HDL, or VHDL design languages. The design creation using *Schematic Capture* will be explained first (Exercise 1) followed by a brief look at design creation of an OR gate using **VHDL**, a hardware definition language (Exercise 2). Finally, you will program the CPLD by downloading your schematic design for the OR gate (Exercise 3).

Exercise 1. OR Gate Schematic Capture Using the Block Editor.

a) In the QUARTUS® II window, select **File | New**. A window appears, shown in Figure D-9, which allows the student to choose the type of file that should be created. The possible file types include AHDL (Altera Hardware Definition Language), block diagram/schematics, EDIF (Electronic Design Interface Format), Verilog HDL, and VHDL.

Appendix D

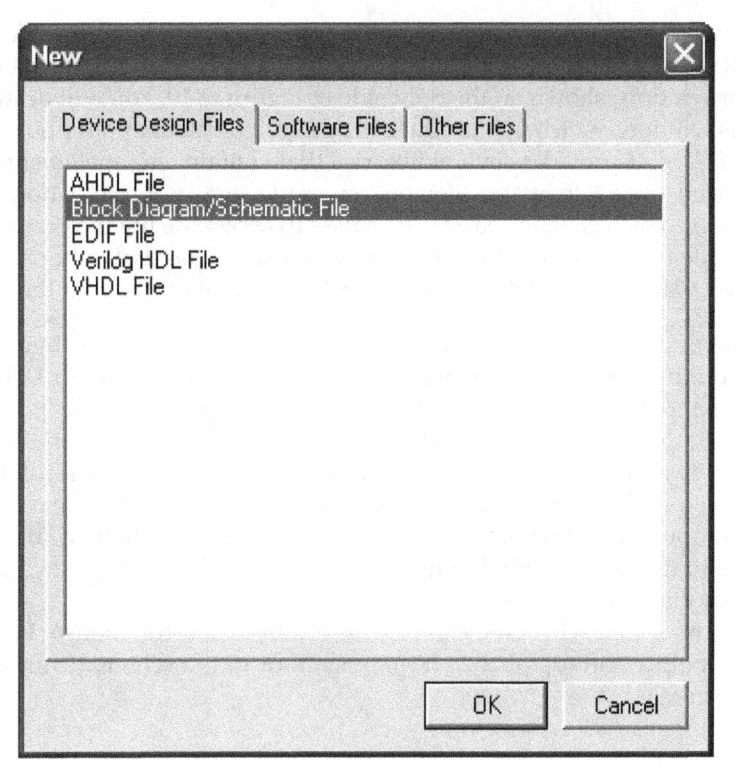

Figure D-9

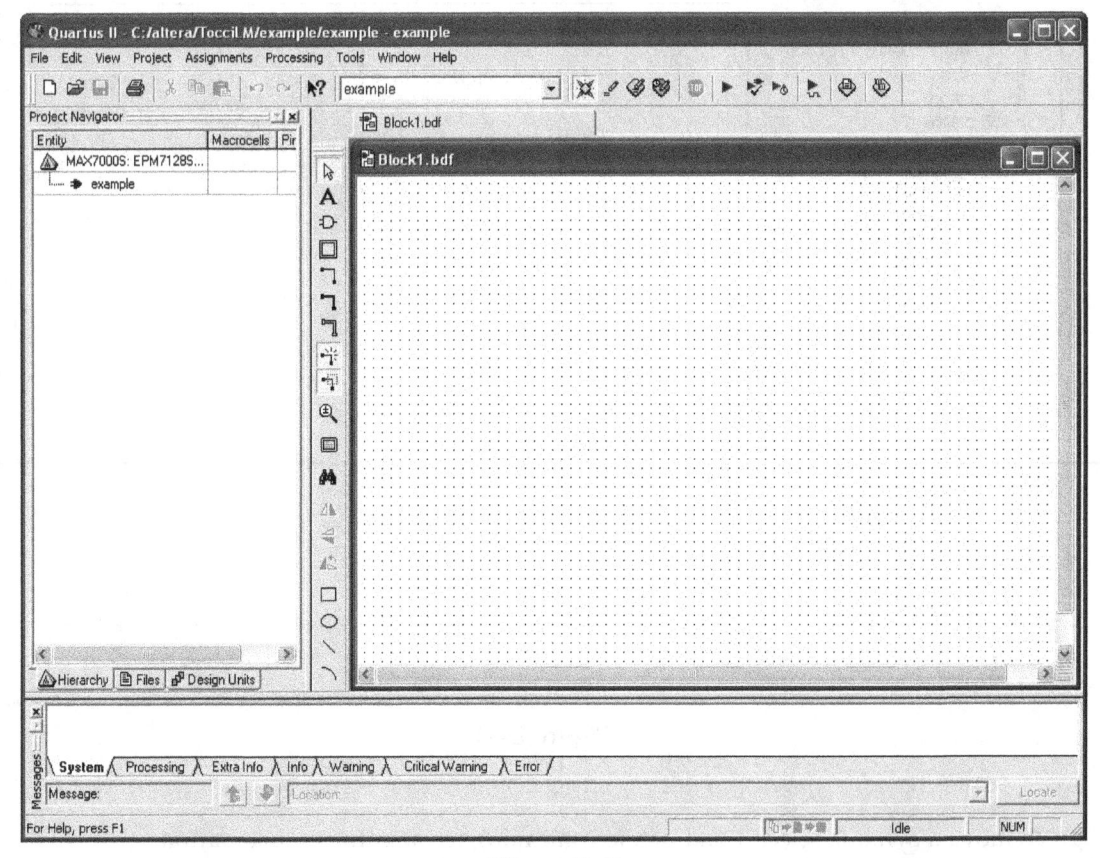

Figure D-10

b) Since we are illustrating the schematic-entry approach in this section, choose **Block Diagram/Schematic File** and click **OK**. This selection opens the Block Editor window, shown on the right side of Figure D-10. You will draw a circuit in this window, which will produce the desired block diagram file.

c) QUARTUS® II provides several libraries that contain circuit elements that can be placed on a schematic. For our example, we will use a library called *primitives*, which contains basic logic gates. To access the library, double-click on the inside of the Block Editor window to open the *Symbol* window shown in Figure D-11 (selecting **Edit | Insert Symbol** or right-clicking on the AND gate symbol in the toolbar will produce the same result). In the figure, the box labeled Libraries lists several libraries that are provided with QUARTUS® II.

d) To expand the list, click on the + symbol next to *c:/altera/quartus50sp1/libraries/*, then click on the + next to *primitives*, and finally click on the + next to *logic*. Now, type in **or2** in the **Name:** box and click on **OK** (or scroll to the or2 entry in the symbol list and click OK) to place it into the schematic.

e) A two-input OR-gate symbol now appears in the Block Editor window. Use the mouse to move the symbol to the desired location in the diagram and place it there by clicking the mouse.

f) Any symbol in a schematic can be selected by using the mouse. Position the mouse pointer on top of the OR-gate symbol in the schematic and click the mouse to select it.

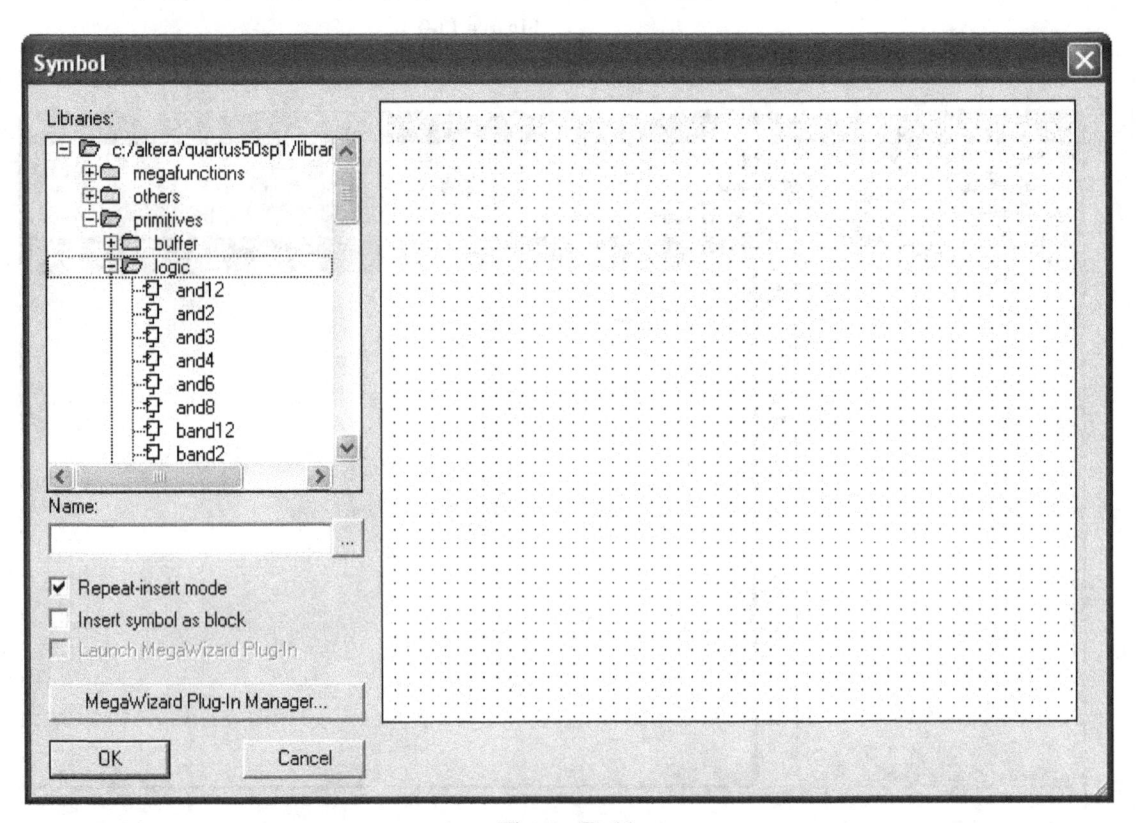

Figure D-11

g) To move a symbol, select it and, while continuing to press the mouse button, drag the mouse to move the symbol. This is called "click and drag." Place the OR-gate symbol as shown in Figure D-12.

Appendix D

Note: To make it easier to position the graphical symbols, a grid of guidelines can be displayed in the Block Editor window by checking the box next to **View | Show Guidelines**.

h) Now that the OR-gate symbol has been placed, you will now place symbols to represent the input and output ports of the circuit. Open the *primitives* library again. Scroll down past the gates until you reach *pin* and then click on the + symbol next to it. Find and place the symbol named *input* into the schematic two times by clicking twice.

i) To represent the output of the circuit, open the *primitives* library, then find and place the symbol named *output*.

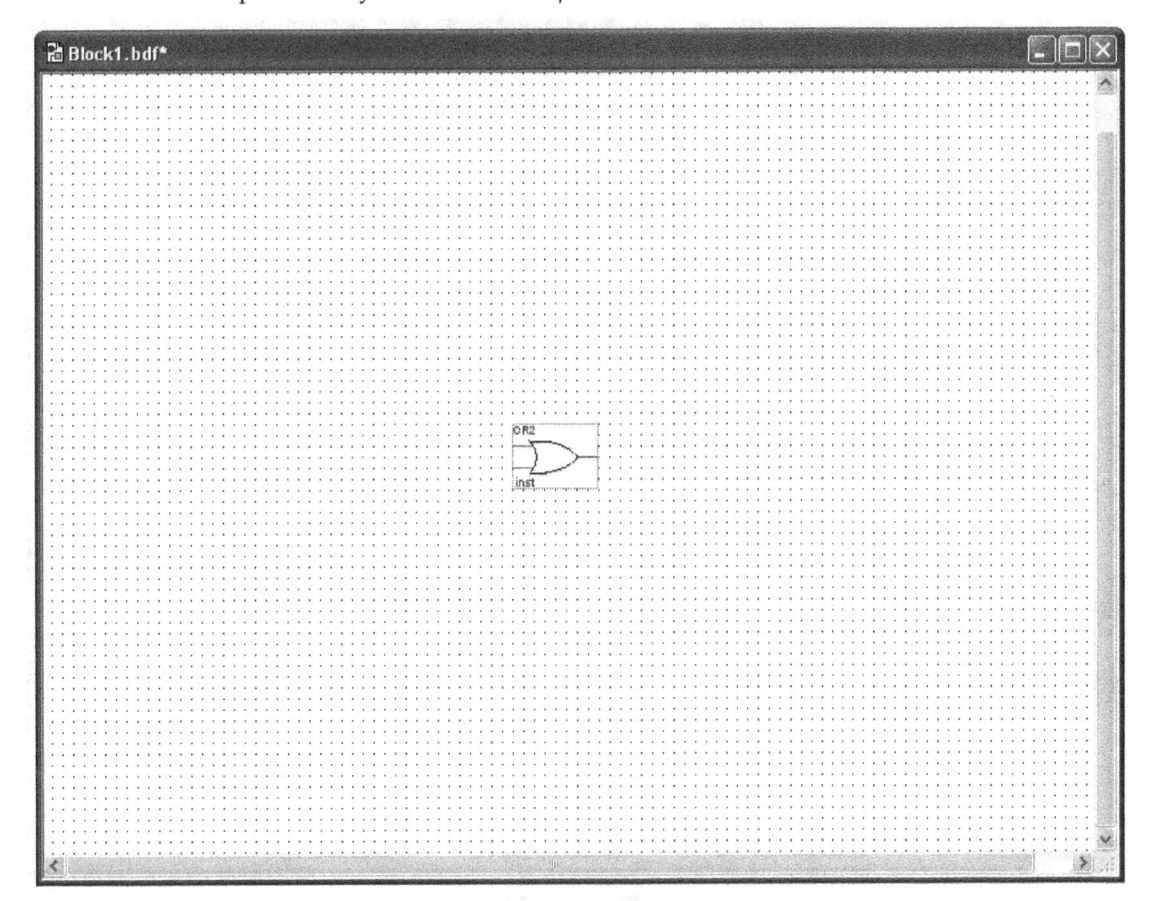

Figure D-12

j) Arrange the symbols to appear as illustrated in Figure D-13.

k) To use the schematic, you must assign names to the input and output symbols. Point to the words *pin name* on the upper input pin symbol in the schematic and double-click the mouse. A dialog box like the one in Figure D-14 named **Pin Properties** appears allowing a new pin name to be typed. Type *A* as the pin name and click **OK**.

l) Use the same procedure as step k to assign the name *B* to the lower input symbol and *X* to the output pin.

m) The next step is to draw lines (wires) to connect the symbols in the schematic together. Click on the icon in the vertical toolbar. This icon is called the *Selection and Smart Drawing* tool, and it allows the Block Editor to change automatically between the modes of selecting a symbol on the screen or drawing wires to interconnect symbols. The appropriate mode is chosen depending on where the mouse is pointing.

n) Move the mouse pointer on top of the *A* input symbol. When pointing anywhere on the symbol except at the right edge, the mouse pointer appears as crossed arrowheads. This indicates that the symbol will be selected if the mouse button is pressed.
o) Move the mouse to point to the small line, called a *pinstub*, on the right edge of the *A* input symbol. The mouse pointer changes to a crosshair, which allows a wire to be drawn to connect the pinstub to another location in the schematic. A connection between two or more pinstubs in a schematic is called a *node*.

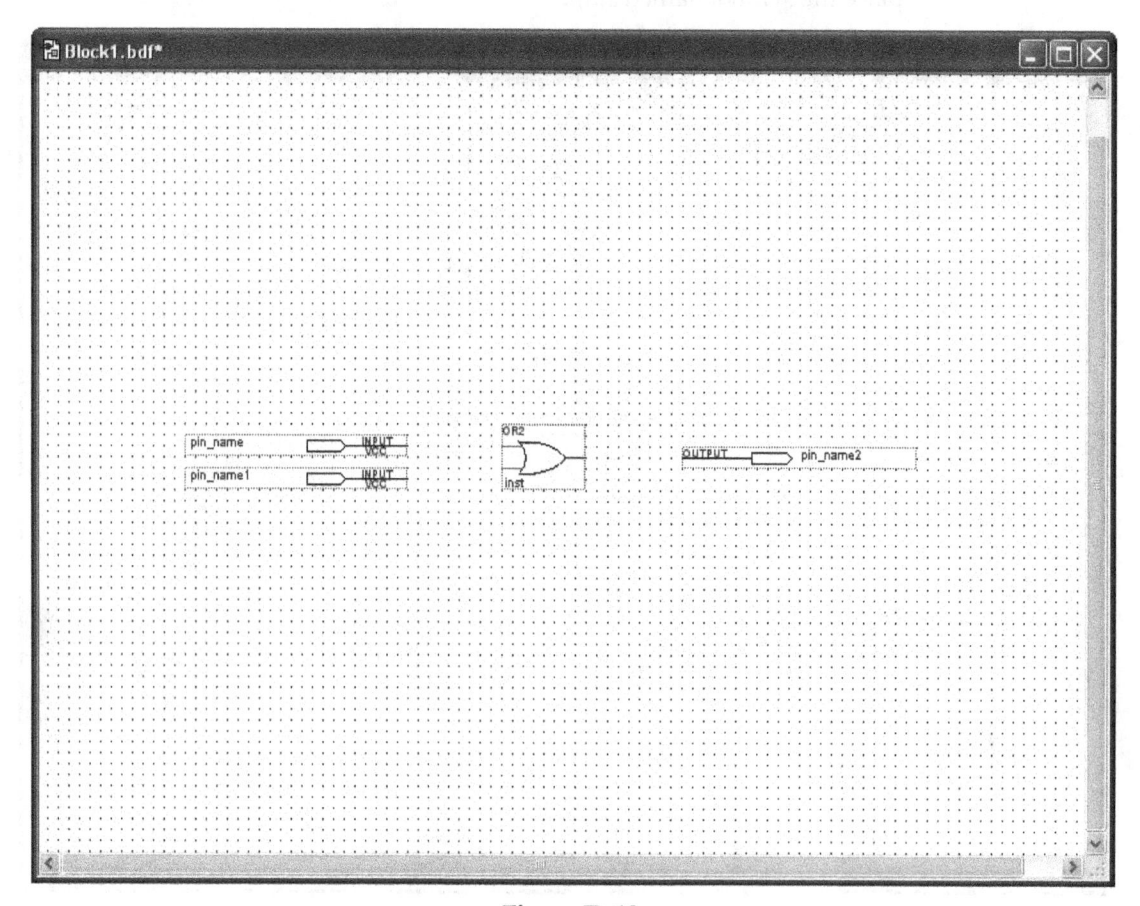

Figure D-13

Appendix D

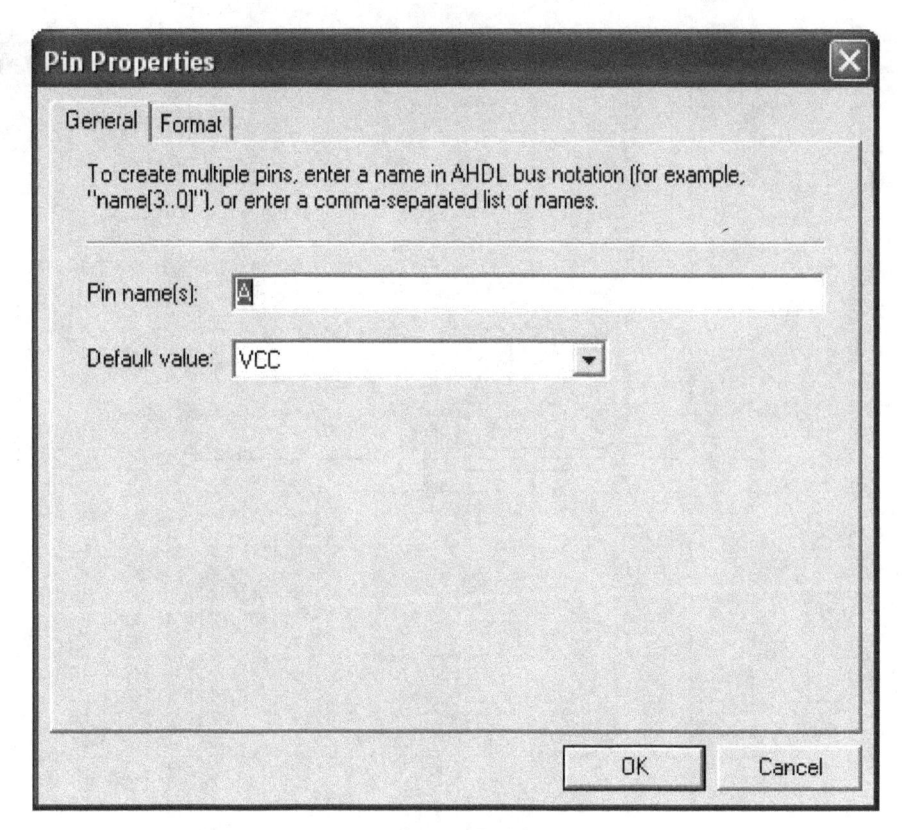

Figure D-14

p) Connect the input symbol for A to the OR gate at the top of the schematic in the following manner:
   1) While the mouse is pointing at the pinstub on the A symbol, click and hold the mouse button.
   2) Drag the mouse to the right until the line (wire) that is drawn reaches the pinstub on the top input of the OR gate; then release the button. The two pinstubs are now connected and represent a single node in the circuit.
q) Use the procedure in p) to draw a wire from the pinstub on the B input symbol to the lower input on the OR gate.
r) Similarly, draw a wire from the pinstub on the output of the OR gate to the pinstub on the X output symbol.
s) The completed circuit should look like the one shown in Figure D-15.

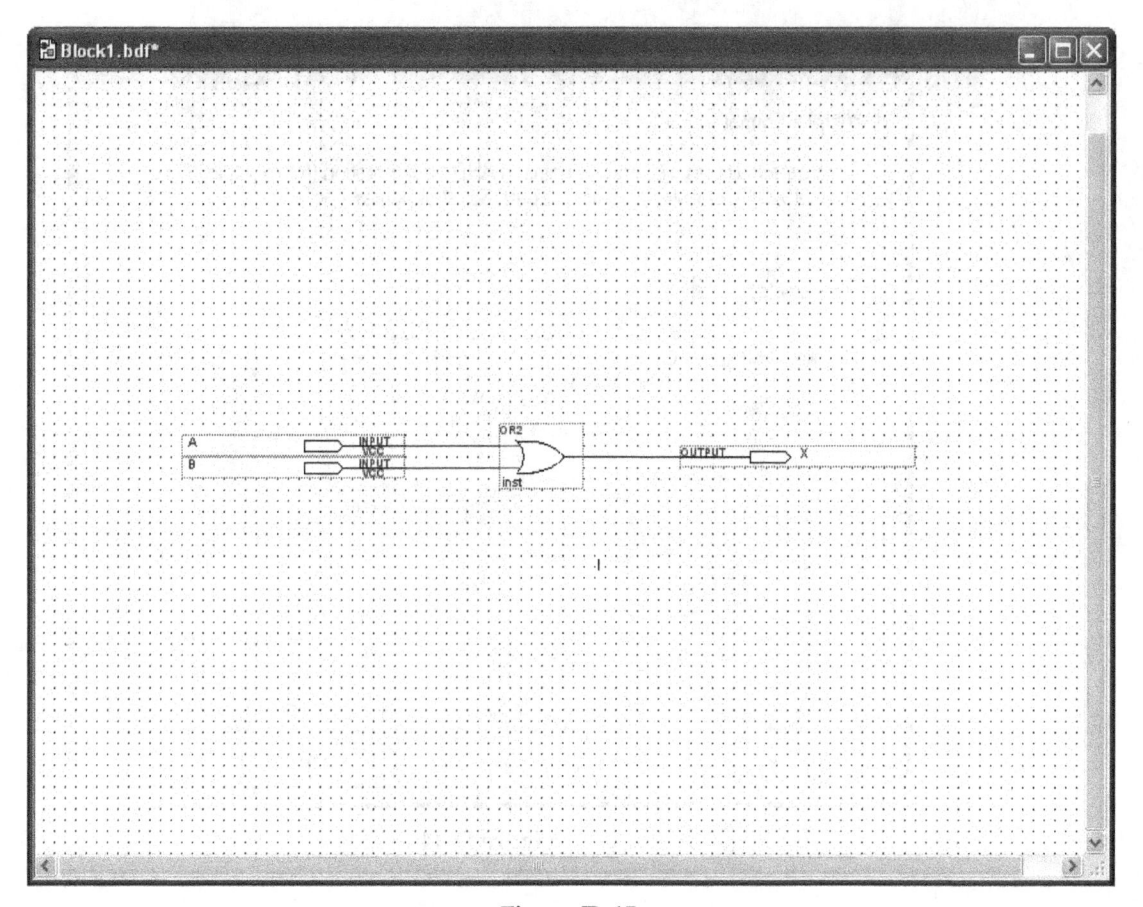

Figure D-15

t) Save the schematic using **File | Save As** using the name *example*. Note that the saved file is called *example.bdf*.

u) To rearrange the layout of the circuit, select the OR-gate and move it. Observe that as you move the gate symbol all connecting wires are adjusted automatically. This takes place because QUARTUS® II has a feature called *rubberbanding*, which was activated by default when you chose to use the *Selection and Smart Drawing* tool.

v) There is a **rubberbanding** icon . Observe that this icon is highlighted to indicate the use of rubberbanding. Turn the icon off and move one of the gates to see the effect of this feature.

Note: Our example schematic is quite simple and it is easy to draw all the wires in the circuit without producing a messy diagram. In larger schematics some nodes that have to be connected may be far apart, in which case it may be difficult to draw wires between them without creating a cluttered look. In such cases the nodes can be connected by assigning labels to them, instead of drawing wires. See the online **Help** for a more detailed description.

Appendix D

The QUARTUS® II Compiler. Several modules make up the CAD tools in QUARTUS®II. Select **Tools | Compiler Tool** to open the window in Figure D-16, which shows four of the main modules. The *Analysis & Synthesis* module produces a circuit of logic elements; each element of the circuit can be directly implemented in the target chip. This is the synthesis step in QUARTUS® II. The *Fitter* module determines the exact location on the chip where each of these elements produced by synthesis will be placed. The *Assembler* module completes project processing by generating a device programming image, in the form of one or more *Programmer Object Files* (**.pof**). The *Timing Analyzer* is used to analyze, debug, and validate the timing performance of all logic in a design. A detailed discussion of CAD modules is found in the QUARTUS® II Manual.

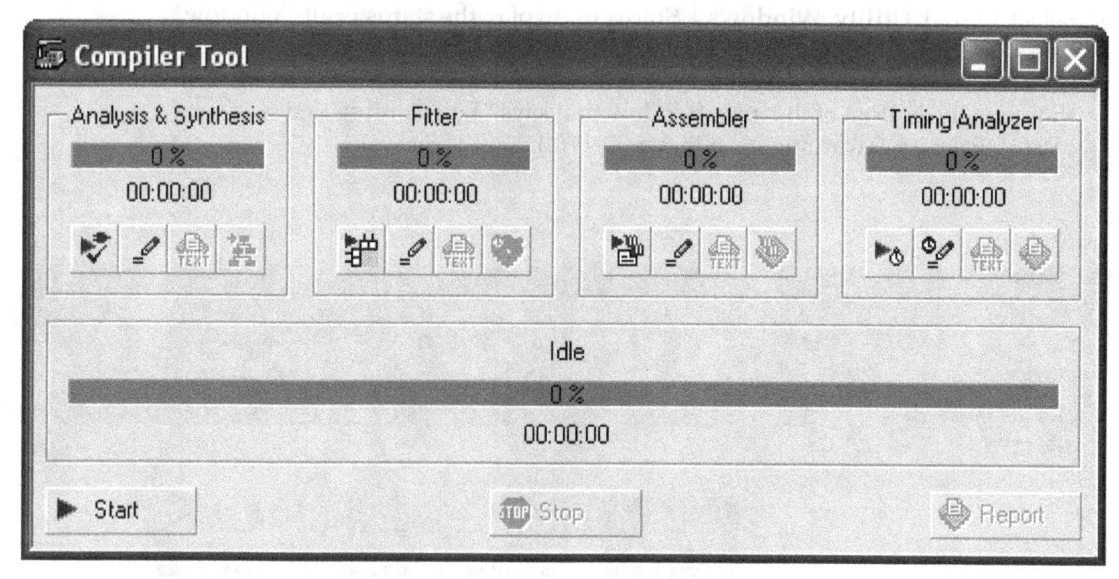

Figure D-16

The QUARTUS® II modules are part of an application program called the *Compiler*. The Compiler can be used to run a single module at a time, or it can run more than one of the modules in sequence. There are several ways to run the Compiler in the QUARTUS® II user interface:

- Pressing the button under Analysis & Synthesis will run this module. Note that the Fitter can also be executed by pressing the button, the Assembler by pressing the button, and the Timing analyzer by pressing the button.
- *Note:* You cannot run a module out of order. For example, you can only run the Assembler after a successful running of Analysis & Synthesis and Fitter modules. Also, an error in one module will prevent the next module in the sequence from being run.

- Pressing the **Start Compilation** button runs the modules in Figure D-16 in the required sequence. An equivalent method is selecting the **Processing | Start Compilation** command.
- Selecting the **Processing | Start** menu will give you access to all of the compiler tools.

The progress of the compilation is indicated by the individual and overall progress bars. The status can also be displayed to the left of the main window by selecting **View | Utility_Windows | Status.**

Detailed information concerning the compilation will be displayed in the Messages window at the bottom. This will include any warnings and errors. While a warning does not necessarily stop the progress of your compilation, an error will. Errors are listed in the QUARTUS® II Users' Manual. You must correct any error detected and re-start the compilation.

Compiling the Design. We now return to the task of compiling the design of the project created in the previous steps.

a) Select **View | Utility_Windows | Status** to display the status utility window.
b) Press the **Start Compilation** button to initiate the compilation process.
c) If there are no errors detected, you will be notified in a message box like the one shown at the bottom of the main window that says "Full Compilation was successful (0 errors, 0 warnings)" in Figure D-17(a).

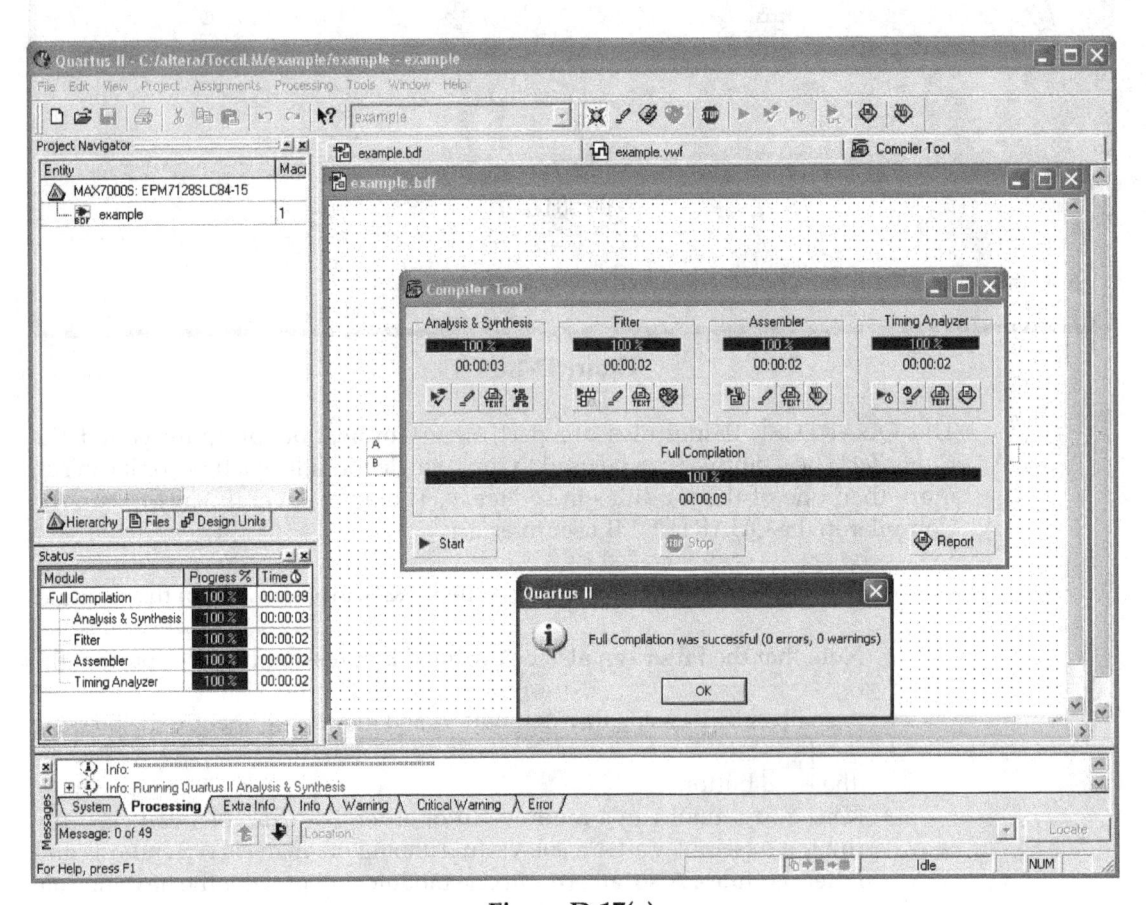

Figure D-17(a)

d) Press the **Report** button to get a *Compilation Report* like the one in Figure D-17(b).
e) To see what happens if an error is made, disconnect input A from the OR gate and compile the modified schematic.

f) The compilation is not successful and one warning and one error message are indicated (see Figure D-18). In the Messages window (see Figure D-19), the warning tells you that pin "A" is not connected and the error tells you that the OR gate is missing a source.

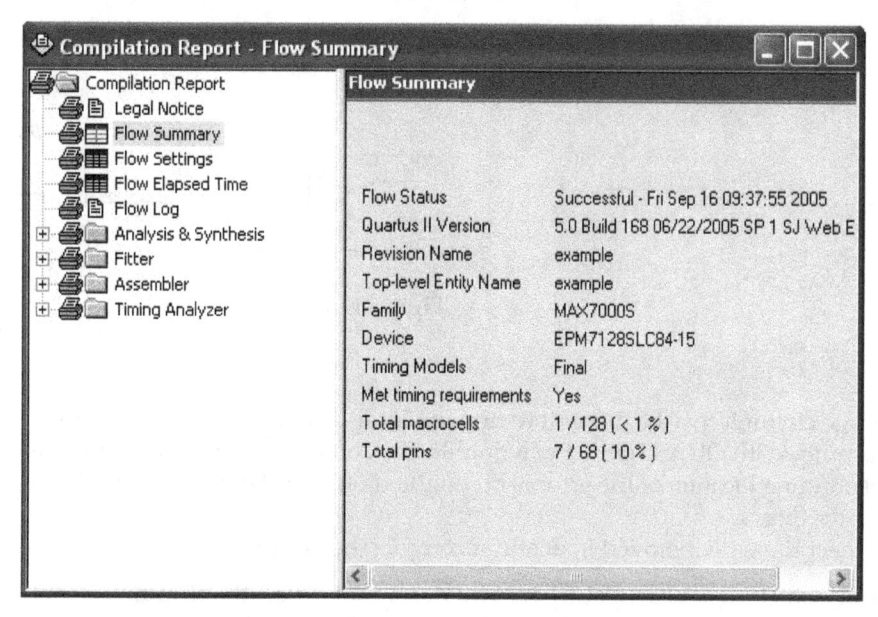

Figure D-17(b)

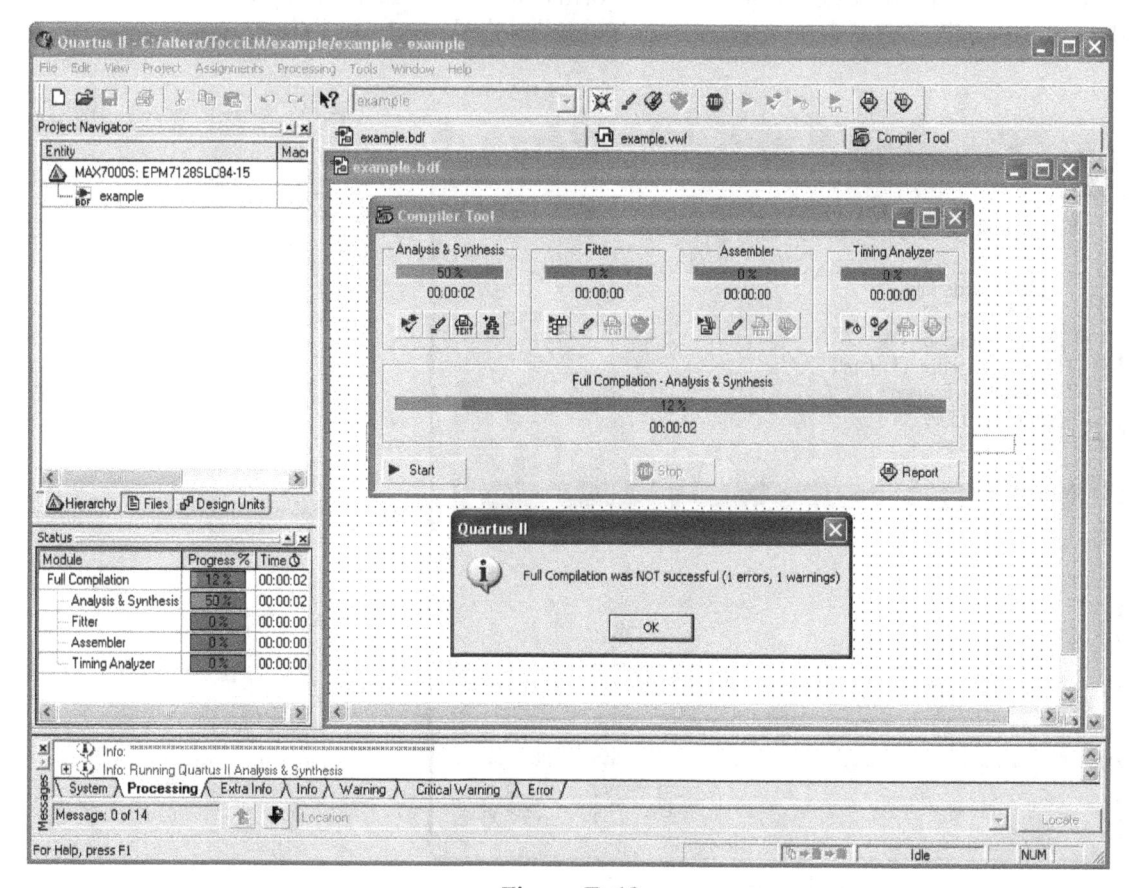

Figure D-18

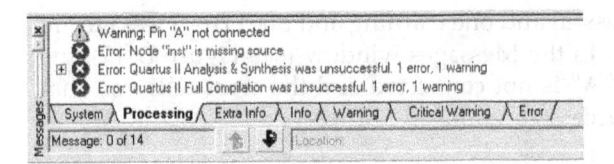

Figure D-19

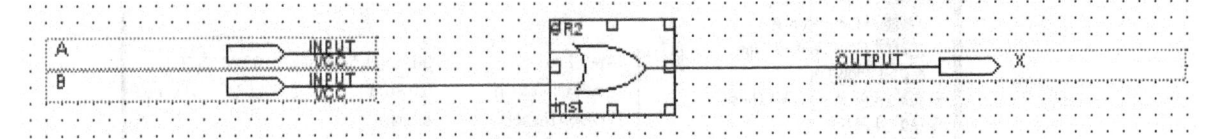

Figure D-20

g) In a large circuit it may be difficult to find the location of the source for an error or warning. With QUARTUS® II you double-click on the error message and the corresponding location of the error is highlighted on the schematic. Figure D-20 illustrates this.

h) Reconnect the wire removed in **e)** and recompile the design.

Simulating the Design. QUARTUS® II provides a tool called the **Simulator** that can be used to simulate the behavior of the designed circuit. In order to use this tool, it is necessary to create the desired waveforms, called *test vectors*, to represent the input signals. This is done with another tool called the **Waveform Editor**.

Figure D-21

Using the Waveform Editor

a) Select **File | New**, which gives the window in Figure D-9.
b) Click on the **Other Files** tab to access the window in Figure D-21.
c) Select **Vector Waveform File** and click OK.
d) The Waveform Editor window is displayed in Figure D-22. Save the file as *example.vwf*.
e) Set the desired simulation to run from 0 to 100 ns by selecting **Edit | End Time** and entering 100 ns in the pop-up dialog.
f) Select **View | Fit in Window** to display the entire simulation range of 0 to 100 ns in the window. It may be necessary to resize the window to its maximum size.

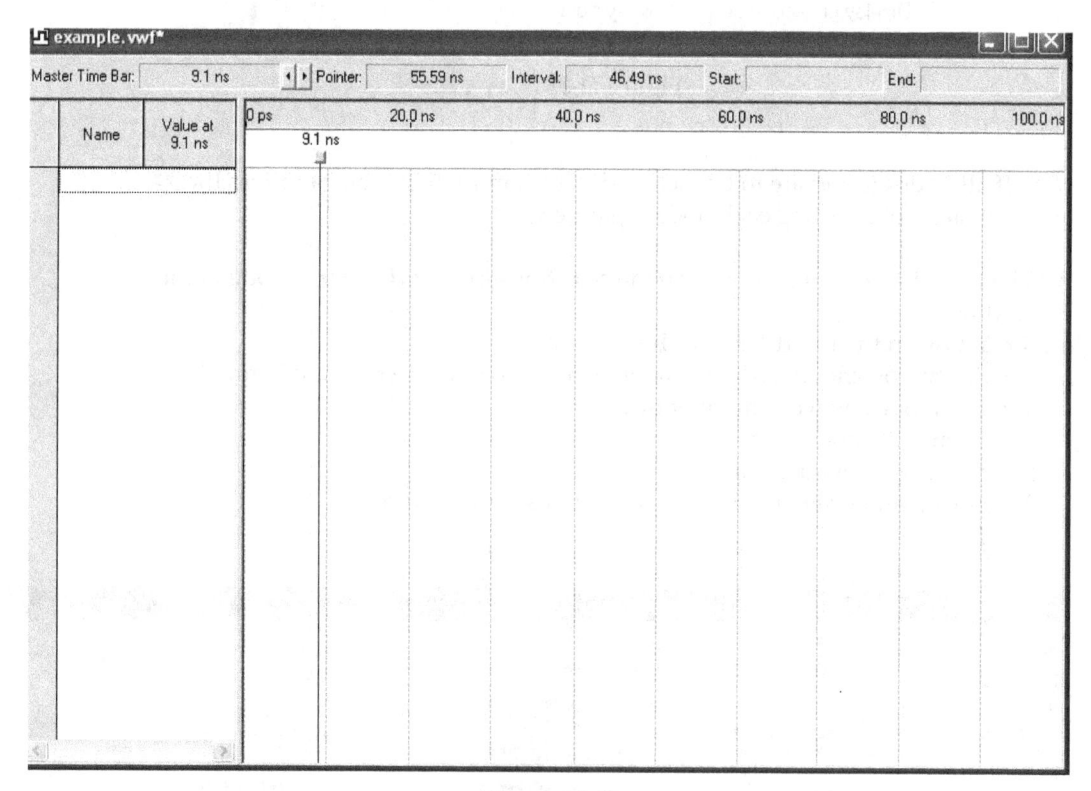

Figure D-22

g) Include the input and output nodes of the circuit to be simulated by using the **Node Finder** utility. Select **Edit | Insert Node or Bus** to open the window in Figure D-23.
h) You can type the name of a signal (pin) into the **Name** box or press the button labeled **Node Finder** to open the window in Figure D-24. The Node Finder utility includes a filter used to select the type of nodes that are to be found. Since we are interested in input and output pins, set the filter to *Pins: all*.
i) Press the **List** button to find the input and output nodes. On the left side of the window, the nodes A, B, and X will be displayed.
j) Select node A in the *Nodes Found* box on the left and press the **>** button to add the node to the Selected Nodes box on the right.
k) Repeat step **j)** for each node you want to include in the simulation.

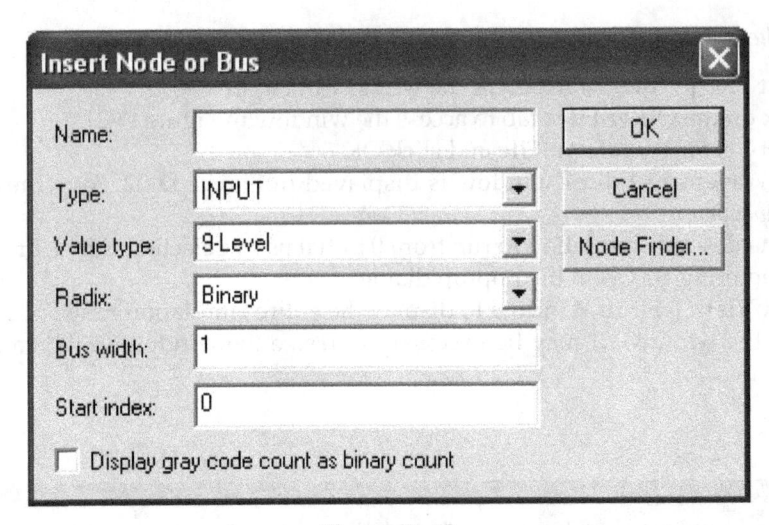

Figure D-23

Note: If all nodes found are to be included in the simulation, you can press the >> button instead of selecting each node separately.

l) When you have completed your node selections, press **OK** on the Node Finder window.
m) Press **OK** on the Insert Node or Bus window.
n) For clarity, you should order the nodes similar to what is shown in Figure D-25. You can move a waveform up or down by
   1) selecting its node name
   2) releasing the mouse button
   3) clicking and dragging the waveform to its desired location

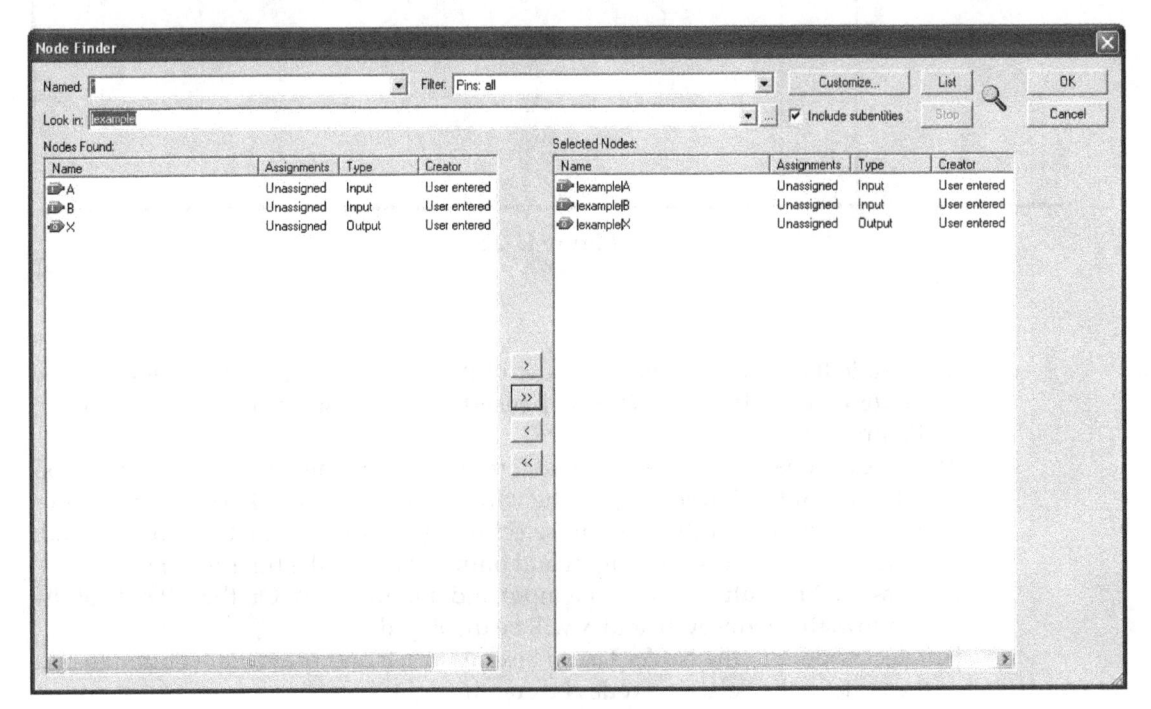

Figure D-24

Appendix D

o) Specify the logic values for all input signals to be used during simulation. Since there are only two inputs, you will only need 4 different values (0 – 3) for the inputs. The values for output X are generated by the simulator.

p) To manually create a waveform for input A using a sequence of 0, 1, 0 ,1
1) Select **View | Snap to Grid**
2) Use the Selection Tool and click the A waveform, which is currently all 0s, at the 20 ns vertical gridline and drag the mouse pointer to the 40 ns gridline. Release the mouse button and you will note that the area of waveform A between 20 ns and 40 ns is highlighted. Pressing the *Forcing High (1)* button at this time will cause the waveform to go HIGH between 20 ns and 40 ns.
3) Repeat step 2 on waveform A between 60 ns and 80 ns.

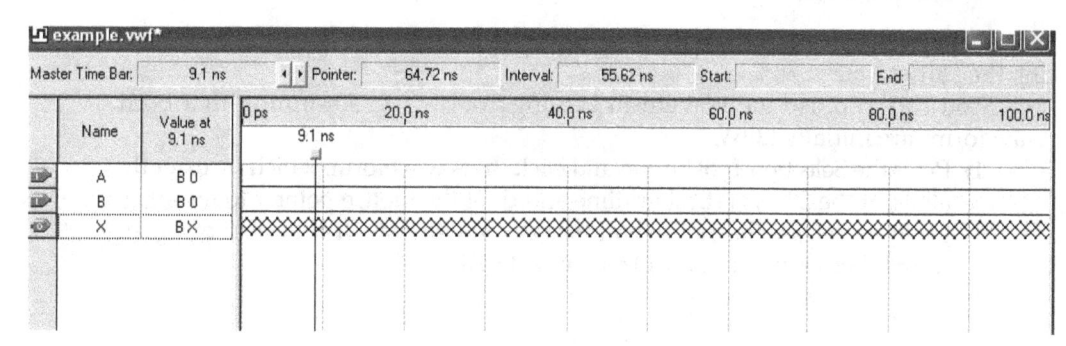

Figure D-25

q) To manually create a waveform for input B using a sequence of 0, 0, 1, 1 click the B waveform, which is currently all 0s, at the 40 ns vertical gridline and drag the mouse pointer to the 80 ns gridline. Release the mouse button and you will note that the area of waveform B between 40 ns and 80 ns is highlighted. Pressing the button at this time will cause the waveform to go HIGH between 40 ns and 80 ns.

r) Leave the area of both waveforms between 80 ns and 100 ns at 0.

s) Your waveforms should be like those in Figure D-26. Your input waveforms together create a sequence of 00, 01, 10, and 11.

Note: To create the A waveform quickly,
1) Select node A and press the Overwrite Clock button. The **Clock** pop-up box in Figure D-27 will appear.

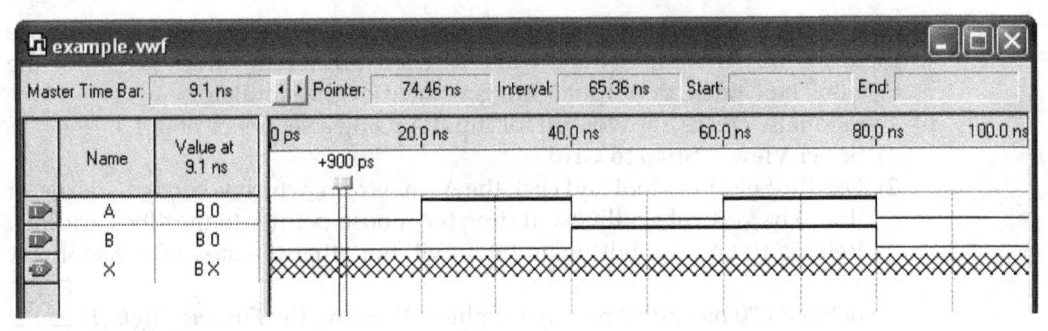

Figure D-26

2) In the Time Period box, enter 40 ns and in the Duty Cycle (%), enter 50 and press **OK**.

Repeat steps 1-2 for the B waveform, using 80 ns for the Time Period while keeping the Duty Cycle at 50%.

You can also use the Waveform Editing Tool . Assuming that both waveforms are initially LOW,

1) Press the Selection Tool button and click the A waveform, which is currently all 0s, at the 20 ns vertical gridline and drag the mouse pointer to the 40 ns gridline. Release the mouse button and you will note that the area of waveform A between 20 ns and 40 ns is now HIGH.

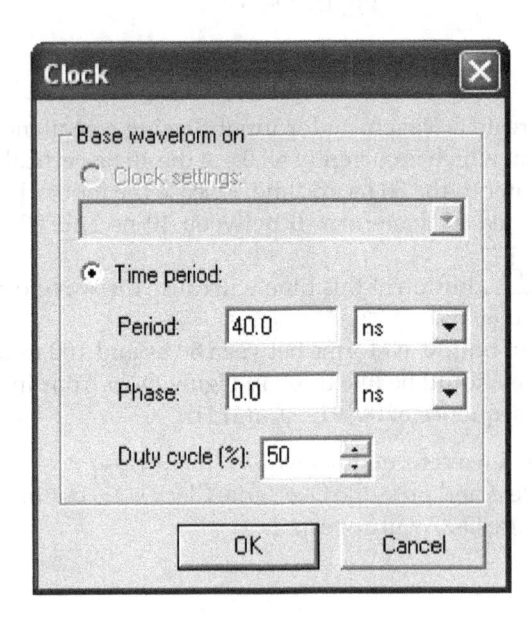

Figure D-27

Appendix D

2) Repeat step 1 on waveform A between 60 ns and 80 ns.
3) Now click the B waveform at the 40 ns vertical gridline and drag the mouse pointer to the 80 ns gridline. Release the mouse button and you will note that the area of waveform B between 40 ns and 80 ns is now HIGH.

Using the Simulator

a) There are two types of simulation that the Quartus Simulator provides. The first is a **Functional Simulation** in which all logic elements and their interconnections have no propagation delay. The second is a **Timing Simulation** in which propagation delays are taken into account. We will perform a functional simulation first.

b) Select **Assignments | Settings** to access the Settings window shown in Figure D-28.

c) Click on *Simulator* at the left.

m) Choose Functional from the *Simulator Mode* pull-down menu.

e) Find and use for the *Simulator Input example.vwf* you just created.

f) Select **Processing | Generate Functional Simulation Netlist.** A pop-up information box like that in Figure D-29 will display indicating the results of the generation. Upon a successful generation, press **OK.**

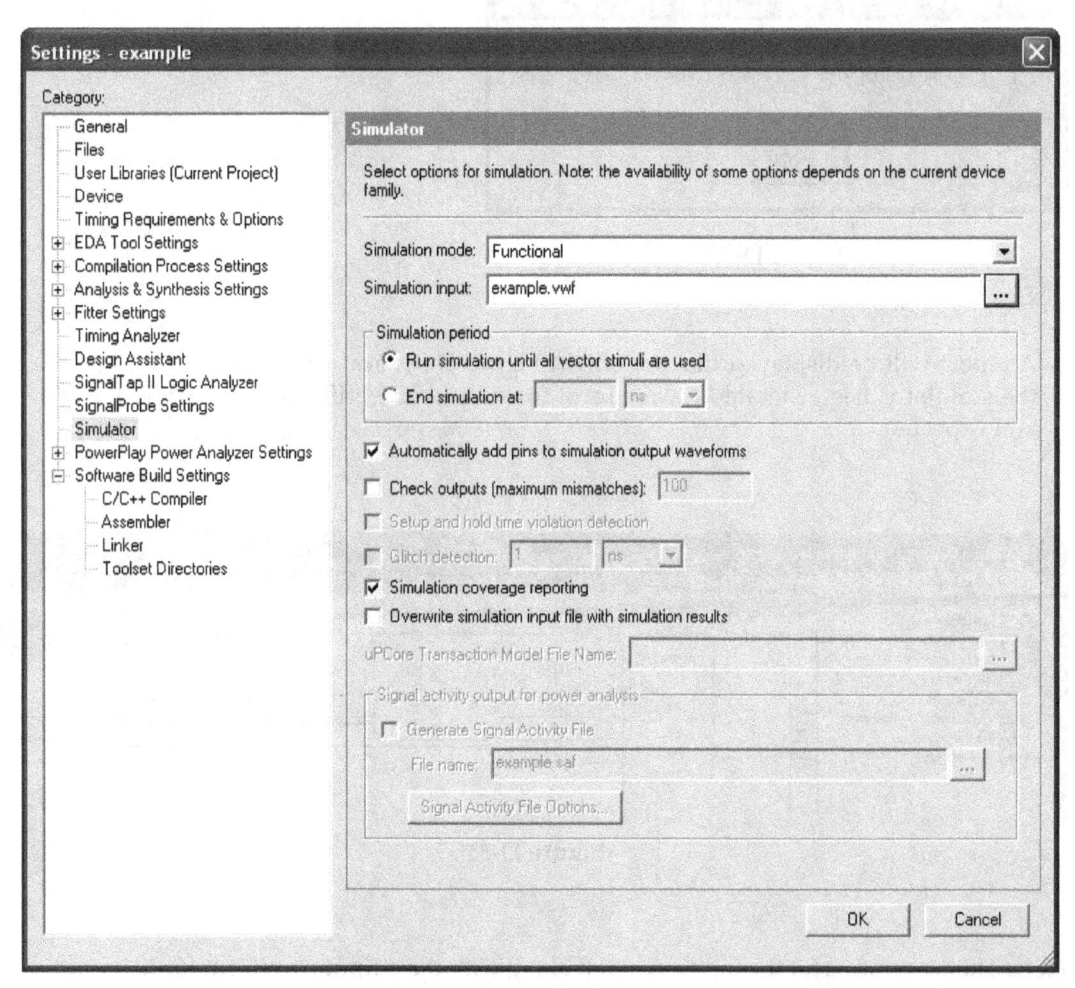

Figure D-28

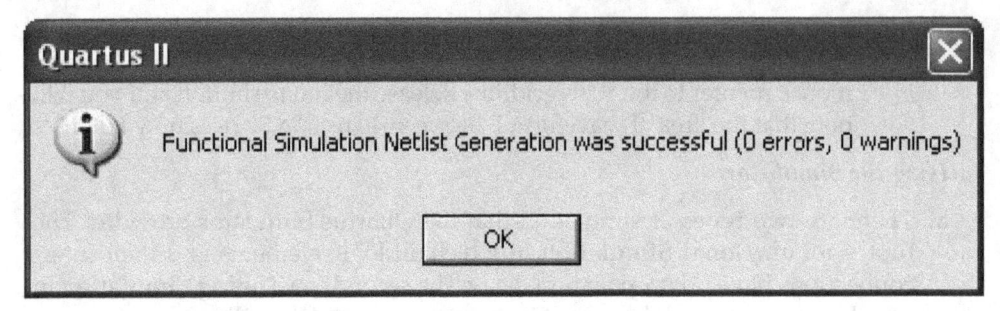

Figure D-29

g) Run the simulation by selecting **Processing | Start Simulation** or pressing the Start Simulation ▶ button.

h) When the simulation is complete, an information box will be displayed like the one in Figure D-30. Press **OK**.

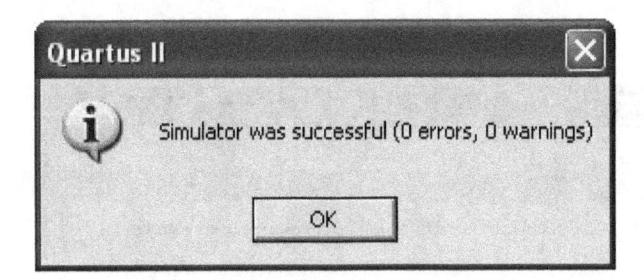

Figure D-30

i) A report will be displayed like the one in Figure D-31. You should note that the simulator has generated a waveform for output X. Verify that it is what you expected.

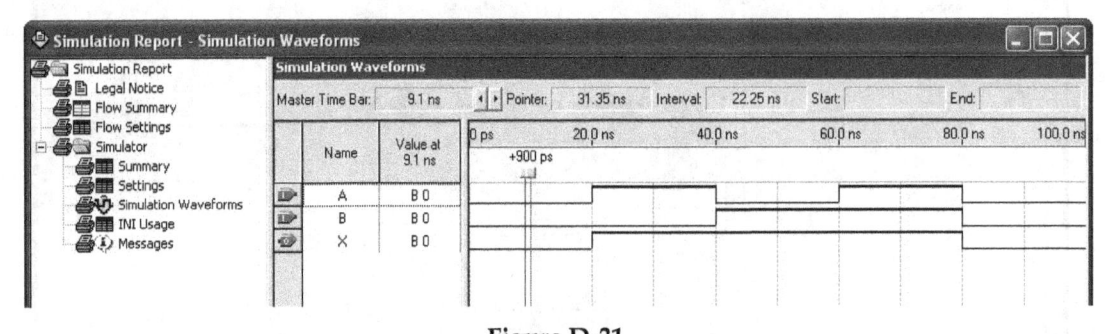

Figure D-31

Note: The waveforms for the inputs A and B along with the output waveform for X represents our design's truth table. To verify the output, use the **Master Time Bar:**

Appendix D

1) Use the Master Time Bar Pointer at the top of the *Simulation Report* window to place the Master Time Bar at 0. Note that in the Value column, A = 0, B = 0, and X = 0.
2) Click on the right arrow of the Master Time Bar Pointer so that the bar moves to 20 ns. Note that now A = 1, B = 0, and X = 1 in the Value column.
3) Click on the right arrow of the Master Time Bar Pointer so that the bar moves to 40 ns. Note that now A = 0, B = 1, and X = 1 in the Value column.
4) Click on the right arrow of the Master Time Bar Pointer so that the bar moves to 60 ns. Note that now A = 1, B = 1, and X = 1 in the Value column.
5) Click on the right arrow of the Master Time Bar Pointer so that the bar moves to 80 ns. Note that now A = 0, B = 0, and X = 0 in the Value column.
6) You can deduce from these results that the waveform values verify the expected truth table for our device.

j) Select **File | Close Project** to close the *example* project. Next, we will show how to use QUARTUS® II to implement circuits specified in VHDL.

k) Have your Instructor or Faculty Assistant initial and record the date completed below.

```
Demonstrated To
Instructor/FA _____ Date _____
```

Exercise 2. VHDL Design using the Text Editor.

a) Select **File | New Project Wizard** to access the features of this wizard. You will get an introduction window, which requires no input.
b) Click **Next** in the introduction window to get the *Directory, Name, Top-Level Entity* window shown in Figure D-3.
c) Set the working directory to be C:/*altera /TocciLM/ example2*. Name the project, which may optionally be the same as the name of the directory. The name used here is *example2*.
d) Click **Next**. A pop-up box, like the one shown in Figure D-4 will appear, indicating that the desired directory needs to be created.
e) Click **Yes**. This will bring up the *Add Files* window, shown in Figure D-5, in which the student can specify which existing files (if any) should be included in the project.
f) Using the New Project Wizard, create a new project for the VHDL design in the directory C:/*altera /TocciLM/ example2*. Choose the same CPLD chip family for implementation. Note that we are creating this project in a different directory than the *example* (schematic) project. It is good practice to create projects in separate directories.
g) You will use the QUARTUS® II Text Editor for entering text-based designs in the VHDL scripting language. You can also use the Text Editor to enter, edit, and view other ASCII text files. In this part, we will use the OR gate again to demonstrate how QUARTUS® II can be used to design and simulate a logic circuit using VHDL.
h) In the QUARTUS® II window, select **File | New**. When the **New** window appears, shown in Figure D-9, select **VHDL** and press **OK**. This will open the Text Editor window as in Figure D-32.

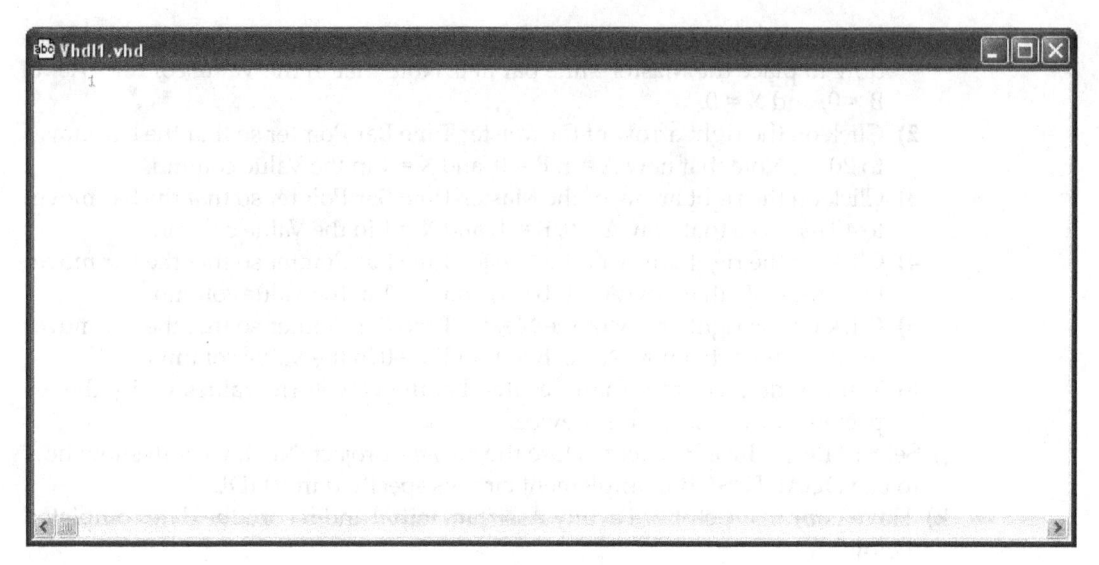

Figure D-32

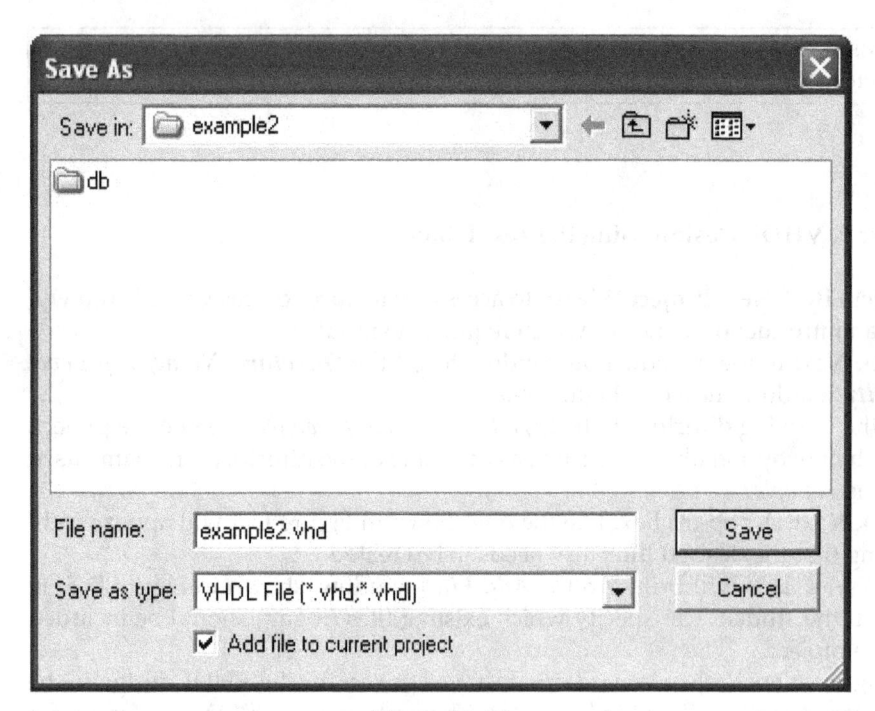

Figure D-33

i) Select **File | Save As** to display the *Save As* pop-up box similar to the one in Figure D-33.
j) Choose *VHDL* as the **Save as type** and use *example2.vhd* as the **File Name**.
k) Leave the **Add file to current project** box checked at the bottom of the figure and click on **Save**.
l) Type in the VHDL program from the listing in Figure D-34. The typed code should be like that shown in Figure D-35.

Appendix D

```
Library IEEE;
use IEEE.std_logic_1164.all;

ENTITY   or_gate IS
    PORT (      a, b       : IN BIT;
                y          : OUT BIT);

END or_gate;

ARCHITECTURE ckt OF or_gate IS
    BEGIN
        y <= a OR b;
    END ckt;
```

Figure D-34

```
example2.vhd
 1  Library IEEE;
 2  use IEEE.std_logic_1164.all;
 3
 4  ENTITY  example2 IS
 5      PORT    (   a, b        :IN BIT;
 6                  y            :OUT BIT);
 7  END example2;
 8
 9  ARCHITECTURE   ckt OF   example2    IS
10      BEGIN
11          y   <=   OR  b;
12      END ckt;
```

Figure D-35

Synthesizing a Circuit from the VHDL Code

m) Select **Processing | Start | Start Analysis & Synthesis** to synthesize a circuit that implements the given VHDL code. If there are no errors detected, you will be notified in a message box like the one shown at the bottom of the main window that says "Full Compilation was successful (0 errors, 0 warnings)" in Figure D-17(a).

n) A summary of the compilation report will be essentially the same as in Figure D-17(b).

o) To see what happens if an error is made, change the "b" in the VHDL text line 5 to a "c" and compile the modified VHDL program.

p) The compilation is not successful and one warning and two error messages are indicated. In the Message window, the first error tells you that object "b" is used but not declared. The second error tells you that the compiler ignored construct "ckt" because of previous errors. If you need to know more about the message, select the message and press the **F1** keyboard key.

q) Double-click on the first error message and the location of the error is highlighted on the VHDL text. In this case, the code "y <= a OR b" will be highlighted (line 11 in our VHDL text). Double-clicking on the second error will highlight the ARCHITECTURE declaration (line 9 in our VHDL text).

r) Correct the VHDL text changed in **o)** above and recompile the design.

Simulating the Design. Functional simulation of the VHDL code is performed in exactly the same way as the simulation described earlier for the design created with schematic capture.

a) Create a new Waveform Editor file with the name *example2.vwf*.
b) Import the nodes in the project into the Waveform Editor.
c) Draw the waveforms for inputs *a and b* shown in Figure D-26.

Note: It is also possible to open the previously drawn waveform file *example.vwf* and then "copy and paste" the waveforms for *a and b*. The procedure for copying waveforms is described in **Help**. It is also possible that since the contents of the two files are identical, we can simply make a copy of the *example.vwf* file and save it under the name *example2.vwf*. However, you will need to change the output name "X" to output name "y" in the *example2.vwf* file.

d) Select the **Functional Simulation** option in the simulator settings and select **Processing | Generate Functional Simulation Netlist**.
e) Start the simulation. The waveform generated by the Simulator for the output "y" should be the same as the waveform "X" in Figure D-31.

Including a VHDL Design in a Schematic

a) To include a circuit from a VHDL design in a schematic, the circuit must have a symbol. To create a symbol that can be imported into the Block Editor, select **File | Create/Update | Create Symbol Files for Current File**.
b) In response, Quartus® II generates a Block Symbol File, *example2.bsf* in the *example2* project folder.
c) To test our symbol, create a new project *example3* in C:/altera /TocciLM/ example3.
d) Create a new schematic and save it as *example3.bdf*.
e) Double-click on the schematic window to access the **Symbol** window.
f) Use the Symbol browser to locate and select the *example2* symbol. Place the symbol in the schematic. Note that the symbol is block-shaped.
g) Place input symbols and connect them to inputs "a" and "b." Place an output symbol on the schematic and connect it to "y." Rename the connectors as you did in the schematic design above. Your schematic should look like Figure D-36.
h) Select **Assignments | Settings | Files**. Use the Files browser on the Settings page to find and **Add** the design file *example2.vhd* to your list of design files for this project.
i) Compile the project.
j) Create a waveform file with the Waveform Editor (or copy one from one of the previous projects) and save it as *example3.vwf*.
k) Simulate the project and verify that the output "y" waveform is as expected.

PART 2 - PROGRAMMING A PLD PROTOTYPING BOARD

As mentioned earlier, one of our objectives for this appendix is to give the student some familiarity with using the QUARTUS® II software to program a CPLD chip such as one from the Altera MAX7000S family. One such chip is the EPM7128SLC84-15, which is found on various PLD prototyping boards such as the DeVry eSOC Board, the Altera UP-1 Board, and the RSR PLDT-2 Board. These

Appendix D

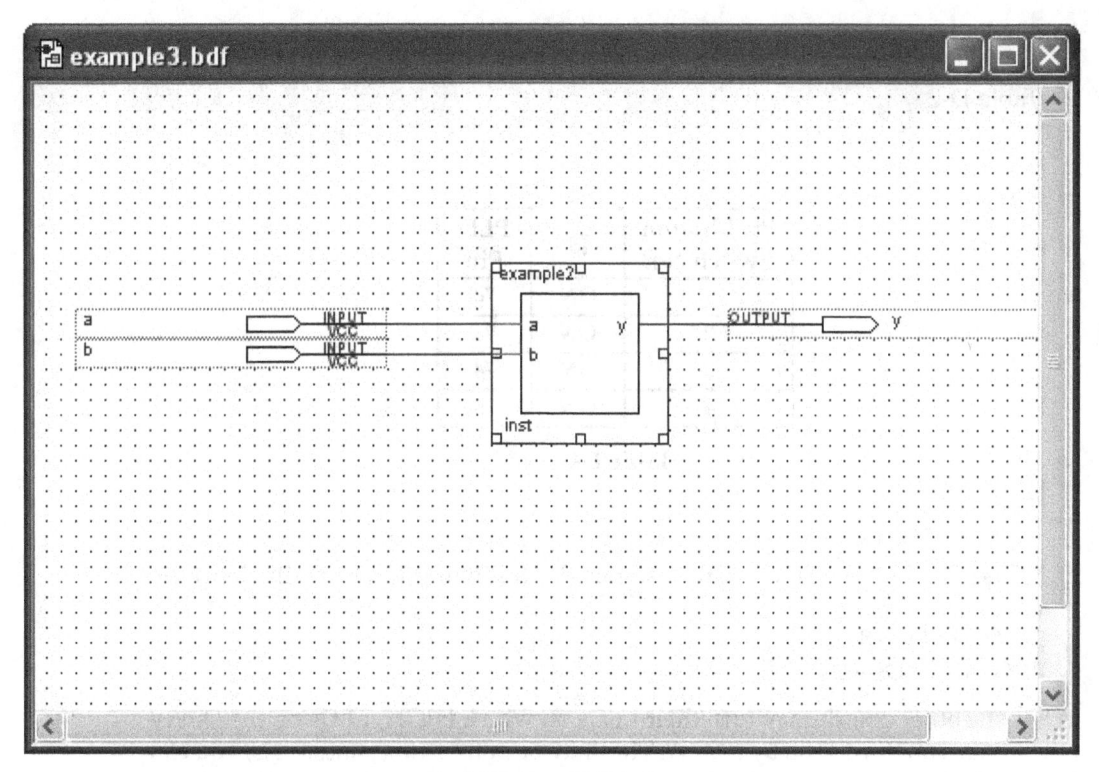

Figure D-36

boards are all programmed the same. Differences, however, appear in the switches and LEDs available and in the available pins on the CPLD that can be programmed. We will begin this section of the appendix by examining these boards and end by learning to use the QUARTUS Programmer tool.

DeVry eSOC Board. The eSOC board has the following switches available to the user:

a) *Toggle Switches.* There are sixteen toggle switches available to the user. The pin assignments for these switches are given in Table D-1.

Toggle Switch No.	J4	PLD Pin	Toggle Switch No.	J5	PLD Pin
1	ON	50	9	ON	60
2	ON	51	10	ON	61
3	ON	52	11	ON	63
4	ON	54	12	ON	64
5	ON	55	13	ON	65
6	ON	56	14	ON	67
7	ON	57	15	ON	68
8	ON	58	16	ON	69

Table D-1

b) *Pushbutton Switches.* There are four user-accessible pushbutton switches on the eSOC board. The pin assignments for these switches are given in Table D-2.

Pushbutton Switch No.	J3	PLD Pin
1	ON	70
2	ON	73
3	ON	74
4	ON	75

Table D-2

c) *LEDs.* The DeVry eSOC board also has the following LEDs available to the user:
   1. Red LEDs – 8
   2. Green LEDs – 8

While the other boards may turn their LEDs on with a HIGH, LEDs on the DeVry eSOC board are turned on by a LOW. This may be bothersome, and if it is, you can invert all of your outputs to LEDs. NOT gates placed in the outputs will do, but if you are using a megafunction or macrofunction such as a counter, adder, or register, you can customize them with the **Symbol Properties** dialog box menu. You can access this by right-clicking on the symbol and selecting **Properties**. You select the outputs you want to invert and click on the Inversion **All** radio button. When you have done this for all the outputs you want to invert, click **OK**.

The pin assignments for both the green and the red LEDs are in Table D-3.

Red LED No.	PLD Pin	Green LED No.	PLD Pin
1	4	9	16
2	5	10	17
3	8	11	18
4	9	12	20
5	10	13	21
6	11	14	22
7	12	15	24
8	15	16	25

Table D-3

7-Segment LEDs. There are two 7-segment LEDs accessible to the user. All segments are turned on by a LOW (see *LEDs* above). Each 7-segment display actually has eight segments, A to G, and a segment referred to as DP (for decimal point). The layout of a 7-segment display is given in Figure D-37.

Appendix D

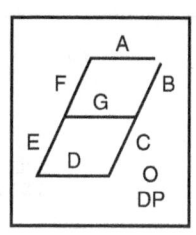

Figure D-37

The pin assignments for each segment of the two 7-segment devices are given in Table D-4.

Left Segment	PLD Pin	Right Segment	PLD Pin
A	31	A	45
B	30	B	44
C	29	C	41
D	28	D	37
E	27	E	36
F	34	F	48
G	33	G	40
DP	35	DP	49

Table D-4

There is also an off-board interface for connecting to other circuits such as those on breadboarded experiments. The connections are available through headers at the bottom edge of the eSOC board. The headers and their on-board connections or PLD pin assignments are illustrated in Figure D-38:

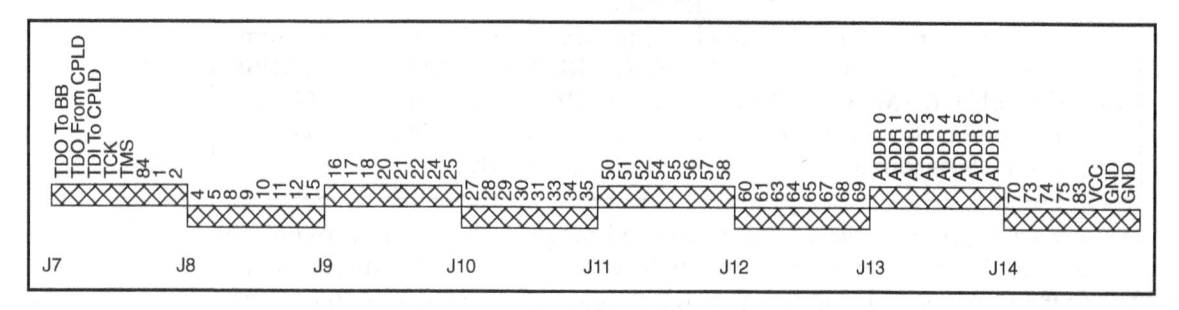

Figure D-38

Your attention should be directed to the J11, J12, and J14 headers. These headers have 8 female pin-jacks each and each numbered 1-8 from left to right. The PLD pins shown on the headers (50, 51, 52, 54–58, 60, 61, 63–65, and 67–69) for J11 and J12 can be accessed externally by turning off the associated DIP switch at J4 and J5, respectively, and by inserting a wire from your external circuit into one of the headers' pin-jacks. For example if you have assigned an output from your external circuit to pin 50, you would

1. Plug the wire from the external circuit into J-11 pin 1 (right below "50 Switch 1").
2. Turn the J4-1 switch to OFF.
3. Assign the connection to pin 50 during the pin assignment procedures, being sure to make the Pin Type Input.

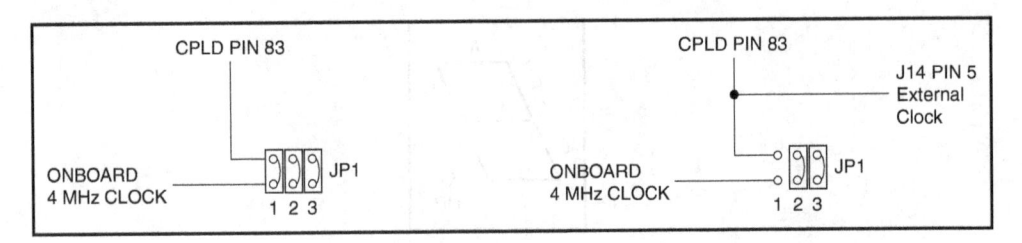

Figure D-39

Electrical and signal grounds for the external connections are located at J14 pins 7 and 8.

If you are going to use an external debounced switch or a TTL-compatible function generator at PLD pin 83 (header J14 pin 5) for your global PLD clock, this connection is through J14 pin 5. The jumper at JP1 must be removed. See Figure D-39.

RSR PLDT-2 Board. The PLDT-2 board has the following devices that are user accessible:

Toggle Switches. There are sixteen toggle switches, numbered 1 to 8, provided by two DIP mounted switch modules, S1 and S2. When a toggle switch is in the position marked H (high), the switch pole is connected via a pull-up resistor to +5 volts. In the position marked L (low), the switch pole is connected to ground.

The eight poles from module S1 are wired to pin-jack strips J1 and J2, and to pins on one side of jumper strip HD1. The opposite side of HD1 is connected to CPLD pins. When no jumpers are installed on HD1, the CPLD is isolated from the switches.

The eight poles from S2 are wired to pin-jack strips J6 and J5. Note that pin-jack strips J2, J5, J6, and J8 are positioned for easy connection to an external solderless breadboard.

> *** WARNING ***
> After testing the board, remove all eight jumpers from HD1 before programming your own design into the CPLD. **FAILURE TO DO SO WILL CAUSE PERMANENT DAMAGE TO THE CPLD CHIP.** When using the SK1 Dip-Switch as inputs in your design, DO NOT program the associated CPLD to be Highs (1s). Those pins are 33, 34, 35, 36, 37, 39, 40, and 41.

Debounced Toggle Switches. Four debounced toggle switches are provided by switch module S5. The outputs of the switches are available at pin-jacks on J12. The four LEDs next to J12 indicate switch status. When a switch is "on," it provides a high TTL level to J12 and the LED lights. When a switch is off, it places a TTL low on J12 and the LED is off.

Momentary Pushbutton Switches. Two momentary push-button switches, S3 and S4, can be used to apply manually generated input pulses to the CPLD. Both switches have pull-up resistors and are normally open (NO), so they will produce a pulse with a falling (+5 V to 0) edge. Neither switch is debounced. The output of S3 is on pin jack J9 while the output of S4 is on pin jack J12.

Oscillator. The PLDT-2 has an on-board 4 MHz oscillator module. The output of this module can be connected to pin 83 of the CPLD by placing a shorting bar across jumper J1. The clock signal is also accessible via single pin socket "CLK"

LEDs. The PLDT-2 has sixteen LEDs, numbered 1 to 16, connected through resistor packs to pin jacks. LEDs 1 to 8 are connected to pin-jack strips J3 and J4 while

Appendix D

LEDs 9 to 16 are connected to pin-jack strips J7 and J8. LEDs 1 to 8 are also connected to one side of jumper-strip HD2. The opposite side of HD2 is connected to CPLD pins as shown in Table D-5. When no jumpers are installed on HD2, the CPLD is isolated from the LEDs.

7-Segment LEDs. A common-anode dual 7-segment display is connected through current-limiting resistors to the CPLD pins as shown in Table D-6 below.

> **NOTE:** The above descriptive paragraphs were extracted from the RSR PLDT-2 user manual with RSR's permission.

SWITCH S1	CPLD PIN	LEDs 1-8	CPLD PIN
1	34	1	44
2	33	2	45
3	36	3	46
4	35	4	48
5	37	5	49
6	40	6	50
7	39	7	51
8	41	8	52

Table D-5

7- SEGMENT DISPLAY			
DISPLAY 1 SEGMENTS	CPLD PIN	DISPLAY 2 SEGMENTS	CPLD PIN
A	58	A	69
B	60	B	70
C	61	C	73
D	63	D	74
E	64	E	76
F	65	F	75
G	67	G	77
DP	68	DP	79

Table D-6

The other devices that are accessible to the user as well as the interface connections are not listed here. Refer to the RSR PLDT-2 user manual.

Altera Project Board. The PLDT-2 board has the following devices that are user accessible:

DIP Switches. The Altera Project Board has sixteen DIP switches (two banks of 8 switches labeled as SW1-1 through SW1-8 and SW2-1 through SW2-8) accessible through pin-jack headers P5 and P6. To use these switches, select a switch and run a wire from its pin-jack to the desired CPLD input pin.

Pushbutton Switches. There are two pushbutton switches, PB1 and PB2, each accessible through its header pin-jack. PB1 has a header pin-jack P9 and PB2 has a header pin-jack P10. To use these switches, select a switch and run a wire from its pin-jack to the desired CPLD input pin.

LEDs. The Altera Project Board has sixteen LEDs (two banks of 8 LEDs labeled as LED 1-8 and LED 9-16) accessible through pin-jack headers P7 and P8. To use these LEDs, select an LED and run a wire from its pin-jack to the desired CPLD output pin.

7-Segment LEDs. The Altera Project Board has two 7-segment LEDs named Digit 1 and Digit 2. The pin assignments for these devices are given in Table D-7.

Pin Assignments. Table D-7 gives a pin assignment for the I/O devices on the Altera UP1 Project Board. The 7-segment LED pin assignments are fixed. The rest are flexible.

7- SEGMENT DISPLAY			
DIGIT 1 SEGMENTS	CPLD PIN	DIGIT 2 SEGMENTS	CPLD PIN
A	58	A	69
B	60	B	70
C	61	C	73
D	63	D	74
E	64	E	76
F	65	F	75
G	67	G	77
DP	68	DP	79
DIP SWITCHES			
SWITCH	CPLD PIN	SWITCH	CPLD PIN
SW1-1	34	SW2-1	28
SW1-2	33	SW2-2	29
SW1-3	36	SW2-3	30
SW1-4	35	SW2-4	31
SW1-5	37	SW2-5	57
SW1-6	40	SW2-6	55
SW1-7	39	SW2-7	56
SW1-8	41	SW2-8	54
LEDS			
SWITCH	CPLD PIN	SWITCH	CPLD PIN
LED1	44	LED9	80
LED2	45	LED 10	81
LED3	46	LED 11	4
LED4	48	LED 12	5
LED5	49	LED 13	6
LED6	50	LED 14	8
LED7	51	LED 15	9
LED8	52	LED 16	10
Pushbuttons			
PB1	11	LED9	1

Table D-7

Appendix D

Special and Unassigned Pins. There are a few unassigned and special function pins. These are listed in Table D-8. You are encouraged to read the UP1 user manual for other pin information.

Function	CPLD PIN	Function	CPLD PIN
GCLRn	1	Unassigned	21
OE2/GCLK2	2	Unassigned	22
Unassigned	12	Unassigned	24
Unassigned	15	Unassigned	25
Unassigned	16	Unassigned	27
Unassigned	17	GLCLK1	83
Unassigned	18	OE1	84
Unassigned	20		

Table D-8

Programming Procedure Using QUARTUS® II. The CPLD of interest to us and used on the various boards listed above is the EPM7128S. (The Altera UP 1 Project Board also has an EPF10K20 installed, but that CPLD will not be discussed here.) The EPM7128S comes in several speeds from 6ns to 15ns. You should examine the CPLD installed on your board and list the device number here: _____

The DeVry eSOC board comes with an EPM7128SLC84-15 CPLD installed. The "-15" refers to the speed of the chip, which is 15ns. The EPM7128SLC84 CPLD contains 128 macrocells with 2500 usable gates arranged in 8 logic array blocks. The chip is available in speeds from 6 ns to 15 ns delay time pin-to-pin. Outputs can drive 25 mA (max) loads.

Regardless of the CPLD used, if it can be programmed using Altera QUARTUS® II the labs in the Supplemental Experiment set should work provided the particular board can facilitate the necessary I/O.

Exercise 3. Implementing a Circuit Using the DeVry eSOC Board.

The CPLD programming procedure will be the same for every experiment with the exception of the pin assignments. Pin assignments are discussed in each experiment where the project is to be downloaded to your board. When the pin assignments have been made and the project has been recompiled successfully, use the following steps to program the CPLD:

a) If you are programming an eSOC board, make sure that all of the DIP switches on J3, J4, and J5 are OFF. Connect the parallel cable for ByteBlaster to the board and the parallel port you are using for downloading. Power up the board.
b) Select **File | Open Project** and browse to the directory *example*, which contains the project file *example.qpf* for the OR gate used earlier.
c) Select the *example.qpf* file and press the **Open** button.
d) This project is already set to use the EPM7128SLC84-15 CPLD. If necessary you can change the CPLD with the **Assignments | Device** window.

Pin Assignments. Suggested pin assignments for this project are: Input A – toggle switch 1 (pin 50), Input B – toggle switch 2 (pin 51), and Output X – LED 1 (pin 4).

e) To make these assignments, select **Assignments | Pins** to access the **Assignment Editor**. If necessary, press the [Pin] button to make the Category indicate Pins.
f) The cursor is now a "solid cross." In the **Edit:** panel, double-click under **To** to pull down a menu of the circuit's input and output labels. Select "**a**," double-click under **Location,** and select "**PIN_50**" from the pull-down menu.
g) Repeat f) above selecting "**b**" and "**PIN_51**."
h) Repeat f) above selecting "**y**" and "**PIN_4**."
i) The pin assignments are now complete. The Assignment Editor window should look like the one in Figure D-40.
j) Recompile the project.

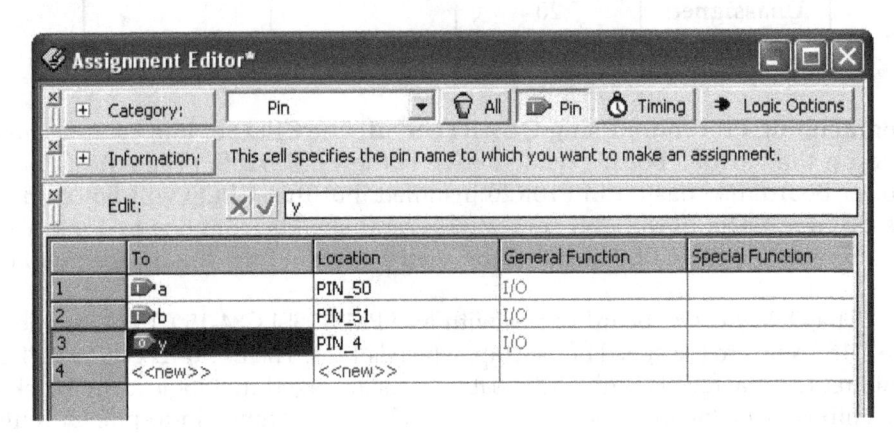

Figure D-40

Downloading to the CPLD. Once pin assignments have been made and the design recompiled, you are ready to download to the CPLD device on the prototyping board.

k) Select **Tools | Programmer.** A **Chain Description** window like that in Figure D-41 will appear.
l) Verify that your hardware is set up correctly. The Programmer window should indicate that **ByteBlaster(MV)** is installed. If it is, go on to **p**).
m) If the Chain Description window appears like the one in Figure D-42 (No Hardware), ByteBlaster(MV) is not installed, the board is not set up correctly, or some other problem has occurred, which requires your instructor's attention.

Appendix D   677

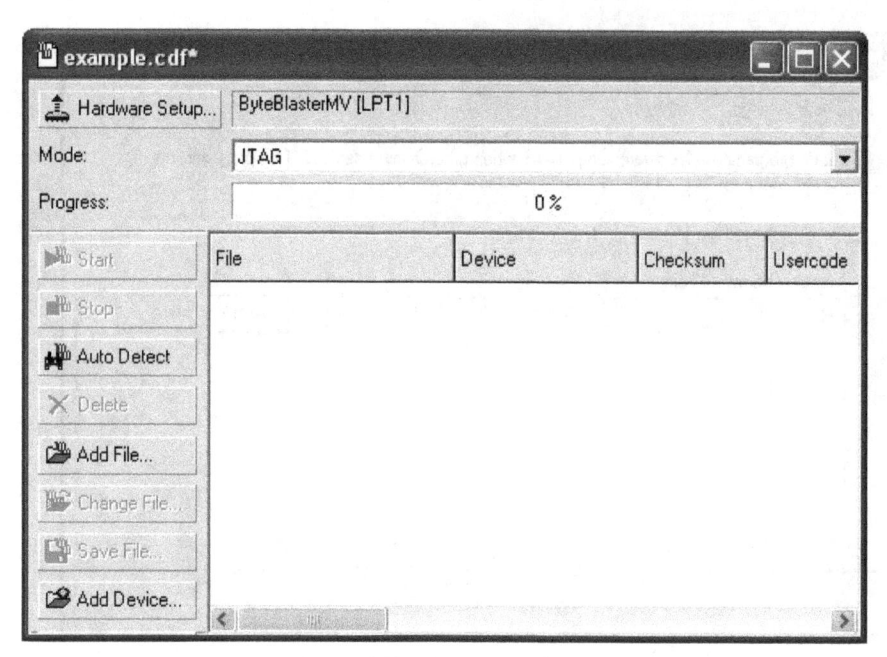

Figure D-41

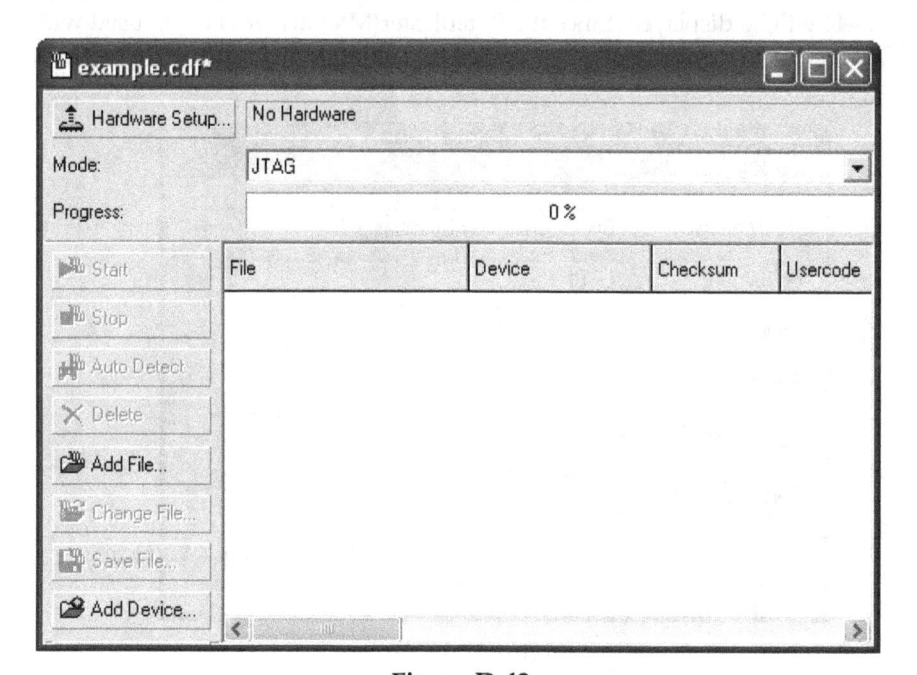

Figure D-42

n) Press the ⟦Hardware Setup...⟧ button. The Hardware Setup window will be displayed like the one in Figure D-43. Note that the **Currently Selected Hardware** box indicates *No Hardware*.

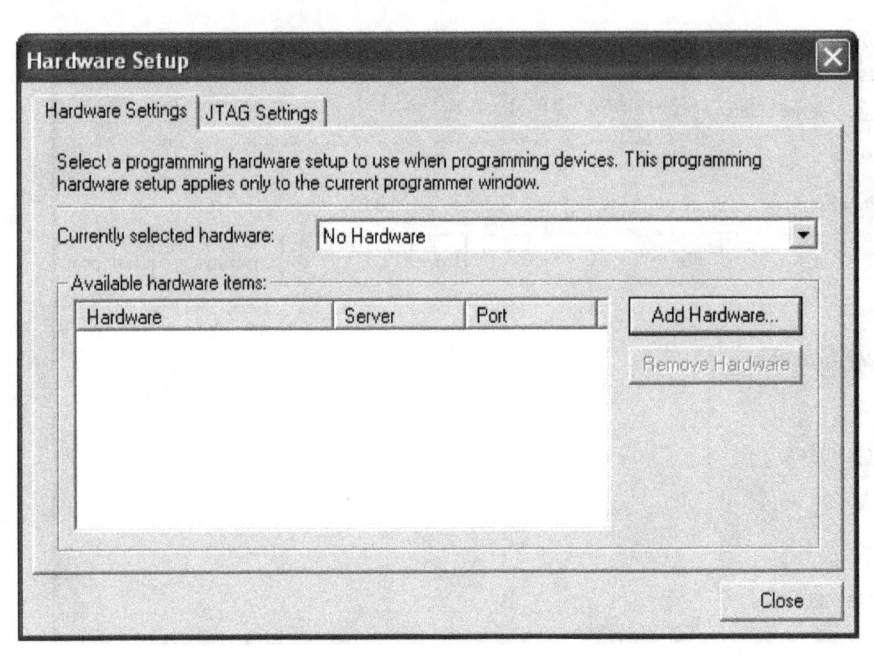

Figure D-43

o) Press the [Add Hardware...] button. A pop-up dialog box like the one in Figure D-44 will be displayed. Since the ByteBlaster(MV) driver is being used with LPT1, click **OK**. The Hardware Setup window should now look like Figure D-45.

Figure D-44

p) Press the [Auto Detect] button. The EPM7128S device should be detected and displayed in the Device column of the Chain Description File. If not, see your instructor.

q) Press the [Add File...] button in the Chain Description File. Locate and Add the *example.pof* file.

Appendix D 679

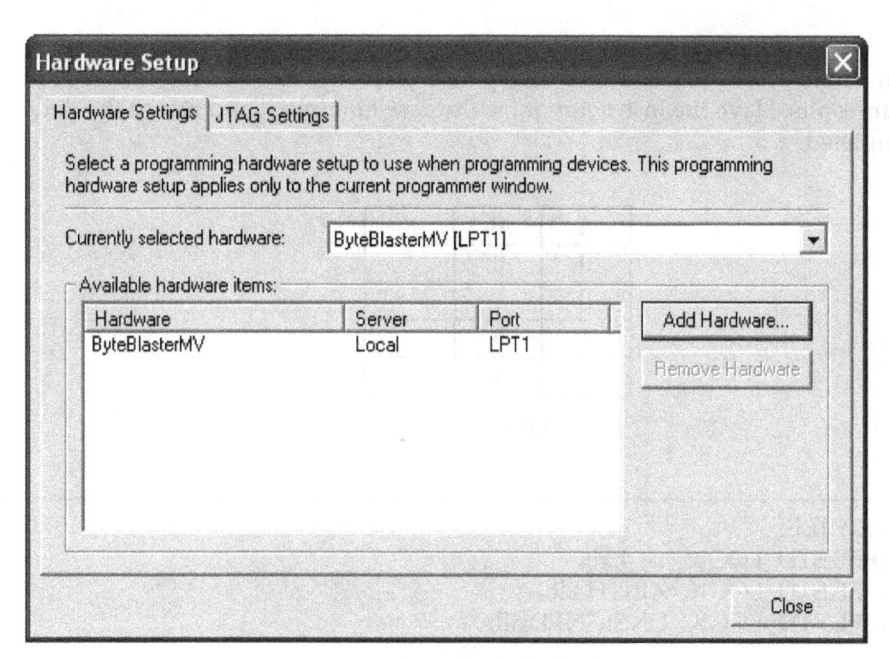

Figure D-45

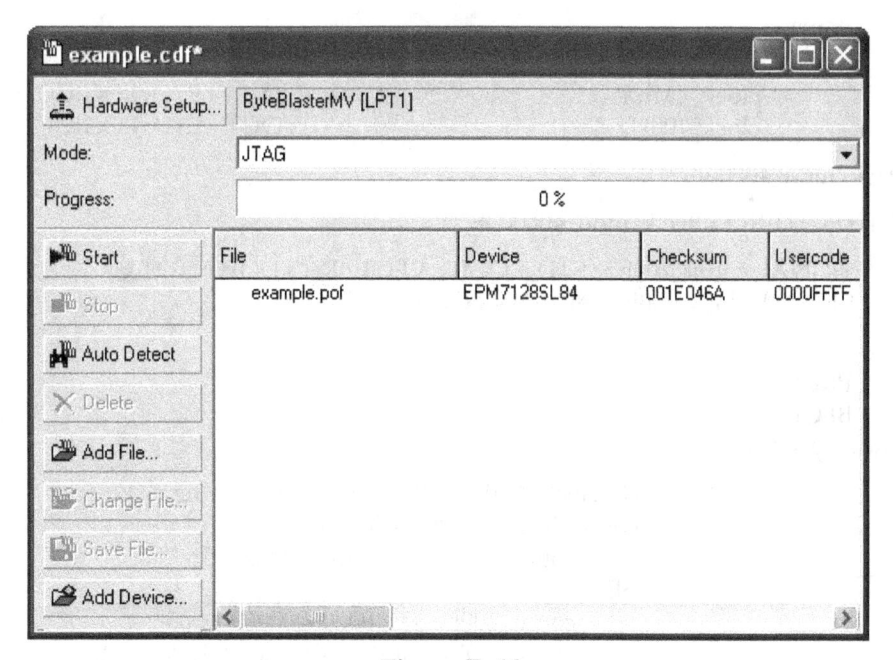

Figure D-46

r) Select the file and device line and scroll (horizontally) to the available programming option. Check the **Program/Configure** and **Verify** boxes.

s) Press the Start button. Observe the progress bar.
t) Check the Messages to see if the programmer detected any problems.
u) On the eSOC board, push the DIP switches 1 and 2 on J4 to ON.

v) Manipulate toggle switches 1 and 2 through the various inputs for your design circuit and verify that LED 1 responds correctly. Use Table D-9 to record your results. Have the instructor or faculty assistant initial and record the date completed.

A	B	X
0	0	
0	1	
1	0	
1	1	

Table D-9

```
LIBRARY IEEE;
USE IEEE.STD_LOGIC_1164.all;
USE IEEE.STD_LOGIC_ARITH.all;
USE IEEE.STD_LOGIC_UNSIGNED.all;

-- This routine receives a 4 MHz clock
-- and outputs a 100 Hz (10 ms)clock required by debouncer

ENTITY clk_mod_40000 IS

        PORT
        (
               clock_4Mhz                          : IN    STD_LOGIC;
               clock_100Hz                         : OUT   STD_LOGIC);

END clk_mod_40000;

ARCHITECTURE a OF clk_mod_40000 IS

        SIGNAL count_100hz : STD_LOGIC_VECTOR(15 DOWNTO 0);
        SIGNAL clock_100hz_int : STD_LOGIC;

BEGIN
        PROCESS
        BEGIN
-- Divide by 40000
               WAIT UNTIL clock_4Mhz'EVENT and clock_4Mhz = '1';
                    IF count_100hz /= 20000 THEN
                         count_100hz <= count_100hz + 1;
                    ELSE
                         count_100hz <= "0000000000000000";
                         clock_100hz_int <= NOT clock_100hz_int;
                    END IF;

               -- Sync output to master clock signal

               clock_100hz <= clock_100hz_int;

        END PROCESS;
END a;
```

Figure D-47

Appendix D

```vhdl
LIBRARY ieee;
USE ieee.STD_LOGIC_1164.all;
USE ieee.STD_LOGIC_UNSIGNED.all;
-- This routine receives active low pb
-- Delays outputing for about 40 ms
-- Outputs active low pulse
ENTITY DEBOUNCER IS
        PORT (
                        input_clock    : IN STD_LOGIC;
                        pb      : IN STD_LOGIC; -- active low input
                        pulse   : OUT STD_LOGIC);
END DEBOUNCER;
ARCHITECTURE filtered_pulse OF DEBOUNCER IS

        SIGNAL cnt      : STD_LOGIC_VECTOR (2 DOWNTO 0);
BEGIN
  PROCESS (input_clock)
  BEGIN
    IF pb = '1' THEN
      cnt <= "000";
    ELSIF (input_clock'EVENT AND input_clock = '1') THEN
      IF (cnt /= "111") THEN
              cnt <= cnt + 1;
        END IF;
      END IF;
    IF (cnt = "101") AND (pb = '0') THEN
              pulse <= '0';
        ELSE pulse <= '1';
        END IF;
    END PROCESS;
END filtered_pulse;
```

Figure D-48

DeVry eSOC VHDL Switch Debouncing Routines

Included in this section are two VHDL routines for software debouncing the DeVry eSOC's active low pushbuttons. These routines can be used for debouncing on the RSR PLDT's pushbutton switches and with modification, the Altera UP1 and UP2 boards. They will be needed for most manual clocking experiments.

The Clock Divider Routine. The eSOC board's master clock is 4 MHz. In order to debounce the pushbutton (or toggle) switches, a clock speed of 100 Hz is needed. A VHDL routine for dividing the master clock to 100 Hz is shown in Figure D-47. The routine effectively divides the clock down by a factor of 40000.

The Debouncing Routine. The VHDL switch debouncer is shown in Figure D-48. It takes an active low pushbutton or toggle switch and delays it for about 40 ms. The output is also active low without any switch bouncing.

Creating Schematic Symbols for the Clock Divider and Debouncing Routines. To debounce the eSOC switches, the routines above must be compiled, a symbol created for each routine, and then the symbols placed and cascaded between a switch input and a device (such as a flip-flop, counter, or register) clock input. A procedure for doing this is as follows:

a) Start the QUARTUS® II program.
b) Select **File | New Project Wizard...** and create a new project *Clock_MOD_40000*.
c) Select **File | New | VHDL File**.
d) Enter the program for the clock divider.
e) Save the program in its own folder for future use.
f) Compile the program.
g) Choose **File | Create/Update > Create Symbol Files for Current File**.
h) The symbol is now available to be placed on a schematic diagram.
i) Repeat the above procedure for the debouncer routine.

Cascading the Symbols for the Clock Divider and Debouncing Routines.
Figure D-49 shows an example for cascading the two switch debouncing routines in a schematic. Figure D-50 shows the Pulse output of the cascaded routines given in a typical undebounced pushbutton input PB3.

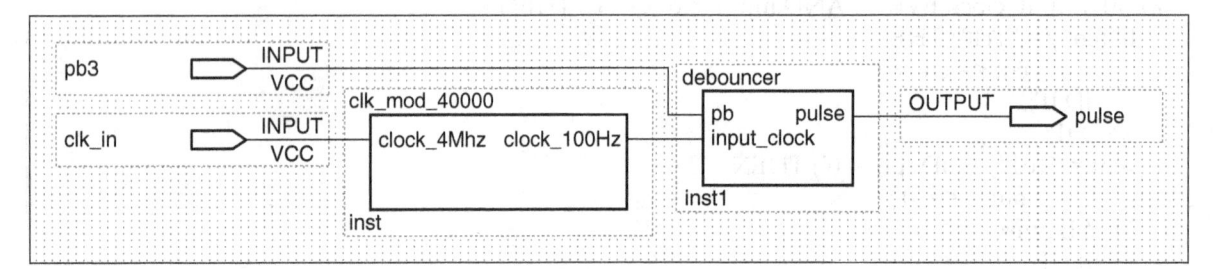

Figure D-49

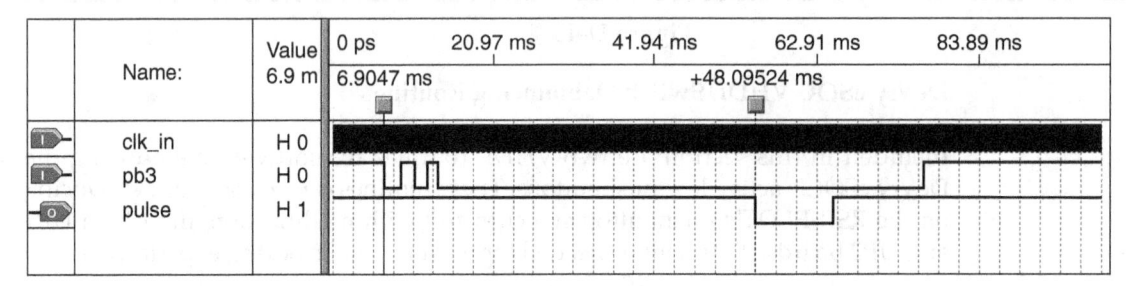

Figure D-50